A TEXTBOOK OF

HYDRAULIC MACHINES

(FLUID POWER ENGINEERING)

SI UNITS

MULTICOLOUR EDITION

A TEXTBOOK OF

HYDRAULIC MACHINES

[For Engineering Students of Various Disciplines and Competitive Examinations]

in

SI UNITS

Er. R.K. RAJPUT

M.E. (Hons.), Gold Medallist; Grad. (Mech.Engg. & Elect. Engg.); M.I.E. (India);

M.S.E.S.I.; M.I.S.T.E.; C.E. (India)

Recipient of:

"Best Teacher (Academic) Award"

"Distinguished Author Award"

"Jawahar Lal Nehru Memorial Gold Medal"

for an outstanding research paper

(Institution of Engineers–India)

Principal *(Formerly)*:

- *Thapar Polytechnic College;*
- *Punjab College of Information Technology,*

PATIALA

S.CHAND PUBLISHING

empowering minds

S. CHAND & COMPANY PVT. LTD.

(AN ISO 9001 : 2008 COMPANY)

RAM NAGAR, NEW DELHI-110 055

S.CHAND PUBLISHING
empowering minds

S. CHAND & COMPANY PVT. LTD.

(An ISO 9001 : 2008 Company)

Head Office: 7361, RAM NAGAR, NEW DELHI - 110 055
Phone: 23672080-81-82, 9899107446, 9911310888 Fax: 91-11-23677446
www.schandpublishing.com; e-mail: helpdesk@schandpublishing.com

Branches

Ahmedabad : Ph: 27541965, 27542369, ahmedabad@schandpublishing.com
Bengaluru : Ph: 22268048, 22354008, bangalore@schandpublishing.com
Bhopal : Ph: 4274723, 4209587, bhopal@schandpublishing.com
Chandigarh : Ph: 2625356, 2625546, chandigarh@schandpublishing.com
Chennai : Ph. 28410027, 28410058, chennai@schandpublishing.com
Coimbatore : Ph: 2625356, 2625446, coimbatore@schandpublishing.com (Marketing Office)
Cuttack : Ph: 2332580; 2332581, cuttack@schandpublishing.com
Dehradun : Ph: 2711101, 2710861, dehradun@schandpublishing.com
Guwahati : Ph: 2738811, 2735640, guwahati@schandpublishing.com
Hyderabad : Ph: 27550194, 27550195, hyderabad@schandpublishing.com
Jaipur : Ph: 2219175, 2219176, jaipur@schandpublishing.com
Jalandhar : Ph: 2401630, 5000630, jalandhar@schandpublishing.com
Kochi : Ph: 2378740, 2378207-08, cochin@schandpublishing.com
Kolkata : Ph: 22367459, 22373914, kolkata@schandpublishing.com
Lucknow : Ph: 4026791, 4065646 lucknow@schandpublishing.com
Mumbai : Ph: 22690881, 22610885, mumbai@schandpublishing.com
Nagpur : Ph: 6451311, 2720523, 2777666, nagpur@schandpublishing.com
Patna : Ph: 2300489, 2302100, patna@schandpublishing.com
Pune : Ph: 64017298, pune@schandpublishing.com
Raipur : Ph: 2443142, raipur@schandpublishing.com (Marketing Office)
Ranchi : Ph: 2361178, ranchi@schandpublishing.com
Siliguri : Ph: 2520750, siliguri@schandpublishing.com (Marketing Office)
Visakhapatnam: Ph: 2782609 visakhapatnam@schandpublishing.com (Marketing Office)

First Edition 1998
Subsequent Editions and Reprints 2002, 2003, 2004, 2006, 2007, 2008 (Twice), 2009, 2010, 2011, 2012
Fully Revised Multicolour Edition 2013; Reprint 2014
Sixth Edition 2016

ISBN : 978-93-856-7631-4 **Code :** 1010C 194

This book is for Sale in India & other SAARC countries only.

PRINTED IN INDIA

By Vikas Publishing House Pvt. Ltd., Plot 20/4, Site-IV, Industrial Area Sahibabad, Ghaziabad-201010 and Published by S.Chand & Company Pvt. Ltd., 7361, Ram Nagar, New Delhi -110 055.

To my wife

Ramesh Rajput

PREFACE TO THE SIXTH EDITION

I am pleased to present the **Sixth Edition (Multicolour)** of this book. The warm reception, which the previous editions and reprints of this book have enjoyed all over India and abroad has been a matter of satisfaction to me.

In this edition, a **New Chapter No. 9** has been added with the following contents.
Chapter 9 : GATE and UPSC Examinations Questions with Answers/Solutions (Latest Selected).

A. Conventional Questions with Answers/Solutions.

B. Multiple-Choice Questions with Answers and "Explanations".

I'm thankful to the Management Team and the Editorial Department of S. Chand & Company Pvt. Ltd., New Delhi for all help and support in the publication of this book.

Any constructive suggestions for the improvement of this book are most welcome.

Er. R.K. Rajput
(Author)

PREFACE TO THE FIRST EDITION

This treatise on ***"Hydraulic Machines"*** (Fluid Power Engineering) assimilates comprehensive treatment of the subject matter in simple, lucid and direct language and envelopes a large number of solved problems properly graded, including typical examples from examination point of view.

The book comprises 6 chapters. All chapters of the book are saturated with much needed text supported by simple and self-explanatory figures and a large number of *Worked Examples* including *Typical Examples* (for competitive examinations). At the end of each chapter *Highlights, Objective Type Questions, Theoretical Questions* and *Unsolved Examples* have been added to make the book a comprehensive and a complete unit in all respects.

The book will prove to be a boon to the students preparing for engineering undergraduate, AMIE Section B (India) and competitive examinations.

The author's thanks are due to his wife Ramesh Rajput for extending all cooperation during preparation of the manuscript and proof reading.

In the end the author wishes to express his gratitude to Shri Ravindra Kumar Gupta, Director, S. Chand & Company Ltd., New Delhi, for taking a lot of pains in bringing out the book, with extremely good presentation, in a short span of time.

Although every care has been taken to make the book free of errors both in the text as well as in solved examples, yet the author shall feel obliged if errors present are brought to his notice. Constructive criticism of the book will be warmly received.

Er. R.K. Rajput
(Author)

NOMENCLATURE

a	Acceleration
A	Area
A_s	Area of suction pipe, surge tank
A_d	Area of delivery pipe
B	Width of wheel (turbine)
b	Width, bed width of rectangular or trapezoidal channel
c_p	Specific heat at constant pressure
CP	Centipoise
C_v	Specific heat at constant volume
C	Chezy's discharge coefficient
C	Celerity of a pressure wave
C_c	Coefficient of contraction
C_d	Discharge coefficient of weirs, orifice plates
C_D	Drag coefficient
C_D	Local drag coefficient
C_v	Coefficient of velocity
d	Diameter of orifice plate, pipe, particle
D	Diameter of pipe, wheel
d_d	Diameter of delivery pipe
d_s	Diameter of suction pipe
e	Linear strain
E	Young's modulus of elasticity of material
f	Darcy Weisbach friction coefficient, frequency
F	Force
F_B	Force exerted by boundary on the fluid
F_D	Drag force on the body
F_L	Lift force
F_r	Froude number
g	Gravitational acceleration
h	Piezometric head, specific enthalpy
h_d	Delivery head
h_f	Frictional loss of head
h_s	Suction head
H_g	Gross head
H	Total energy head, net head
h_{ad}	Acceleration head for delivery pipe
h_{as}	Acceleration head for suction pipe
I	Moment of inertia (of area), moment of inertia (of mass)
l_d	Length of delivery pipe
l_s	Length of suction pipe
l_d'	Length of delivery pipe between cylinder to air vessel
l_s'	Length of suction pipe between cylinder and air vessel

k	Roughness height
K	Conveyance
K	Head loss coefficient, bulk modulus of elasticity, blade friction coefficient
K_t	Vane thickness factor
K_u	Speed ratio
K_f	Flow ratio
m	Mass
M	Momentum, Mach number
n	ratio B/D
N	Manning's roughness coefficient, revolutions per minute
N_s	Specific speed
p, p_s	Pressure, stagnation pressure
P	Power, shaft power (turbine), Poise, force
q	Discharge per unit width, discharge per jet
Q	Discharge, heat
r	Distance from the centre
R	Radius of pipe, hydraulic radius, radius of pipe bend
R_o	Universal gas constant
Re	Reynolds number
S	Specific gravity, bed slope of channel
t	Thickness, time
T	Absolute temperature in Kelvins
T	Torque, water surface width
u	Instantaneous velocity at a point in X-direction
u_f	Shear friction velocity
U	Free stream velocity
V_d	Velocity of flow in delivery pipe
V	Velocity of flow in the cylinder
V_s	Velocity of flow in suction pipe
v	Instantaneous velocity at a point in Y-direction
v	Specific volume
v_c	Critical velocity
V_a	Velocity of approach
v	Time averaged velocity at a point in Y-direction
V_r	Relative velocity
V_f	Velocity of flow (in turbines and pumps)
V_w	Velocity of swirl (in turbines and pumps)
V	Volume
w	Weight density, Instantaneous velocity at a point in Z-direction
W	Weight of fluid, workdone
x	Distance in X-direction
y	Distance in Y-direction, depth of flow
y_c	Critical depth
$\bar{x}$	Depth of centroid of area below water surface
Z	Number of buckets/vanes
z	elevation

Greek Notations

α	Energy correction factor, Mach angle, angle
β	Momentum correction factor, angle
γ	Ratio of specific heats
δ	Boundary layer thickness
δ'	Laminar sub-layer thickness
δ	Displacement thickness of boundary layer
$*\Delta s$	Change in entropy
η	Efficiency, dimensionless distance (y/δ)
θ	Angle, momentum thickness of boundary layer
μ	Coefficient of dynamic viscosity
ν	Kinematic viscosity
ρ	Mass density of fluid
σ	Coefficient of surface tension, cavitation number (Thoma number)
τ	Shear stress
τ_0	Bottom shear stress
ϕ	Angle, velocity potential
ψ	Stream function
ω	Angular velocity
Γ	Circulation
Ω	Vorticity

Subscript 0	refer to any quantity at reference section
Subscripts 1, 2	refer to any quantity at section 1 or 2
Subscripts x, y, z	refer to any quantity in x, y, z direction
Subscripts m, p	refer to any quantity in model and prototype
Subscript r	refer to the ratio of any quantity in model to that in prototype.

CONTENTS

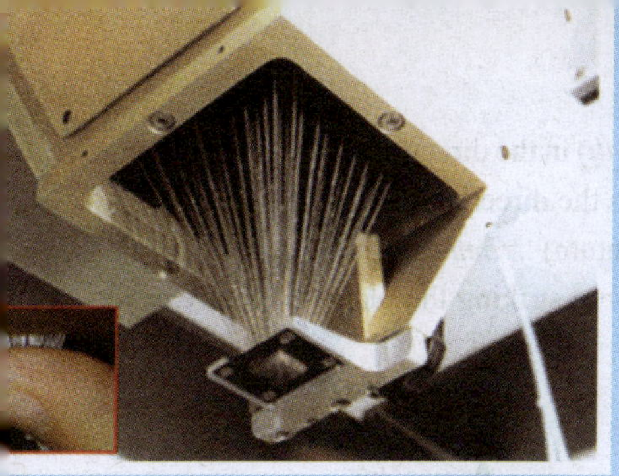

Chapter 1

IMPACT OF FREE JETS

1.1. INTRODUCTION

A fluid **jet** *is a stream of fluid issuing from a nozzle with a high velocity and hence a high kinetic energy.* When a jet impinges on a plate or vane, it exerts a force on it (due to change in momentum). This force (hydrodynamic) can be evaluated by using *'Impulse-momentum principle'*. This chapter deals with the application of the impulse-momentum equation for evaluating the hydrodynamic force on the stationary and moving vanes. The following cases of impact of jet will be considered:

A. Force exerted by the jet on the stationary plate:

1. When flat plate is held normal to the jet;
2. When flat plate is held inclined to the jet;
3. When plate is curved.

B. Force exerted by the jet on the moving plate:

1. When plate is held normal to the jet;
2. When plate is held inclined to the jet;
3. When plate is curved.

A. FORCE EXERTED BY THE JET ON THE STATIONARY PLATE

1.2. FORCE EXERTED ON A STATIONARY FLAT PLATE HELD NORMAL TO THE JET

Fig. 1.1 shows a fluid jet striking a stationary flat plate held perpendicular to the flow direction. Let a and V be cross-sectional area and velocity of the jet respectively. The jet, after striking this plate (vertical), will get its direction changed through 90°; but, it will move on and off the plate with velocity V, if we *neglect* the friction between the jet and the plate as is possible when the plate is *smooth*. If the *friction is considered*, the velocity of liquid coming off the plate will be *slightly less* than V.

The force exerted by the jet on the plate (assuming it *smooth*) in the direction of jet (*X*-direction),

F_x = Rate of change of momentum (in the direction of force)

= (Initial momentum – final momentum) ... *Impulse-momentum principle.*

= (Mass/sec) × [velocity of jet before striking the plate – velocity of jet after striking the plate]

$= \rho a V (V - 0)$ [$\because$ Mass/second = $\rho a V$]

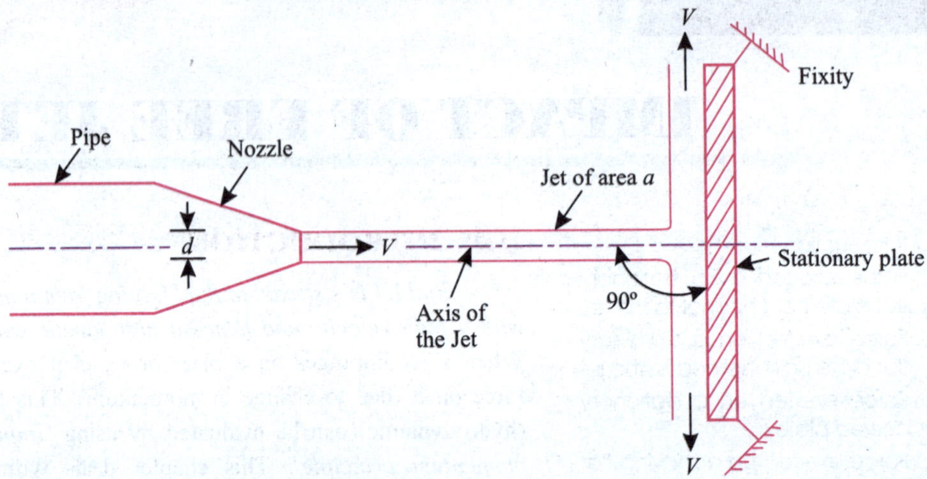

Fig. 1.1. Fluid jet striking a stationary plate.

or, $$F_x = \rho a V^2 \quad ...(1.1)$$

(where, ρ = mass density of liquid; a = area of jet $= \frac{\pi}{4} d^2$, d being diameter of the jet).

It may be noted that a jet leaves in the direction normal to X-axis, the final velocity in the X-direction is *zero.*

Since the plate is *stationary*, therefore, the *work done on the plate is zero.*

1.3. FORCE EXERTED ON A STATIONARY FLAT PLATE HELD INCLINED TO THE JET

Fig. 1.2 shows the stationary flat plate inclined at θ° to the direction of the horizontal jet. If a and V are the cross-sectional area and velocity of the jet respectively, then the mass of liquid per second striking the plate

$= \rho \times aV$ (where $a = \frac{\pi}{4} d^2$, d being diameter of the jet)

After striking the plate (assuming it *smooth*), the jet leaves the plate with a velocity equal to initial velocity (V).

Let us apply the *impulse- momentum equation* in the direction normal to the plate.

Force in normal direction, $F_n = \rho a V (V \sin\theta - 0)$

$$= \rho a V^2 \sin\theta \quad ...(1.2)$$

This normal force can be resolved into two components; component F_x, *parallel* to the direction of jet and component F_y, *normal* to the direction of jet.

$$F_x = F_n \sin\theta = \rho a V^2 \sin\theta \times \sin\theta = \rho a V^2 \sin^2\theta \quad ...(1.3)$$

$$F_y = F_n \cos\theta = \rho a V^2 \sin\theta \times \cos\theta = \rho a V^2 \sin\theta \cos\theta \quad \text{...(1.4)}$$

In this case also, the work done is *zero* as the plate is stationary. When the fluid strikes the plate, it gets divided into two streams Q_1 and Q_2. Since the frictional resistance has been assumed to be negligible, the resultant force in the direction tangential to the plate is zero. Applying the impulse-momentum equation in the direction tangential to the plate, we obtain:

$$(\rho Q_1 V - \rho Q_2 V) - \rho Q V \cos\theta = 0$$

$$Q_1 - Q_2 - Q \cos\theta = 0 \quad \text{...(i)}$$

Also, $$Q_1 + Q_2 = Q \quad \textit{(Continuity equation)} \text{ ...(ii)}$$

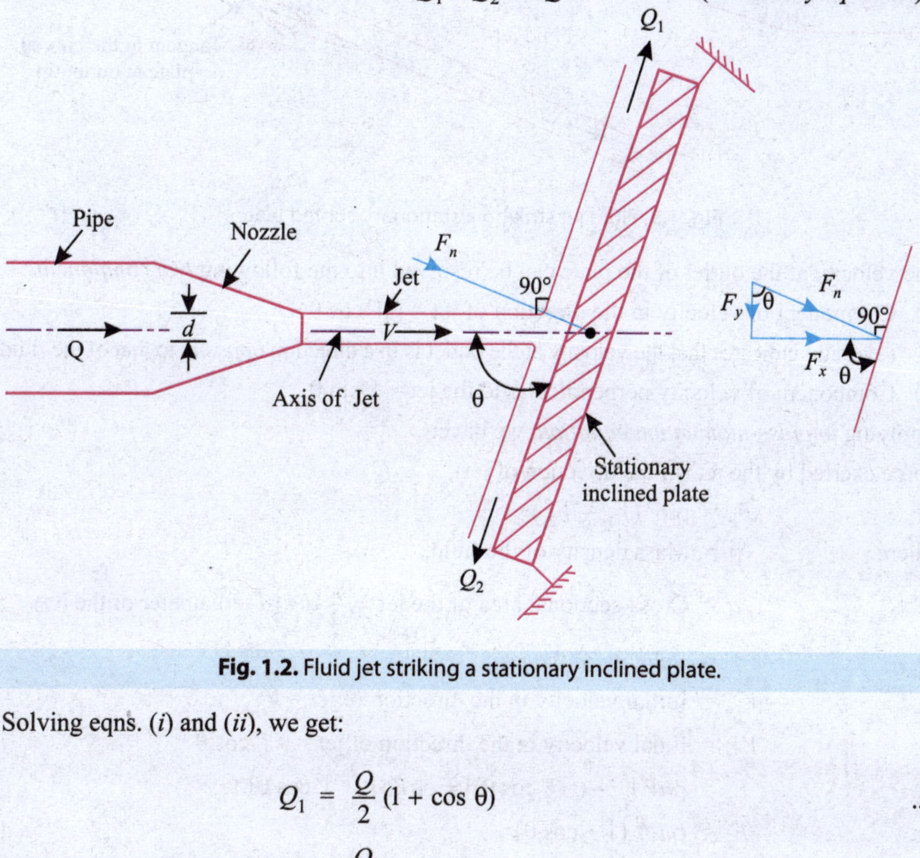

Fig. 1.2. Fluid jet striking a stationary inclined plate.

Solving eqns. (*i*) and (*ii*), we get:

$$Q_1 = \frac{Q}{2}(1 + \cos\theta) \quad \text{...(1.5)}$$

and, $$Q_2 = \frac{Q}{2}(1 - \cos\theta) \quad \text{...(1.6)}$$

Ratio of discharges, $$\frac{Q_1}{Q_2} = \frac{1 + \cos\theta}{1 - \cos\theta} \quad \text{...(1.7)}$$

1.4. FORCE EXERTED ON STATIONARY CURVED PLATE

Case I. Jet strikes the curved plate at the centre:

Cosider a fluid jet striking a stationary curved plate (*smooth*) at the centre as shown in Fig. 1.3. The jet after striking the plate comes out with the same velocity, in the tangential direction of the curved plate.

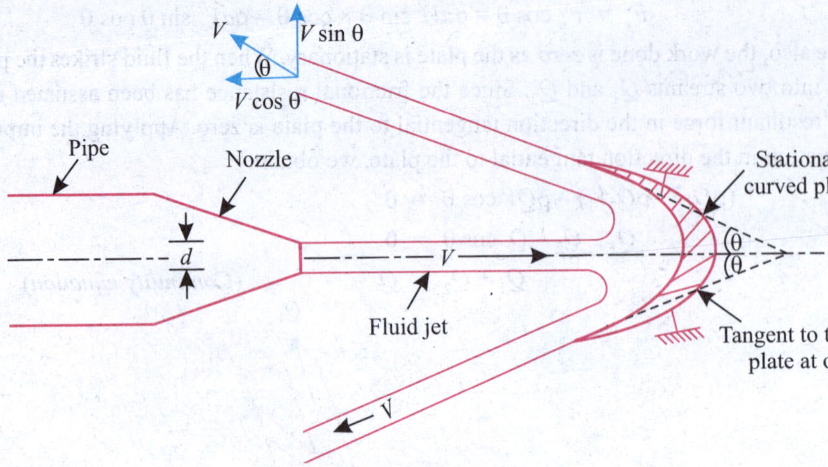

Fig. 1.3. Fluid jet striking a stationary curved plate.

The velocity at the outlet of the plate can be resolved into the following *two components*:

(*i*) Component of velocity in the direction of jet = $-V\cos\theta$

(–ve sign indicates that the velocity at the outlet is in a direction *opposite* to that of the fluid jet)

(*ii*) Component of velocity perpendicular to the jet = $V\sin\theta$

Applying *impulse-momentum equation*, we have:

Force exerted by the jet (in the direction of jet),

$$F_x = \rho a V(V_{1x} - V_{2x})$$

where, ρ = Mass density of the fluid,

a = Cross-sectional area of the jet $= \frac{\pi}{4}d^2$ (d = diameter of the jet),

V = Velocity of the jet,

V_{1x} = Initial velocity in the direction of jet = V

V_{2x} = Final velocity in the direction of jet = $-V\cos\theta$

$\therefore \quad F_x = \rho a V[V - (-V\cos\theta)] = \rho a V[V + V\cos\theta]$

or, $$F_x = \rho a V^2(1+\cos\theta) \qquad ...(1.8)$$

Similarly, $F_x = \rho a V(V_{1y} - V_{2y})$

where, V_{1y} = Initial velocity in the direction of $y = 0$

V_{2y} = Final velocity in the direction of $y = V\sin\theta$

$$\therefore \quad F_y = \rho a V(0 - V\sin\theta) = -\rho a V^2\sin\theta \qquad ...(1.9)$$

–ve sign indicates that force is acting in the *downward* direction.

Note: The angle of defection of the jet = (180° – θ).

Case II. Jet strikes the curved plate at one end tangentially when the plate is symmetrical:

Fig. 1.4 shows a fluid jet striking a stationary symmetrical curved plate (*smooth*) at one end tangentially, the plate is symmetrical about X-axis. Let V be the velocity of the jet and θ be the angle made the jet with X-axis at inlet tip of the curved plate. The velocity of fluid at outlet tip of the curved plate will be equal to V (since the curved plate is assumed smooth).

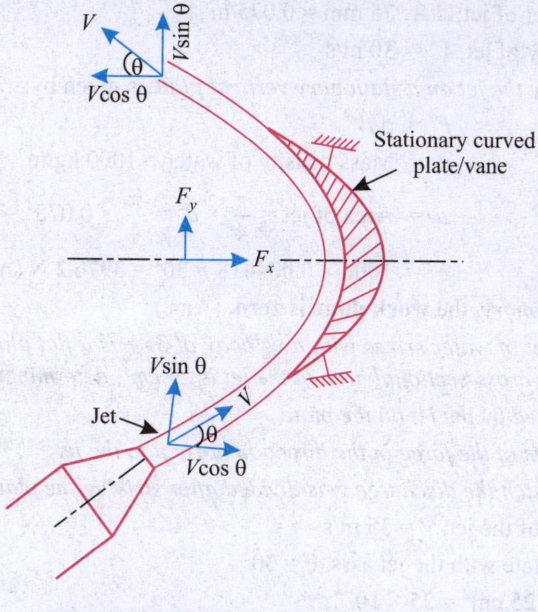

Fig. 1.4. Fluid jet striking stationary curved plate/vane.

The forces exerted by the fluid jet in X and Y directions are:

$$F_x = \rho a V(V_{1x} - V_{2x})$$
$$= \rho a V[V\cos\theta - (-V\cos\theta)]$$
$$= \rho a V(V\cos\theta + V\cos\theta) = \rho a V \times 2V\cos\theta$$

or,
$$F_x = 2\rho a V^2 \cos\theta \quad ...(1.10)$$
$$F_y = \rho a V(V_{1y} - V_{2y})$$
$$= \rho a V[V\sin\theta - V\sin\theta] = 0$$

i.e.,
$$F_y = 0$$

Case III. Jet strikes the curved plate or vane at one end tangentially when the plate is unsymmetrical:

In this case, as the plate is *unsymmetrical* about X-axis, therefore, the angles made by the tangents drawn at inlet and outlet tips of the plate with X-axis will be *different*. Let θ and ϕ be the angles made by the tangents at inlet tip and outlet tip respectively with X-axis.

∴ Components of velocity at inlet: $V_{1x} = V\cos\theta$; $V_{1y} = V\sin\theta$

Components of velocity at outlet: $V_{2x} = -V\cos\phi$; $V_{2y} = V\sin\phi$

Now, the forces exerted by the fluid jet in X and Y direction are:

$$F_x = \rho a V(V_{1x} - V_{2x})$$
$$F_x = \rho a V[V\cos\theta - (-V\cos\phi)] = \rho a V(V\cos\theta + V\cos\phi)$$

or
$$F_x = \rho a V^2(\cos\theta + \cos\phi) \quad ...(1.11)$$

and,
$$F_y = \rho a V(V_{1y} - V_{2y})$$
$$= \rho a V(V\sin\theta - V\sin\phi)$$

or
$$F_y = \rho a V^2(\sin\theta - \sin\phi) \quad ...(1.12)$$

Example 1.1. *A jet of water, 75 mm in diameter, issues with a velocity of 30 m/s and impinges on a stationary flat plate which destroys its forward motion. Flnd the force exerted by the jet on the plate and work done.*

Solution. Diameter of jet, $d = 75 \text{ mm} = 0.075 \text{ m}$

Velocity of jet, $V = 30 \text{ m/s}$

The *force exerted by the jet on a stationary vertical plate* is given by:

$$F = \rho a V^2$$

where, ρ = Mass density of water = 1000 kg/m^3

$$a = \text{Area of jet} = \frac{\pi}{4} \times d^2 = \frac{\pi}{4} \times 0.075^2 = 0.004418 \text{ m}^2$$

$$\therefore \quad F = 1000 \times 0.004418 \times 30^2 = \mathbf{3976.2 \text{ N}} \textbf{ (Ans.)}$$

As the plate is *stationary*, the **work done is zero. (Ans.)**

Example 1.2. *A jet of water srikes with a velocity of 35 m/s a flat plate inclined at 30° with the axis of the jet. If the cross-sectional area of the jet is 25 cm^2, determine:*

(i) The force exerted by the jet on the plate,

(ii) The components of the force in the direction normal to the jet

(iii) The ratio in which the discharge gets divided after striking the plate.

Solution. Velocity of the jet, $V = 35$ m/s

Inclination of the plate with the jet axis, $\theta = 30°$

Area of the jet, $a = 25 \text{ cm}^2 = 25 \times 10^{-4} \text{ m}^2$

(i) The force exerted by the jet, F :

$$F = \rho a V^2 \sin\theta \qquad \text{[Eqn. (1.3)]}$$

$$= 1000 \times (25 \times 10^{-4}) \times 35^2 \times \sin 30° = \mathbf{1531.25 \text{ N}} \textbf{ (Ans.)}$$

(ii) The components of the force, F :

$$F_x = F \sin\theta = 1531.25 \times \sin 30° = \mathbf{765.625 \text{ N}} \textbf{ (Ans.)}$$

$$F_y = F \cos\theta = 1531.25 \times \cos 30° = \mathbf{1326.1 \text{ N}} \textbf{ (Ans.)}$$

(iii) The ratio in which the discharge gets divided :

$$\frac{Q_1}{Q_2} = \frac{1 + \cos\theta}{1 - \cos\theta} \qquad \text{[Eqn. (1.7)]}$$

or

$$\frac{Q_1}{Q_2} = \frac{1 + \cos 30°}{1 - \cos 30°} = \frac{1 + 0.866}{1 - 0.866} = \mathbf{13.925} \textbf{ (Ans.)}$$

Example 1.3. *A jet of water of diameter 40 mm moving with a velocity of 30 m/s, strikes a curved fixed symmetrical plate at the centre. Find the force exerted by the jet water in the direction of the jet if the jet is deflected through an angle of 120 ° at the outlet of the curved plate.*

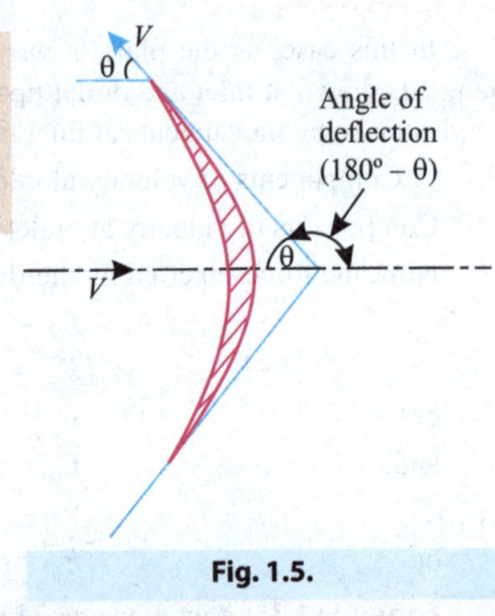

Fig. 1.5.

Solution. Diameter of jet, $d = 40 \text{ mm} = 0.04 \text{ m}$

$$\therefore \quad \text{Area of jet, } a = \frac{\pi}{4} \times d^2 = \frac{\pi}{4} \times 0.04^2$$

$$= 0.001256 \text{ m}^2$$

Velocity of jet, $V = 30$ m/s

Angle of deflection = 120°, or, $180 - \theta = 120°$

$$\therefore \quad \theta = 180° - 120° = 60°$$

Force exerted by the jet of water in the direction of jet,

$$F_x = \rho a V^2 (1 + \cos\theta) \qquad \text{[Eqn. (1.8)]}$$

$$= 1000 \times 0.001256 \times 30^2 (1 + \cos 60°)$$

$$= \mathbf{1695.6\ N\ (Ans.)}$$

Example 1.4. *A jet of water of 20 mm diameter and moving at 15 m/s, strikes upon the centre of a symmetrical vane. After impingement, the jet gets deflected through 160° by the vane. Presuming vane to be smooth determine:*

(i) The force exerted by jet on the vane, and

(ii) The ratio of velocity at outlet to that at inlet if actual reaction of the vane is 127 N.

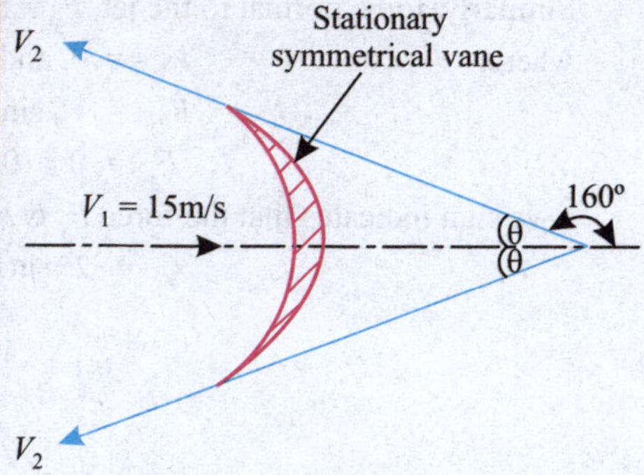

Fig. 1.6.

Solution. Diameter of the jet, $d = 20$ mm $= 0.02$ m

Velocity of jets, V = 15 m/s

Angle of deflection = 160°.

(i) The force exerted by the jet on the vane, F :

Refer to Fig 1.6.

$$160° = 180 - \theta, \text{ or, } \theta = 180 - 160 = 20°$$

For *smooth vane*, the theoretical force (or thrust) exerted by the jet on the vane is,

$$F = \rho a V^2 (1 + \cos\theta) \quad ...(i)$$

$$= 1000 \times \left(\frac{\pi}{4} \times 0.02^2\right) \times 15^2 (1 + \cos 20°)$$

$$= \mathbf{137.1\ N\ (Ans.)}$$

(ii) $\dfrac{V_2}{V_1}$:

Actual reaction of the vane = 127 N (*Given*)

If the vane is *not smooth*, then outgoing velocity at the vane tip is less than the incoming velocity, *i.e.*, $\dfrac{V_2}{V_1} = K$ where $K < 1$. The eqn. (*i*) gets modified to

$$F = \rho a V^2 (1 + K\cos\theta)$$

$$127 = 1000 \times \left(\frac{\pi}{4} \times 0.02^2\right) \times 15^2 (1 + K\cos 20°)$$

or, $$1 + K\cos 20° = \frac{127}{1000 \times \left(\frac{\pi}{4} \times 0.02^2\right) \times 15^2} = 1.796$$

or, $$K = \frac{1.796 - 1}{\cos 20°} = \mathbf{0.847\ (Ans.)}$$

Example 1.5. *A jet of water from a nozzle is deflected through 60° from its original direction by curved plate which it enters tangentially without shock with a velocity of 30 m/s and leaves with a mean velocity of 25 m/s. If the discharge form the nozzle is 0.8 kg/s, calculate the magnitude and direction of the resultant force on the vane, if the vane is stationary.* **[UPTU]**

Solution. Velocity at inlet, $V_1 = 30$ m/s

Velocity at outlet, V_2 = 25 m/s

Mass per second = 0.8 kg/s

Force in the direction of jet, F_x = Mass per sec. × $(V_{1x} - V_{2x})$

where, $V_{1x} = V_1 = 30$ m/s

$$V_{2x} = V_2 \cos 60° = 25 \cos 60° = 12.5 \text{ m/s}$$

$$\therefore \quad F_x = 0.8\,(30 - 12.5) = 14 \text{ N}$$

Similarly, force normal to the jet, F_y = mass per sec. × $(V_{1y} - V_{2x})$

where, $V_{1y} = 0$, and

$$V_{2y} = V_2 \sin 60° = 25 \sin 60° = 21.65 \text{ N}$$

$$\therefore \quad F_y = 0.8\,(0 - 21.65) = -17.32 \text{ N}$$

(–ve sign indicates that the force F_y *is acting in the vertically downward direction*)

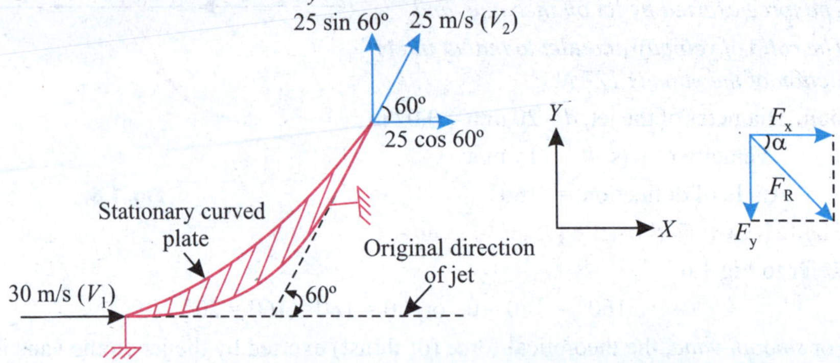

Fig. 1.7.

∴ Resultant force in the curved plate,

$$F_R = \sqrt{F_x^2 + F_y^2} = \sqrt{14^2 + 17.32^2} = \mathbf{22.27\ N\ (Ans.)}$$

The angle made by the resultant force with X-axis (Refer to Fig. 1.7),

$$\tan \alpha = \frac{F_y}{F_x} = \frac{17.32}{14} = 1.237$$

$$\therefore \quad \alpha = \tan^{-1} 1.237 \simeq \mathbf{51°\ (Ans.)}$$

Example 1.6. *A jet of water of diameter 60 mm moving with a velocity of 40 m/s, strikes a curved fixed plate tangentially at one end at an angle of 30° to horizontal. The jet leaves the plate at an angle of 20° to the horizontal. Find the force exerted by the jet on the plate in the horizontal and vertical directions.*

Solution. Diameter of jet, d = 60 mm = 0.06 m

$$\therefore \quad \text{Area, } a = \frac{\pi}{4} \times d^2 = \frac{\pi}{4} \times 0.06^2 = 0.002827 \text{ m}^2$$

Velocity of the jet, V = 40 m/s

Angle made by the jet at inlet tip with horizontal, $\theta = 30°$

Angle made by the jet at outlet tip with horizontal, $\phi = 20°$.

Force exerted by the jet :

Force exerted by the jet in X-direction,

$$F_x = \rho a V^2 (\cos\theta + \cos\phi) \qquad \text{[Eqn. (1.11)]}$$

$$= 1000 \times 0.002827 \times 40^2 (\cos 30° + \cos 20°) = \mathbf{8167.6\ N\ (Ans.)}$$

Force exerted by the jet in Y-direction,

$$F_y = \rho a V^2 (\sin\theta - \sin\phi) \qquad \text{[Eqn. (1.12)]}$$

$$= 1000 \times 0.002827 \times 40^2 (\sin 30° - \sin 20°) = \mathbf{714.57\ N\ (Ans.)}$$

Example 1.7. *A rectangular plate, weighing 60 N is suspended vertically by a hinge on the top horizontal edge. The centre of gravity of the plate is 100 mm from the hinge. A horizontal jet of* *water 20 mm diameter, whose axis is 150 mm below the hinge impinges normally on the plate with a velocity of 5 m/s. Determine:*

(i) The horizontal force applied at the centre of gravity to maintain the plate in its vertical portion.

(ii) The corresponding velocity of the jet, if the plate is deflected through 30° and the same force continues to act at the centre of gravity of the plate. **[Delhi University]**

Solution. Weight of plate, $W = 60$ N

Distance of weight W from hinge, $x = 100$ mm $= 0.1$ m

Diameter of the jet, $d = 20$ mm $= 0.02$ m

$$\therefore \quad \text{Area, } a = \frac{\pi}{4} \times d^2 = \frac{\pi}{4} \times (0.02)^2 = 0.000314 \text{ m}^2$$

Distance of the axis of the jet from hinge = 150 mm = 0.15 m

Velocity of jet, $V = 15$ m/s

Refer to Fig. 1.8 for the plate and its orientation in the vertical and deflected positions.

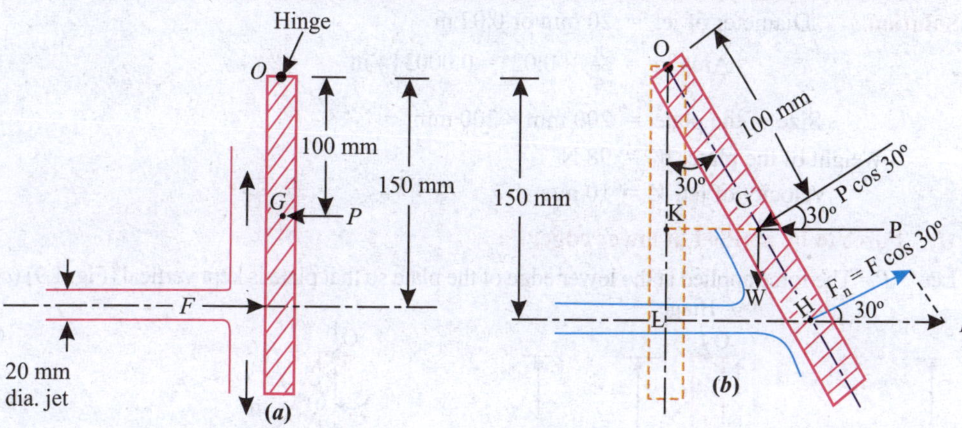

Fig. 1.8.

(i) Horizontal force, P :

Normal force exerted by the jet on plate,

$$F = \rho a V^2 = 1000 \times 0.000314 \times 5^2 = 7.85 \text{ N}$$

Let P be the force to be applied at the centre of gravity G [Fig. 1.8 (a)] to keep the plate in its vertical position. Then taking moments about the hinge point O, we get:

$$F \times 150 = P \times 100$$

$$\text{or,} \quad P = \frac{F \times 150}{100} = \frac{7.85 \times 150}{100} = \mathbf{11.77 \ (Ans.)}$$

(ii) Velocity of the jet :

Let, V = Velocity of the jet, if the plate is deflected through 30° and the same force continues to act at the centre of gravity of the plate.

Refer to Fig. 1.8. (b). Taking moments about the hinge point O, we get

$$W \times GK + P \cos 30° \times OG = F \cos 30° \times OH \quad ...(i)$$

But, $GK = 0.1 \sin 30° = 0.05$ m, $OG = 0.1$ m; $OH = \dfrac{OL}{\cos 30°} = \dfrac{0.15}{\cos 30°} = 0.1732$ m

Substituting the various values, we get

$$60 \times 0.05 + 11.77 \times 0.866 \times 0.1 = F \times 0.1732$$

$$3 + 1.02 = 0.15\,F$$

$$\therefore \quad F = 26.8 \text{ N}$$

Now, jet force, $F = \rho a V^2$, or, $26.8 = 1000 \times 0.000314 \times V^2$

$$\therefore \quad V = \left(\frac{26.8}{1000 \times 0.000314}\right)^{1/2} = \mathbf{9.24 \text{ m/s}} \textbf{ (Ans.)}$$

Example 1.8. *A jet of water of diameter 20 mm strikes a 200 mm × 200 mm square plate of uniform thickness with a velocity of 10 m/s at the centre of the plate which is suspended vertically by a hinge on its top horizontal edge. The weight of the plate is 98 N. The jet strikes normal to the plate.*

(i) What force must be applied at the lower edge of the plate so that plate is kept vertical ?

(ii) If the plate is allowed to deflect freely, what will be the inclination of the plate with vertical due to the force exerted by jet water?

Solution. Diameter of jet = 20 mm or 0.02 m

$$\therefore \quad \text{Area, } a = \frac{\pi}{4} \times 0.02^2 = 0.000314 \text{ m}^2$$

Size of the plate = 200 mm × 200 mm

Weight of the plate, W = 98 N

Velocity of jet, V = 10 m/s

(*i*) Force to be applied at lower edge, P :

Let, P = The force applied at the lower edge of the plate so that plate is kept vertical [Fig. 1.9) (*a*)].

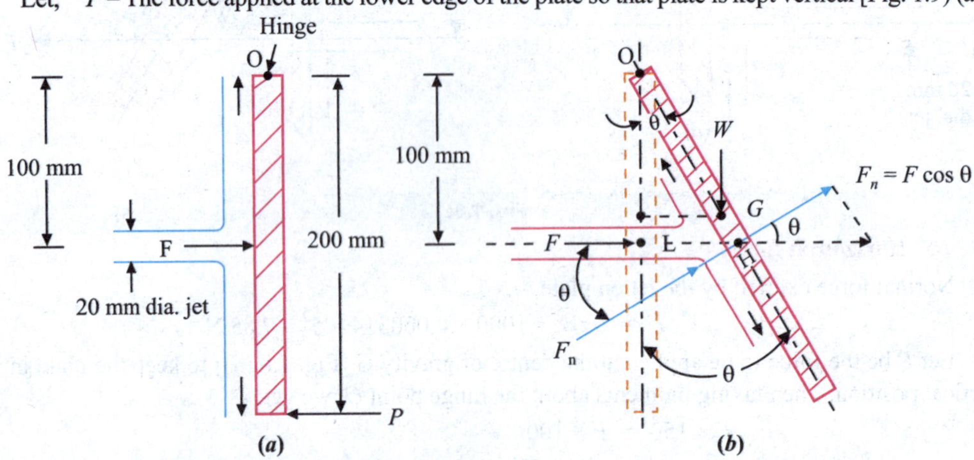

Fig. 1.9.

Force exerted by the jet on the plate,

$$F = \rho a V^2 = 1000 \times 0.000314 \times 10^2 = 31.4 \text{ N}$$

Now, taking moments about the hinge point O, we get

$$P \times 200 = F \times 100$$

or,

$$P = \frac{F \times 100}{200} = \frac{31.4 \times 100}{200} = \mathbf{15.7 \text{ N}} \textbf{ (Ans.)}$$

(*ii*) Inclination of the plate :

Let, θ = Inclination of the plate with vertical due to force exerted by the jet [Fig. 1.9 (*b*)].

Force exerted by water normal to the plate is given by:

$$F_n = F \cos\theta = \rho a V^2 \cos\theta = 31.4 \cos\theta$$

[$\because$ $\rho a V^2 = 31.4$ N, calculated earlier]

Taking moments about the hinge point O, we get:

$$F_n \times OH = W \times GK \quad ...(i)$$

where, $OH = \dfrac{OL}{\cos\theta} = \dfrac{0.1}{\cos\theta}$ m, $GK = 0.1 \sin\theta$ m

Substituting the values in eqn. (*i*), we have

$$31.4 \cos\theta \times \frac{0.1}{\cos\theta} = 98 \times 0.1 \sin\theta$$

$$\therefore \quad \sin\theta = \frac{31.4}{98} = 0.32$$

Hence, $\theta = \sin^{-1} 0.32 =$ **18.66° (Ans.)**

B. FORCE EXERTED BY THE JET ON THE MOVING PLATE

1.5. FORCE EXERTED ON MOVING FLAT PLATE HELD NORMAL TO JET

Fig. 1.10 shows a fluid jet striking a flat vertical plate moving with a uniform velocity away from the jet.

Let, V = Absolute velocity of the jet,

a = Cross-sectional area of the jet, and

u = Velocity of the flat plate held normal to the jet.

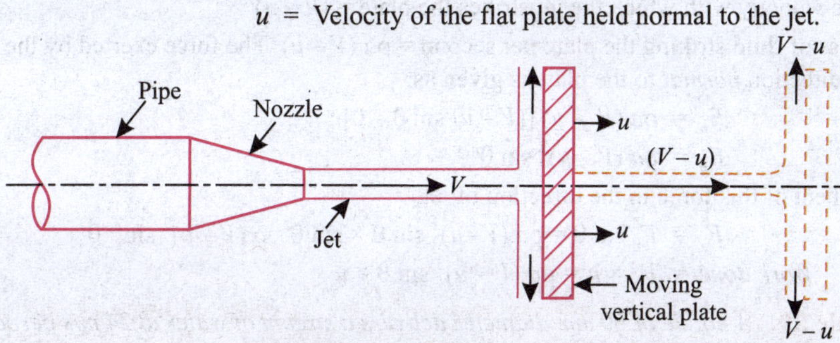

Fig. 1.10. Fluid jet striking a moving plate.

The relative velocity with which the jet strikes the plate is $(V - u)$.

Mass of water striking the plate per second $= \rho a (V - u)$

$\therefore$ Force exerted by the jet on the plate in the direction of jet,

F_x = Mass of water striking the plate/sec × (initial velocity with which water strikes–final velocity)

$= \rho a (V - u) [(V - u) - 0]$ ($\because$ Final velocity in the diction of jet = 0)

or, $$F_x = \rho a (V - u)^2 \quad ...(1.13)$$

The work done = Force × the distance through which the body moves in the direction of force.

$$\therefore \quad \textit{Work done} = \rho a (V - u)^2 \times u \quad ...(1.14)$$

Note : This case is not practically feasible because the distance between the nozzle and the plate will go on increasing. However, if a series of plates is fitted on the wheel, there is always a plate facing the jet. Thus, the entire fluid issuing from the jet strikes the plates.

1.6. FORCE EXERTED ON A MOVING PLATE HELD INCLINED TO THE DIRECTION OF JET

Fig. 1.11 shows a fluid jet striking an inclined plate, which is moving with uniform velocity in the direction of the jet.

Let, V = Absolute velocity of the jet,

u = Velocity of plate in the direction of jet

a = Cross-sectional area of jet, and

θ = Angle between jet and the plate.

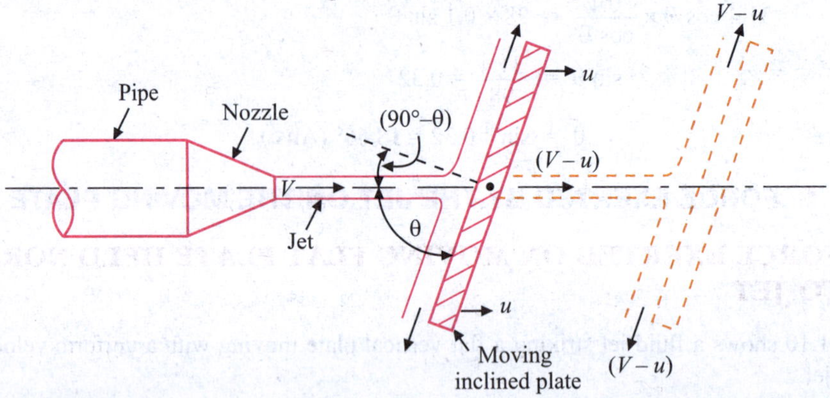

Fig. 1.11. Fluid jet striking a moving plate held inclined to the direction of jet.

Relative velocity with which the jet strikes the plate = $(V - u)$

The mass of fluid striking the plate per second = $\rho a\,(V - u)$. The force exerted by the jet on the plate in the direction *normal* to the plate is given as:

$$F_n = \rho a\,(V - u)\,[(V - u)\sin\theta - 0]$$

or, $$F_n = \rho a\,(V - u)^2 \sin\theta \qquad ...(1.15)$$

Component of this force in the direction of jet,

$$F_x = F_n \sin\theta = \rho a\,(V - u)^2 \sin\theta \times \sin\theta = \rho\,(V - u)^2 \sin^2\theta \qquad ...(1.16)$$

$$\therefore \quad \textit{Work done} = F_x \times u = \rho a\,(V - u)^2 \sin\theta \times u \qquad ...(1.17)$$

Example 1.9. *A nozzle of 60 mm diameter delivers a stream of water at 24 m/s perpendicular to a plate that moves away from the jet at 6 m/s.*

Find: (i) The force on the plate,

(ii) The work done, and

(iii) The efficiency of the jet.

Solution. Diameter of the nozzle/jet, d = 60 mm = 0.06 m

Velocity of water, V = 24 m/s

Velocity of the plate, u = 6 m/s

(*i*) The force on the plate, F :

$$F = \rho a\,(V - u)^2 \qquad \text{[Eqn. (1.13)]}$$

$$= 1000 \times \left(\frac{\pi}{4} \times 0.06^2\right) \times (24 - 6)^2 = \mathbf{916\ N\ (Ans.)}$$

(*ii*) The work done :

$$\textit{Work done} = F \times u = 916 \times 6 = \mathbf{5496\ Nm/s\ (Ans.)}$$

(iii) The efficiency of jet, η_{jet} :

$$\text{Kinetic energy of issuing jet} = \frac{1}{2} mV^2 = \frac{1}{2}(\rho aV) \times V^2 \qquad (\because \text{ Mass, } m = \rho aV)$$

$$= \frac{1}{2}\left(1000 \times \frac{\pi}{4} \times 0.06^2 \times 24\right) \times 24^2 = 19543.2 \text{ Nm/s}$$

$$\therefore \quad \eta_{jet} = \frac{\text{Work done}}{\text{Kinetic energy of issuing jet}}$$

$$= \frac{5496}{19543.2} = 0{\cdot}281 \text{ or } \mathbf{28.1\%} \textbf{ (Ans.)}$$

Example 1.10. *A 75 mm diameter jet having a velocity of 30 m/s strikes a flat plate, the normal of which is inclined at 45° to the axis of the jet. Find the normal pressure on the plate,*

(i) When the plate is stationary;

(ii) When the plate is moving with a velocity of 15 m/s in the direction of jet, away from the jet.

Also determine the power and efficiency of the jet when the plate is moving.

[Allahabad University]

Solution. Diameter of the jet, $d = 75 \text{ mm} = 0.075 \text{ m}$

$$\therefore \quad \text{Area, } a = \frac{\pi}{4} \times 0.075^2 = 0.004418 \text{ m}^2$$

Angle between the jet and the plate, $\theta = 90° - 45° = 45°$

Velocity of the jet, $V = 30 \text{ m/s}$

(i) *When the plate is stationary,* the normal force on the plate is:

$$F_n = \rho a V^2 \sin\theta \qquad \text{[Eqn. (1.2)]}$$

$$= 1000 \times 0.004418 \times 30^2 \times \sin 45° = \mathbf{2811.6 \text{ N}} \textbf{ (Ans.)}$$

(ii) *When the plate is moving with a velocity of 15 m/s and moving away from the jet,* the normal force on the plate is

$$F_n = \rho a (V - u)^2 \sin\theta \quad (\text{where, } u = 15 \text{ m/s}) \qquad \text{[Eqn. 1.15]}$$

$$= 1000 \times 0.004418 (30 - 15)^2 \times \sin 45° = \mathbf{702.9 \text{ N}} \textbf{ (Ans.)}$$

Work done per second by the jet

= Force in the direction of jet × distance moved by the plate in the direction of jet/sec.

$$= F_n \times u$$

where, $F_x = F_n \sin\theta = 702.9 \times \sin 45° = 497 \text{ N}$

$$\therefore \quad \text{Work done} = 497 \times 15 = 7455 \text{ Nm/s}$$

$\therefore$ *Power of the jet* $= 7455 \text{ J/s} = 7455 \text{ W} = \mathbf{7.455 \text{ kW}}$ **(Ans.)**

$$\textit{Efficiency of the jet} = \frac{\text{Work done on the plate}}{\text{Kinetic energy supplied by the jet}}$$

$$= \frac{7455}{\frac{1}{2}(\rho aV) \times V^2} = \frac{7455}{\frac{1}{2} \times (1000 \times 0{\cdot}004418 \times 30) \times 30^2}$$

$$= 0.125 \text{ or } \mathbf{12.5\%} \textbf{ (Ans.)}$$

Example 1.11. *A 75 mm diameter water jet having a velocity of 12 m/s impinges on a plane, smooth plate at an angle of 60° to the normal to the plate. What will be the impact when (i) the plate is stationary, and (ii) the plate is moving in the direction of the jet at 6 m/s? Estimate the work done per unit time by the jet on the plate in each case. Take the density of water as 998 kg/m³.* **[N.U.]**

Solution. Given: $d = 75 \text{ mm} = 0.075 \text{ m}$; $V = 12 \text{ m/s}$; $\theta = 90 - 60° = 30°$; $u = 6 \text{ m/s}$; $\rho = 998 \text{ kg/m}^3$.

Force on the plate and work done :

(*i*) *When the plate is stationary :*

Force on the plate, $F = (\rho a V) V \sin 30°$

$$= \left[998 \times \frac{\pi}{4} \times (0.075)^2 \times 12 \times 12 \sin 30°\right] = \mathbf{317.45\ N}\ \textbf{(Ans.)}$$

Work done = **0**, since the plate is stationary **(Ans.)**

(*ii*) *When the plate is moving :*

Normal force on the plate, $F_n = [\rho a (V - u)] \times (V - u) \sin 30°$

$= \rho a (V - u)^2 \sin 30° = 998 \times \frac{\pi}{4} \times (0.075)^2 \times (12 - 6)^2 \times 0.5 = \mathbf{79.36\ N}$ **(Ans.)**

Work done on the plate $= F_n \sin \theta \times u$

$= 79.36 \sin 30° \times 6 = \mathbf{238.08\ Nm/s}$ **(Ans.)**

Example 1.12. (*a*) *Derive the following expression for force F exerted by a jet of area 'a' which strikes a flat plate at an angle θ to the normal to the plate with velocity V. The plate itself is moving with velocity u in the direction of normal to the plate surface.*

$$F = \rho a \frac{(V \cos\theta - u)^2}{\cos\theta}$$

(*b*) *A 40 mm diameter jet having a velocity of 20 m/s strikes a flat plate, the normal of which is inclined at 30° to the axis of jet. If the plate itself is moving with a velocity of 8 m/s parallel to itself and in the direction of normal to its surface, calculate:*

(*i*) *Normal force exerted on the plate,*

(*ii*) *Work done per second, and*

(*iii*) *Efficiency of the jet* **[Punjab University]**

Solution. (*a*) Fig 1.12 shows the orientation of the jet and the arrangement of the plate.

Component of plate velocity in the direction of jet $= \dfrac{u}{\cos\theta}$

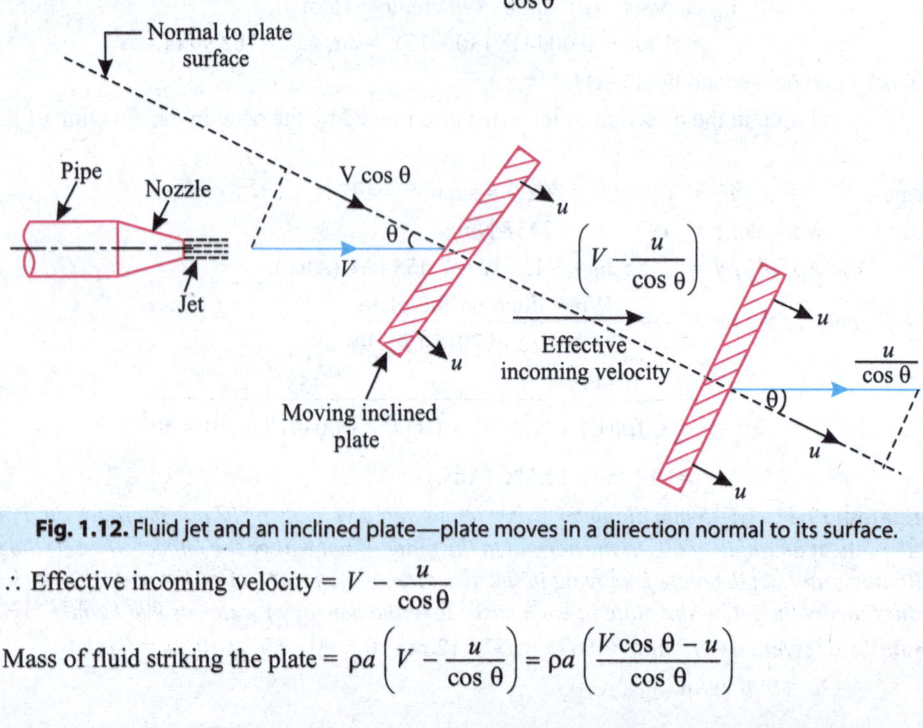

Fig. 1.12. Fluid jet and an inclined plate—plate moves in a direction normal to its surface.

∴ Effective incoming velocity $= V - \dfrac{u}{\cos\theta}$

Mass of fluid striking the plate $= \rho a \left(V - \dfrac{u}{\cos\theta}\right) = \rho a \left(\dfrac{V \cos\theta - u}{\cos\theta}\right)$

Component of absolute velocity of jet *normal* to plate = $V \cos\theta$

∴ Change of velocity *normal* to plate = $(V \cos\theta - u)$

Force exerted on the plate (F) = Mass × change of velocity

or, $$F = \rho a \left(\frac{V \cos\theta - u}{\cos\theta}\right) \times (V \cos\theta - u)$$

or, $$F = \rho a \frac{(V \cos\theta - u)^2}{\cos\theta}$$ **Proved.**

(*b*) Diameter of the jet, $d = 40$ mm = 0.04 m

∴ $$\text{Area, } a = \frac{\pi}{4} \times d^2 = \frac{\pi}{4} \times 0.04^2 = 0.001256 \text{ m}^2$$

Velocity of jet (absolute), $V = 20$ m/s

Velocity of the plate, $u = 8$ m/s

The inclination of the normal to the plate with the axis of jet, $\theta = 30°$

(*i*) Normal force exerted on the plate, F :

$$F = \rho a \frac{(V \cos\theta - u)^2}{\cos\theta} = 1000 \times 0{\cdot}001256 \times \frac{(20 \cos 30° - 8)^2}{\cos 30°} = \mathbf{125{\cdot}98\ N}$$ **(Ans.)**

(*ii*) Work done per second :

$$\text{Work done/sec} = \text{Normal force exerted on the plate} \times \text{distance moved by the plate}$$
$$= 125.98 \times 8 = \mathbf{1007.84\ Nm/s}$$ **(Ans.)**

(*iii*) Efficiency of the jet, η_{jet} :

$$\eta_{jet} = \frac{\text{Work done}}{\text{Kinetic energy of issuing jet}} = \frac{1007.84}{\frac{1}{2} mV^2}$$

$$= \frac{1007.84}{\frac{1}{2}(\rho a V)V^2} = \frac{1007.84}{\frac{1}{2}(1000 \times 0.001256 \times 20) \times 20^2} = 0.20 \text{ or } \mathbf{20\%}$$ **(Ans.)**

1.7. FORCE EXERTED ON A CURVED PLATE (OR VANE) WHEN THE PLATE (OR VANE) IS MOVING IN THE DIRECTION OF JET

A. Single vane:

Fig. 1.13 shows a fluid jet striking at the centre of curved vane moving with a uniform velocity in the direction of jet.

Let, V = Absolute velocity of the jet,

a = Area of jet, and

u = Velocity of the plate in the direction of the jet.

When the jet strikes the moving vane, the effective velocity is the relative velocity $(V - u)$. The component of this velocity [*i.e.*, $(V - u)$] in the direction of jet

$$= -(V - u) \cos\theta$$

(– ve sign indicates that the component is in the direction *opposite* to that of the jet.)

Mass of fluid striking the plate = $\rho a (V - u)$

∴ Force exerted by the jet on the vane,

F_x = Mass striking per sec × (initial velocity with which the jet strikes the vane in the direction of jet – final velocity)

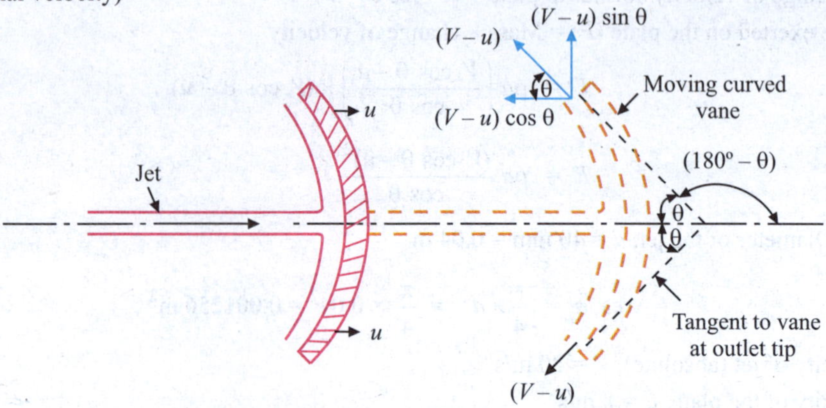

Fig. 1.13. Jet striking a curved moving vane.

$$= \rho a (V-u)\,[(V-u) - \{-(V-u)\cos\theta\}]$$

$$= \rho a (V-u)\,[(V-u) + (V-u)\cos\theta]$$

or, $$F_x = \rho a (V-u)^2 (1+\cos\theta) \qquad ...(1.18)$$

Work done on the vane per second

$$= F_x \times u = \rho a (V-u)^2 (1+\cos\theta) \times u \qquad ...(1.19)$$

Kinetic energy of issuing jet $= \frac{1}{2} mV^2 = \frac{1}{2}(\rho a V)\, V^2 = \frac{1}{2}\rho a V^3$

$$\text{Efficiency, } \eta = \frac{\text{Work done}}{\text{Kinetic energy supplied by the jet}}$$

$$= \frac{\rho a (V-u)^2 (1+\cos\theta) \times u}{\frac{1}{2}\rho a V^3}$$

or, $$\eta = \frac{2(V-u)^2 (1+\cos\theta)\, u}{V^3} \qquad ...(1.20)$$

For a given jet velocity, the efficiency will be *maximum* when

$$\frac{d\eta}{du} = 0$$

or, $$\frac{d}{du}\left[\frac{2(V-u)^2 (1+\cos\theta)\, u}{V^3}\right] = 0$$

or, $$\frac{2}{V^3}(1+\cos\theta)\frac{d}{du}[(V-u)^2 \times u] = 0$$

or, $$\frac{2}{V^3}(1+\cos\theta)\frac{d}{du}[(V^2+u^2-2Vu)u] = 0$$

or, $$\frac{2}{V^3}(1+\cos\theta)\frac{d}{du}(V^2u+u^3-2Vu^2) = 0$$

or, $$\frac{2}{V^3}(1+\cos\theta)\,(V^2+3u^2-4Vu) = 0$$

For a given system $\frac{2}{V^3}(1 + \cos\theta) \neq 0$

$$\therefore \quad V^2 + 3u^2 - 4Vu = 0$$

$$\text{or,} \quad (V - u)(V - 3u) = 0$$

$$\text{i.e.,} \quad u = V \text{ or } u = \frac{V}{3} \quad \text{...(1.21)}$$

When $u = V$, the work done becomes zero. Thus, for maximum efficiency $u = V/3$, *i.e., vane velocity is 1/3 rd of the jet velocity.*

B. Series of vanes:

In this case, there is always one vane or the other facing the jet (e.g., curved vanes mounted equidistantly around the periphery of a wheel) and the entire fluid is utilized. The mass of fluid striking the vanes per second equals ρaV.

Thrust exerted by the jet on the vane,

$$F_x = \rho aV[(V - u) - \{-(V - u)\cos\theta)\}]$$

$$= \rho aV[(V - u) + (V - u)\cos\theta)]$$

$$\text{or,} \quad F_x = \rho aV(V - u)(1 + \cos\theta) \quad \text{...(1.22)}$$

$$\text{Work done per second} = F_x \times u$$

$$= \rho aV[(V - u)(1 + \cos\theta) \times u \quad \text{...(1.23)}$$

Kinetic energy supplied by the jet

$$= \frac{1}{2}(\rho aV)V^2 = \frac{1}{2}\rho aV^3$$

$$\therefore \quad \text{Efficiency, } \eta = \frac{\text{Work done on the wheel}}{\text{Kinetic energy supplied by the jet}}$$

$$= \frac{\rho aV\ (V - u)\ (1 + \cos\theta) \times u}{\frac{1}{2}\rho aV^3}$$

$$\text{or,} \quad \eta = \frac{2u\ (V - u)\ (1 + \cos\theta)}{V^2} \quad \text{...(1.24)}$$

The efficiency is *maximum* if

$$\frac{d\eta}{du} = 0$$

$$\text{or,} \quad \frac{d}{du}\left[\frac{2u\ (V - u)\ (1 + \cos\theta)}{V^2}\right] = 0$$

$$\text{or,} \quad \frac{2\ (1 + \cos\theta)}{V^2} \times \frac{d}{du}(uV - u^2) = 0$$

$$\text{or,} \quad \frac{2\ (1 + \cos\theta)}{V^2}(V - 2u) = 0$$

For a given system, $\frac{2\ (1 + \cos\theta)}{V^2} \neq 0$ and therefore

$$V - 2u = 0 \text{ or } u = \frac{V}{2} \quad \text{...(1.25)}$$

Hence for the *maximum efficiency of the wheel, the peripheral speed (u) is one-half the velocity of jet.*

$$\text{Maximum efficiency, } \eta_{max} = \frac{2\left(\frac{V}{2}\right)\left(V - \frac{V}{2}\right)(1 + \cos\theta)}{V^2}$$

$$\left[\text{Substituting } u = \frac{V}{2} \text{ in eqn. (1.24)}\right]$$

or,
$$\eta_{max} = \frac{1 + \cos\theta}{2} \qquad \text{...(1.26)}$$

When $\theta = 0$, *i.e.* the vanes are *semicircular,* $\eta_{max} = \frac{1+1}{2} = 1$ or 100%. Thus, the theoretical maximum efficiency of the wheel is 100%. This concept (of having vanes of semicircular configuration) is utilized in the design of buckets for Pelton wheel. The buckets are, however not exactly semicircular. The angle of deflection is limited to 160° – 170° depending upon a particular design. This *ensures that the jet coming out of a bucket does not interfere with the jet striking the bucket.*

Example 1.13. *A jet of water, 60 mm in diameter, strikes a curved vane at its centre with a velocity of 18 m/s. The curved vane is moving with a velocity of 6 m/s in the direction of the jet. The jet is deflected through an angle of 165°. Assuming the plate to be smooth, find:*

(i) Thrust on the plate in the direction of jet,

(ii) Power of the jet, and

(iii) Efficiency of the jet.

Solution. Diameter of the jet, $d = 60 \text{ mm} = 0.06 \text{ m}$

$$\therefore \quad \text{Area, } a = \frac{\pi}{4} \times d^2 = \frac{\pi}{4} \times 0.06^2 = 0.002827 \text{ m}^2$$

Velocity of the jet, $V = 18$ m/s

Velocity of the vane, $u = 6$ m/s

Angle of deflection of the jet = 165°

∴ Angle made by the relative velocity at the outlet of the vane,

$$\theta = 180 - 165 = 15°$$

(i) Thrust on the plate :

Thrust on the plate in the direction of the jet,

$$F_x = \rho a (V - u)^2 (1 + \cos\theta) \qquad \text{...[Eqn. 1.18]}$$

$$= 1000 \times 0.002827 (18 - 6)^2 (1 + \cos 15°) = \mathbf{800.3\ N\ (Ans.)}$$

(ii) Power of the jet :

Work done by the jet on the vane per second

$$= F_x \times u = 800.3 \times 6 = 4801.8 \text{ Nm/s}$$

$$\therefore \quad \textit{Power of the jet} = 4801.8 \text{ J/s} = 4801.8 \text{ W} \simeq \mathbf{4.8\ kW\ (Ans.)}$$

(iii) Efficiency of the jet, η_{jet} :

$$\eta_{jet} = \frac{\text{Work done by the jet/sec.}}{\text{Kinetic energy of the jet/sec.}}$$

$$= \frac{4801.8}{\frac{1}{2}(\rho a V) \times V^2} = \frac{4801.8}{\frac{1}{2}(1000 \times 0.002827 \times 18) \times 18^2} = 0.5825 \text{ or } \mathbf{58.25\ \%\ (Ans.)}$$

Example 1.14. **(*a*)** *A stationary vane having an inlet angle of zero degree and an outlet angle of 25° receives water at velocity of 50 m/s. Determine the components of force acting on it in the direction of the jet velocity and normal to it. Also find the resultant force in magnitude and direction per kg of flow.*

(*b*) *If the vane stated above is moving with a velocity of 20 m/s in the direction of the jet, calculate the force components in the direction of the vane velocity and across it, also the resultant force in magnitude and direction. Calculate the work done and power developed per kg of flow.*

[UPTU]

Solution.

(*a*) Stationary vane :

Inlet angle of the vane = 0°

Outlet angle of the vane = 25°

Velocity of water jet, V = 50 m/s

The force in the direction of jet per kg of flow,

$$F_x = 1 \times (V_{1x} - V_{2x})$$

where, V_{1x} = 50 m/s

$$V_{2x} = -V\cos 25° = -50 \cos 25° = -45.315 \text{ m/s}$$

$$\therefore \quad F_x = 1 \times [50 - (-45.315)] = 95.315 \text{ N}$$

Similarly, the force exerted by the jet in the direction perpendicular to the direction of the jet per kg of flow,

$$F_y = 1 \times (V_{1y} - V_{2y})$$

where, V_{1y} = 0, and

$$V_{2y} = V\sin 25° = 50 \sin 25° = 21.13 \text{ m/s}$$

$$\therefore \quad F_y = 1\,(0 - 21.13) = -21.13 \text{ N}$$

(– ve sign indicates that F_y is acting in the *downward direction.*)

∴ Resulting force per N of water (Refer to Fig. 1.14),

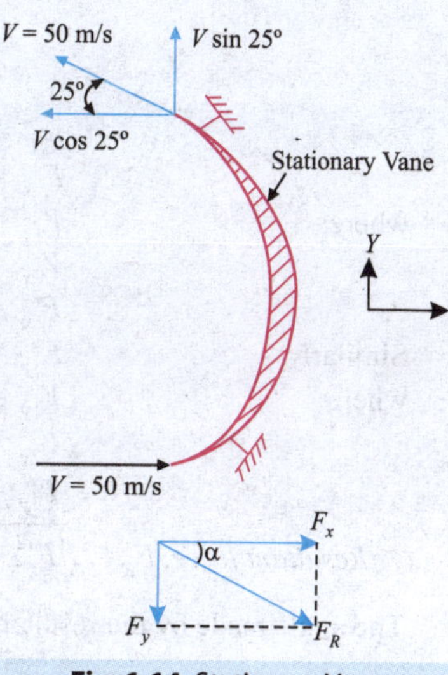

Fig. 1.14. Stationary Vane.

$$F_R = \sqrt{F_x^2 + F_y^2} = \sqrt{95.315^2 + 21.13^2} = \mathbf{97.63\ N\ (Ans.)}$$

The angle made by the resultant with X-axis,

$$\tan\alpha = \frac{F_y}{F_x} = \frac{21.13}{95.315} = 0.2217$$

$$\therefore \quad \alpha = \tan^{-1} 0.2217 = \mathbf{12.5°\ (Ans.)}$$

(*b*) Moving vane :

Velocity of the vane, u = 20 m/s

The force exerted by the jet on the moving vane in the direction of jet,

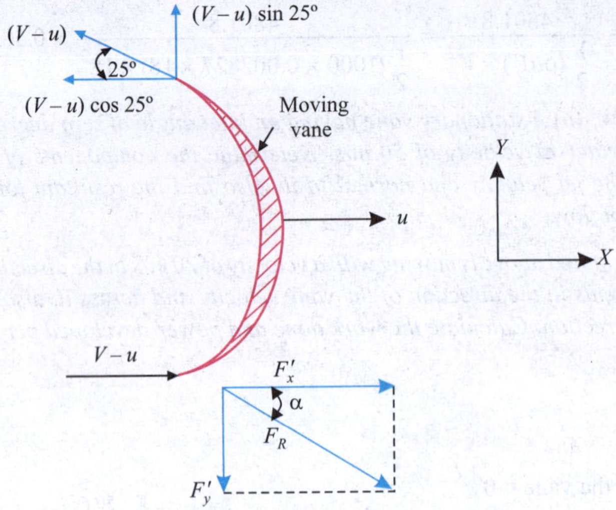

Fig. 1.15. Moving Vane.

$$F'_x = 1 \times (V_{1x} - V_{2x})$$

where, $V_{1x} = V - u = 50 - 20 = 30$ m/s

$V_{2x} = -(V-u)\cos 25° = -(50-20)\cos 25° = -27.19$ m/s

$\therefore$ $\boldsymbol{F'_x = 1 \times [30 - (-27.19)] = 57.19}$ **N (Ans.)**

Similarly, $F'_y = 1 \times (V_{1y} - V_{2y})$

where $V_{1y} = 0,$

$V_{2y} = (V-u)\sin 25° = (50-20)\sin 25° = 12.67$ m/s

$\boldsymbol{F'_y} = 1 \times (0 - 12.67) = \boldsymbol{-12.67}$ **N (Ans.)**

$\therefore$ *Resultant force,* $F_R = \sqrt{F_x'^2 + F_y'^2} = \sqrt{57.19^2 + 12.67^2} = \mathbf{58.57}$ **N (Ans.)**

The angle made by the resultant with X-axis,

$$\tan \alpha = \frac{F_y'}{F_x'} = \frac{12.67}{57.19} = 0.2215$$

or, $\alpha = \tan^{-1} 0.2215 = \mathbf{12.5°}$ **(Ans.)**

Work done per second $= F'_x \times u = 57.19 \times 20 =$ **1143.8 Nm/s (Ans.)**

$\therefore$ *Power developed* = 1143.8 J/s or W or **1.1438 kW (Ans.)**

Example 1.15. *A jet of water 80 mm diameter and having a velocity of 20 m/s impinges at the centre of hemispherical vane. The linear velocity of vane is 10 m/s in the direction of jet. Find the force exerted on the vane. How this force would change if the jet impinges on a series of vanes attached to the circumference of wheel ?*

Solution. Diameter of the jet, $d = 80$ mm $= 0.08$ m

$\therefore$ Area, $a = \frac{\pi}{4} \times 0.08^2 = 0.005026$ m^3

Velocity of the jet, $V = 20$ m/s

Velocity of the vane, $u = 10$ m/s

When the jet strikes at the centre of a moving vane, then effective incoming velocity = $V - u$

Force exerted on the vane, $F = \rho a (V-u)^2 (1 + \cos\theta)$ ...[Eqn. (1.18)]

For a hemispherical vane, $\theta = 0$

$$\therefore \quad F = \rho a (V-u)^2 (1 + \cos 0°) = \rho a (V-u)^2 \times 2 = 2\rho a (V-u)^2$$
$$= 2 \times 1000 \times 0.005026 (20-10)^2 = \mathbf{1005.2\ N\ (Ans.)}$$

When the jet impinges on a *series of vanes,* the entire fluid mass issuing from the jet ($m = \rho aV$) hits the vane and the force exerted by the jet on the vane system is

$$F = \rho aV (V-u)(1 + \cos\theta) \qquad \text{...[Eqn. (1.22)]}$$
$$= \rho aV (V-u)(1 + \cos 0°)$$
$$= 2\rho aV (V-u) \qquad (\because \theta = 0°)$$
$$= 2 \times 1000 \times 0.005026 \times 20 \times (20-10) = \mathbf{2010.4\ N\ (Ans.)}$$

Example 1.16. *A jet strikes tangentially a smooth curved vane moving in the same direction as the jet, and the jet gets reversed in the direction. Show that the maximum efficiency is slightly less than 60 per cent.*

Solution. Refer to Fig 1.16.

Let, V = Velocity of the jet, and

u = Velocity of the vane.

($V > u$ for impact)

The force exerted by jet in the direction of motion of the vane,

$$F_x = \text{Mass/sec} \times \text{change in velocity}$$
$$= \rho a (V-u)[(V-u) - \{-(V-u)\}]$$

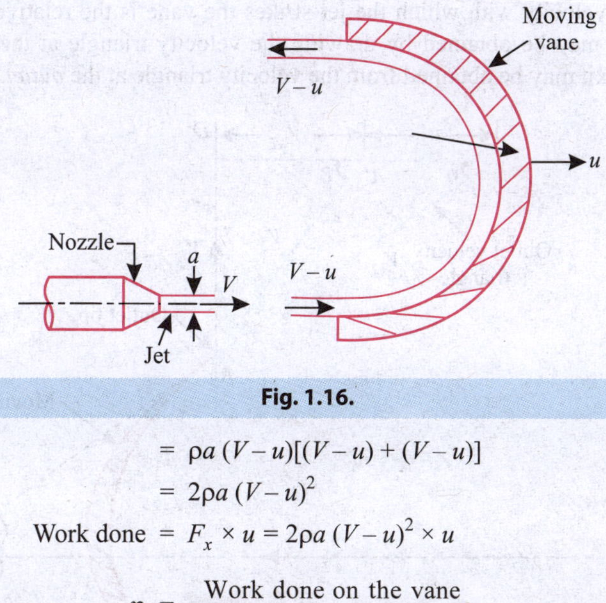

Fig. 1.16.

$$= \rho a (V-u)[(V-u) + (V-u)]$$
$$= 2\rho a (V-u)^2$$

$$\text{Work done} = F_x \times u = 2\rho a (V-u)^2 \times u$$

$$\eta = \frac{\text{Work done on the vane}}{\text{Kinetic energy of the jet}}$$

$$= \frac{2\rho a (V-u)^2 \times u}{\frac{1}{2}(\rho a V) V^2} = 4 \times \frac{u (V-u)^2}{V^3} \qquad \text{...}(i)$$

For *maximum efficiency,* $\dfrac{d\eta}{dV} = 0$

or, $$\frac{d}{dV}\left[\frac{4u (V-u)^2}{V^3}\right] = 4u\left[\frac{V^3 \times 2 (V-u) - (V-u)^2 \times 3V^2}{V^6}\right] = 0$$

or, $$\frac{2V^3(V-u)-(V-u)^2\times 3V^2}{V^6} = 0 \qquad (\because u \neq 0)$$

or, $$2V(V-u)-3(V-u)^2 = 0$$

or, $$(V-u)[2V-3(V-u)] = 0$$

or, $$(V-u)(2V-3V+3u) = 0$$

or, $$(V-u)(3u-V)=0,\ (V-u) \neq 0, \text{ for impact,}$$

$$\therefore \quad 3u-V = 0 \text{ or } u=\frac{V}{3}$$

Substituting this value of $u\left(=\frac{V}{3}\right)$ in expression (*i*), we get

$$\eta_{max} = 4\times\frac{V/3\left(V-\frac{V}{3}\right)^2}{V^3} = \frac{16}{27} = 0.593 \text{ or } \mathbf{59.3\%}$$

which is slightly less than 60%. **(Ans.)**

1.8. JET STRIKING A MOVING CURVED VANE TANGENTIALLY AT ONE TIP AND LEAVING AT THE OTHER

A. Single vane:

Fig. 1.17, shows a jet striking a moving curved vane tangentially at one tip and leaving at the other. The effective velocity with which the jet strikes the vane is the relative velocity (V_{r1}). The relative velocity V_{r1} may be obtained by drawing the velocity triangle at the *inlet*. The absolute velocity (V_2) at the exit may be obtained from the velocity triangle at the *outlet*. Fig. 1.17 shows the velocity triangles.

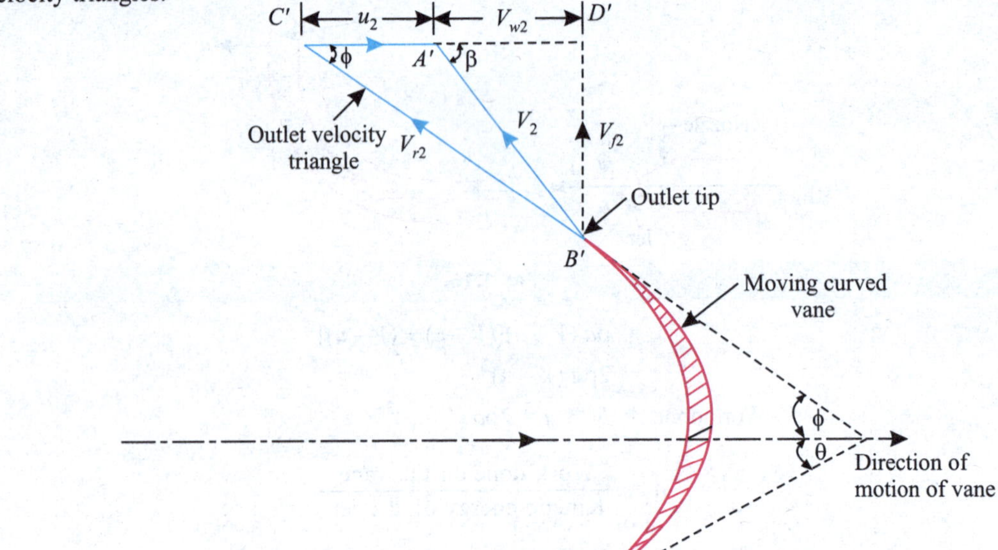

Fig. 1.17. Jet striking a moving curved vane tangentially at one tip and leaving at the other.

Let, V_1, V_2 = Absolute velocities of the jet at the inlet and outlet respectively,
u_1, u_2 = Peripheral velocities of the vane at the inlet and outlet respectively,
V_{r1}, V_{r2} = Relative velocities at the inlet and outlet respectively,
V_{f1}, V_{f2} = Velocities of the flow at the inlet and outlet respectively,
V_{w1}, V_{w2} = Velocities of the whirl at the inlet and outlet respectively,
θ, ϕ = Tip angles at the inlet and outlet respectively,
α, β = Angles which the absolute velocities make at the inlet and outlet respectively.

It may be *noted* that:

- All angles are measured with the direction of motion of vane,
- The velocity of whirl is the component of the absolute velocity in the direction of motion.
- The velocity of flow is the component of the absolute velocity normal to the direction of motion.

The triangles *ABD* and *B′C′D′* are called *inlet* and *outlet velocity triangles* and are drawn as follows:

1. Inlet velocity triangle:

- Take any point *A* and draw a line $AB = V_1$ (in magnitude), making an angle α with the horizontal line *AD*.
- Draw a line $AC = u_1$ and join *C* to *B*, *CB* then represents relative velocity (V_{r1}) of the jet at inlet. If the loss of energy at inlet due to *impact is zero,* then *CB must be in the tangential direction* to the vane at inlet.
- From *B* draw a perpendicular *BD* meeting the horizontal line *AC* produced at *D*. Then *BD* represents the velocity of flow at Inlet (V_{f1}). *AD* represents the velocity of whirl at inlet (V_{w1}). $\angle BCD = \theta$ = vane angle at inlet.

2. Outlet velocity triangle:

If the vane surface is assumed *smooth,* the energy loss due to friction will be zero and thus $V_{r1} = V_{r2}$ will be in tangential direction to the vane at outlet.

- Draw *B′C′* in the tangential direction of the vane at outlet and cut $B'C' = V_{r2}$.
- From *C′* draw a line *C′ A′* in the direction of vane at outlet and equal to u_2 (the velocity of vane at outlet). Join *B′A′*. Then *B′ A′* represents the absolute velocity of the jet (V_2) at outlet in magnitude and direction.
- From *B′* draw a perpendicular *B′ D′* to meet the line *C′ A′* produced at *D′*. Then *B′ D′* and *A′ D′* represent the velocity of flow (V_{f2}) and velocity of whirl (V_{w2}) at outlet respectively.
- ϕ = angle of vane at outlet, β = angle made by V_2 with the direction of motion of vane at outlet.

If vane is *smooth* and is having velocity in the direction of motion at inlet and outlet *equal,* then

$$u_1 = u_2 = u \text{ (velocity in the direction of motion), and } V_{r1} = V_{r2}$$

Mass of water striking the vane per second = $\rho a V_{r1}$

(where, a = area of jet water)

∴ Force exerted in the direction of motion,

F_x = Mass of water striking the vane per second × (Initial velocity with which the jet strikes in the *direction of motion* – final velocity)

$$= \rho a V_{r1}[V_{r1} \cos\theta - (-V_{r2}\cos\phi)]$$

But, $V_{r1}\cos\theta = (V_{w1} - u_1)$, and $V_{r2}\cos\phi = (u_2 + V_{w2})$ (Refer to Fig. 1.17)

$$\therefore \quad F_x = \rho a V_{r1}[(V_{w1} - u_1) - \{-(u_2 + V_{w2})\}]$$

$$= \rho a V_{r1}[V_{w1} - u_1 + u_2 + V_{w2}]$$

or, $$F_x = \rho a V_{r1}(V_{w1} + V_{w2}) \qquad (\because u_1 = u_2) \qquad ...(i)$$

The eqn. (*i*) is *true only when* β *is an actue angle* (See Fig. 1.17), When $\beta = 90°$, $V_{w2} = 0$ the eqn. (*i*) redues to

$$F_x = \rho a V_{r1}(V_{w1})$$

If β is an *obtuse angle,* the expression for F_x will become

$$F_x = \rho a V_{r1}(V_{w1} - V_{w2})$$

Thus, in general F_x is written as:

$$F_x = \rho a V_{r1}(V_{w1} \pm V_{w2}) \qquad ...(1.27)$$

Work done per second by the jet on the vane

$$= F_x \times u = \rho a V_{r1}(V_{w1} \pm V_{w2}) \times u \qquad ...(1.28)$$

$\therefore$ Work done per second per unit weight of fluid striking

$$= \frac{\rho a V_{r1}(V_{w1} \pm V_{w2}) \times u}{\text{Weight of fluid striking}}$$

$$= \frac{\rho a V_{r1}(V_{w1} \pm V_{w2}) \times u}{\rho a V_{r1} \times g}$$

$$= \frac{1}{g}(V_{w1} \pm V_{w2}) \times u \qquad ...(1.29)$$

B. Force exerted on a series of radial curved vanes :

Consider a series of radial curved vanes mounted on a wheel as shown in Fig. 1.18. The jet water impinges upon the vanes and the wheel starts rotating at a constant angular speed.

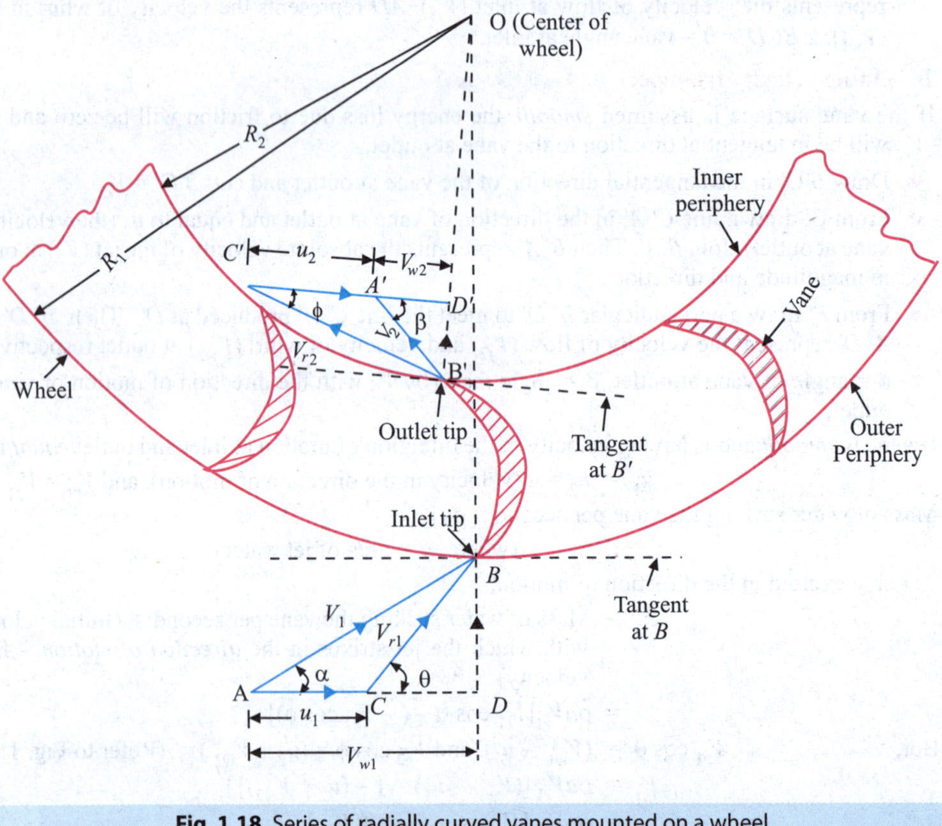

Fig. 1.18. Series of radially curved vanes mounted on a wheel.

Let, ω = Angular speed of the wheel, and

R_1, R_2 = Radii of the wheel at the inlet and outlet of the vane respectively.

As the vanes are situated radially round the wheel the blade velocities at the inlet and outlet tips of the vane would be different, *i.e.,*

$$u_1 = \omega R_1 \quad \text{and} \quad u_2 = \omega R_2$$

The flow system is inward or outward, depending upon whether the jet enters the outer periphery or the inner periphery.

The inlet and outlet triangles are shown in Fig. 1.18.

The mass of water striking per second (for a series of vanes)

= Mass of water issuing from the nozzle per second

= $\rho a V_1$ (where, a = area of jet, V_1 = velocity of jet).

Momentum of water striking the vanes (in tangential direction) per second *at inlet*

$$= (\rho a V_1) \times V_{w1}$$

[where, V_{w1} (= $V_1 \cos \alpha$) = component of V_1 in the *tangential direction.*]

Similarly, momentum of water per second at outlet

$$= \rho a V_1 \times (-V_{w2}) = -\rho a V_1 \times V_{w2}$$

[where, $V_{w2} = (V_2 \cos \beta)$ = component of V_2 in tangential direction]

–ve sign is taken as V_2 at outlet is in opposite direction.

Now, angular momentum per second at *inlet* = momentum at inlet × radius at inlet.

$$= (\rho a V_1) \times V_{w1} \times R_1$$

Angular momentum per second at outlet = $-(\rho a V_1) \times V_{w2} \times R_2$

Torque exerted by water on the wheel,

T = Rate of change of angular momentum

= (initial angular momentum per second – final angular momentum per second)

$$= \rho a V_1 \times V_{w1} \times R_1 - \{-(\rho a V_1 \times V_{w2} \times R_2)\}$$

$$= \rho a V_1 (V_{w1} \times R_1 + V_{w2} \times R_2)$$

Work done per second on the wheel = $T \times \omega$

$$= \rho a V_1 (V_{w1} \times R_1 + V_{w2} \times R_2) \times \omega$$

$$= \rho a V_1 (V_{w1} \times \omega R_1 + V_{w2} \times \omega R_2)$$

$$= \rho a V_1 (V_{w1} \times u_1 + V_{w2} \times u_2)$$

$$(\because u_1 = \omega R_1 \quad \text{and} \quad u_2 = \omega R_2)$$

In case β is an *obtuse angle* (Fig. 1.18) then work done per second is

$$= \rho a V_1 (V_{w1} u_1 - V_{w2} u_2)$$

$\therefore$ The general expression for the work done per second on the wheel

$$= \rho a V_1 (V_{w1} u_1 \pm V_{w2} u_2) \qquad \text{...(1.30)}$$

If the discharge is radial at outlet, then $\beta = 90°$ and work done is

$$= \rho a V_1 (V_{w1} u_1) \qquad (\because V_{w_2} = 0)$$

Efficiency of the radial curved vane, η_{vane}:

$$\eta_{\text{vane}} = \frac{\text{Work done per second}}{\text{Kinetic energy per second}}$$

$$= \frac{\rho a V_1 [V_{w1} u_1 \pm V_{w2} u_2]}{\frac{1}{2} (\rho a V_1) \times V_1^2}$$

$$= \frac{2\,[V_{w1}u_1 \pm V_{w2}u_2]}{V_1^2} \qquad ...(1.31)$$

Example 1.17. *A jet of water having a velocity of 40 m/s strikes a curved vane, which is moving with a velocity of 20 m/s. The jet makes an angle of 30° with the direction of motion of vane at inlet and leaves at an angle of 90° to the direction of motion of vane at outlet. Draw the velocity triangles at inlet and outlet and determine the vane angles at inlet and outlet so that the water enters and leaves the vane without shock.*

[Delhi University]

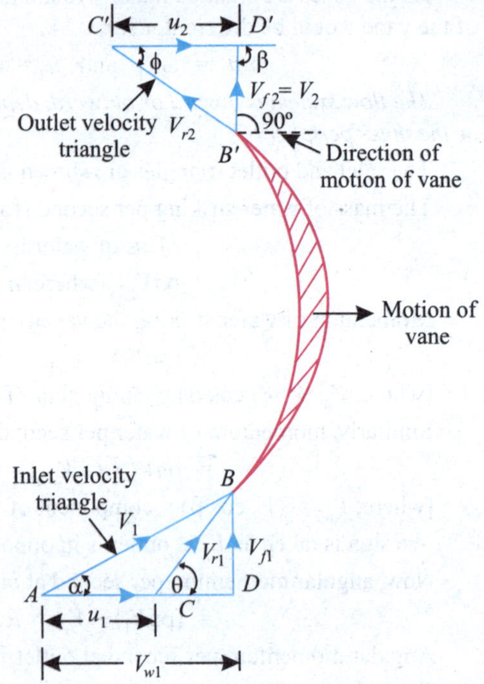

Fig. 1.19.

Solution. Velocity of jet, $V_1 = 40$ m/s

Velocity of vane, $u_1 = 20$ m/s

Angle made by the jet at inlet, $\alpha = 30°$

Angle made by the jet at outlet = 90°

$\therefore \quad \beta = 180° - 90° = 90°$

Here, we have $u_1 = u_2 = u = 20$ m/s

Vane angles at inlet and outlet:

Vane angles at inlet and outlet are θ and ϕ respectively.

From $\Delta\,BCD$, we have

$$\tan\theta = \frac{BD}{CD} = \frac{BD}{AD - AC} = \frac{V_{f1}}{V_{w1} - u_1}$$

where,

$$V_{f1} = V_1 \sin\alpha = 40 \times \sin 30° = 20 \text{ m/s}$$

$$V_{w1} = V_1 \cos\alpha = 40 \times \cos 30° = 34.64 \text{ m/s}$$

$$u_1 = 20 \text{ m/s} \qquad (Given)$$

$$\because \quad \tan\theta = \frac{20}{34.64 - 20} = 1.366$$

or, $\theta = \tan^{-1}(1.366) = $ **53.79° (Ans.)**

Again form $\Delta\,BCD$, we have:

$$\sin\theta = \frac{V_{f1}}{V_{r1}}, \text{ or, } V_{r1} = \frac{V_{f1}}{\sin\theta} = \frac{20}{\sin 53.79°} = 24.78 \text{ m/s}$$

But, $V_{r2} = V_{r1} = 24.78$ m/s

Hence from $\Delta\,B'C'D'$, we have:

$$\cos\phi = \frac{u_2}{V_{r2}} = \frac{20}{24.78} = 0.8071$$

or, $\phi = \cos^{-1}(0.8071) = $ **36.18° (Ans.)**

Example 1.18. *A jet of water having a velocity of 45 m/s impinges without shock on a series of vanes moving at 15 m/s. The direction of motion of the vanes is inclined at 20° to that of jet. The relative velocity at outlet is 0.9 of that at inlet, and absolute velocity of water at exit is to be normal to the motion of vanes. Find:*

(*i*) *Vane angles at inlet and outlet,*

(ii) Work done on vanes per N (newton) of water supplied by the jet, and

(iii) Hydraulic efficiency. **[UPTU]**

Solution. Velocity of the jet, $V_1 = 45$ m/s

Velocity of vane, $u_1 = u_2 = (u =)$ 15 m/s

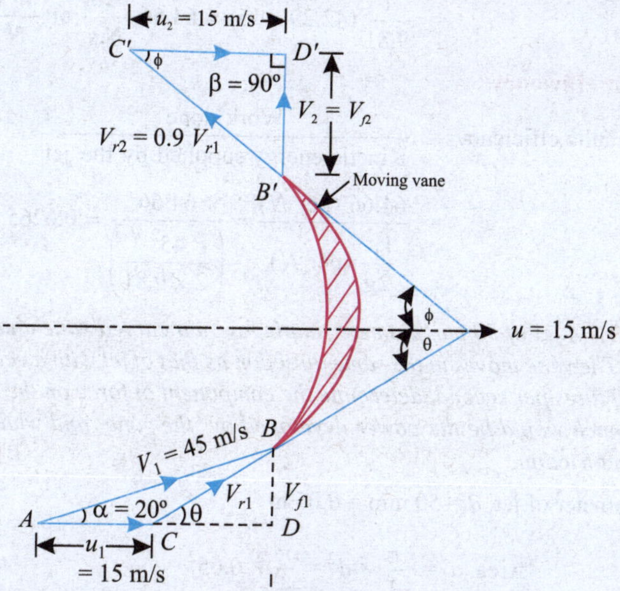

Fig. 1.20.

Angle made by jet at inlet, with direction of motion of vane, $\alpha = 20°$

Relative velocity at outlet, $V_{r2} = 0.9V_{r1}$ (relative velocity at inlet).

(i) Vane angles at inlet and outlet :

Vane angles at inlet and outlet are θ and ϕ .

$$\text{In } \Delta\, BCD,\ \tan\theta = \frac{V_{f1}}{(V_{w1} - u_1)}$$

where,

$$V_{f1} = V_1 \sin\alpha = 45° \sin 20° = 15.39 \text{ m/s}$$

$$V_{w1} = V_1 \cos\alpha = 45 \cos 20° = 42.29 \text{ m/s}$$

$$\therefore \quad \tan\theta = \frac{15.39}{42.29 - 15} = 0.564$$

or,

$$\theta = \tan^{-1} 0.564 = \mathbf{29.42°} \textbf{ (Ans.)}$$

Again from $\Delta\ BCD$, $\dfrac{V_{f1}}{V_{r1}} = \sin\theta$ or $V_{r1} = \dfrac{V_{f1}}{\sin\theta} = \dfrac{15.39}{\sin 29.42°} = 31.33$ m/s

$$\therefore \quad V_{r2} = 0.9\ V_{r1} = 0.9 \times 31.33 = 28.2 \text{ m/s}$$

From $\Delta\ B'C'D'$, we have:

$$\cos\phi = \frac{u_2}{V_{r2}} = \frac{15}{28.2} = 0.5319$$

$$\therefore \quad \phi = \cos^{-1} 0.5319 = \mathbf{57.87°} \textbf{ (Ans.)}$$

(ii) Work done on vanes per N of water :

Work done per second per N of water striking

$$= \frac{1}{g}(V_{w1}u_1 + V_{w2}u_2) = \frac{1}{g}(V_{w1}u_1) \qquad [\because V_{w2} = 0]$$

$$= \frac{1}{9.81}(42.29 \times 15) = \mathbf{64.66}\ \frac{\textbf{N.m}}{\textbf{N.s}}\ \textbf{or}\ \frac{\textbf{J/s}}{\textbf{N}}\ \textbf{(Ans.)}$$

(iii) Hydraulic efficiency :

$$\text{Hydraulic efficiency} = \frac{\text{Work done}}{\text{Kinetic energy supplied by the jet}}$$

$$= \frac{64.66\ (\text{per } N)}{\frac{V_1^2}{2g}\ (\text{per } N)} = \frac{64.66}{\left(\frac{45^2}{2 \times 9.81}\right)} = 0.6265 \text{ or } \mathbf{62.65\%\ (Ans.)}$$

Example 1.19. *A jet of 50 mm diameter impinges on a curved vane and is deflected through an angle of 175°. The vane moves in the same direction as that of jet with a velocity of 35 m/s. If the rate of flow is 170 litres per second, determine the component of force on the vane in the direction of motion. How much would be the power developed by the vane and what would be the water efficiency? Neglect friction.* **[Punjab University]**

Solution. Diameter of jet, $d = 50$ mm $= 0.05$ m

$$\therefore \quad \text{Area, } a = \frac{\pi}{4} \times d^2 = \frac{\pi}{4} \times 0.05^2$$

$$= 0.001963 \text{ m}^2$$

Angle of deflection = 175°

Velocity of the vane, $u_1 = u_2 (= u) = 35$ m/s.

Rate of flow, Q = 170 litres = 0.17 m³/s

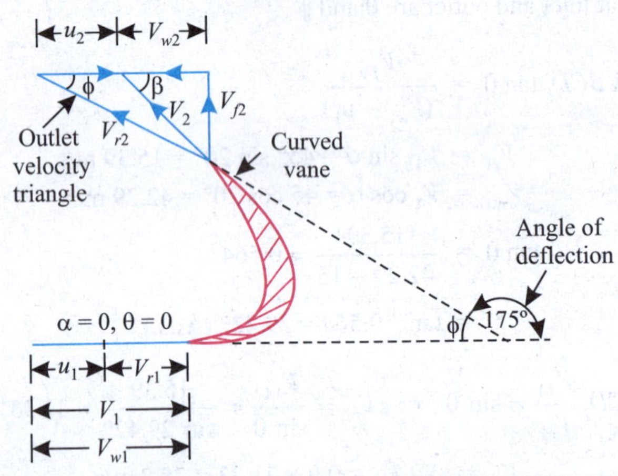

Fig. 1.21.

Since the jet of water moves in the same direction as that of vane, $\alpha = \theta = 0$ and, therefore, the inlet velocity triangle will be a *straight line* with,

$$V_1 = \frac{Q}{a} = \frac{0.17}{0.001963} = 86{\cdot}6 \text{ m/s}$$

$$V_{r1} = V_1 - u_1 = 86.6 - 35 = 51.6 \text{ m/s}$$

and, $$V_{w1} = V_1 = 86.6 \text{ m/s}$$

Corresponding to outlet velocity triangle,

$$\phi = 180° - 175° = 5°$$

Further, since the vane is *smooth*, therefore,

$$V_{r2} = V_{r1} = 51.6 \text{ m/s}$$

$$V_{w2} = V_{r2} \cos \phi - u_2$$

$$= 51.6 \times \cos 5° - 35 = 16.4 \text{ m/s}$$

Power developed by the vane :

Force exerted by the jet on the vane in the direction of motion,

$$F = \rho a V_{r1} (V_{w1} + V_{w2})$$

$$= 1000 \times 0.001963 \times 51.6 (86.6 + 16.4) = 10432.9 \text{ N}$$

$$\text{Work done} = \text{Force} \times \text{velocity}$$

$$= 10432.9 \times 35 = 365151 \text{ Nm/s or J/s}$$

Hence, *power developed by the vane* = 365151 J/s or W or **365.151 kW (Ans.)**

Efficiency of vane (water efficiency) = $\dfrac{\text{Work done on the vane}}{\text{Kinetic energy supplied by the jet}}$

$$= \frac{365151}{\frac{1}{2} \rho Q V_1^2} = \frac{365151}{\frac{1}{2} \times 1000 \times 0.17 \times 86.6^2} = 0.573 \text{ or } \mathbf{57.3\%} \text{ (Ans.)}$$

Example 1.20. *A jet of water moving at 12 m/s impinges on a concave shaped vane to deflect the jet through 120° when stationary. The vane is moving at 5 m/s. Find:*

(i) The angle of jet so that there is no shock at inlet,

(ii) The absolute velocity of the jet at exit both in magnitude and direction, and

(iii) The work done per second per N of water.

Assume that vane is smooth. **[Anna University]**

Solution. Velocity at jet, $V_1 = 12$ m/s

Velocity of vane, $u_1 = u_2 (= u) = 5$ m/s

Angle of deflection of the jet = 120°

(*i*) The angle of the jet at inlet of the vane, α :

Assuming vane to be symmetrical, we have

$$\theta = \phi$$

Now, $$120° = 180 - (\theta + \phi)$$

∴ $$\theta + \phi = (180° - 120°) = 60° \quad \therefore \theta = \phi = 30°$$

Applying sine rule to Δ *ABC*, we have:

$$\frac{AB}{\sin (180° - \theta)} = \frac{AC}{\sin (\theta - \alpha)}, \text{ or, } \frac{V_1}{\sin \theta} = \frac{u_1}{\sin (\theta - \alpha)}$$

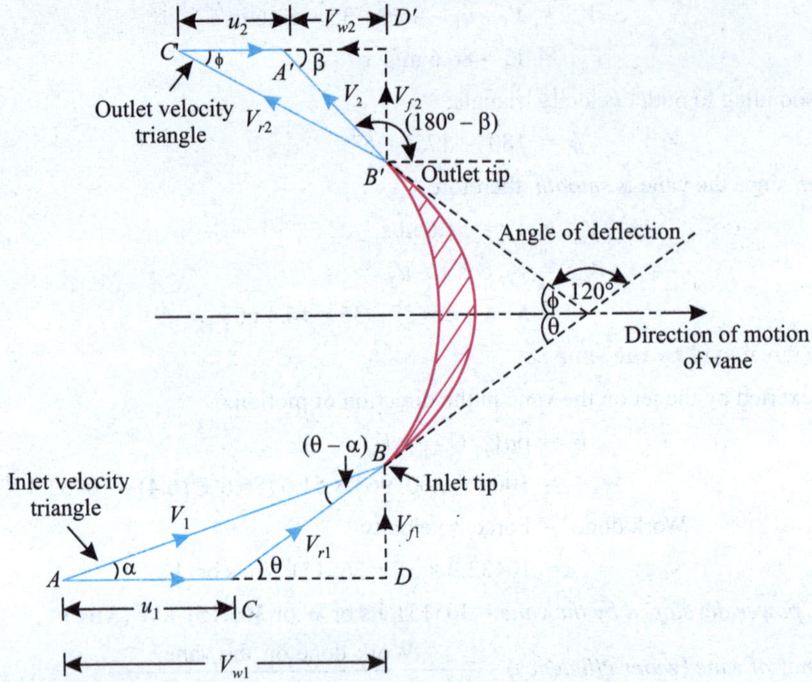

Fig. 1.22.

or, $$\frac{12}{\sin 30°} = \frac{5}{\sin (30° - \alpha)}, \text{ or, } \sin (30° - \alpha) = \frac{5 \times \sin 30°}{12} = 0.2083$$

∴ $$30° - \alpha = \sin^{-1} 0.2083 = 12°$$

or, $$\alpha = 30° - 12° = \mathbf{18°} \text{ (Ans.)}$$

(ii) The absolute velocity of the jet at exit; V_2:

Again applying sine rule to ΔABC, we have:

$$\frac{V_1}{\sin (180° - \theta)} = \frac{V_{r1}}{\sin \alpha}, \quad \text{or,} \quad \frac{12}{\sin \theta} = \frac{V_{r1}}{\sin 18°}$$

∴ $$V_{r1} = \frac{12 \times \sin 18°}{\sin \theta} = \frac{12 \times \sin 18°}{\sin 30°} = 7.42 \text{ m/s}$$

In ΔABD: $V_{w1} = V_1 \cos \alpha = 12 \cos 18° = 11.41$ m/s

Now, since the vane is *smooth,* therefore,

$$V_{r2} = V_{r1} = 7.42 \text{ m/s}$$

At *outlet,* from $\Delta B'C'D'$, we have:

$$V_{r2} \cos \phi = u_2 + V_{w2}$$

∴ $$V_{w2} = V_{r2} \cos \phi - u_2 = 7.42 \cos 30° - 5 = 1.42 \text{ m/s}$$

Also, $$V_{f2} = V_{r2} \sin \phi = 7.42 \sin 30° = 3.71 \text{ m/s}$$

Now, $$\tan \beta = \frac{V_{f2}}{V_{w2}} = \frac{3.71}{1.42} = 2.613$$

∴ Angle of jet at outlet, $\beta = \tan^{-1} 2.613 = 69.06°$

Hence, angle made by V_2 at oulet with direction of motion of vane is

$$= 180° - \beta = 180° - 69.06° = \mathbf{110.94°} \text{ (Ans.)}$$

Absolute velocity of jet at exit, $V_2 = \sqrt{V_{w2}^2 + V_{f2}^2}$

$$= \sqrt{1.42^2 + 3.71^2} = \mathbf{3.97\ m/s} \text{ (Ans.)}$$

(iii) The work done per second per N of water :

The work done per second per N of water

$$= \frac{1}{g}(V_{w1}u_1 + V_{w2}u_2) = \frac{1}{g}(V_{w1} + V_{w2}) \times u \quad [\because \beta < 90°]$$

$$= \frac{9}{9.81}(11.41 + 1.42) \times 5 = \mathbf{6.539\ Nm} \text{ (Ans.)}$$

Example 1.21. *A jet of water having a velocity of 35 m/s impinges on a series of vanes moving with a velocity of 20 m/s. The jet makes an angle of 30° to the direction of motion of vanes when entering and leaves at an angle of 120°. Draw the triangles of velocities at inlet and outlet and find:*

(i) The angles of vanes tips so that water enters and leaves without shock,

(ii) The work done per N of water entering the vanes, and

(iii) The efficiency. **[AMIE, Fluid Power Engg.]**

Solution. Velocity of jet, $V_1 = 35$ m/s

Velocity of vane, $u_1 = u_2 (= u) = 20$ m/s

Angle made by jet at inlet, $\alpha = 30°$

Angle made by the jet at outlet $= 120°$ (with the direction of motion of vane)

$\therefore \quad \beta = 180 - 120 = 60°$ (See Fig. 1.23)

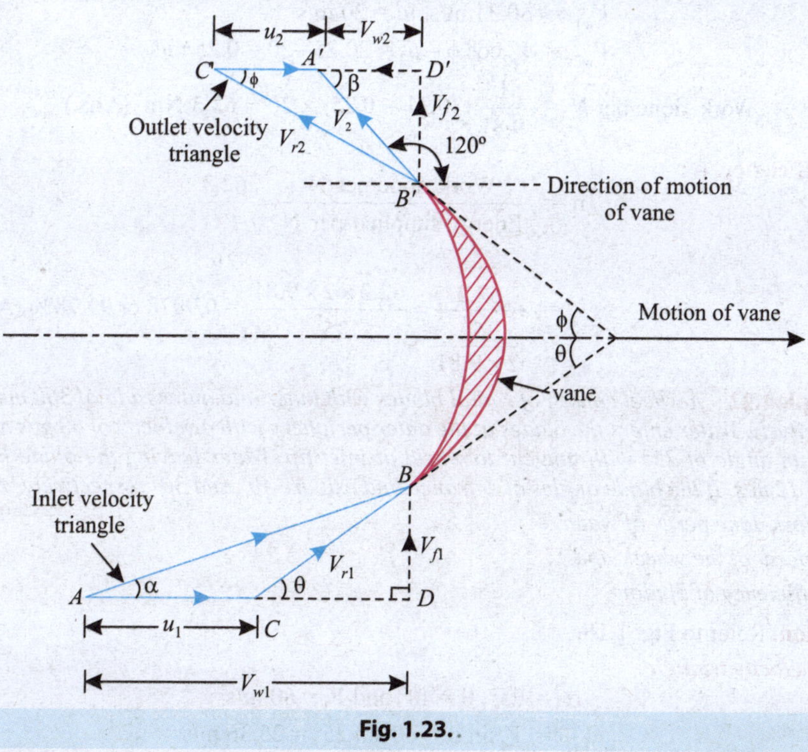

Fig. 1.23.

(i) Angles of vane tips :

From inlet triangle, we have:

$$V_{w1} = V_1 \cos\alpha = 35\cos 30° = 30.31 \text{ m/s}$$

$$V_{f1} = V_1 \sin\alpha = 35\sin 30° = 17.50 \text{ m/s}$$

$$\tan\theta = \frac{V_{f1}}{V_{w1} - u_1} = \frac{17.5}{30.31 - 20} = 1.697$$

$$\therefore \quad \theta = \tan^{-1} 1.697 \simeq \mathbf{60°} \textbf{ (Ans.)}$$

Applying sine rule to inlet velocity triangle, we have:

$$\frac{V_{r1}}{\sin 90°} = \frac{V_{f1}}{\sin\theta} \quad \text{or} \quad \frac{V_{r1}}{1} = \frac{17.50}{\sin 60°} \quad \text{or} \quad V_{r1} = 20.25 \text{ m/s}$$

Now, $V_{r2} = V_{r1} = 20.25$ m/s

Applying sine rule to outlet velocity triangle, we have:

$$\frac{V_{r2}}{\sin 120°} = \frac{u_2}{\sin(60° - \phi)}, \quad \text{or,} \quad \frac{20.25}{\sin 120°} = \frac{20}{\sin(60° - \phi)}$$

$$\text{or,} \quad \sin(60° - \phi) = \frac{20 \times \sin 120°}{20.25} = 0.855$$

$$\text{or,} \quad 60° - \phi = \sin^{-1}(0.855) \quad \text{or} \quad \phi = 58.75°$$

$$\therefore \quad \phi = 60° - 58.75° = \mathbf{1.25°} \textbf{ (Ans.)}$$

(ii) Work done per N of water entering the vanes :

Work done per N of water entering the vanes

$$= \frac{1}{g}(V_{w1} + V_{w2}) \times u_1 \qquad [\because \beta < 90°]$$

$$V_{w1} = 30.31 \text{ m/s}, u_1 = 20 \text{ m/s}$$

$$V_{w2} = V_{r2}\cos\phi - u_2 = 20.25 - 20 = 0.25 \text{ m/s}$$

$$\therefore \quad \text{Work done per } N = \frac{1}{9.81}(30.31 + 0.25) \times 20 = \mathbf{62.3 \text{ Nm}} \textbf{ (Ans.)}$$

(iii) Efficiency, η :

$$\eta = \frac{\text{Work done per N}}{\text{Energy supplied per N}} = \frac{62.3}{\frac{V_1^2}{2g}}$$

$$= \frac{62.3}{\frac{35^2}{2 \times 9.81}} = \frac{62.3 \times 2 \times 9.81}{35^2} = 0.9978 \text{ or } \mathbf{99.78\%} \textbf{ (Ans.)}$$

Example 1.22. *A wheel consist of radial blades with inner and outer radii of 360 mm and 720 mm respecitvely. Water enters the blades at the outer periphery with a velocity of 60 m/s and supply jet makes an angle of 25° with tangent to wheel at inlet tip. Water leaving the blade has a flow velocity of 12 m/s. If the blade angles at entrance and exit are 40° and 30° respectively, determine:*

(i) Work done per N of water,

(ii) Speed of the wheel, and

(iii) Efficiency of blading.

Solution. Refer to Fig. 1.18.

Inlet velocity triangle :

$$\alpha = 25°, \theta = 40° \text{ and } V_1 = 60 \text{ m/s}$$

$$V_{f1} = V_1 \sin\alpha = 60 \sin 25° = 25.36 \text{ m/s}$$

$$V_{w1} = V_1 \cos\alpha = 60 \cos 25° = 54.38 \text{ m/s}$$

$$u_1 = V_{w1} - \frac{V_{f1}}{\tan\theta} = 54.38 - \frac{25.36}{\tan 40°} = 24.16 \text{ m}$$

Outlet velocity triangle :

$$\phi = 30°, V_{f2} = 12 \text{ m/s}$$

Since, $u_1 = \omega R_1$ and $u_2 = \omega R_2$

$$\therefore \quad u_2 = u_1 \times \frac{R_2}{R_1} = 24.16 \times \frac{0.36}{0.72} = 12.08 \text{ m/s}$$

$$(\because R_2 = 360 \text{ mm} = 0.36 \text{ m} \quad \text{and} \quad R_1 = 720 \text{ mm} = 0.72 \text{ m})$$

$$V_{w2} = \frac{V_{f2}}{\tan 30°} - u_2 = \frac{12}{\tan 30°} - 12.08 = 8.70 \text{ m/s}$$

(*i*) Work done per N of water :

The work done per second per N of water

$$= \frac{1}{g}(V_{w1} u_1 + V_{w2} u_2) \qquad [\because \beta < 90°]$$

$$= \frac{1}{9.81}(54.38 \times 24.16 + 8.70 \times 12.08) = \mathbf{144.6 \text{ Nm}} \text{ (Ans.)}$$

(*ii*) Speed of the wheel :

$$\text{Angular velocity, } \omega = \frac{u_1}{R_1} = \frac{24.16}{0.72} = 38.55 \text{ rad/s}$$

But, $\omega = \frac{2\pi N}{60}$ where N is the speed of the wheel.

$$\therefore \quad N = \frac{\omega \times 60}{2\pi} = \frac{33.55 \times 60}{2\pi} \mathbf{320.38 \text{ r.p.m.}} \text{ (Ans.)}$$

(*iii*) Efficiency of blading, η :

$$\eta = \frac{\text{Work done}}{\text{Energy supplied by jet}} = \frac{144.6}{\frac{V_1^2}{2g} \text{(per N of water)}}$$

$$= \frac{144.6}{\frac{60^2}{2 \times 9.81}} = \frac{144.6 \times 2 \times 9.81}{60^2} = 0.788 \text{ or } \mathbf{78.8\%} \text{ (Ans.)}$$

Example 1.23. *A jet of water having a velocity of 18 m/s strikes a curved vane which is moving with a velocity of 6 m/s. The vane is symmetrical and so shaped that the jet is deflected through 120°. Determine:*

(*i*) *The angle of the jet at inlet of the vane so that there is no shock,*

(*ii*) *The absolute velocity of the jet at outlet in magnitude and direction, and*

(*iii*) *The work done per N of water.*

Solution. Velocity of jet, $V_1 = 18$ m/s

Velocity of vane, $u_1 = u_2 (= u) = 6$ m/s

Refer to Fig 1.24.

(i) The angle of jet at inlet of the vane, α :

As the vane is *symmetrical*, hence $\theta = \phi$

Angle of deflection of jet = 120° *(Given)*

$$120° = 180° - (\theta + \phi) \quad \text{or} \quad \theta + \phi = 180° - 120° = 60°$$

$$\therefore \qquad \theta = \phi = 30°$$

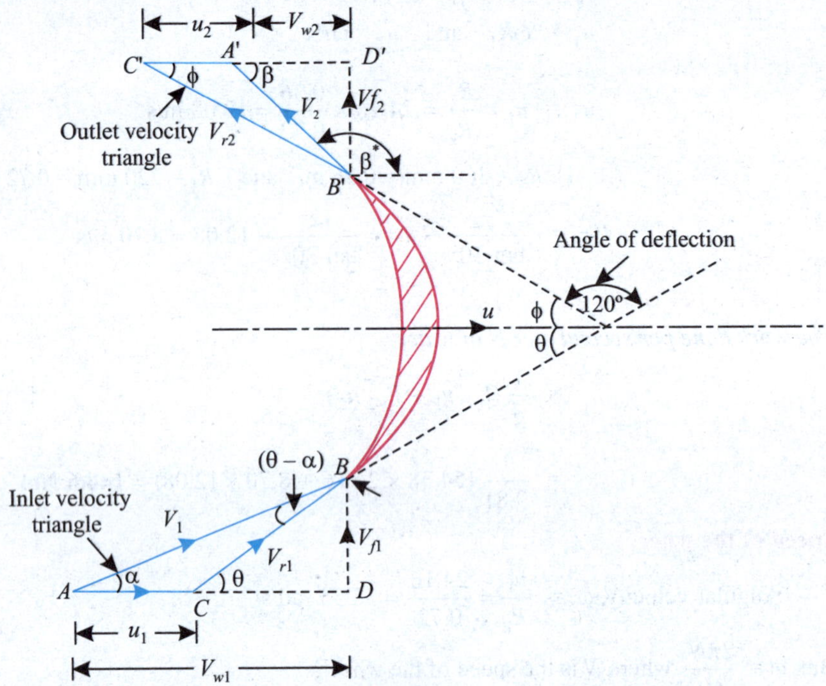

Fig. 1.24.

Applying sine rule to ΔABC, we have:

$$\frac{AC}{\sin(\theta - \alpha)} = \frac{AB}{\sin(180° - 30°)}$$

or, $$\frac{u_1}{\sin(30° - \alpha)} = \frac{V_1}{\sin 30°}$$

or, $$\frac{6}{\sin(30° - \alpha)} = \frac{18}{\sin 30°}$$

or, $$\sin(30° - \alpha) = \frac{6 \times \sin 30°}{18} = 0.1667$$

or, $$30° - \alpha = \sin^{-1}(0.1667) = 9.6°$$

$$\therefore \qquad \alpha = 30° - 9.6° = \mathbf{20.4°} \text{ (Ans.)}$$

(ii) The absolute velocity of jet at outlet, V_2 :

$$V_{r1} = V_{r2} \qquad (\because \text{ Vane is } smooth.)$$

Again, applying sine rule to ΔABC, we have:

$$\frac{AB}{\sin(180° - \theta)} = \frac{BC}{\sin \alpha}, \quad \text{or,} \quad \frac{V_1}{\sin(180° - 30°)} = \frac{V_{r1}}{\sin 20.4°}$$

or, $$\frac{18}{\sin 30°} = \frac{V_{r1}}{\sin 20.4°} \quad \text{or} \quad V_{r1} = \frac{18 \times \sin 20.4°}{\sin 30°} = 12.55 \text{ m/s}$$

$\therefore$ $$V_{r2} = V_{r1} = 12.55 \text{ m/s}$$

From outlet velocity triangle $B'C'D'$, we have:

$$V_{r2} \cos \phi = u_2 + V_{w2}$$

$$12.55 \cos 30° = 5 + V_{w2}$$

or, $$V_{w2} = 12.55 \cos 30° - 6 = 4.87 \text{ m/s}$$

Also, $$V_{f2} = V_{r2} \sin \phi = 12.55 \times \sin 30° = 6.27 \text{ m/s}$$

From $\Delta B'A'D'$, we have:

$$V_2 = \sqrt{V_{w2}^2 + V_{f2}^2} = \sqrt{(4.87)^2 + (6.27)^2} = \mathbf{7.94 \text{ m/s}} \text{ (Ans.)}$$

$$\tan \beta = \frac{V_{f2}}{V_{w2}} = \frac{6.27}{4.87} = 1.287$$

$\therefore$ $$\beta = \tan^{-1} 1.287 = 52.15°$$

$\therefore$ *Angle made by absolute velocity at outlet with the direction of motion,*

$$\beta^* = 180 - \beta = 180° - 52.15° = \mathbf{127.85°} \text{ (Ans.)}$$

(*iii*) The work done per N of water :

The work done per *N* of water

$$= \frac{1}{g}(V_{w1} + V_{w2}) \times u \qquad (\because \beta < 90°)$$

$$= \frac{1}{9.81}(V_1 \cos \alpha + V_{w2}) \times u \qquad (\because V_{w1} = V_1 \cos \alpha)$$

$$= \frac{1}{9.81}(18 \cos 20.4° + 4.87) \times 6 = \mathbf{13.3 \text{ Nm/s}} \text{ (Ans.)}$$

Example 1.24. *A jet of water having a velocity of 36 m/s strikes a series of radial curved vanes mounted on a wheel which is rotating at 240 r.p.m. The jet makes an angle of 20° with the tangent to the wheel at inlet and leaves the wheel with a velocity of 6 m/s at an angle of 130° to the tangent to the wheel at outlet. Water is flowing from outward in a radial direction. The outer and inner radii of the wheel are 500 mm and 250 mm respectively. Determine:*

(*i*) *Vane angles at inlet and outlet,*

(*ii*) *Work done per second per N of water, and*

(*iii*) *Efficiency of the wheel.*

Solution. Velocity of jet, V_1 = 36 m/s

Speed of wheel, N = 240 r.p.m.

$\therefore$ Angular speed, $$\omega = \frac{2\pi N}{60} = \frac{2\pi \times 240}{60} = 25.13 \text{ rad./s}$$

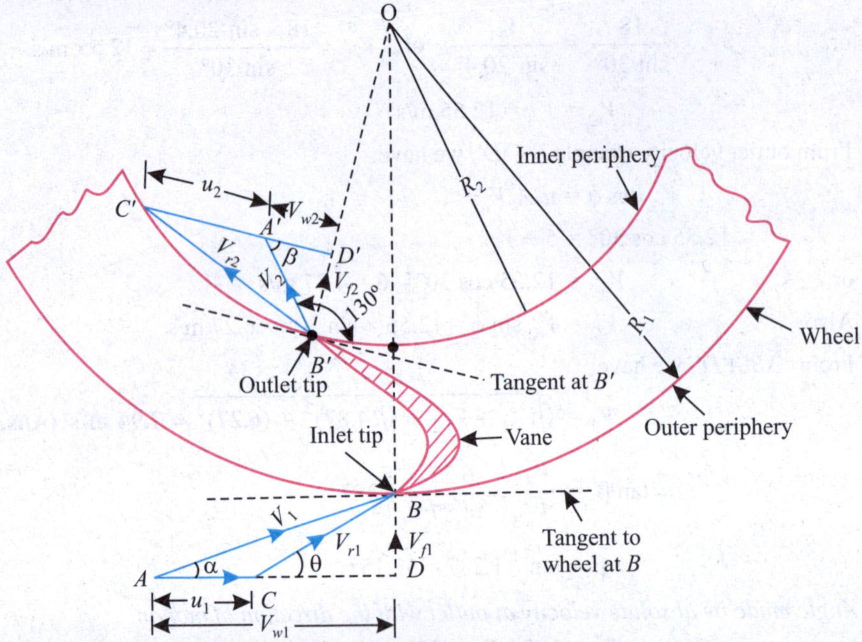

Fig. 1.25.

Angle of jet inlet, $\alpha = 20°$

Velocity of jet at outlet, $V_2 = 6$ m/s

Angle made by the jet at outlet with the tangent to wheel = 130°

$\therefore$ Angle, $\beta = 180 - 130° = 50°$

Outer radius, $R_1 = 500$ mm $= 0.5$ m

Inner radius, $R_2 = 250$ mm $= 0.25$ m

$\therefore$ Velocity, $u_1 = \omega R_1 = 25.13 \times 0.5 = 12.56$ m/s

and, $u_2 = \omega R_2 = 25.13 \times 0.25 = 6.28$ m/s

(*i*) Vane angles at inlet and outlet, θ and ϕ :

From ΔABD, we have:

$$V_{w1} = V_1 \cos\alpha = 36 \times \cos 20° = 33.83 \text{ m/s}$$

$$V_{f1} = V_1 \sin\alpha = 36 \times \sin 20° = 12.31 \text{ m/s}$$

In ΔCBD, $\tan\theta = \dfrac{BD}{CD} = \dfrac{BD}{AD - AC} = \dfrac{V_{f1}}{V_{w1} - u_1} = \dfrac{12.31}{33.83 - 12.56} = 0.578$

$\therefore$ $\theta = \tan^{-1} 0.578 =$ **30° (Ans.)**

From outlet velocity triangle, we have:

$$V_{w2} = V_2 \cos\beta = 6 \cos 50° = 3.86 \text{ m/s}$$

$$V_{f2} = V_2 \sin\beta = 6 \sin 50° = 4.59 \text{ m/s}$$

In $\Delta B'C'D'$, $\tan\phi = \dfrac{V_{f2}}{u_2 + V_{w2}} = \dfrac{4.59}{6.28 + 3.86} = 0.4527$

$\therefore$ $\phi = \tan^{-1}(0.4527) =$ **24.35° (Ans.)**

(ii) Work done per second per N of water :

$$\text{Work done per second} = \frac{1}{g}(V_{w1}u_1 + V_{w2}u_2) \qquad (\because \beta < 90°)$$

$$= \frac{1}{9.81}(33.83 \times 12.56 + 3.86 \times 6.28) = \mathbf{45.78\ Nm}\ \textbf{(Ans.)}$$

(iii) Efficiency of the wheel, η :

$$\eta = \frac{2(V_{w1}u_1 + V_{w2}u_2)}{V_1^2} \qquad \text{[Eqn. (1.31)]}$$

$$= \frac{2(33.83 \times 12.56 + 3.86 \times 6.28)}{36^2} = 0.693 \text{ or } \mathbf{69.3\%}\ \textbf{(Ans.)}$$

Example 1.25. *A 30 mm diameter jet strikes without shock on a series of vanes. The jet velocity is 60 m/s and the vanes move in the same direction as the jet. The shape of each vane is such that, when stationary, it would deflect the jet through an angle of 150°. The friction reduces the relative velocity by 10% as water flows across the vanes and there is a further windage loss given by $\frac{u^2}{2}$ Nm/kg of water, where u is the vane speed. Determine:*

(i) The velocity of vanes corresponding to maximum efficiency, and
(ii) The corresponding thrust on the vanes in the direction of motion.

Solution. Diameter of the jet, $d = 30 \text{ mm} = 0.03 \text{ m}$

$\therefore$ Area, $a = \frac{\pi}{4} \times d^2 = \frac{\pi}{4} \times 0.03^2 = 0.0007068 \text{ m}^2$

Velocity of jet, $V_1 = 60$ m/s.

Reduction in relative velocity due to friction = 10%

$\therefore$ $V_{r2} = (1 - 0.1)V_{r1} = 0.9\,V_{r1}$

Deflection of jet = 150°

Windage loss $= \frac{u^2}{2}$ Nm/kg of water, where u is the vane speed.

(i) The velocity of vanes corresponding to maximum efficiency :

Refer to Fig. 1.26 for inlet and outlet velocity tringles.

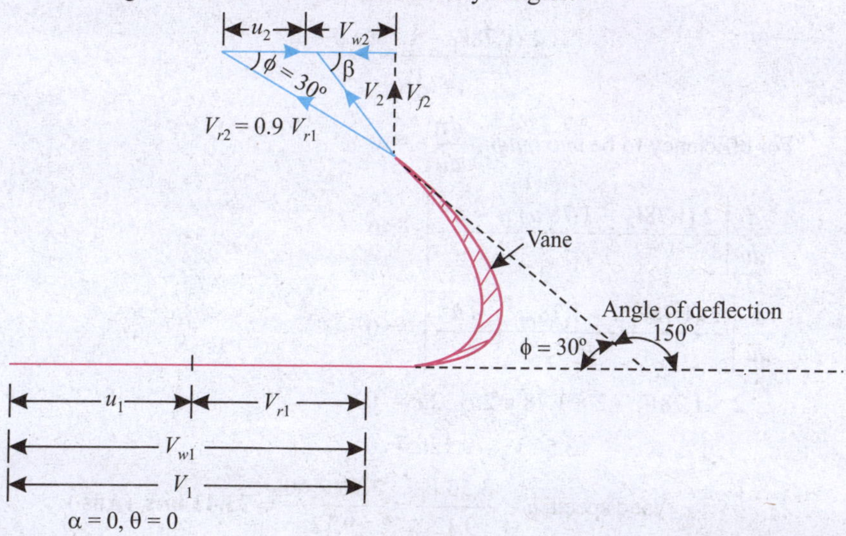

Fig. 1.26.

$$V_{w1} = V_1 = 60 \text{ m/s}; u_1 = u_2 = u$$
$$V_{r1} = V_1 - u; \phi = 180° - 150° = 30°$$
$$V_{r2} = 0.9 \ V_{r1} = 0.9 \ (V_1 - u)$$
$$V_{w2} = V_{r2} \cos \phi - u$$
$$= 0.9 \ (V_1 - u) \cos 30° - u = 0.78 \ (V_1 - u) - u$$

When series of such vanes are fixed to the wheel, the entire fluid of mass $\rho a V_1$ issuing from the jet is utilized in striking the vanes.

Thrust or force on the vane in the direction of its motion,

$$F = \rho a V_1 (V_{w1} + V_{w2}) \quad (\because \beta < 90°)$$
$$= \rho a V_1 [V_1 + 0.78 \ (V_1 - u) - u]$$
$$= \rho a V_1 (1.78 \ V_1 - 1.78 \ u) \quad ...(i)$$

$$\text{Work done per second} = F \times u = \rho a V_1 (1.78 \ V_1 - 1.78 \ u) \ u$$

$$\text{Windage loss} = \frac{u^2}{2} \text{ Nm per unit mass of water}$$
$$= \rho a V_1 \times \frac{u^2}{2}$$

$$\therefore \quad \text{Useful work done} = \rho a V_1 (1.78 \ V_1 - 1.78 \ u) \ u - \rho a V_1 \times \frac{u^2}{2}$$

Energy supplied by the water jet

$$= \frac{1}{2} m V_1^2 = \frac{1}{2} \rho a V_1 \times V_1^2$$

Now, $$\text{efficiency, } \eta = \frac{\text{Useful work done}}{\text{Energy supplied}}$$

$$= \frac{\rho a V_1 \ (1.78 \ V_1 - 1.78 \ u) \ u - \rho a V_1 \times \frac{u^2}{2}}{\frac{1}{2} \rho a V_1 \times V_1^2}$$

$$= \frac{2 \ (1.78 V_1 - 1.78 u) \ u - u^2}{V_1^2}$$

For efficiency to be *maximum*, $\frac{d\eta}{du} = 0$

$$\therefore \quad \frac{d}{du}\left[\frac{2 \ (1.78 V_1 - 1.78 \ u) \ u - u^2}{V_1^2}\right] = 0$$

or, $$\frac{d}{du}\left[\frac{2 \ (1.78 V_1 u - 1.78 \ u^2) - u^2}{V_1^2}\right] = 0$$

or, $$2 \times 1.78 V_1 - 2 \times 1.78 \times 2u - 2u = 0$$

or, $$3.56 \ V_1 - 9.12 u = 0$$

or, $$\text{Vane speed, } u = \frac{3.56 \ V_1}{9.12} = \frac{3.56 \times 60}{9.12} = \mathbf{23.42 \ m/s \ (Ans.)}$$

(ii) The corresponding thrust on the vanes, F :

Substituting the relavent data in eqn (*i*), we get

$$F = 1000 \times 0.0007068 \times 60\,(1.78 \times 60 - 1.78 \times 23.42) = \mathbf{2761.2\ N\ (Ans.)}$$

Example 1.26. *The rotor of inward flow hydraulic turbine has a diameter over the tips of the moving vanes of 1.2 m. The diameter at the bottom of the vanes is 0.72 m. The speed is 300 r.p.m. The water is supplied through fixed vanes at 10° to the tangent to the outer circumference of the rotor, the velocity of water being 12 m/s. If the water leaves the moving vanes with the velocity entirely radial and equal to 4.2 m/s, determime:*

(i) The blade angles at entry and exit, so that the water may enter and leave the moving blades without shock.

(ii) The velocity of water relative to the vanes at the exit.

Solution. Radius of the wheel at inlet of the vane, $R_1 = \dfrac{1.2}{2} = 0.6$ m

Radius of the wheel at outlet of the vane, $R_2 = \dfrac{0.72}{2} = 0.36$ m

Speed of the wheel, $N = 300$ r.p.m.

$\therefore$ Angular speed, $\omega = \dfrac{2\pi N}{60} = \dfrac{2\pi \times 300}{60} = 10\pi$ rad./s

Then, tangential velocity at inlet tip of this blade, $u_1 = \omega R_1 = 10\pi \times 0.6 = 18.85$ m/s,

and, tangential velocity at outlet tip of the blade, $u_2 = \omega R_2 = 10\pi \times 0.36 = 11.31$ m/s

Angle at inlet, $\alpha = 10°$

Absolute velocity of jet at inlet, $V_1 = 12$ m/s

Absolute velocity of jet at outlet, $V_2 = 4.2$ m/s

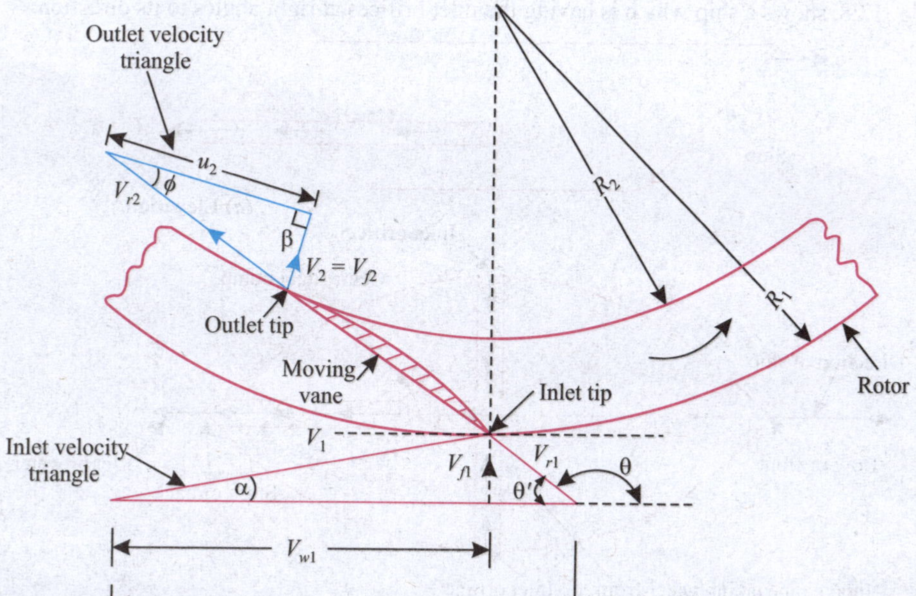

Fig. 1.27.

(i) The blade angles at entry and exit, θ and ϕ :

From inlet velocity triangle, we have :

$$V_{w1} = V_1 \cos\alpha = 12 \cos 10° = 11.82 \text{ m/s}$$

$$V_{f1} = V_1 \sin\alpha = 12 \sin 10° = 2.08 \text{ m/s}$$

$$\therefore \quad \tan\theta' = \frac{V_{f1}}{u_1 - V_{w1}} = \frac{2.08}{18.85 - 11.82} = 0.295$$

or, $$\theta' = \tan^{-1} 0.295 = 16.43°$$

∴ *Blade angle at inlet,* $\theta = 180° - \theta' = 180° - 16.43° =$ **163.57° (Ans.)**

From outlet triangle, we have:

$$\tan\phi = \frac{V_2}{u_2} = \frac{4.2}{11.31} = 0.3713$$

or, $$\phi = \tan^{-1} 0.3713 = \mathbf{20.37°}$$ **(Ans.)**

(ii) The velocity of water relative to the moving vanes, V_{r2} :

From outlet velocity triangle, we have:

$$\frac{V_2}{V_{r2}} = \sin\phi, \text{ or, } V_{r2} = \frac{V_2}{\sin\phi} = \frac{4.2}{\sin 20.37°} = \mathbf{12.06 \text{ m/s}}$$ **(Ans.)**

1.9. JET PROPULSION OF SHIPS

One of the applications of the *impulse-momentum equation* is **'jet propulsion'** wherein the reaction of a high velocity jet issuing from a nozzle provides the necessary thrust. This principle is employed in propelling the ships, aircrafts and missiles.

Case I. When the inlet orifices are at right angles to the direction of the motion of the ship.

Fig. 1.28. shows a ship which is having the inlet orifices at right angles to its direction.

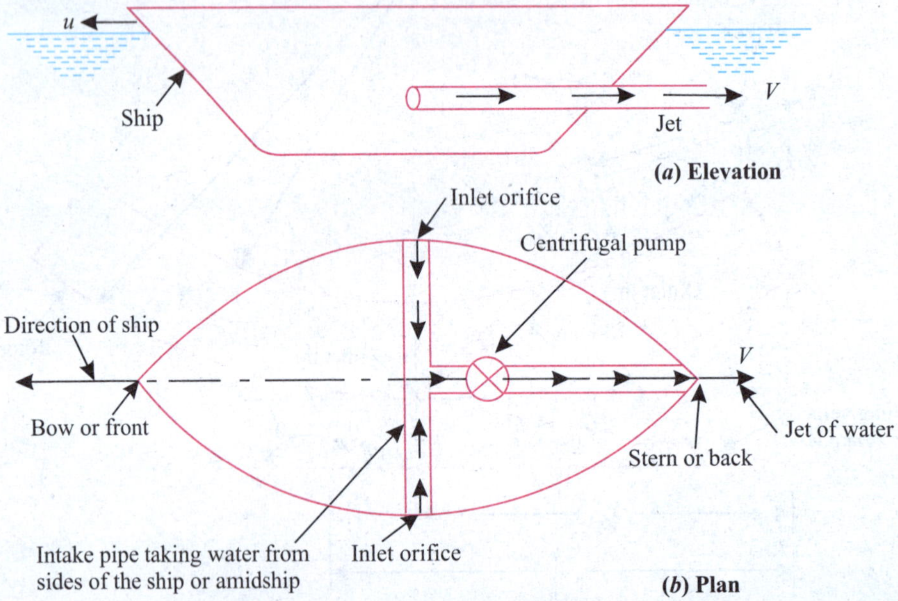

Fig. 1.28. Inlet orifices at right angles to the ship motion.

The ship carries a centrifugal pump which takes water from the sea and discharges it through nozzle at the rear of the ship.

Let, V = Absolute velocity of the issuing jet,

u = Velocity of the moving ship,

V_r = Relative velocity of the jet with respect to ship

$= V - (-u) = V + u$ (since the ship velocity is in a direction opposite to that of absolute jet velocity), and

a = Area of orifices.

Mass of water issuing from the orifice at the back of the ship $= \rho a V_r = \rho a (V + u)$

∴ Propulsive force exerted on the ship,

F = Mass flow rate of water × change in jet velocity

$$= \rho a (V+u)(V_r - u) = \rho a (V+u)[(V+u) - u] = \rho a (V+u) \times V \qquad ...[1.32]$$

Thrust or propulsive work per second = Forward thrust × speed of ship

$$= F \times u = \rho a (V+u) \times V \times u \qquad ...[1.33]$$

As the intake is at right angles to the direction of motion, inlet velocity of water is zero; hence the outlet velocity equals V_r and hence the kinetic energy supplied by the jet per second is

$$= \frac{1}{2}(\rho a V_r) \times V_r^2 = \frac{1}{2}\rho a V_r^3 = \frac{1}{2}\rho a (V+u)^3 \qquad ...[1.34]$$

$$\therefore \text{ Efficiency of propulsion, } \eta = \frac{\text{Propulsive work}}{\text{Kinetic energy supplied by the jet}}$$

$$= \frac{\rho a (V+u) \times V \times u}{\frac{1}{2}\rho a (V+u)^2} = \frac{2Vu}{(V+u)^2} \qquad ...[1.35]$$

For a given jet velocity V, the condition for *maximum efficiency of propulsion* is given by, $\frac{d\eta}{du} = 0$

i.e.,
$$\frac{d}{du}\left(\frac{2Vu}{(V+u)^2}\right) = 0$$

or,
$$\frac{(V+u)^2 \times 2V - 2Vu \times 2(V+u)}{(V+u)^4} = 0$$

or,
$$(V+u)^2 \times 2V - 2Vu \times 2(V+u) = 0$$

or,
$$(V+u)[(V+u)2V - 4Vu] = 0$$

$$(V+u)2V - 4Vu = 0 \qquad (\because V \neq -u)$$

or,
$$2V(V+u-2u) = 0 \qquad (\because V \neq -u)$$

or,
$$V - u = 0$$

$$\therefore \quad u = V \qquad ...[1.36]$$

Hence for *maximum efficiency of propulsion*, $u = V$ and by substitution,

$$\eta_{max} = \frac{2 \times u \times u}{(u+u)^2} = \frac{2u^2}{(2u)^2} = 05 \text{ or } \mathbf{50\%}$$

(Neglecting loss of head due to friction, etc. in the intake and ejecting pipes.)

Case II. When the inlet orifices face the direction of motion of the ship.

Fig. 1.29 shows a ship which is having the inlet orifices *facing the direcion of the motion of the ship.* The water is drawn in by a longitudinal (intake) pipe from bow (front) of the ship and is discharged at the stern (back).

For this case the expressions for propulsive force and work done per second will be the *same* as those for the case I *i.e.,*when the inlet orifices are at right angles to the ship motion. But the energy supplied by the jet will be different, as in this case the water enters with a velocity equal to the velocity of the ship (u).

$\therefore$ Kinetic energy supplied by the jet $= \frac{1}{2}$ mass of water supplied per second $\times (V_r^2 - u^2)$

$$= \frac{1}{2} \rho a V_r (V_r^2 - u^2)$$

But, $V_r = V + u$ (as in case I)

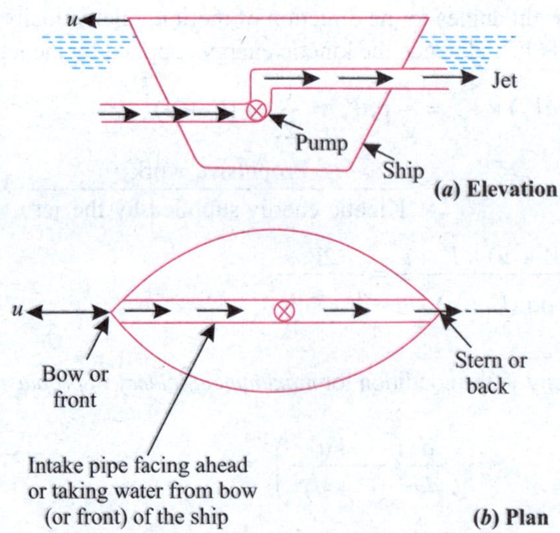

Fig. 1.29. Inlet orifices facing the direction of ship.

$\therefore$ Kinetic energy supplied by the jet $= \frac{1}{2} \rho a (V + u) [(V + u)^2 - u^2]$ ...(1.37)

$\therefore$ Efficiency of propulsion,

$$\eta = \frac{\text{Work done per second by jet}}{\text{Energy supplied by the jet}}$$

$$= \frac{\rho a (V + u) \times V \times u}{\frac{1}{2} \rho a (V + u) [(V + u)^2 - u^2]}$$

[Work done $= \rho a (V + u) \times V \times u$] ...from eqn. (1.33)

$$= \frac{2Vu}{(V + u)^2 - u^2} = \frac{2Vu}{V^2 + u^2 + 2Vu - u^2} = \frac{2Vu}{V^2 + 2Vu} = \frac{2u}{V + 2u} \quad ...(1.38)$$

In this case, however it is not possible to derive a practical condition for maximum efficiency. But for $u = V$, which is the condition for the maximum efficiency in the previous case (*i.e.,* case I), corresponding value of the efficiency for this case will be:

$$\eta = \frac{2u}{V + 2u} = \frac{2u}{u + 2u} = \frac{2}{3} \text{ or } 0.6667 = \mathbf{66.67\%}$$

In actual practice, since the velocity of ship u will normally be *less* than the velocity of jet V, and therefore the *limiting value of u is equal to V*. Hence the above obtained value of the efficiency may be *cosidered as the maximum possible efficiency for this case*.

Note: The propulsion of ships is also known as ***water rocket.*** In practice, jet propulsion is used only in small life boats. For large ships, the screw propulsion is generally perferred because of *high overall efficiency of screw propulsion*.

Example 1.27. *A ship driven by reaction jets and discharging astern is found to have resistance to motion of 2950 N when moving at 30 km/h. The velocity of jet relative to ship is 18 m/s. Determine:*

(*i*) *The number of jets if each jet has an area of 85 cm²;*

(*ii*) *The power required to work the pump and the propulsive efficiency for the following cases:*

(*a*) *Inlet orifices at right angles to ship motion;*

(*b*) *Inlet orifices face the direction of ship motion.*

Solution. Resistance to motion = 2950 N

$$\text{Velocity of the ship, } u = 30 \text{ km/h} = \frac{30 \times 1000}{60 \times 60} = 8.33 \text{ m/s}$$

Velcoity of the jet relative to the ship, $V_r = (V + u) = 18$ m/s

$\therefore$ Absolute velocity of the jet, $V = V_r - u = 18 - 8.33 = 9.67$ m/s

Area of each jet, $a = 85 \text{ cm}^2 = 85 \times 10^{-4} \text{ m}^2$

(*i*) **Number of jets required :**

Thrust force exerted on the ship, $F = \rho a\,(V + u) \times V$ ...[Eqn. (1.32)]

Since thrust equals the resistance to motion of ship, therefore,

$$2950 = 1000 \times a \times 18 \times 9.67, \text{ or, } a = \frac{2950}{1000 \times 18 \times 9.67} = 0.01695 \text{ m}^2$$

$$\therefore \quad \text{Number of jets} = \frac{\text{Total area of jets}}{\text{Area of each jet}} = \frac{0.01695}{85 \times 10^{-4}} = \mathbf{2} \textbf{ (Ans.)}$$

(*ii*) **Power required to drive the pump and propulsive efficiency :**

Propulsive work = Thrust × forward velocity = 2950 × 8.33 = 24573.5 Nm/s

$\therefore$ *Power required* = 24573.5 J/s or W or **24.573kW (Ans.)**

This will be same for both the cases (*a*) and (*b*).

(*a*) When intake orifices are at right angles to motion of ship, then kinetic energy supplied by jets

$$= \frac{1}{2} \rho\, a\, (V + u)^3 \qquad \text{...[Eqn. 1.34]}$$

$$= \frac{1}{2} \times 1000 \times 0.01695 \times 18^3 = 49426.2 \text{ Nm/s}$$

$\therefore$ *Propulsive efficiency*

$$= \frac{\text{Propulsive work}}{\text{Kinetic energy supplied by the jet}} = \frac{24573.5}{49426.2} = 0.497 \text{ or } \mathbf{49.7\%} \textbf{ (Ans.)}$$

$$\left[\begin{array}{l}\text{Alternatively,}\\ \textit{Propulsive efficiency} = \dfrac{2Vu}{(V+u)^2} = \dfrac{2\times 9.67\times 8.33}{18^2} = 0.497 \text{ or } \mathbf{49.7\%}\end{array}\right]$$

(*b*) When intake orifices face the direction of ship motion, then kinetic energy supplied by the jet,

$$= \frac{1}{2}\rho a\,(V+u)\,[(V+u)^2 - u^2] \qquad \text{...[Eqn. (1.37)]}$$

$$= \frac{1}{2}\times 1000\times 0.016695\times 18\,(18^2 - 8.33^2) = 38840.9 \text{ Nm/s}$$

$$\therefore \text{ Propulsive efficiency} = \frac{\text{Propulsive work}}{\text{Kinetic energy supplied by the jet}}$$

$$= \frac{24573.5}{38840.9} = 0.6326 \text{ or } \mathbf{63.26\%} \textbf{ (Ans.)}$$

Example 1.28. *A small ship is fitted with jets of total area 0.65 m². The velocity through the jet is 9 m/s and speed of the ship is 18 km/h in sea water. The efficiencies of the engine and the pump are 85% and 65% and respectively. If the water is taken amidship, determine:*

(*i*) *Propelling force, and*

(*ii*) *Overall efficiency.*

Assume the pipe losses to be 10% of the kinetic energy of the jets.

[MDU Haryana]

Solution. Total area of jets, $a = 0.65 \text{ m}^2$

Velocity of jet relative to ship, $V_r = 9$ km/s

Speed of the ship, $u = 18 \text{ km/h} = \dfrac{18\times 1000}{60\times 60} = 5$ m/s

Efficiency of the engine, $\eta_E = 85\%$

Efficiency of the pump, $\eta_P = 65\%$

Pipe losses, $h_f = 10\%$ of K.E. of jets

$$= \frac{10}{100}\times\frac{V_r^2}{2} = \frac{V_r^2}{20}$$

Now, $V_r = V + u$

or, $V = V_r - u = 9 - 5 = 4$ m/s

(where, V = absolute velocity of jet)

(*i*) Propelling force, F :

$$F = \rho a\,(V+u)\,V \qquad \text{...[Eqn. (1.32)]}$$

$$= 1000\times 0.65\,(4+5)\times 4 = \mathbf{23400\ N\ (Ans.)}$$

(*ii*) Overall efficiency, $\eta_{overall}$:

Work done by the jets per second

$$= F\times u = 23400\times 5 = 117000 \text{ Nm/s}$$

The output of the pump should be such as to give the jet a relative velocity V_r and also overcome the pipe losses.

∴ Output of the pump per kg of water

$$= \text{K.E. of jet} + \text{pipe losses}$$

$$= \frac{V_r^2}{2} + \frac{V_r^2}{20} = \frac{V_r^2}{2}(1 + 0.1) = 1.1\frac{V_r^2}{2}$$

$$\text{Input to the pump per kg of water} = \frac{\text{Output of the pump}}{\text{Efficiency of the pump}(\eta_p)}$$

$$= \frac{1.1V_r^2}{2 \times 0.65}$$

The input to the pump = Output of the engine

$$\therefore \text{ Input to the engine per kg of water} = \frac{1.1V_r^2}{2 \times 0.65 \times \eta_E}$$

$$= \frac{1.1 \times 9^2}{2 \times 0.65 \times 0.85} = 80.63 \text{ Nm}$$

$\therefore$ Total input to the engine = Mass of water × input per kg of water

$$= (\rho a V_r) \times 80.63$$

$$= 1000 \times 0.65 \times 9 \times 80.63 = 471685.5 \text{ Nm}$$

$$\therefore \quad \eta_{overall} = \frac{\text{Work done by jets}}{\text{Total input to engine}} = \frac{117000}{471685.5} = 0.248 \text{ or } \mathbf{24.8\%} \textbf{ (Ans.)}$$

Example 1.29. *A jet propelled boat, moving with a velocity of 6 m/s, draws water amidship. The total area of the jet is 424 cm². If the total resistance offered to the motion of the boat is 5890 N, determine:*

(i) Volume of water drawn by the pump per second, and

(ii) Efficiency of the jet propulsion.

Solution. Velocity of boat, $u = 6$ m/s

Total area of jets, $a = 424 \text{ cm}^2 = 424 \times 10^{-4} = 0.0424 \text{ m}^2$

Total resistance to motion = 5890 N

Since the propelling force must be equal to the resistance to the motion, therefore,

Propelling force, $F = 5890$ N

But propelling force, $F = \rho a (V + u) V$ ...[Eqn. (1.32)]

$$\therefore \quad \rho a (V + u) V = 5890$$

or, $1000 \times 0.0424 (V + 6) \times V = 5890$

$$\text{or,} \quad (V + 6) V = \frac{5890}{1000 \times 0.0424} = 138.9$$

$$\text{or,} \quad V^2 + 6V - 138.9 = 0$$

$$V = \frac{-6 \pm \sqrt{6^2 + 4 \times 138.9}}{2} \cdot \frac{-6 \pm 24.32}{2} = 916 \text{ m/s}$$

(i) Volume of water drawn by the pump per second :

= Volume of water discharged through orifices at the back in the form of the jets

$= aV_r = a(V + u) = 0.0424 (9.16 + 6) = \mathbf{0.643 \ m^3/s}$ **(Ans.)**

(ii) Efficiency of the jet propulsion :

$$\eta = \frac{2Vu}{(V+u)^2} \qquad \text{...Eqn. (1.35)}$$

$$= \frac{2 \times 9.16 \times 6}{(9.16+6)^2} = 0.478 \text{ or } \mathbf{47.8\%} \text{ (Ans.)}$$

Example 1.30. (*a*) *A small ship driven by reaction jets and discharging astern is estimated to have a relative velocity of 12 m/s when moving at 30 km/h. The cross-sectional area of the jets at discharge is 240 cm². Find resistance to the motion of ship and propulsive work.*

(*b*) *If water enters through orifices facing the direction of motion of ship and pump is 85 per cent efficient and frictional losses in the pipe are equivalent to 3.6 m of water head, find:*

(*i*) *Power required to drive the pump, and*

(*ii*) *Overall efficiency of propulsion.*

Solution. Relative velocity of ship, $V_r = 12$ m/s

$$\text{Speed of the ship, } u = 30 \text{ km/h} = \frac{30 \times 1000}{60 \times 60} = 8.33 \text{ m/s}$$

The cross-sectional area of the jet at discharge, $a = 240 \text{ cm}^2 = 0.024 \text{ m}^2$

Efficiency of pump, $\eta_p = 85\%$

Frictional losses, $h_f = 3.6$ m of water head.

(*a*) Resistance to motion and propulsive work :

Thrust or propulsive force, $F = \rho a (V + u) V$ ...[Eqn. (1.32)]

Now, $V_r = V + u$ or $V = V_r - u = 12 - 8.33 = 3.67$ m/s

(where, V = absolute velocity of the jet)

$\therefore \quad F = 1000 \times 0.024 \times 12 \times 3.67 = 1057$ N

Hence, *resistance to motion* = F = **1057 N (Ans.)**

Propulsive work = Thrust × forward velocity

= 1057 × 8.33 = 8804.8 Nm/s

= 8804.8 J/s or W or **8.8 kW (Ans.)**

(*b*) When the intake of water is at the front end and the water enters with the boat speed u and discharges astern (back) with relative velocity V_r, then,

$$\text{Energy supplied by the jet} = \frac{1}{2} m (V_r^2 - u^2)$$

Actual energy supplied to water, considering frictional loss in the pipes

$$= \frac{1}{2} m (V_r^2 - u^2) + mgh_f = \frac{1}{2} \rho a V_r (V_r^2 - u^2) + (\rho a V_r) g h_f$$

$$= \rho a V_r \left[\frac{1}{2}(V_r^2 - u^2) + gh_f\right]$$

$$= 1000 \times 0.024 \times 12 \left[\frac{1}{2}(12^2 - 8.33^2) + 9.81 \times 3.6\right] = 20913 \text{ Nm/s} \simeq 20.9 \text{ kW}$$

This is equal to the *output of the pump.*

(*i*) Power required to drive the pump :

$$\text{Power required ro drive the pump} = \frac{\text{Output of the pump}}{\eta_p} = \frac{20.9}{0.85} = \mathbf{24.58 \text{ kW}} \text{ (Ans.)}$$

(ii) Overall efficiency of propulsion, $\eta_{overall}$:

$$\eta_{overall} = \frac{\text{Propulsive power}}{\text{Power supplied to the pump}}$$

$$= \frac{8.8}{24.58} = 0.358 \text{ or } \mathbf{35.8\%} \text{ (Ans.)}$$

Example 1.31. *A jet-propelled boat discharges water through a jet of area 240 cm^2; the water being drawn from inlet openings facing the direction of motion. The total drag is estimated to be 21.2 u^2 N where u is the speed of the boat in m/s. If the boat moves at 64.8 km/h, determine:*

(i) Relative velocity of jet,

(ii) Energy supplied by the jet,

(iii) Power of the motor required to work the pumps, and

(iv) Efficiency of propulsion. **[Anna University]**

Take efficiency of pump set as 80% and density of water 1020 kg/m^3.

Solution. Area of jet, $a = 240 \text{ cm}^2 = 0.024 \text{ m}^2$

Total drag $= 21.2\, u^2 N$, where u is the speed of boat in m/s.

Velocity of the boat, $u = 64.8$ km/h

$$= \frac{64.8 \times 1000}{60 \times 60} = 18 \text{ m/s}$$

Efficiency of the pump, $\eta_p = 80\%$

(i) Relative velocity of jet, V_r :

Thrust force on the boat $= \rho a V_r (V_r - u)$

Also, resistance to the motion of ship (drag) equals the thrust force,

$$\therefore \quad 21.2u^2 = \rho a V_r (V_r - u)$$

or, $$21.2 \times 18^2 = 1020 \times 0.024\, V_r (V_r - 18)$$

or, $$6868.8 = 24.48\, V_r (V_r - 18)$$

or, $$V_r (V_r - 18) = \frac{6868.8}{24.48} = 280.6$$

or, $$V_r^2 - 18V_r - 280.6 = 0$$

or, $$V_r = \frac{18 \pm \sqrt{18^2 + 4 \times 280.6}}{2}$$

$$= \frac{18 + 38.03}{2} = \mathbf{28 \text{ m/s}} \text{ (Ans.)}$$ (–ve value is not possible.)

(ii) Energy supplied by the jet :

Mass of water discharged $= \rho a V_r$

$\therefore$ Kinetic energy supplied by the jet

$$= \frac{1}{2} (\rho a V_r)\, (V_r^2 - u^2)$$

$$= \frac{1}{2}(1020 \times 0.024 \times 28)\, (28^2 - 18^2) = 157651.2 \text{ N m/s}$$

***(iii)* Power of motor :**

Power of the motor required to work the pumps

$$= \frac{157651.2}{0.8} = 197064 \text{ Nm/s or } 197064 \text{ J/s or W or} \simeq \mathbf{197\ kW}\ \textbf{(Ans.)}$$

***(iv)* Efficiency of propulsion, η_{prop} :**

$$\eta_{prop.} = \frac{2u}{V + 2u} \qquad \text{...[Eqn. (1.38)].}$$

$$= \frac{2u}{V_r + u} = \frac{2 \times 18}{28 + 18} = 0.7826 \text{ or } \mathbf{78.26\%}\ \textbf{(Ans.)} (\because V + u = V_r)$$

HIGHLIGHTS

1. A fluid jet is a stream of fluid issuing from a nozzle with a high velocity and hence a high kinetic energy.
2. The force exerted by a jet of water on a "*stationary plate*" (F_x):
   $F_x = \rho a V^2$ ...for a vertical plate,
   $= \rho a V^2 \sin^2 \theta$ ...for an inclined plate,
   $= \rho a V^2 (1 + \cos \theta)$ ...for a curved plate and jet strikes at the centre,
   $= 2\, \rho a V^2 \cos \theta$ ...for a curved plate and jet strikes at one of the tips of the jet.
   where, V = Velocity of the jet,
   θ = Angle between the jet and the plate for inclined plate, and
   = Angle made by the jet with the direction of motion for curved plates.
3. The force exerted by a jet of water on a *moving plate* in the direction of the motion of the plate (F_x):
   $F_x = \rho a (V - u)^2$ ...for a moving vertical plate,
   $= \rho a (V - u)^2 \sin^2 \theta$ ...for an inclined moving plate, and
   $= \rho a (V - u)^2 (1 + \cos \theta)$ ...when jet strikes the curved plate at the centre.
4. When a jet of water strikes a curved moving vane at one of its tips and comes curved out at the other tip, the force exerted and work done are given by (from inlet and outlet velocity triangles):
   Force exerted, $F_x = \rho a V_{r1} (V_{w1} \pm V_{w2})$
   Work done per second $= \rho a V_{r1} (V_{w1} \pm V_{w2}) \times u$
   + ve sign is taken ...when $\beta < 90°$ (*i.e.*, β is an acute angle)
   – ve sign is taken ...when $\beta > 90°$ (*i.e.*, β is an obtuse angle)
   $V_{w2} = 0$ ...when $\beta = 90°$
   Work done per second per N of fluid

   $$= \frac{1}{g} (V_{w1} + V_{w2}) \times u$$

   For series of vanes:

   Force exerted, $F_x = \rho a V_1 (V_{w1} \pm V_{w2})$

   Work done per sec $= \rho a V_1 (V_{w1} \pm V_{w2}) \times u$

   Work done per sec per N of fluid $= \frac{1}{g} (V_{w1} + V_{w2}) \times u$

where, V_1 = Absolute velocity of jet at inlet,
V_{w1} = Velocity of whirl at inlet,
V_{w2} = Velocity of whirl at outlet, and
u = Velocity of the vane.

5. ***For seires of radial curved vanes:***

Work done per second on the wheel

$$= \rho a V_1 (V_{w1} \times u_1 \pm V_{w2} \times u_2)$$

Efficiency of the radial curved vane,

$$\eta_{vane} = \frac{\rho a V_1 (V_{w1} u_1 \pm V_{w2} u_2)}{\frac{1}{2}(\rho a V_1) \times V_1^2} = \frac{2(V_{w1} u_1 \pm V_{w2} u_2)}{V_1^2}$$

where, u_1 = Tangential velocity of vane at inlet, and
u_2 = Tangential velocity of vane at outlet.

6. ***Jet propulsion of ships:***

Case I. When the inlet orifices are at right angles to the direction of motion of the ships.

Efficiency of propulsion, $\eta = \frac{2Vu}{(V+u)^2}$

Conditions for maximum efficiency, $\frac{d\eta}{du} = 0$, *i.e.*, $u = V$

η_{max} = 50% (neglecting loss of head due to friction, etc. in the intake and ejecting pipes)

Case II. When the inlet orifices face the direction of motion of the ship.

Efficiency of propulsion, $\eta = \frac{2u}{V + 2u}$

OBJECTIVE TYPE QUESTIONS

Choose the Correct Answer:

1. The force exerted by a jet of water on a stationary vertical plate in the direction of jet is given by
(*a*) $\rho a V$ (*b*) $\rho a V^2$
(*c*) $\rho a^2 V$ (*c*) $\rho a V^3$.

2. The force exerted by a jet of water on a moving vertical plate, in the direction of motion of plate is given by
(*a*) $\rho a V^2$ (*b*) $\rho a V^3$
(*c*) $\rho a (V - u)^2$ (*d*) $\rho a (V - u)^3$.

3. When a steady jet impinges on a fixed inclined surface
(*a*) the flow is divided into parts proportional to the angle of inclination of the surface
(*b*) no force is exerted by the jet on the vane
(*c*) the momentum component is unchanged parallel to the surface
(*d*) none of the above.

4. For maximum efficiency of a series of curved vanes, the speed is
(*a*) equal to the jet speed
(*b*) $\frac{3}{4}$ of the jet speed
(*c*) $\frac{1}{2}$ of the jet speed
(*d*) $\frac{1}{3}$ of the jet speed.

5. The efficiency of jet propulsion with inlet orifices at right angles to the direction of motion of ship is given by
(*a*) $\frac{2u}{V+u}$ (*b*) $\frac{2V}{(V+u)^2}$
(*c*) $\frac{2Vu}{(V+u)^2}$ (*d*) $\frac{2u(V-u)}{V^3}$.

6. The efficiency of jet propulsion when the inlet orifices face the direction of motion of the ship is given by

(a) $\frac{2V}{V+u}$ (b) $\frac{2u}{V+2u}$.

(c) $\frac{2Vu}{V+u}$ (d) $\frac{2V}{V+u}$.

ANSWERS

1. (b) 2. (c) 3. (a) 4. (c) 5. (c) 6. (b).

THEORETICAL QUESTIONS

1. Derive an expression for the force exerted by a jet of water on a fixed vertical plate in the direction of the jet.
2. Show that the force exerted by a jet of water on moving inclined plate in the direction of jet is given by

$$F_x = \rho a (V-u)^2 \sin^2 \theta$$

where, a = Area of jet,

V = Velocity of the jet, and

θ = Inclination of the plate with the jet.

3. Prove that for a curved radial vane the efficiency is given by

$$\eta = \frac{2 (V_{w1}\, u_1 \pm V_{w2}\, u_2)}{V_1^2}$$

4. Show that efficiency of propulsion when the inlet orifices face the direction of motion of the ship is given by, $\eta = \frac{2u}{V+2u}$, where V is absolute velocity of issuing jet and u is the velocity of the ship.

UNSOLVED EXAMPLES

1. A jet of water, 50 mm in diameter, issues with a velocity of 10 m/s and impinges on a stationary flat plate which destroys its forward motion. Find the force exerted by the jet on the plate and the work done. **[Ans.** 196.35 N]
2. A jet of water of diameter 75 mm moving with a velocity of 20 m/s strikes a fixed plate in such a way that the angle between the jet and the plate is 60°. Find the force exerted by the jet on the plate.
   (*i*) in the direction normal to the plate, and
   (*ii*) in the direction of jet.
   [Ans. (*i*) 1530 N; (*ii*) 1325 N]
3. A jet of water of diameter 60 mm moving with a velocity of 40 m/s, strikes a curved fixed symmetrical plate at the centre. Find the force exerted by the jet of water in the direction of the jet, if the jet is deflected through an angle of 120° at the outlet of the curved plate. **[Ans.** 6785.8 N]
4. A jet of water of diameter 70 mm moving with a velocity of 40 m/s, strikes a curved fixed plate tangentially at one end at an angle of 30° to the horizontal. The jet leaves the plate at an angle 20° to the horizontal. Find the force exerted by the jet on the plate in the horizontal and vertical directions. **[Ans.** 11117 N; 972.61 N]
5. A rectangular plate weighing 60 N is suspended vertically by a hinge on the top horizontal edge. The centre of gravity of the plate is 100 mm from the hinge. A horizontal jet of water of 25 mm diameter, whose axis is 150 mm below the hinge, impinges normally to the plate with a velocity of 6 m/s. Find:
   (*i*) The horizontal force applied at the centre of gravity to maintain the plate in vertical position, and
   (*ii*) The change in velocity of jet if the plate is deflected through 30° and the same horizontal force continues to act at the centre of gravity of the plate.
   [Ans. (*i*) 26.49 N, (*ii*) 2.48 m/s (increase)]
6. A square plate weighing 117.72 N and of uniform thickness and 300 mm edge is hung so that horizontal jet 20 mm dimater and having a velocity of 15 m/s impinges on the plate. The centre line of the jet is 150 mm below the upper edge of the plate, and when the plate is vertical the jet strikes the plate normally and at its centre.
   (*i*) Find what force must be applied at the lower edge of the plate in order to keep the plate vertical.

(*ii*) If the plate is allowed to swing freely, find the inclination to the vertical which the plate will assume under the action of jet.

[**Ans.** (*i*) 35.32 N, (*ii*) 36.87°]

7. A nozzle of 50 mm diameter delivers a stream of water at 20 m/s prependicular to a plate that moves away from the jet at 5 m/s. Find:

(*i*) The force on the plate,

(*ii*) The work done, and

(*iii*) The efficiency of jet.

[**Ans.** (*i*) 441.45 N; (*ii*) 2207 Nm/s (*iii*) 28.1%]

8. A jet of water of 75 mm diameter strikes a curved vane at its centre with a velocity of 20 m/s. The curved vane is moving with a velocity of 8 m/s in the direction of jet. Find the force exerted on the plate in the direction of the jet, power and efficiency of the jet.

Assume the plate to be smooth.

[**Ans.** 1250.4 N; 10 kW (app.), 56.4%]

9. A jet of water 100 mm diameter and having a velocity of 30 m/s strikes tangentially on a wheel which deflects the jet through an angle of 120°. Calculate the thrust on the vane when

(*i*) The axis of symmetry of the vane is horizontal.

(*ii*) The tangent at inlet tip is horizontal.

[**Ans.** 1958 N; 30°, (*ii*) 1958 N; 30°]

10. A jet of water 100 mm diameter and having a velocity of 15 m/s impinges at the centre of a hemispherical vane. The linear velocity of vane is 5 m/s in the direction of the jet. Find the force exerted on the vane. How this force would change if the jet impinges on a series of vanes attached to the circumference of a wheel ?

[**Ans.** 1569.6 N, 2354.4 N]

11. A jet of water moving with a velocity of 30 m/s impinges on series of curved vanes moving with a velocity of 15 m/s. The jet makes an angle of 20° with the direction of motion of vanes. Assuming $K = 0{\cdot}9$, and that the absolute velocity of water at exit is normal to the direction of motion of vanes, determine:

(*i*) Vane angles at entrance and exit,

(*ii*) Work done on the vanes per N of water, and

(*iii*) Efficiency of the system.

[**Ans.** (*i*) 37.85°, 3.8°; (*ii*) 43.2 Nm/s; (*iii*) 94%]

12. A jet of water moving with a velocity of 20 m/s impinges on a curved vane, which is moving with a velocity of 10 m/s. The jet makes an angle of 20° with the direction of motion of vane at inlet and leaves at angle of 130° to the direction of motion of vane at outlet. Determine:

(*i*) The angles of curved vane tips so that water enters and leaves without shock;

(*ii*) The work done per N of water entering the vane.

[**Ans.** (*i*) $\theta = 37.87°$; $\phi = 6.56°$; (*ii*) 20.24 Nm/s or W]

13. A wheel consists of radial blades with inner and outer radii of 300 mm and 600 mm respectively. Water enters the blades at the outer periphery with a velocity of 50 m/s and the supply jet makes an angle of 25° with tangent to wheel at inlet tip. Water leaving the blade has a flow velocity of 10 m/s. If the blade angles at entrance and exit are 40° and 30° respectively, determine:

(*i*) Work done per N of water,

(*ii*) Speed of the wheel, and

(*iii*) Efficiency of blading.

[**Ans.** (*i*) 100.43 Nm; (*ii*) 320.38 r.p.m., (*iii*) 78.8%]

14. In a jet propelled boat water is drawn amidship and discharged at the back with an absolute velocity of 20 m/s. If the cross-sectional area of the jet is 200 cm^2 and the boat is moving in sea water with a speed of 8·33 m/s determine: (*i*) The propelling force on the boat, (*ii*) Power required to drive the pump, and (*iii*) Efficiency of jet propulsion.

[**Ans.** 11332 N; (*ii*) 94.3 kW; (*iii*) 41.5%]

15. A jet propelled boat is discharging water at a speed of 10 m/s relative to the ship in a jet of 0·02 m^2 corss-sectional area. If the boat moves through water with a velocity of 20 km/h, determine:

(*i*) The resistance of the vessel and power exerted by the jet, and

(*ii*) The efficiency of jet apart from losses in pumping machinery.

Assume that inlet orifices face the direction of flow.

[**Ans.** (*i*) 890 N; (*ii*) 6.91 kW; 71.4%]

16. A jet-propelled boat discharges water through a jet of area 200 cm^2; the water being drawn form inlet openings facing the direction of motion. The total drag is estimated to be $17{\cdot}66\,u^2$ N where u is the speed of the boat in m/s. If the boat moves at 54 km/h, determine:

(*i*) Relative velocity of jet,

(*ii*) Energy supplied by the jet,

(*iii*) Power of motor required to work the pumps, and

(*iv*) Efficiency of propulsion.

Assume: Efficiency of pump-set = 75 per cent, and density of water = 1020 kg/m^3.

[**Ans.** (*i*) 23.34 m/s; (*ii*) 76125 Nm/s; (*iii*) 101.4 kW; (*iv*) 78.2%]

Chapter 2

HYDRAULIC TURBINES

2.1. INTRODUCTION

A **hydraulic turbine** is a *prime mover* (a machine which uses the raw energy of a substance and converts it into mechanical energy) *that uses the energy of flowing water and converts it into the mechanical energy* (*in the form of rotation of the runner*). This mechanical energy is used in running an electric generator which is directly coupled to the shaft of the hydraulic turbine; from this electric generator, we get electric power which can be transmitted over long distances by means of transmission lines and transmission towers. The hydraulic turbines are also known as '*water turbines*' since the fluid medium used in them is water.

First hydroelectric station was probably started in America in 1882 and thereafter development took place very rapidly. In India, the first major hydroelectric power plant of 4.5 MW capacity named as Sivasamudram Scheme in Mysore was commissioned in 1902.

Hydro (water) power is a conventional renewable source of energy which is clean, free from pollution and generally has a good environment effect. However, the following factors are major obstacles in the utilisation of hydropower resources:

(*i*) Large investments,

(*ii*) Long gestation period, and

(*iii*) Increased cost of power transmission.

Fig. 2.1 shows the flow sheet of hydroelectric power plant.

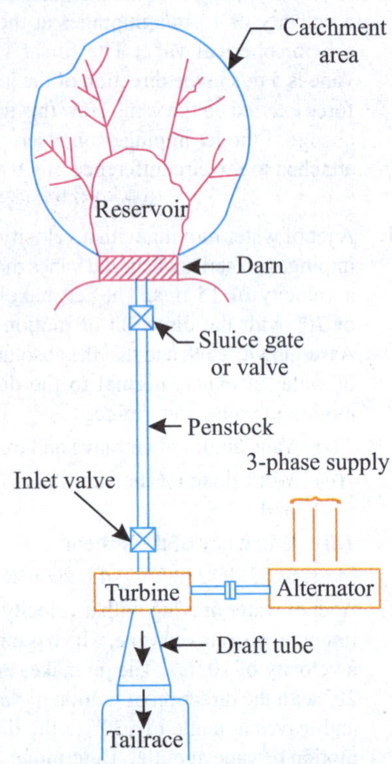

Fig. 2.1. Flow sheet of hydroelectric power plant.

2.2. CLASSIFICATION OF HYDRAULIC TURBINES

The hydraulic turbines are *classified* as follows :

1. According to the head and quantity of water available.
2. According to the name of the originator.
3. According to the action of water on moving blades.
4. According to the direction of flow of water in the runner.
5. According to the disposition of the turbine shaft.
6. According to the specific speed N.

1. According to the head and quantity of water avaliable :

(*i*) *Impulse turbine* ... requires *high head* and *small quantity of flow*.

(*ii*) *Reaction turbine* ... requires *low head* and *high rate of flow*.

Actually there are two types of reaction turbines, one for medium head and medium flow and the other for low head and large flow.

2. According to the name of the originator :

(*i*) *Pelton turbine* ... named after Lester Allen Pelton of California (U.S.A.). It is an impulse type of turbine and is used for *high head and low discharge*.

(*ii*) *Francis turbine* ... named after James Bichens Francis. It is a reaction type of turbine from *medium high to medium low heads and medium small to medium large quantites of water*.

(*iii*) *Kalpan turbine* ... named after Dr. Victor Kaplan. It is a reaction *type of turbine for low heads and large quanties of flow*.

3. According to action of water on the moving blades :

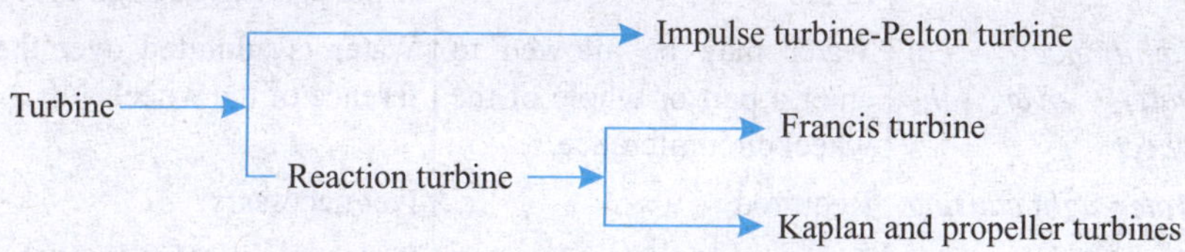

4. According to direction of flow of water in the runner :

(*i*) Tangential flow turbines (Pelton turbine)

(*ii*) Radial flow turbine (no more used)

(*iii*) Axial flow turbine (Kaplan turbine)

(*iv*) Mixed (radial and axial) flow turbine (Francis turbine).

In *tangential flow* turbine of Pelton type the water strikes the runner tangential to the path of rotation.

In *axial flow* turbine water flows parallel to the axis of the turbine shaft. Kaplan turbine is an axial flow turbine. In Kaplan turbine the runner blades are *adjustable and can be rotated* about pivots fixed to the boss of the runner. If the runner blades of the axial flow turbines are *fixed,* these are called "*propeller turbines*".

In *mixed flow* turbines the water enters the blades radially and comes out axially, parallel to the turbine shaft. *Modern Francis turbines have mixed flow runners.*

5. According to the disposition of the turbine shaft :

Turbine shaft may be either vertical or horizontal. In modern practice, Pelton turbines usually have horizontal shafts whereas the rest, especially the large units, have vertical shafts.

6. **According to specific speed :**

The *specific speed* of a turbine is defined as the speed of a geometrically similar turbine that would develop *1 kW* under *1 m* head. All geometrically similar turbines (irrespective of the sizes) will have the same specific speeds when operating under the same head.

$$\text{Specific speed, } N_s = \frac{N\sqrt{P}}{H^{5/4}}$$

where,
N = The normal working speed,
P = Power output of the turbine, and
H = The net or effective head in metres.

Turbines with low specific speeds work under high head and low discharge conditions, while high specific speed turbines work under low head and high discharge conditions.

The following table gives the comparison between the impulse and reaction turbines with regard to their operation and application.

Table 2.1. Comparison between Impulse and Reaction Turbines

S. No.	Aspects	Impulse turbine	Reaction turbine
1.	*Conversion of fluid energy*	The available fluid energy is converted into K.E. by a nozzle.	The energy of the fluid is partly transformed into K.E. before it (fluid) enters the runner of the turbine.
2.	*Changes in pressure and velocity*	The pressure remains same (atmospheric)throughout the action of water on the runner.	After entering the runner with an excess pressure, water undergoes changes both in velocity and pressure while passing through the runner.
3.	*Admittance of water over the wheel*	Water may be allowed to enter a part or whole of the wheel circumference.	Water is admitted over the circumference of the wheel.
4.	*Water-tight causing*	Required	Not necessary.
5.	*Extent to which the water fills the wheel/ turbine*	The wheel/turbine does not run full and air has a free access to the buckets.	Water completely fills all the passages between the blades and while flowing between inlet and outlet sections does work on the blades.
6.	*Installation of unit*	Always installed above the tail race. No draft tube is used.	Unit may be installed above or below the tail race, use of a draft tube is made.
7.	*Relative velocity of water*	Either remaining constant or reduces slightly due to friction.	Due to continuous drop in pressure during flow through the blade, the relative velocity increases.
8.	*Flow regulation*	— By means of a needle valve fitted into the nozzle. — Impossible without loss.	— By means of a guide-vane assembly. — Always accompanied by loss.

2.3. IMPULSE TURBINES—PELTON WHEEL

In an impulse turbine the *pressure energy of water is converted into kinetic energy* when passed through the nozzle and forms the high velocity jet of water. The formed water jet is used for driving the wheel.

Pelton wheel (named after the American engineer Lester Allen Pelton), among the various impulse turbines that have been designed and utilized, is by far the important. The Pelton wheel or Pelton turbine is a *tangential flow impulse turbine.*

Important Pelton turbine installations in India :

S. No.	Scheme/Project	Location (State)	Source of water
1.	Koyana hydroelectric project	Koyana (Maharashtra)	Koyana river
2.	Mahatama Gandhi hydroelectric works	Sharavathi (Karnataka)	Sharavathi river
3.	Mandi hydroelectric scheme	Joginder Nagar (Himachal Pradesh)	Uhl river
4.	Pallivasal power station	Pallivasal (Kerala)	Mudirapuzle river
5.	Pykara hydroelectric scheme	Pykara (Tamil Nadu)	Pykara river.

2.3.1. Construction and working of Pelton Wheel/Turbine

A Pelton wheel/turbine consists of a **rotor**, at the periphery of which are mounted equally spaced *double hemispherical or double ellipsoidal buckets.* Water is transferred from a high head source through penstock which is fitted with a **nozzle**, through which the water flows out at a *high speed jet.* **A needle spear** moving inside the nozzle *controls the water flow through the nozzle* and the same time, provides a smooth flow with negligible energy loss. All the available *potential energy is thus converted into kinetic energy* before the jet strikes the **buckets** of the **runner**. *The pressure all over the wheel is constant and equal to atmosphere, so that energy transfer occurs due to purely impulse action.*

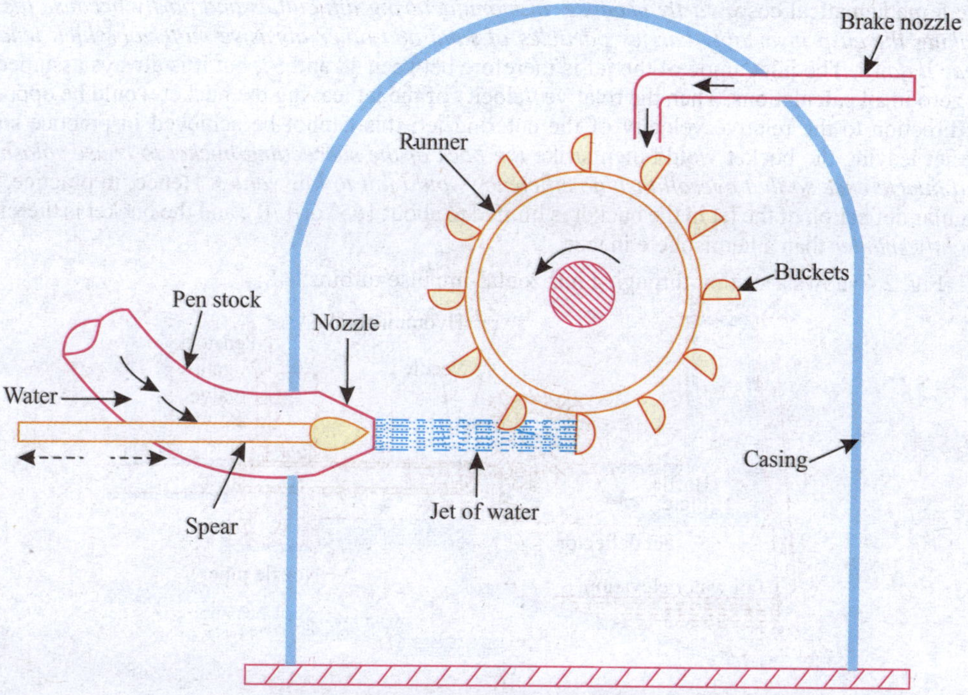

Fig. 2.2. Pelton wheel.

The Pelton turbine is provided with a **casing** the function of which is to *prevent the splashing of water and to discharge water to the tail race.*

When the nozzle is completely closed by moving the spear in the forward direction the amount of water striking the runner is reduced to zero but the runner due to inertia continues revolving for a long time. In order to bring the runner to rest in a short time, a nozzle (brake) is provided which directs the jet of water on the back of buckets; this jet of water is called **braking jet.**

Speed of the turbine runner is kept constant by a **governing mechanism** *that automatically regulates the quantity of water flowing through the runner in accordance with any variation of load.*

Fig. 2.2 shows a schematic diagram of a Pelton wheel, while Fig. 2.3 shows two views of its bucket.

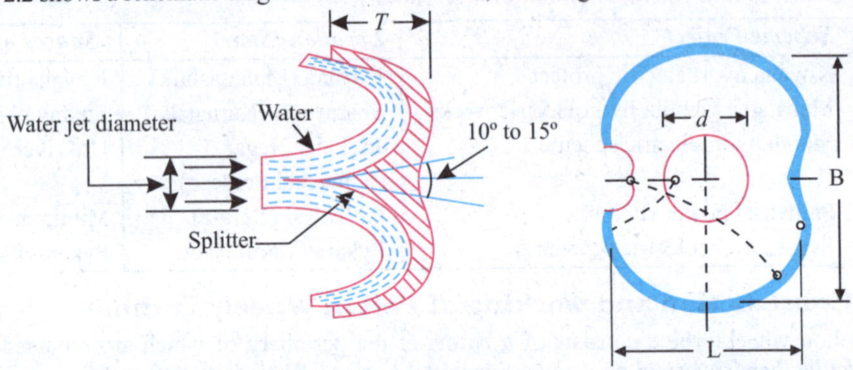

$$\frac{L}{d} = 2 \text{ to } 3; \quad \frac{B}{d} = 3 \text{ to } 4; \quad \frac{D}{d} = 11 \text{ to } 16; \quad \frac{T}{d} = 0.8 \text{ to } 1.2; \quad \text{notch (width)} = 1.1d + 5 \text{ mm}$$

Fig. 2.3. The bucket dimensions.

The jet emerging from the nozzle hits the splitter symmetrically and is equally distributed into the two halves of hemispherical bucket as shown. The bucket centre line cannot be made exactly like a mathematical cusp, *partly because of manufacturing difficulties and partly because the jet striking the cusp invariably carries particles of sand and other abrasive material which tend to wear it down.* The inlet angle of the jet is therefore between 1° and 3°, but it is always assumed to be zero in all calculations. Then the relative velocity of the jet leaving the bucket would be opposite in direction to the relative velocity of the entering jet; this cannot be achieved in practice since the jet leaving the bucket would then strike *the back of the succeeding bucket to cause splashing and interference so that overall turbine efficiency would fall to low values.* Hence, in practice, the angular deflection of the jet in the bucket is limited to about 165° or 170°, and the bucket is therefore *slightly smaller* than a hemisphere in size.

Fig. 2.4 shows a section through a horizontal-impulse turbine.

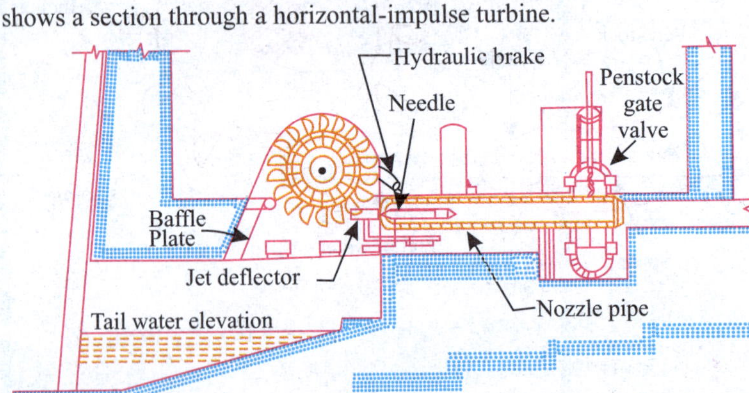

Fig. 2.4. Section through a horizontal-impulse turbine.

2.3.2. Work done and Efficiency of a Pelton Wheel

Fig. 2.5 shows the velocity triangles.

Let, N = Speed of wheel in r.p.m.,

D = Diameter of the wheel,

d = Diameter of the jet,

u = Peripheral (or circumferential) velocity of runner. It will be same at inlet and outlet of the runners at the mean pitch. (*i.e.* $u = u_1 = u_2$)

$$= \frac{\pi DN}{60},$$

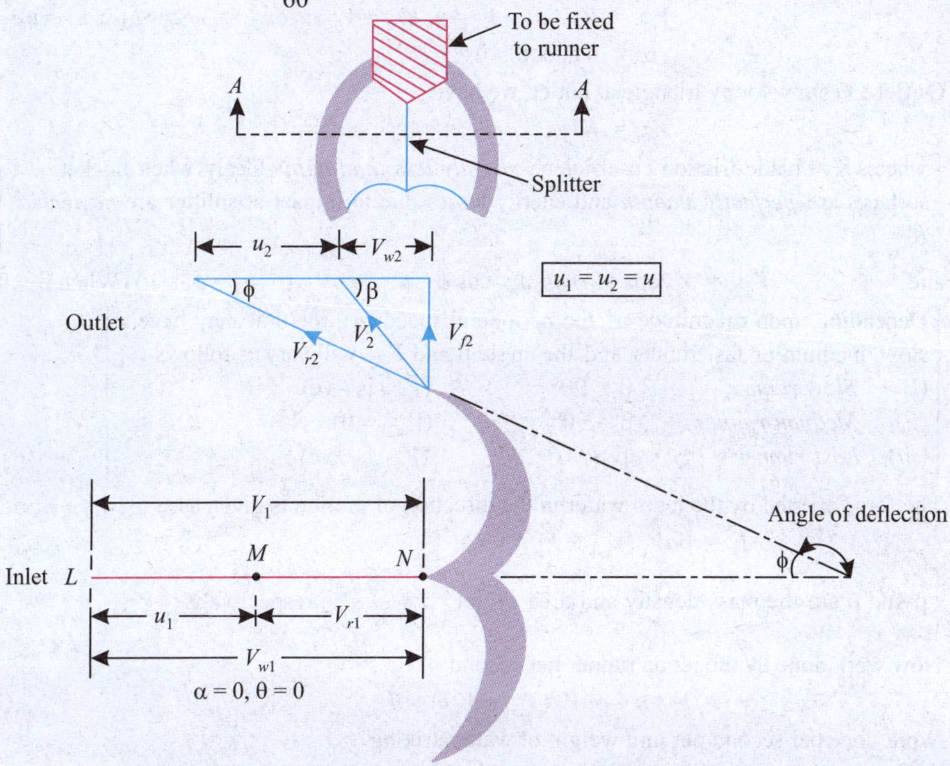

Fig. 2.5. Velocity triangles.

V_1 = Absolute velocity of water at inlet,

V_{r1} = Jet velocity relative to vane/bucket at inlet,

α = Angle between the direction of the jet and direction of motion of the vane/bucket (also called *guide angle*),

θ = Angle made by the relative velocity (V_{r1}) with the direction of motion at inlet (also called *vane angle at inlet*),

V_{w1} and V_{f1} = The components of the velocity of the jet V_1, in direction of motion and perpendicular to the direction of motion of the vane respectively;

V_{w1} is also known as *velocity of whirl* at inlet,

V_{f1} is also known as *velocity of flow* at inlet,

V_2 = Velocity of jet, leaving the vane or velocity of jet at outlet of the vane,

V_{r2} = Relative velocity of the jet with respect to the vane at outlet,

ϕ = Angle made by the relative velocity V_{r2} with the direction of motion of the vane at outlet and also called *vane angle at outlet*,

β = Angle made by the velocity V_2 with the direction of motion of the vane at outlet, and

V_{w2} and V_{f2} = Components of the velocity V_2, in the direction of motion of vane and perpendicular to the direction of motion of vane at outlet;
V_{w2} is also called the *velocity of whirl at outlet*, and
V_{f2} is also called the *velocity of flow at outlet*.

Inlet. The velocity triangle at *inlet* will be a *straight line* where

$$V_{r1} = V_1 - u_1 = V_1 - u,\ V_{w1} = V_1 \qquad (\because u_1 = u_2 = u)$$

$$\alpha = 0 \text{ and } \theta = 0$$

Outlet : From velocity triangle at outlet, we have

$$V_{r2} = KV_{r1},$$

[where, K = blade friction co-efficient, *slightly less than unity*. Idealy when bucket surfaces are *perfectly smooth* and energy losses due to impact at splitter are *neglected*, $K = 1$]

and, $$V_{w2} = V_{r2}\cos\phi - u_2 = V_{r2}\cos\phi - u \qquad (\because u_1 = u_2 = u) \text{ (When } \beta < 90°)$$

[Depending upon magnitude of the peripheral speed (u), the unit may have a slow, medium or fast runner and the angle β and V_{w2} will vary as follows :

(*i*)	*Slow runner*	$\beta < 90°$	(V_{w2} is –ve)
(*ii*)	*Medium runner*	$\beta = 90°$	($V_{w2} = 0$)
(*iii*)	*Fast runner*	$\beta > 90°$	(V_{w2} is +ve)

]

The force exerted by the jet of water in the direction of motion is given as:

$$F = \rho a V_1 (V_{w1} + V_{w2}) \qquad ...(2.1)$$

[ρ and a are the mass density and area of jet $\left(a = \frac{\pi}{4} d^2\right)$ respecitvely.]

Now work done by the jet on runner per second

$$= F \times u = \rho a V_1 (V_{w1} + V_{w2}) \times u \qquad ...(2.2)$$

Work done per second per unit weight of water striking

$$= \frac{\rho a V_1 (V_{w1} + V_{w2}) \times u}{\text{Weight of water striking}} = \frac{\rho a V_1 (V_{w1} + V_{w2}) \times u}{\rho a V_1 \times g}$$

$$= \frac{1}{g}(V_{w1} + V_{w2})\, u \qquad ...[2.2\,(a)]$$

The energy supplied to the jet at inlet is in the form of K.E. and is equal to $\frac{1}{2} m V_1^2$.

$\therefore$ Kinetic energy (K.E.) of jet per second $= \frac{1}{2}(\rho a V_1) \times V_1^2$

$\therefore$ Hydraulic efficiency, $$\eta_h = \frac{\text{Work done per second}}{\text{K.E. of jet per second}} = \frac{\rho a V_1 (V_{w1} + V_{w2}) \times u}{\frac{1}{2}(\rho a V_1) \times V_1^2}$$

or, $$\eta_h = \frac{2(V_{w1} + V_{w2}) \times u}{V_1^2} \qquad ...(2.3)$$

From inlet and outlet velocity triangles, we have:

$$V_{w1} = V_1, V_{r1} = V_1 - u_1 = V_1 - u$$

$$V_{w2} = V_{r2} \cos\phi - u_2 = V_{r2}\cos\phi - u = KV_{r1}\cos\phi - u = K(V_1 - u)\cos\phi - u$$

Substituting the values of V_{w1} and V_{w2} in eqn (2.3), we have:

$$\eta_h = \frac{2[V_1 + K(V - u)\cos\phi - u]u}{V_1^2} = \frac{2[(V_1 - u)(1 + K\cos\phi)]u}{V_1^2} \quad ...(2.4)$$

The hydraulic efficiency will be *maximum* for given value of V_1 when,

$$\frac{d}{du}(\eta_h) = 0$$

i.e.,
$$\frac{d}{du}\left[\frac{2(V_1 - u)(1 + \cos\phi)u}{V_1^2}\right] = 0$$

or,
$$\frac{2(1 + K\cos\phi)}{V_1^2} \times \frac{d}{du}(V_1 u - u^2) = 0$$

Since,
$$\frac{2(1 + K\cos\phi)}{V_1^2} \neq 0, \therefore \frac{d}{du}(V_1 u - u^2) = 0$$

or,
$$V_1 - 2u = 0, \text{ or, } u = \frac{V_1}{2} \quad ...(2.5)$$

The above equation states that *hydraulic efficiency of a Pelton wheel is maximum when the velocity of the wheel is half the velocity of jet of water at inlet.* The maximum efficiency can be obtained by substituting the value of $u = \frac{V_1}{2}$ in eqn. (2.4).

$$(\eta_h)_{max} = \frac{2\left(V_1 - \frac{V_1}{2}\right)(1 + K\cos\phi)\frac{V_1}{2}}{V_1^2} = \frac{2 \times \frac{V_1}{2}(1 + K\cos\phi) \times \frac{V_1}{2}}{V_1^2}$$

or,
$$(\eta_h)_{max} = \frac{(1 + K\cos\phi)}{2} \quad ...(2.6)$$

If friction factor, $K = 1$ (*i.e.,* assuming *no friction*), we have

$$(\eta_h)_{max} = \frac{1 + \cos\phi}{2} \quad ...[2.6(a)]$$

2.3.3. Definitions of Heads and Efficiencies

Fig. 2.6 shows a general layout of a hydroelectric power plant using an impulse turbine (Pelton wheel).

1. Gross head. The gross (total) head is the difference between the water level at the reservoir (also known as the *head race*) and the water level at the tail race. It is denoted by H_g.

2. Net or effective head. The head available at the inlet of the turbine is known as net or effective head. It is denoted by H and is given by:

$$H = H_g - h_f - h$$

where, h_f = Total loss of head between the head race and entrance of the turbine

$$= \frac{4fLV^2}{D \times 2g} \quad (L = \text{length of penstock}, D = \text{diameter of penstock},$$

V = velocity of flow in penstock), and

h = Height of nozzle above the tail race.

3. Efficiencies. The following are the important *efficiencies of turbine* :

***(i)* Hydraulic efficiency (η_h).** It is defined as the *ratio of power developed by the runner to the power supplied by the jet at entrance to the turbine.*

Mathematically,
$$\eta_h = \frac{\text{Power developed by the runner}}{\text{Power supplied at the inlet of turbine}}$$

$$= \frac{\rho Q_a (V_{w1} \pm V_{w2}) u}{wQ_a H} = \frac{\left(\frac{w}{g}\right) Q_a (V_{w1} \pm V_{w2}) u}{wQ_a H}$$

$$= \frac{(V_{w1} \pm V_{w2}) u}{gH} = \frac{H_r}{H} \quad \text{...(2.7)}$$

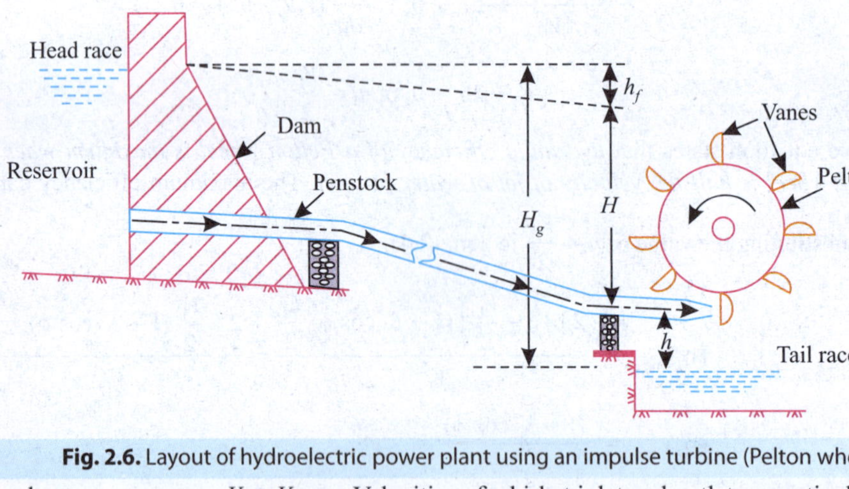

Fig. 2.6. Layout of hydroelectric power plant using an impulse turbine (Pelton wheel).

where, V_{w1}, V_{w2} = Velocities of whirl at inlet and outlet respectively,

u = Tangential velocity of vane,

H = Net head on the turbine, and

Q_a = Actual flow rate to turbine runner (bucket).

The parameter, $H_r = \frac{1}{g}(V_{w1} + V_{w2})\, u$ represents the energy transfer per unit weight of water and is referred to as the '*runner head*' or '*Euler head*'.

$H - H_r = \Delta H$ = Hydraulic losses within the turbine.

***(ii)* Mechanical efficiency (η_m).** It is defined as the *ratio of the power obtained from the shaft of the turbine to the power developed by the runner.* These two powers differ by the amount of mechanical losses, *viz.,*bearing friction, etc.

Mathematically,
$$\eta_h = \frac{\text{Power available at the turbine shaft}}{\text{Power developed by turbine runner}} = \frac{\text{Shaft power}}{\text{Bucket power}}$$

$$= \frac{P}{wQ_a\left(\dfrac{V_{w1}+V_{w2}}{g}\right)u} = \frac{P}{wQ_aH_r} \quad ...(2.8)$$

Values of mechanical efficiency for a Pelton wheel usually lie between *97 to 99 percent depending on size and capacity of the unit.*

***(iii)* Volumetric efficiency (η_v).** The volumetric efficiency is the *ratio of the volume of water actually striking the runner to the volume of water supplied by the jet to the turbine.* That is,

$$\eta_v = \frac{\text{Volume of water actually striking the runner } (Q_a)}{\text{Total water supplied by the jet to the turbine } (Q)} \quad ...(2.9)$$

For Pelton turbines, $\eta_v \simeq 0.97$ to 0.99.

***(iv)* Overall efficiency (η_o).** It is defined as the *ratio of power available at the turbine shaft to the power supplied by the water jet.* That is

$$\eta_0 = \frac{\text{Power available at the turbine shaft}}{\text{Power available from the water jet}} = \frac{\text{Shaft power}}{\text{Water power}} = \frac{P}{wQH} \quad ...(2.10)$$

(where, Q = the total discharge in m^3/s supplied by the jet.)

The values of overall efficiency for a Pelton wheel lie between 0.85 to 0.90.

The individual efficiencies may be combined to give,

$$\eta_0 = \eta_h \times \eta_m \times \eta_v$$

$$= \frac{H_r}{H} \times \frac{P}{wQ_aH_r} \times \frac{Q_a}{Q} = \frac{P}{wQH}, \text{ which is the same as defined vide eqn. (2.10)}$$

If η_g is the efficiency of a generator, then power output of hydrounit (turbine + hydrogenerators)

$$= (wQH) \times \eta_0 \times \eta_g$$

The product $\eta_0 \times \eta_g$ is known as *hydroelectric plant efficiency.*

2.3.4. Design Aspects of Pelton wheel

The following points should be considered while *designing a Pelton wheel* :

1. Velocity of jet. The velocity of jet at inlet is given by,

$$V_1 = C_v\sqrt{2gH} \quad ...(2.11)$$

where, C_v = Co-efficient of velocity (= 0·98 or 0·99), and

H = Net head on turbine.

2. Velocity of wheel. The velocity of wheel (u) is given by,

$$u = K_u\sqrt{2gH} \quad ...(2.12)$$

where, K_u = Speed ratio. It varies from 0.43 to 0.48.

3. Angle of deflection of the jet. The angle of deflection of the jet through the buckets is taken as 165° if no angle of deflection is given.

4. Mean diameter of the wheel (*D*). The mean diameter or pitch diameter D of the Pelton wheel is given by,

$$u = \frac{\pi DN}{60} \text{ or } D = \frac{60\,u}{\pi N} \quad ...(2.13)$$

5. Jet ratio (m). It is defined as the *ratio of the pitch diameter* (D) *of the Pelton wheel to the diameter of the jet* (d). It is denoted by 'm' and is given as :

$$m = \frac{D}{d} \text{ (lies between 11 and 16 for maximum hydraulic efficiency)} \quad ...(2.14)$$

Normally, the jet ratio is adopted as 12 in practice.

6. **Bucket dimensions.** Some of the main dimensions of the bucket of a Pelton wheel are as follows :

Refer to Fig. 2.3 : $B = 3$ to $4d$; L = 2 to $3d$; T = 0.8 to $1.2d$.

7. **Number of jets.** Normally a Pelton wheel has one nozzle or one jet. However, a number of nozzles may be employed when more power is to be produced with the same wheel. Theoretically *six nozzles* can be used on Pelton wheel. However, practical considerations limit the use of not more than two jets per runner for a vertical runner and not more than four jets per runner if it is of horizontal configuration.

Number of jets is obtained by *dividing the total rate of flow through the turbine by the rate of flow of water through a single jet.*

8. **Number of buckets (Z).** The number of buckets for a Pelton wheel should be such that the jet is always completely intercepted by the buckets so that volumetric efficiency of the turbine is very close to unity. Number of buckets on a runner is given by,

$$Z = 15 + \frac{D}{2d} = 15 + 0.5\text{ m} \quad ...(2.15)$$

Example 2.1. *A Pelton wheel is receiving water from a penstock with a gross head of 510 m. One-third of gross head is lost in friction in the penstock. The rate of flow through the nozzle fitted at the end of the penstock is 2·2 m³/s. The angle of deflection of the jet is 165°. Determine :*

(*i*) *The power given by water to the runner, and*

(*ii*) *Hydraulic efficiency of the Pelton wheel.*

Take C_v (co-efficient of velocity) = 1·0 and speed ratio = 0·45.

Solution. Gross head, $H_g = 510$ m

Head lost in friction, $h_f = \frac{H_g}{3} = \frac{510}{3} = 170$ m

∴ Net head, $H = H_g - h_f = 510 - 170 = 340$ m

Discharge, $Q = 2{\cdot}2\text{ m}^3/\text{s}$

Angle of deflection = 165°

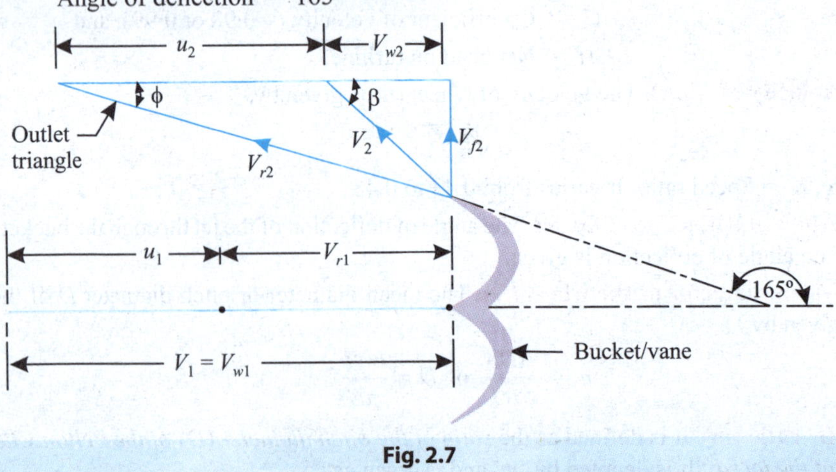

Fig. 2.7

$\therefore$ Angle, $\phi = 180° - 165° = 15°$

Co-efficient of velocity, $C_v = 1.0$

Speed ratio, $K_u = 0.45$

(i) The power given by water to the runner :

Velocity of jet, $V_1 = C_v\sqrt{2gH} = 1.0\sqrt{2 \times 9.81 \times 340} = 81.67$ m/s

Velocity of wheel, $u = K_u\sqrt{2gH} = 0.45\sqrt{2 \times 9.81 \times 340} = 36.75$ m/s

Refer to fig. 2.7. $V_{r1} = V_1 - u_1 = V_1 - u = 81.67 - 36.75 = 44.92$ m/s $(\because u_1 = u_2 = u)$

Also, $V_{w1} = V_1 = 81.67$ m/s

From outlet velocity triangle, we have:

$$V_{r2} = V_{r1} = 44.92 \text{ m/s}$$

Also, $V_{r2}\cos\phi = u_2 + V_{w2} = u + V_{w2}$

or, $V_{w2} = V_{r2}\cos\phi - u = 44.92\cos 15° - 36.75 = 6.64$ m/s

Work done by the jet on the runner per second

$$= \rho Q(V_{w1} + V_{w2}) \times u \quad \text{...[Eqn (2.2)]}$$

$$= 1000 \times 2.2\,(81.67 + 6.64) \times 36.75 = 7139863 \text{ Nm/s}$$

$\therefore$ Power given by water to the runner = 7139863 J/s

or, $W \simeq$ **7139.8 kW (Ans.)**

(ii) Hydraulic efficiency of the Pelton wheel, η_h :

$$\eta_h = \frac{2\,(V_{w1} + V_{w2}) \times u}{V_1^2} \quad \text{...[Eqn (2.3)]}$$

$$= \frac{2\,(81.67 + 6.64) \times 36.75}{(81.67)^2} = 0.973 \text{ or } \mathbf{97.3\ \%\ (Ans.)}$$

Alternatively :

$$\eta_h = \frac{(V_{w1} + V_{w2})\,u}{gH} \quad \text{...[Eqn (2.7)]}$$

$$= \frac{(18.67 + 6.64) \times 36.75}{9.81 \times 340} = 0.973 \text{ or } \mathbf{97.3\ \%\ (Ans.)}$$

Example 2.2. *A Pelton wheel having a mean bucket diameter of 1·2 m is running at 1000 r.p.m. The net head on the Pelton wheel is 840 m. If the side clearance angle is 15° and discharge through the nozzle is 0·12 m³/s, determine :*

(i) Power available at the nozzle, and

(ii) Hydraulic efficiency of the turbine.

Solution. Mean diameter of Pelton wheel, $D = 1.2$ m

Speed of wheel, $N = 1000$ r.p.m.

$\therefore$ Tangential velocity of the wheel, $u = \dfrac{\pi DN}{60} = \dfrac{\pi \times 1.2 \times 1000}{60} = 62.83$ m/s

Net head on the turbine, $H = 840$ m

Side clearance angle, $\theta = 15°$

Discharge $Q = 0.12\ m^3/s$

(i) Power available at the nozzle :

Velocity of jet at inlet, $V_1 = C_v\sqrt{2gH}$

$$= 1.0\sqrt{2 \times 9.81 \times 840} = 128.38 \text{ m/s}$$

(Assume $C_v = 1.0$ if not given)

∴ Power available at nozzle

$= wQH = 9810 \times 0.12 \times 840 = 988848$ Nm/s or J/s or W = **988.85 kW (Ans.)**

(ii) Hydraulic efficiency, η_h :

$$\eta_h = \frac{2(V_1 - u)(1 + K\cos\phi)\,u}{V_1^2} \quad \text{...[Eqn. (2.4)]}$$

$$= \frac{2(128.38 - 62.83)(1 + \cos 15°) \times 62.83}{(128.38)^2} = \frac{131.1(1 + 0.966) \times 62.83}{(128.38)^2}$$

(Assume $K = 1$)

$= 0.982$ or **98.2 % (Ans.)**

Example 2.3. *A Pelton wheel is to be designed for the following specifications :*

Power (brake or shaft) ... *9560 kW*
Head ... *350 metres*
Speed ... *750 r.p.m.*
Overall efficiency ... *85%*
Jet diameter ... *not to exceed 1/6 th of the wheel diameter*

Determine the following :

(i) The wheel diameter, *(ii) Diameter of the jet, and*
(iii) The number of jets required.

Take $C_v = 0{\cdot}985$, Speed ratio = 0.45. **[UPTU]**

Solution. Shaft or brake power = 9560 kW

Head, $H = 350$ m

Speed, $N = 750$ r.p.m.

Overall efficiency, $\eta_0 = 85\%$

Ratio of jet diameter to wheel, $\dfrac{d}{D} = \dfrac{1}{6}$

Co-efficient of velocity, $C_v = 0.985$

Speed ratio, $K_u = 0.45$

(i) The wheel diameter, D :

Velocity of jet, $V_1 = C_v\sqrt{2gH} = 0.985\sqrt{2 \times 9.81 \times 350} = 81.62$ m/s

The velocity of wheel, $u = u_1 = u_2$

$$= K_u \times \sqrt{2gH} = 0.45\sqrt{2 \times 9.81 \times 350} = 37.3 \text{ m/s}$$

But, $$u = \frac{\pi DN}{60}$$

∴ $$37.3 = \frac{\pi D \times 750}{60}, \quad \text{or,} \quad D = \frac{37.3 \times 60}{\pi \times 750} = \mathbf{0.95\ m}\ \textbf{(Ans.)}$$

(ii) Diameter of the jet, d :

$$\frac{d}{D} = \frac{1}{6}$$

$$\therefore \quad d = \frac{D}{6} = \frac{0.95}{6} = \mathbf{0.158\ m}\ \textbf{(Ans.)}$$

(iii) The number of jets required :

$$\text{Discharge of one jet, } q = \text{Area of jet} \times \text{velocity of jet}$$

$$= \frac{\pi}{4} d^2 \times V_1 = \frac{\pi}{4} \times 0.158^2 \times 81.62 = 1.6\ \text{m}^3/\text{s}$$

Now, overall efficiency, $\eta_0 = \dfrac{\text{Shaft power}}{\text{Water power}} = \dfrac{9560}{wQH}$

or, $0.85 = \dfrac{9560}{9.81 \times Q \times 350}$ $\quad (\because w = 9.81\ \text{kN/m}^3)$

$\therefore$ Total discharge, $Q = \dfrac{9560}{0.85 \times 9.81 \times 350} = 3.27\ \text{m}^3/\text{s}$

$\therefore$ Number of jets $= \dfrac{\text{Total discharge}}{\text{Discharge of one jet}} = \dfrac{Q}{q} = \dfrac{3.27}{1.6} = \mathbf{2\ jets}$ **(Ans.)**

Example 2.4. *A Pelton wheel nozzle, for which $C_v = 0{\cdot}97$, is 400 m below the water surface of a lake. The jet diameter is 80 mm, the pipe diameter is 0·6 m, its length is 4 km, and $f = 0.032$ in the formula $h_f = \dfrac{fLV^2}{D \times 2g}$. The buckets deflect the jet through 165° and they run at 0·48 times the jet speed, bucket friction reducing the velocity at outlet by 15 per cent of the relative velocity at inlet. Mechanical efficiency = 90%. Determine :*

(i) The flow rate, and

(ii) The shaft power developed by the turbine. **[MDU Haryana]**

Solution. Co-efficient of velocity, $C_v = 0.97$

Gross head, $H_g = 400$ m

Diameter of jet, $d = 80$ mm $= 0.08$ m

Diameter of pipe, $D = 0.6$ m

Length of pipe, $L = 4$ km $= 4000$ m

Friction factor, $f = 0.32$

Angle, $\phi = 180° - 165° = 15°$

Bucket speed, $u = 0.48$ times the jet speed

Relative velocity at the outlet (V_{r2}) = 0.85 times the relative velocity at inlet V_{r1}.

(i) The flow rate, Q :

Let, V = Velocity of water in pipe, and

V_1 = Velocity of jet of water.

Also, $AV = aV_1$ ...*Continuity equation*

(where, A = area of pipe, and a = area of jet)

or,
$$\frac{\pi}{4} \times D^2 \times V = \frac{\pi}{4} \times d^2 \times V_1$$

$$V = \frac{d^2}{D^2} \times V_1 = \left(\frac{0.08}{0.6}\right)^2 \times V_1 = 00177\ V_1 \qquad ...(i)$$

Applying Bernoulli's equation to free surface of water in the reservoir and the outlet of the nozzle, we have:

Head at reservoir = Kinetic head of jet of water + head lost due to friction in pipe + head lost in nozzle

$$= \frac{V_1^2}{2g} + \frac{fLV^2}{D \times 2g} + \text{head lost in nozzle,} \qquad ...(ii)$$

Let, $(V_1)_{th}$ = Theoretical velocity at outlet of nozzle, and

V_1 = Actual velocity of jet of water.

Then,
$$\frac{V_1}{(V_1)_{th}} = C_v \text{ or, } (V_1)_{\text{th}} = \frac{V_1}{C_v}$$

Now, Head lost in nozzle = Head corresponding to $(V_1)_{th}$ – head corresponding to V_1

$$= \frac{(V_1)_{th}^2}{2g} - \frac{V_1^2}{2g} = \left(\frac{V_1}{C_v}\right)^2 \times \frac{1}{2g} - \frac{V_1^2}{2g}$$

$$= \frac{V_1^2}{2g}\left(\frac{1}{C_v^2} - 1\right)$$

Substituting this value in eqn. (*ii*), we get:

$$\text{Head at reservoir} = \frac{V_1^2}{2g} + \frac{fLV^2}{D \times 2g} + \frac{V_1^2}{2g}\left(\frac{1}{C_v^2} - 1\right)$$

or,
$$400 = \frac{V_1^2}{2g} + \frac{0.032 \times 4000V^2}{0.6 \times 2 \times 9.81} + \frac{V_1^2}{2 \times 9.81} + \frac{V_1^2}{2g} \times \frac{1}{C_v^2} - \frac{V_1^2}{2g}$$

$$= \frac{0.032 \times 4000 \times (0.0177V_1)^2}{0.6 \times 2 \times 9.81} + \frac{V_1^2}{2 \times 9.81} \times \frac{1}{(0.97)^2} \quad (\because V = 0.0177\text{V}_1)$$

$$= 0.0034\ V_1^2 + 0.054\ V_1^2 = 0.0574\ V_1^2$$

$$V_1 = \left(\frac{400}{0.0574}\right)^{1/2} = 83.48 \text{ m/s}$$

∴ Flow rate = Area of jet × velocity of jet

$$= \frac{\pi}{4} \times (0.08)^2 \times 83.48 = \mathbf{0.419\ m^3/s}\ \textbf{(Ans.)}$$

(*ii*) The shaft power :

Velocity of bucket, $u_1 = 0{\cdot}48V_1 = 0{\cdot}48 \times 83{\cdot}48 = 40{\cdot}07$ m/s

Refer to Fig. 2.8. $V_{r1} = V_1 - u_1 = 83.47 - 40.07 = 43.4$ m/s

$V_{w1} = V_1 = 83.48$ m/s

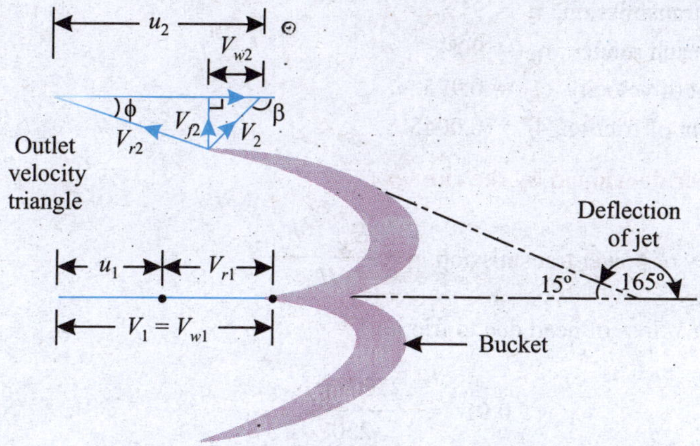

Fig. 2.8

From the outlet velocity triangle, we have:

$$V_{r2} = 0.85 V_{r1} = 0.85 \times 43.4 = 36.89 \text{ m/s}$$

$$V_{w2} = u_2 - V_{r2} \cos \phi$$

$$= 40.07 - 36.89 \times \cos 15° = 4.44 \text{ m/s}$$

Mechanical efficiency,

$$\eta_m = \frac{\text{Shaft power}}{\text{Power given to runner}}$$

$\therefore$ Shaft power, $= \eta_m \times$ power given to runner

But power given to runner

$$= \frac{w}{g} Q (V_{w1} - V_{w2}) \times u$$

[Here –ve sign is taken since $\beta > 90°$ (*i.e.,* V_{w1} and V_{w2} are in the *same direction*)]

$$\therefore \quad \text{Shaft power} = \eta_m \times \frac{w}{g} Q (V_{w1} - V_{w2}) \times u_1$$

$$= 0.9 \times \frac{9.81}{9.81} \times 0.419 (83.48 - 4.44) \times 40.07 \text{ kW} \qquad (\because w = 9.81 \text{ kN/m}^3)$$

$$= \mathbf{1194.3 \text{ kW}} \text{ (Ans.)}$$

Example 2.5. *The water available for a Pelton wheel is 4 m³/s and the total head from the reservoir to the nozzle is 250 m. The turbine has two runners with two jets per runner. All the four jets have the same diameters. The pipe is 3 km long. The efficiency of transmission through the pipeline and the nozzle is 91 % and efficiency of each runner is 90 %. The velocity co-efficient of each nozzle is 0·975 and co-efficient of friction '4f' for the pipe is 0·0045. Determine :*

(i) The power developed by the turbine,

(ii) The diameter of the jet, and

(iii) The diameter of the pipeline. **[M.U]**

Solution. Rate of flow, $Q = 4 \text{ m}^3/\text{s}$

Total or gross head, $H_g = 250$ m

Total number of jets $= 2 \times 2 = 4$

Length of pipe, $L = 3 \text{ km} = 3000$ m

Efficiency of transmission, η = 91%

Efficiency of each runner, η_h = 90%

Co-efficient of velocity, C_v = 0.975

Co-efficient of friction, $4f$ = 0.0045

(i) The power developed by the runner :

Efficiency of power transmission, $\eta = \dfrac{H_g - h_f}{H_g}$

(where, h_f = loss of head due to friction)

or, $$0.91 = \frac{250 - h_f}{250}$$

$$\therefore \quad h_f = 250 - 250 \times 0.91 = 22.5 \text{ m}$$

$\therefore$ Net head on the turbine, $H = H_g - h_f = 250 - 22.5 = 227.5$ m

Velocity of jet, $V_1 = C_v\sqrt{2gH} = 0.975\sqrt{2 \times 9.81 \times 227.5} = 65.14$ m/s

Now, Water power = Kinetic energy of the jet

$$\frac{1}{2}mV_1^2 = \frac{1}{2}\rho QV_1^2 = \frac{1}{2} \times 1000 \times 4 \times 65.14^2 = 8486439 \text{ Nm/s}$$

$$= 8486439 \text{ J/s or } W = 8486.44 \text{ kW}$$

But, hydraulic efficiency, $\eta_h = \dfrac{\text{Power developed by the turbine}}{\text{Water power}}$

or, $$0.9 = \frac{\text{Power developed by the turbine}}{8486.44}$$

$\therefore$ Power developed by the turbine

$= 0.9 \times 8486.44 =$ **7637.8 kW (Ans.)**

(ii) The diameter of the jet, *d* :

Discharge per jet, $q = \dfrac{\text{Total discharge}}{\text{No. of jets}} = \dfrac{4}{4} = 1.0 \text{ m}^3\text{/s}$

But, $$q = \frac{\pi}{4} \times d^2 \times V_1$$

$$\therefore \quad 1.0 = \frac{\pi}{4} \times d^2 \times 65.14 \text{ or } d = \left(\frac{1.0 \times 4}{\pi \times 65.14}\right)^{1/2}$$ = **0.14 m (Ans.)**

(iii) The diameter of the pipeline, *D* :

Head lost due to friction, $h_f = \dfrac{4fLV^2}{D \times 2g}$

where, V = velocity through pipe $= \dfrac{Q}{\text{Area}} = \dfrac{Q}{(\pi/4) \times D^2} = \dfrac{4Q}{\pi D^2}$

$$\therefore \quad h_f = \frac{0.0045 \times 3000 \times \left(\frac{4Q}{\pi D^2}\right)^2}{D \times 2g}$$

$$\text{or,} \quad 22.5 = \frac{0.0045 \times 3000 \times 16Q^2}{D \times 2 \times 9.81 \times \pi^2 \times D^4}$$

$$= \frac{0.0045 \times 3000 \times 16 \times (4)^2}{D^5 \times 2 \times 9.81 \times \pi^2} = \frac{17.85}{D^5}$$

$$\text{or,} \quad D^5 = \frac{12.85}{22.5} \quad \text{or} \quad D = \left(\frac{17.85}{22.5}\right)^{1/5} = 0.955 \text{ m}$$

Hence, *the diameter of the pipeline, D* = **0.955 m (Ans.)**

Example 2.6. *A single jet Pelton wheel runs at 300 r.p.m. under a head of 510 m. The jet diameter is 200 mm, its deflection inside the bucket is 165° and its relative velocity is reduced by 15% due to friction. Determine :*

(i) Water power,

(ii) Resultant force on the bucket, and

(iii) Overall efficiency.

Take: Mechanical losses = 3%, co-efficient of velocity = 0.98, and speed ratio = 0.46.

[PTU]

Solution. Speed of the wheel, N = 300 r.p.m.

Diameter of jet, d = 200 mm = 0.2 m

Net head, H = 510 m

Angle of deflection of jet = 165°

Reduction of relative velocity due to friction = 15%

Mechanical losses = 3%

Co-efficient of velocity, C_v = 0.98

Speed ratio, K_u = 0.46.

(i) Water power :

$$\text{Velocity of jet, } V_1 = C_v\sqrt{2gH} = 0.98\sqrt{2 \times 9.81 \times 510} = 98 \text{ m/s}$$

∴ Discharge through the Pelton wheel,

$$Q = \text{Area of jet } (a) \times \text{velocity } (V_1)$$

$$= \frac{\pi}{4} \times (0.2)^2 \times 98 = 3.078 \text{ m}^3\text{/s}$$

Water power = wQH = 9.81 × 3.078 × 510 kW = **15399.5 kW (Ans.)**

(ii) Resultant force on the bucket :

$$\text{Peripheral speed of the wheel, } u = K_u\sqrt{2gH} = 0.46\sqrt{2 \times 9.81 \times 510} = 46 \text{ m/s}$$

Refer to fig. 2.8. *At inlet to turbine* :

$$V_{w1} = V_1 = 98 \text{ m/s}$$

$$V_{r1} = (V_1 - u_1) = 98 - 46 = 52 \text{ m/s}$$

At exit from the turbine :

The blade angle at exit, $\phi = 180° - 165° = 15°$

$$V_{r2} = 0.85\, V_{r1} \qquad \text{...(Given)}$$

or, $$V_{r2} = 0.85 \times 52 = 44.2 \text{ m/s}$$

As $V_{r2} \cos \phi$ is *less* than blade speed u, the velocity triangle at outlet will be as shown in Fig. 2.8 ($\beta > 90°$)

$$V_{w2} = u_2 - V_{r2} \cos \phi = 46 \qquad (\because u_1 = u_2 = u)$$
$$= 46 - 44.2 \cos 15° = 3.31 \text{ m/s}$$

∴ Resultant force on the bucket,

$$F = \rho Q (V_{w1} - V_{w2}) \qquad (\because \beta > 90°)$$
$$= 1000 \times 3.078\,(98 - 3.31) = \mathbf{291455.8\ N\ (Ans.)}$$

(*iii*) Brake power, P :

Power developed by the wheel $= F \times u = 291455.8 \times 46$ Nm/s or J/s or W

$$= 291455.8 \times 46 \times 10^{-3} \text{ kW}$$
$$= 13406.97 \text{ kW}$$

∴ Brake power (power produced at the shaft),

$$P = 13406.97 \times (1 - 0.03) = \mathbf{13004.76\ kW\ (Ans.)}$$

(*iv*) Overall efficiency, η_o :

$$\eta_0 = \frac{\text{Brake power}}{\text{Water power}}$$

$$= \frac{13004.76}{15399.5} = 0.844 \text{ or } \mathbf{84.4\%\ (Ans.)}$$

Example 2.7. *A Pelton wheel running at 480 r.p.m. and operating under an available head of 420 m is required to develop 4800 kW. There are two equal jets and the bucket deflection angle is 165°. The overall efficiency is 85 percent when the water is discharged from the wheel in a direction parallel to the axis of rotation. The co-efficient of velocity of nozzle = 0·97 and blade speed ratio = 0·46. The relative velocity of water at exit from the bucket is 0·86 times the relative velocity at inlet. Calculate the following :*

(*i*) *Cross-sectional area of each jet,*

(*ii*) *Bucket pitch circle diameter, and*

(*iii*) *Hydraulic efficiency of the turbine.*

Solution. Speed of the wheel, N = 480 r.p.m.

Available head, H = 420 m

Shaft power, P = 4800 kW

Angle of deflection of jet = 165°

Overall efficiency, η_0 = 85%

Co-efficient of velocity of nozzle, C_v = 0.97

Blade speed ratio, K_u = 0.46

Relative velocity of water at exit = 0·86 times the relative velocity at inlet

(*i*) Cross-sectional area of each jet, a :

$$\text{Shaft power, } P = wQH \times \eta_0$$
$$4800 = 9.81 \times Q \times 420 \times 0.85$$

$\therefore$ Total discharge through the wheel, $Q = \dfrac{4800}{9.81 \times 420 \times 0.85} = 1.37 \text{ m}^3/\text{s}$

$$\text{Velocity of jet, } V_1 = C_v \sqrt{2gH} = 0.97\sqrt{2 \times 9.81 \times 420} = 88.05 \text{ m/s}$$

Now, total discharge Q = No. of jets × area of each nozzle (a) × velocity of jet (V_1)

or, $1.37 = 2 \times a \times 88.05$

$\therefore$ $a = \dfrac{1.37}{2 \times 88.05} = \mathbf{7.779 \times 10^{-3}\ m^2}$ **(Ans.)**

(*ii*) Bucket pitch circle diameter, D :

Velocity of bucket, $u = K_u \sqrt{2gH} = 0.46\sqrt{2 \times 9.81 \times 420} = 41.76$ m/s

Also, $u = \dfrac{\pi DN}{60}$; $41.76 = \dfrac{\pi D \times 480}{60}$;

$\therefore$ $D = \dfrac{41.76 \times 60}{\pi \times 480} = \mathbf{1.66\ m}$ **(Ans.)**

(*iii*) Hydraulic efficiency of the turbine, η_h :

$$\eta_h = \frac{2(V_1 - u)(1 + K\cos\phi)\,u}{V_1^2} \quad \text{...[Eqn. (2.4)]}$$

The blade angle at exit; $\phi = 180° - 165° = 15°$

Substituting the relevant data in the above eqn. we get:

$$\eta_h = \frac{2 \times (88.05 - 41.76)(1 + 0.86 \times \cos 15°) \times 41.76}{(88.05)^2} \quad (\because K = 0.86)$$

$$= \frac{2 \times 46.29(1 + 0.86 \times 0.966) \times 41.76}{7752.8} = 0.913 \text{ or } \mathbf{91.3\%} \textbf{ (Ans.)}$$

Example 2.8. *The following data relate to a Pelton wheel :*

Head at the base of the nozzle = 82 m; diameter of the jet = 100 mm; discharge of the nozzle = 0.30 m³/s; shaft power = 206 kW; power absorbed in mechanical resistance = 4.5 kW. Determine :

(*i*) *Power lost in nozzle, and*

(*ii*) *Power lost due to hydraulic resistance in water.*

Solution. Head at the base of the nozzle, $H_1 = 82$ m

Diameter of the jet, $d = 100$ mm $= 0.1$ m

$\therefore$ Area of the jet, $a = \dfrac{\pi}{4} \times 0.1^2 = 0.007854 \text{ m}^2$

Discharge of the nozzle, $Q = 0.30 \text{ m}^3/\text{s}$

Shaft power, $P = 206$ kW

Power absorbed in mechanical resistance = 4.5 kW.

(*i*) Power lost in nozzle :

Discharge, Q = Area of jet × velocity of jet

$0.30 = 0.007854 \times V_1$

$$\therefore \qquad V_1 = \frac{0.30}{0.007854} = 38.2 \text{ m/s}$$

Power available at the base of the nozzle

$$wQH_1 = 9.81 \times 0.30 \times 82 = 241.3 \text{ kW} \qquad [\because w = 9.81 \text{ kN/m}^3]$$

Power corresponding to K.E. of the jet,

$$\frac{1}{2}\left(\frac{wQ}{g}\right)V_1^2 = \frac{1}{2} \times \left(\frac{9.81 \times 0.30}{9.81}\right) \times 38.2^2 = 218.9 \text{ kW}$$

∴ Power lost in nozzle = Power available at the nozzle base – power corresponding to K.E. of the jet

= 241.3 – 218.9 = **22.4 kW (Ans.)**

(*ii*) Power lost due to hydraulic resistance in water :

Power at the base of the nozzle = Shaft power + power lost in runner + power lost in nozzle

241.3 = 206 + power lost in runner + 22.4

∴ Power lost in runner = 241.3 – 206 – 23.4 = **12.9 kW (Ans.)**

Example 2.9. *A single jet Pelton turbine is required to drive a generator to develop 10000 kW. The available head at the nozzle is 760 m. Assuming electric generation efficiency 95 percent, Pelton wheel efficiency 87 percent, co-efficient of velocity for nozzle 0·97, mean bucket velocity 0·46 of jet velocity, outlet angle of bucket 15° and the relative velocity of the water leaving the buckets 0·85 of that inlet, find :*

(*i*) *The flow in* m^3/s,

(*ii*) *The diameter of jet,*

(*iii*) *The force exerted by the jet on the buckets, and*

(*iv*) *The best synchronous speed for generation at 50 Hz and the corresponding mean diameter if the ratio of the mean bucket circle diameter to the jet diameter is not to be less than 10.*

[N.U.]

Solution. Output of generator = 10000 kW

Generator efficiency = 95%

Availavble head at the nozzle, H = 760 m

Pelton wheel efficiency = 87%

Co-efficient of velocity, C_v = 0.97

Bucket velocity, u = 0.46 × jet velocity

Relative velocity of water at outlet = 0.85 × relative velocity at inlet

Outlet angle of bucket, ϕ = 15°

(*i*) The flow in m^3/s :

Output of turbine, P_t = Input of generator

$$= \frac{\text{Output of the generator}}{\text{Generator efficiency}}$$

$$= \frac{10000}{0.95} = 10526.3 \text{ kW}$$

$$\text{Available power of the turbine} = wQH = \frac{P_t}{\text{Pelton wheel efficiency}}$$

or, $$9.81 \times Q \times 760 = \frac{10526.3}{0.87}$$

$$\therefore \quad Q = \frac{10526.3}{9.81 \times 760 \times 0.87} = \mathbf{1.62\ m^3/s}\ \textbf{(Ans.)}$$

(*ii*) The diameter of jet *d* :

$$Q = \frac{\pi}{4} \times d^2 \times V_1$$

where, $$V_1 = \text{Velocity of jet} = C_v \times \sqrt{2gH}$$

$$= 0.97 \times \sqrt{2 \times 9.81 \times 760} = 118.45 \text{ m/s}$$

$$\therefore \quad 1.62 = \frac{\pi}{4} \times d^2 \times 118.45$$

or, $$d = \left(\frac{1.62 \times 4}{\pi \times 118.45}\right)^{1/2} = 0.132 \text{ m or } \mathbf{132\ mm}\ \textbf{(Ans.)}$$

(*iii*) The force exerted by the jet on the buckets :

Bucket velocity, $u = 0.46V_1 = 0.46 \times 118.45 = 54.5$ m/s

At inlet :

$$V_{w1} = V_1 = 118.45 \text{ m/s}$$

$$u_1 = u_2 = 54.5 \text{ m/s}$$

$$V_{r1} = V_1 - u_1 = 118.45 - 54.5 \simeq 64 \text{ m/s}$$

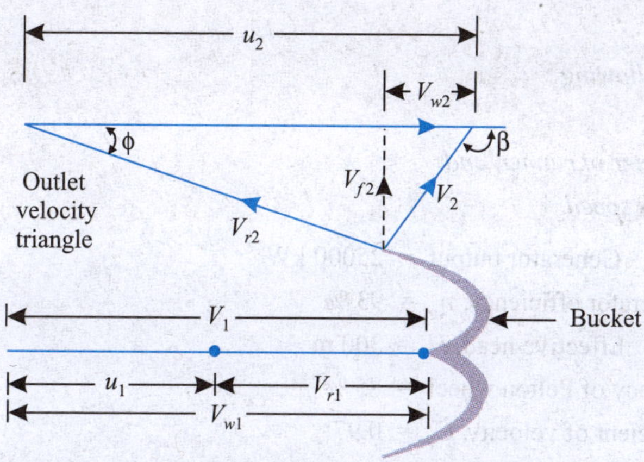

Fig. 2.9

At outlet :

$$V_{r2} = 0.85\, V_{r1} = 0.85 \times 64 = 54.4 \text{ m/s}$$

$$V_{w2} = u_2 - V_{r2} \cos\phi = 54.5 - 54.4 \times \cos 15° = 1.95 \text{ m/s}$$

$\therefore$ Force exerted by jet on water,

$$F = \rho Q (V_{w1} - V_{w2}) \qquad (\because \beta > 90°)$$

$$= 1000 \times 1.62\,(118.5 - 1.95)$$

$$= 188811 \text{ N} \simeq \mathbf{188.8\ kN}\ \textbf{(Ans.)}$$

(iv) Best synchronous speed (N_{syn}); mean bucket diameter (D) :

Now, $\frac{D}{d} = 10$, where D is the mean bucket diameter

$\therefore \quad D = 10 \times d = 10 \times 132 = 1320$ mm or 1.32 m

$$u = \frac{\pi DN}{60} \text{ or } N = \frac{60u}{\pi D} = \frac{60 \times 54.5}{\pi \times 1.32} \simeq 788 \text{ r.p.m.}$$

Frequency of generator, $f = \frac{N_{syn.} P}{120}$, where P = no. of poles.

If $\quad P = 8, N_{syn.} = \frac{120 f}{P} = \frac{120 \times 50}{8} = 750$ r.p.m. which is nearest to 788 r.p.m.

$\therefore \quad D_{(revised)} = \frac{1320 \times 788}{750} = 1387$ mm or **1.387 m (Ans.)**

Example 2.10. *The following data relate to a double overhung Pelton unit :*

Output of generator	*...25000 kW*
Generator efficiency	*...93 %*
Effective head at the base of nozzle	*...300 m*
Pelton wheel efficiency	*...85 %*
Co-efficient of velocity	*...0·97*
Speed ratio	*...0·46*
Jet ratio	*...12*

Determine the following :

(i) Size of jet,

(ii) Mean diameter of runner, and

(iii) Synchronous speed.

Solution. Generator output = 25000 kW

Generator efficiency, η_g = 93 %

Effective head, H = 300 m

Efficiency of Pelton wheel = 85 %

Co-efficient of velocity, C_v = 0.97

Speed ratio, K_u = 0.46

Jet ratio, $m = \frac{D}{d} = 12$

(i) Size of jet, *d* :

There are two runners keyed on the two ends of the shaft, and the generator lies between them. Each runner is to be considered as one complete turbine. Thus, two Pelton turbines are feeding the generator.

$$\therefore \quad \text{Output of each turbine, } P_t = \frac{25000}{2 \times \eta_g} = \frac{25000}{2 \times 0.93} = 13440.8 \text{ kW}$$

Power developed by each turbine $= \dfrac{P_t}{\text{Pelton wheel efficiency}} = \dfrac{13440.8}{0.85} = 15812.7$ kW

But, $\quad 15812.7 = wQH$

$$\therefore \quad Q = \frac{15812.7}{wH} = \frac{15812.7}{9.81 \times 300} = 5.37 \text{ m}^3\text{/s}$$

$$\text{Velocity of jet, } V_1 = C_v\sqrt{2gH} = 0.97\sqrt{2 \times 9.81 \times 300} = 74.4 \text{ m/s}$$

Also, $\quad Q = \text{Area of jet} \times \text{velocity of jet} = a \times V_1 = \dfrac{\pi}{4} \times d^2 \times V_1$

or, $\quad 5.37 = \dfrac{\pi}{4} \times d^2 \times 74.4$

$$\therefore \quad d = \left(\frac{5.37 \times 4}{\pi \times 74.4}\right)^{1/2} = 0.3 \text{ m or } \mathbf{300\ mm\ (Ans.)}$$

(*ii*) Mean diameter of runner, D :

$$m = \frac{D}{d} = 12$$

or, $\quad D = 12d = 12 \times 0.3 = \mathbf{3.6\ m\ (Ans.)}$

(*iii*) Synchronous speed ($N_{syn.}$) :

$$\text{Peripheral speed, } V = K_u \times \sqrt{2gH} = 0.46\sqrt{2 \times 9.81 \times 300} = 35.3 \text{ m/s}$$

But, $\quad u = \dfrac{\pi DN}{60} \quad \therefore N = \dfrac{60u}{\pi D} = \dfrac{60 \times 35.3}{\pi \times 3.6} = 187.3$ r.p.m.

$$\text{Frequency of generator, } f = \frac{N_{syn.} \times P}{120}$$

$$\therefore \quad N_{syn.} = \frac{120f}{P} = \frac{120 \times 50}{P} = \frac{6000}{P}$$

[where, f = frequency = 50 Hz (given), and P = no. of poles]

Assuming $P = 32$, we have:

$$D_{syn.} = \frac{6000}{32} = \mathbf{187.5\ r.p.m.\ (Ans.)}$$

$$D_{(revised)} = \frac{3.6 \times 187.3}{187.5} \simeq \mathbf{3.6\ m\ (Ans.)}$$

Example 2.11. *The following data relate to a Pelton wheel :*

Head	*... 72 m*
Speed of the wheel	*... 240 r.p.m.*
Shaft power of the wheel	*...115 kW*
Speed ratio	*... 0·45*
Co-efficient of velocity	*... 0·98*
Overall efficiency	*... 85%*

Design the Pelton wheel.

Solution. Effective head, H = 72 m

Speed of the wheel, N = 240 r.p.m.

Shaft power, P = 115 kW

Speed ratio, K_u = 0·45

Co-efficient of velocity, C_v = 0·98

Overall efficiency, η_0 = 85%

Design the Pelton wheel; it means to find diameter of the wheel D, diameter of jet (d), width and depth of buckets and number of buckets on the wheel.

(i) Diameter of wheel, D:

$$\text{Velocity of jet, } V_1 = C_v \sqrt{2gH} = 0.98\sqrt{2 \times 9.81 \times 72} = 36.8 \text{ m/s}$$

$$\therefore \text{ Bucket velocity, } u\ (= u_1 = u_2) = K_u \times V_1 = 0.45 \times 36.8 = 16.56 \text{ m/s}$$

But,
$$u = \frac{\pi DN}{60}, \text{ or, } D = \frac{60u}{\pi N} = \frac{60 \times 16.56}{\pi \times 240} = \mathbf{1.32\ m} \ \textbf{(Ans.)}$$

(ii) Diameter of jet, d:

$$\text{Overall efficiency, } \eta_0 = \frac{\text{Shaft power}}{\text{Water power}} = \frac{P}{wQH}$$

or,
$$0.85 = \frac{115}{9.81 \times Q \times 72}$$

or,
$$Q = \frac{115}{0.85 \times 9.81 \times 72} = 0.1915 \text{ m}^3/\text{s}$$

But,
$$Q = \text{Area of jet} \times \text{velocity of jet}$$

$$0.1915 = \frac{\pi}{4} \times d^2 \times V_1 = \frac{\pi}{4} d^2 \times 36.8$$

$$\therefore \quad d = \left(\frac{0.1915 \times 4}{\pi \times 36.8}\right)^{1/2} = 0.0814 \text{ m or } \mathbf{81.4\ mm} \ \textbf{(Ans.)}$$

(iii) Size of buckets:

Width of the bucket, B = 3 to 4 times jet diameter (d)

$\simeq 3.5\,d = 3.5 \times 81.4 =$ **285 mm (Ans.)**

Radial length of bucket, L = 2 to 3 times jet diameter (d)

$\simeq 2.5\,d = 2.5 \times 81.4 =$ **203.5 mm (Ans.)**

Depth of bucket, T = 0.8 to 1.2 times jet diameter (d)

$\simeq 1.0\,d =$ **81.4 mm (Ans.)**

(iv) Number of buckets on the wheel, Z :

$$Z = 15 + \frac{D}{2d} = 15 + \frac{1.32 \times 1000}{2 \times 81.4} = \mathbf{23} \ \textbf{(Ans.)}$$

Example 2.12. *A Pelton wheel of 1.1 m mean bucket diameter works under a head of 500 m. The deflection of jet is 165° and its relative velocity is reduced over the bucket by 15 per cent due to friction. If the diameter of jet is 100 mm and the water is to leave the bucket without any whirl, determine :*

(*i*) *Rotational speed of wheel,*
(*ii*) *Ratio of bucket speed to jet velocity,*
(*iii*) *Impulsive force and power developed by the wheel,*
(*iv*) *Available power (water power),*
(*v*) *Power input to buckets, and*
(*vi*) *Efficiency of the wheel with power input to bucket as reference input.*
Take $C_v = 0.97$. **[UPSC]**

Solution. Mean bucket diameter, $D = 1.1$ m

Net head, $H = 500$ m

Deflection of jet $= 165°$

Reduction of relative velocity due to friction = 15%

Diameter of jet, $d = 100$ mm = 0.1 m

Co-efficient of velocity, $C_v = 0.97$

(*i*) Rotational speed of wheel, N :

$$\text{Velocity of jet, } V_1 = C_v\sqrt{2gH} = 0.97\sqrt{2 \times 9.81 \times 500} = 96.07 \text{ m/s}$$

Let, bucket speed $u_1 = u_2 = u$

Relative velocity at inlet, $V_{r1} = V_1 - u_1 = (96.07 - u)$

Relative velocity at outlet, $V_{r2} = 0.85\, V_{r1}$

$= 0.85\,(96.07 - u)$ ...(*i*)

The blade angle at exit, $\phi = 180° - 165° = 15°$

As the jet leaves the bucket *without any whirl,* the velocity triangle at outlet will be as shown in Fig. 2.10 ($\beta = 90°$)

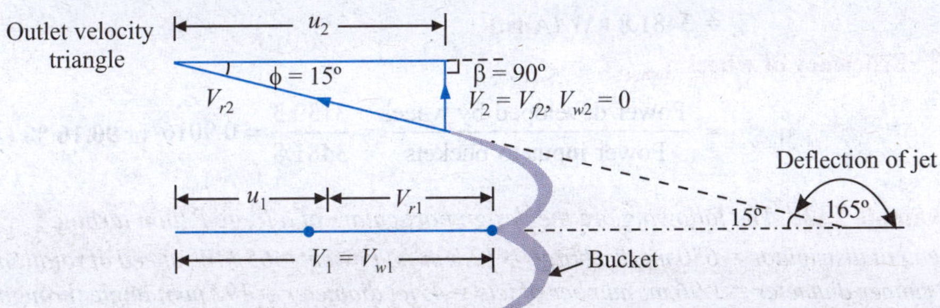

Fig. 2.10

$V_{r2}\cos\phi = u$, or, $V_{r2}\cos 15° = u$ ...(*ii*)

From (*i*) and (*ii*), we get:

$$0.85\,(96.07 - u)\cos 15° = u$$

or, $\quad 0.85\,(96.07 - u) \times 0.966 = u$

or, $\quad 78.88 - 0.821\,u = u$

or, $\quad u = 43.31$ m/s

Also, $\quad u = \dfrac{\pi DN}{60}$, or, $43.31 = \dfrac{\pi \times 1.1 \times N}{60}$

$\therefore$ Rotational speed of wheel, $N = \dfrac{43.31 \times 60}{\pi \times 1.1} \simeq$ **752 r.p.m. (Ans.)**

(ii) Ratio of bucket speed to jet velocity :

$$\frac{u}{V_1} = \frac{43.31}{96.07} = \mathbf{0.4508}\ \textbf{(Ans.)}$$

(iii) Impulsive force and power developed by the wheel :

Discharge through the wheel,

$$Q = \frac{\pi}{4} \times d^2 \times V_1 = \frac{\pi}{4} \times (0.1)^2 \times 96.07 = 0.7545 \text{ m}^3/\text{s}$$

Impulsive force on the buckets,

$$F = \rho Q (V_{w1} \pm V_{w2}) = \rho Q (V_{w1}) \qquad (\because V_{w2} = 0)$$

$$= 1000 \times 0.7545 \times 96.07 \qquad (\because V_{w1} = V_1)$$

$$= \mathbf{72484.8\ N}\ \textbf{(Ans.)}$$

Power developed by the wheel,

$$= F \times u = 72484.8 \times 43.31 = 3139316.7 \text{ Nm/s or J/s or W}$$

$$= \mathbf{3139.3\ kW}\ \textbf{(Ans.)}$$

(iv) Available power (water power) :

$$\text{Available power (water power)} = wQH$$

$$= 9.81 \times 0.7545 \times 500 = \mathbf{3700.8\ kW}\ \textbf{(Ans.)}$$

(v) Power input to buckets :

$$\text{Power input to buckets} = \frac{1}{2} m V_1^2 = \frac{1}{2} (\rho Q) \times V_1^2$$

$$= \frac{1}{2} \times 1000 \times 0.7545 \times (96.07)^2 = 3481808 \text{ Nm/s or J/s or W}$$

$$= \mathbf{3481.8\ kW}\ \textbf{(Ans.)}$$

(vi) Efficiency of wheel η_{wheel} :

$$\eta_{wheel} = \frac{\text{Power developed by wheel}}{\text{Power input to buckets}} = \frac{3139.3}{3481.8} = 0.9016 \text{ or } \mathbf{90.16\ \%}\ \textbf{(Ans.)}$$

Example 2.13. *The following are the design particulars of a large Pelton turbine :*

Head at distributor = 630 m; discharge = 12.5 m³/s; power = 65 MW; speed of rotation = 500 r.p.m; runner diameter = 1.96 m; number of jets = 4; jet diameter = 192 mm; angle through which the jet is defleted by the buket = 165°; and mechanical efficiency of the turbine = 96%.

Determine the hydraulic power losses in the distributor nozzle assembly and the buckets.

[UPSC]

Solution. Head at distributor, H = 630 m

Discharge, Q = 12.5 m^3/s

Shaft power, P = 65 MW = 65000 kW

Speed of rotation, N = 500 r.p.m.

Runner diameter, D = 1.96 m

Number of jets = 4

Jet diameter, d = 192 mm = 0.192 m

Angle of deflection = 165°

Mechanical efficiency of the turbine, η_m = 96%

Hydraulic power losses in nozzle assembly and buckets :

$$\text{Mechanical efficiency, } \eta_m = \frac{\text{Shaft power}}{\text{Power developed by the runner}}$$

or, $$0.96 = \frac{65000}{\text{Power developed by the runner}}$$

or, $$\text{Power developed by the runner} = \frac{65000}{0.96} = 67708 \text{ kW}$$

$\therefore$ $$\text{Power developed per jet} = \frac{67708}{4} = 16927 \text{ kW}$$

$$\text{Discharge per jet, } q = \frac{Q}{4} = \frac{12.5}{4} = 3.125 \text{ m}^3/\text{s}$$

But $$q\,(= 3.125) = \text{Area of jet } (a) \times \text{velocity of jet } (V_1)$$

or, $$3.125 = \frac{\pi}{4} \times d^2 \times V_1 = \frac{\pi}{4} \times 0.192^2 \times V_1$$

$\therefore$ $$= \frac{3.125 \times 4}{\pi \times 0.192^2} = 107.93 \text{ m/s}$$

Also, $$\text{peripheral speed of runner, } u = \frac{\pi D N}{60} = \frac{\pi \times 1.96 \times 500}{60} = 51.31 \text{ m/s}$$

$$\text{Blade angle at outlet, } \phi = 180° - 165° = 15°$$

$$\text{Work developed per jet} = \frac{wQ}{g}(V_1 - u)(1 + K \cos \phi)\, u$$

(where, K = blade friction co-efficient)

$\therefore$ $$\frac{wq}{g}(V_1 - u)(1 + K \cos \phi)\, u = 16927$$

or, $$\frac{9.81 \times 3.125}{9.81}(107.93 - 51.31)(1 + K \cos 15°) \times 51.31 = 16927$$

or, $$9078.7\,(1 + K \cos 15°) = 16927$$

or, $$K \cos 15° = \frac{16927}{9078.7} - 1 = 0.8644$$

$\therefore$ $$K = \frac{0.8644}{\cos 15°} = 0.895$$

Also, $$V_{r2} = K V_{r1}$$

But, $$V_{r1} = (V_1 - u_1) = (V_1 - u)$$

$\therefore$ $$V_{r2} = K(V_1 - u)$$

$$\text{Theoretical velocity of jet } (V_1)_{\text{th}} = \sqrt{2gH}$$

$$\text{Actual velocity of jet } V_1 = C_v \times (V_1)_{th} = C_v \sqrt{2gH}$$

$\therefore$ $$\text{Head lost in nozzle } V_1 = \frac{(V_1)_{th}^2 - V_1^2}{2g} = \frac{2gH - C_v^2\,(2gH)}{2g}$$

$$= (1 - C_v^2)H$$

But, $$C_v = \frac{V_1}{\sqrt{2gH}} = \frac{107.93}{\sqrt{2 \times 9.81 \times 630}} = 0.97$$

$\therefore$ Head lost in nozzle $= (1 - 0.97^2) \times 630 = 37.23$ m

$$\text{Head lost in buckets} = \frac{V_{r1}^2 - V_{r2}^2}{2g} = \frac{V_{r1}^2 - K^2 V_{r1}^2}{2g} = \frac{V_{r1}^2}{2g}(1 - K^2)$$

$$= \frac{(V_1 - u)^2}{2g}(1 - K^2) = \frac{(107.93 - 51.31)^2}{2 \times 9.81} \times (1 - 0.895^2) = 32.5 \text{ m}$$

$\therefore$ The head lost in nozzle and buckets,

$$H_L = 37.23 + 32.5 = 69.73 \text{ m}$$

Power loss $= wQH_L = 9.81 \times 12.5 \times 69.73 =$ **8550.6 kW (Ans.)**

Example 2.14. *Show that in Pelton wheel, where the buckets deflect the water through* (180° – ϕ), *the hydraulic efficiency of the wheel is given by:*

$$\eta_h = \frac{2u\,(V - u)\,(1 + \cos\phi)}{V^2}$$

where, V is the velocity of the jet and u is the velocity of the wheel at the pitch radius.

[UPSC]

Solution. Refer to Article 2.3.2.

Example 2.15. *Prove that the maximum efficiency of Pelton wheel occurs when the ratio of bucket velocity u to the velocity V is given by the expression.*

$$\frac{u}{V} = \frac{1 - \cos\theta + K_1}{2\,(1 - \cos\theta) + K_1 + K_2}$$

where K_1 and K_2 are the constants.

Loss due to bucket friction and shock $= \dfrac{K_1\,(V - u)^2}{2g}$

Loss due to bearing friction and windage losses $= K_2 \dfrac{u^2}{2g}$

θ = *bucket angle at outlet* = (180° – ϕ) *i.e., angle of deflection of jet.*

Volumetric losses to be considered negligible.

Solution. The net amount of work done per unit weight of water (taking into account the losses given)

$$= \frac{1}{g}[(V - u)(1 - \cos\theta)]\,u - K_1 \frac{(V - u)^2}{2g} - K_2 \frac{u^2}{2g}$$

$$\therefore \quad \text{Efficiency, } \eta = \frac{\frac{1}{g}[(V - u)(1 - \cos\theta)]\,u - K_1 \frac{(V - u)^2}{2g} - K_2 \frac{u^2}{2g}}{(V^2/2g)}$$

$$= \frac{2\,[(V-u)\,u \times (1-\cos\theta)] - K_1\,(V-u)^2 - K_2 u^2}{V^2}$$

The efficiency to be *maximum*, $\frac{d\eta}{du} = 0$.

$$\therefore \quad \frac{d\eta}{du} = \frac{2\,(V-2u)\,(1-\cos\theta) + 2K_1\,(V-u) - 2K_2 u}{V^2} = 0$$

or, $(V-2u)\,(1-\cos\theta) + K_1\,(V-u) - K_2 u = 0$

or, $V(1-\cos\theta) - 2u\,(1-\cos\theta) + K_1 V - K_1 u - K_2 u = 0$

or, $V(1-\cos\theta + K_1) = u\,[2\,(1-\cos\theta) + K_1 + K_2]$

$$\therefore \quad \frac{u}{V} = \frac{1-\cos\theta + K_1}{2\,(1-\cos\theta) + K_1 + K_2}$$ ...(Proved)

2.4. REACTION TURBINES

In reaction turbines, the *runner utilizes both potential and kinetic energies*. As the water flows through the stationary parts of the turbine, whole of its pressure energy is *not* transformed to kinetic energy and when the water flows through the moving parts, there is a *change both in pressure and in the direction and velocity of flow of water*. As the water gives up its energy to the runner, both its pressure and absolute velocity get reduced. The water which acts on the runner blades is under a pressure *above atmospheric and the runner passages are always completely filled with water.*

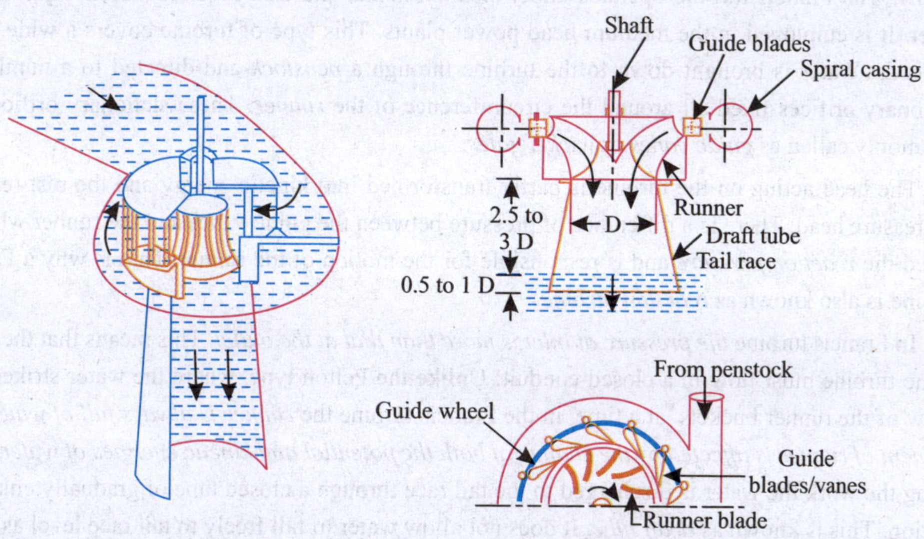

Fig. 2.11. Schematic diagram of a Francis turbine.

Important reaction turbines are *Francis*, *Kaplan* and *Propeller*.

2.4.1. Francis Turbine

Fig. 2.11 shows a schematic diagram of a Francis turbine. The *main parts of a Francis turbine are:*

1. *Penstock* ... It is a large size conduit which conveys water from the upstream of the dam/reservoir to the turbine runner.

2. *Spiral/scroll casing* ... It constitutes a closed passage whose cross-sectional area gradually decreases along the flow direction, area is maximum at inlet and nearly zero at exit.

3. *Guide vanes/wicket gates* ... These vanes direct the water onto the runner at an angle appropriate to the design. The motion to them is given by means of a hand wheel or automatically by a governor.

4. *Governing mechanism* ... It changes the position of the guide blades/vanes to affect a variation in water flow rate, when the load conditions on the turbine change.

5. *Runner and runner blades* ... — The driving force on the runner is both due to impulse and reaction effects;

   — The number of runner blades usually varies between 16 to 24.

6. *Draft tube* ... It is a gradually expanding tube which discharges water, passing through the runner, to the tail race.

The modern Francis turbine is an *inward mixed flow reaction turbine* (in the earlier stages of development, Francis turbine had a purely radial flow runner), *i.e. water under pressure, enters the runner from the guide vanes towards the centre in radial direction and discharges out of the runner axially.* The Francis turbine operates under *medium heads* and also requires *medium quantity* of water. It is employed in the medium head power plants. This type of turbine covers a wide range of heads. Water is brought down to the turbine through a *penstock* and directed to a number of stationary orifices fixed all around the circumference of the *runner*. These stationary orifices are commonly called as *guide vanes* or *wicket gates*.

The head acting on the turbine is partly transformed into kinetic energy and the rest remains as pressure head. There is a difference of pressure between the guide vanes and the runner which is called the *reaction pressure* and is responsible for the motion of the runner. That is why a Francis turbine is also known as *reaction turbine*.

In Francis turbine *the pressure at inlet is more than that at the outlet*. This means that the water in the turbine must flow in a closed conduit. Unlike the Pelton type, where the water strikes only a few of the runner buckets at a time, in the Francis turbine the *runner is always full of water. The moment of runner is affected by the change of both the potential and kinetic energies of water.* After doing the work the water is discharged to the tail race through a closed tube of gradually enlarging section. This is known as *draft tube*. It does not allow water to fall freely to tail race level as in the Pelton turbine. The free end of the draft tube is submerged deep in tail water making, thus, the entire water passage, right from the head race up to the tail race, *totally enclosed*.

Fig. 2.12 shows general layout of a hydroelectric power plant *using a reaction turbine*.

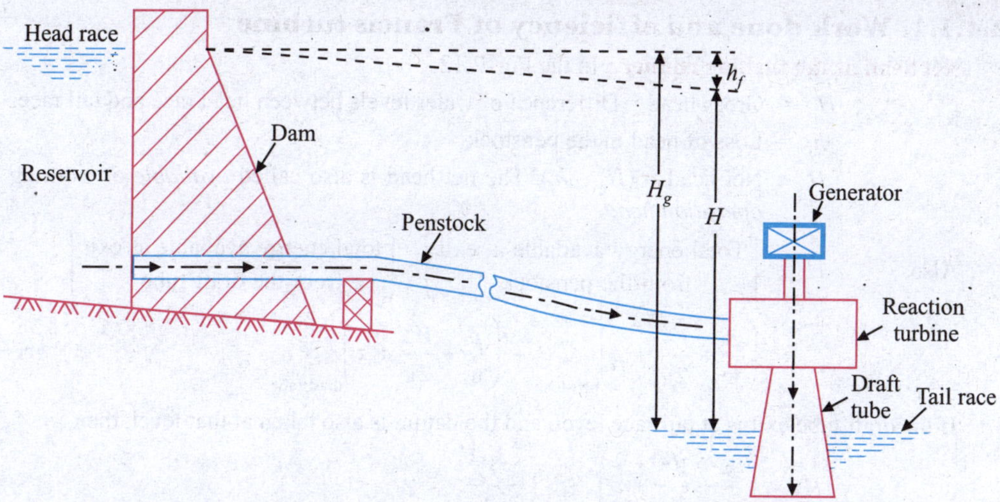

Fig. 2.12. General layout of a hydroelectric power plant using a reaction turbine.

Important Francis Turbine Installations in India :

S.No.	Scheme/Project	Location (State)	Source of water
1.	Bhakhra dam project	Bhakra (Punjab)	Sutlej river
2.	Cauvery hydroelectric scheme	Siva Samudram (Karnataka)	Cauvery river
3.	Chambal hydroelectric scheme	Gandhi sagar (Rajasthan)	Chambal river
4.	Hirakud dam project	Hirakud (Orissa)	Hirakud river
5.	Rihand dam project	Rihand (Uttar Pradesh)	Rihand river

Important differences between Inward and Outward Flow Reaction Turbines :

The following are the important differences between inward and outward flow reaction turbines :

S.No.	Aspects	Inward flow reaction turbine	Outward flow reaction turbine
1	*Entry of water*	Water enters at the outer periphery, flows inward and towards the centre of the turbine and discharges at the outer periphery.	Water enters at the inner periphery flows outward and discharges at the outer periphery.
2	*Centrifugal head imparted*	Negative (negative centrifugal head reduces the relative velocity of water at the outlet).	Positive (Positive centrifugal head increases the relative velocity of water at the outlet).
3	*Discharge*	Does not increase.	The discharge increases.
4	*Speed control*	Easy and effective.	Very difficult.
5	*Tendency of the wheel to race*	Nil. The turbine adjusts the speed by itself.	If the turbine speed increases the wheel tends to race; the turbine cannot adjust the speed by itself.
6	*Suitability*	Quite suitable for medium high heads; best suitable for large outputs and units.	Quite suitable for low or medium heads.
7	*Application*	For power projects.	Practically obsolete.

2.4.1.1. Work done and efficiency of Francis turbine

Net head at the turbine runner : In the Fig. 2.12,

H_g = Gross head = Difference of water levels between head race and tail race;

h_f = Loss of head in the penstock;

H = Net head = $(H_g - h_f)$. The net head is also called *available* or *working* or *operation head.*

Also,

$$H = \begin{bmatrix}\text{Total energy available at exit} \\ \text{from the penstock}\end{bmatrix} - \begin{bmatrix}\text{total energy available at exit} \\ \text{from the draft tube}\end{bmatrix}$$

$$= \left(\frac{p}{w} + \frac{V^2}{2g} + z\right)_{\text{penstock}} - \left(\frac{p}{w} + \frac{V^2}{2g} + z\right)_{\text{draft tube}}$$

If the draft tube exit is at tail race level, and the datum is also taken at that level, then,

$$H = \left(\frac{p}{w} + \frac{V^2}{2g} + z\right)_{\text{penstock}} - \frac{V_d^2}{2g}$$

(where, V_d = velocity at the exit of the draft tube)

Neglecting the velocity at the draft tube exit (V_d), we have:

$$H = \left(\frac{p}{w} + \frac{V^2}{2g} + z\right) \quad \text{...(2.16)}$$

Work done by the runner :

Fig. 2.13 shows the runner and the velocity diagrams (inlet and outlet) for an inward flow reaction turbine. The general expression for the work done with usual notations according to the Euler momentum equation, is given by,

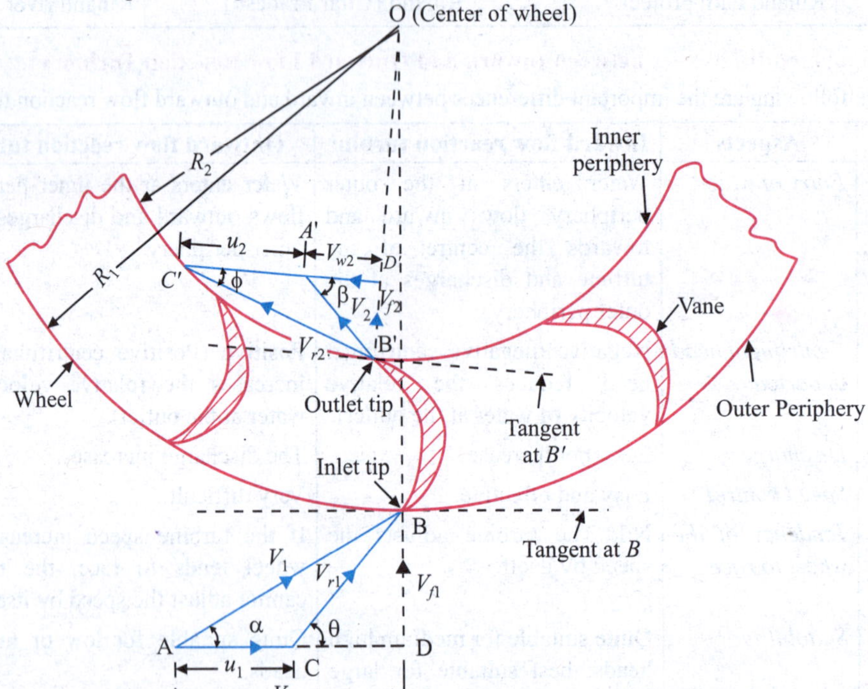

Fig. 2.13. Velocity diagrams for an inward flow reaction turbine.

$$\text{Work done} = \rho Q (V_{w1} u_1 \pm V_{w2} u_2)$$

$$= \frac{wQ}{g}(V_{w1} u_1 \pm V_{w2} u_2) \quad ...(2.17)$$

where, Q = Discharge through the runner, m^3/s.

The maximum output under given conditions is obtained when $V_{w2} = 0$.

Thus, the maximum work done is given by,

$$\text{Work done} = \frac{wQ}{g}(V_{w1} u_1) \quad ...(2.18)$$

This *discharge in this case is radial*. For radial discharge, the *absolute velocity at exit is radial.*

Hydraulic efficiency, η_h :

If H is the net head, then input to the turbine = wQH.

$$\eta_h = \frac{\text{Power developed by the runner}}{\text{Power supplied to the turbine (water power)}} = \frac{\frac{wQ}{g}(V_{w1} u_1)}{wQH}$$

or,

$$\eta_h = \frac{V_{w1} u_1}{gH} \quad ...(2.19)$$

[However, if the velocity of whirl at the exit is *not* zero, then

$$\eta_h = \frac{V_{w1} u_1 \pm V_{w2} u_2}{gH} \quad ...[2.19(a)]$$
]

The hydraulic efficiency of the Francis turbine varies from 85 to 90 percent.

Mechanical efficiency, η_m :

The mechanical efficiency is given by:

$$\eta_m = \frac{\text{Shaft power (P)}}{\text{Power developed by the runner}} \quad ...(2.20)$$

Overall efficiency, η_0 :

The overall efficiency is given as:

$$\eta_0 = \frac{\text{Shaft water}}{\text{Water power}} = \frac{P}{wQH} \quad ...(2.21)$$

and,

$$\eta_0 = \eta_h \times \eta_m \quad [2.21(a)]$$

The overall efficiency varies from 80 to 90 percent.

2.4.1.2. Working proportions of a Francis turbine

The following working proportions pertain to a Francis turbine :

1. Ratio of width to diameter $\left(\frac{B}{D}\right)$:

The ratio of width (B_1) to the diameter of the wheel (D_1) at inlet is represented by n. Thus,

$$n = \frac{B_1}{D_1}$$

The value of n varies from 0.10 to 0.45.

2. Flow ratio (K_f) :

Flow ratio is the *ratio of the velocity of flow at inlet to the theoretical jet velocity*. Thus,

$$\text{Flow ratio, } K_f = \frac{V_{f1}}{\sqrt{2gH}} \quad \text{...(2.23)}$$

The value of K_f varies from 0.15 to 0.30.

3. Speed ratio (K_u)

Speed ratio is the ratio of the peripheral speed at inlet to the theoretical jet velocity. Thus,

$$\text{Speed ratio, } K_u = \frac{u}{\sqrt{2gH}}$$

The value of K_u ranges from 0.6 to 0.9.

2.4.1.3. Design of Francis turbine runner

The runner of a Francis turbine is required to be designed to develop a known power P, when running at a known speed N r.p.m. under a known head H. The design of the runner involves the determination of its *size* and the *vane angles*.

The design of a Francis turbine runner is carried out as follows :

1. Assume suitable values of η_0, η_h, n, K_f and K_t.
2. Determine the required discharged Q from the relation:

$$P = \eta_0 \times wQH \quad \text{...(2.24)}$$

3. Obtain the velocity of flow from the discharge and flow area.

Let B_1, D_1 and t_1 respectively be the width, diameter and thickness of runner vane at inlet (Fig. 2.14).

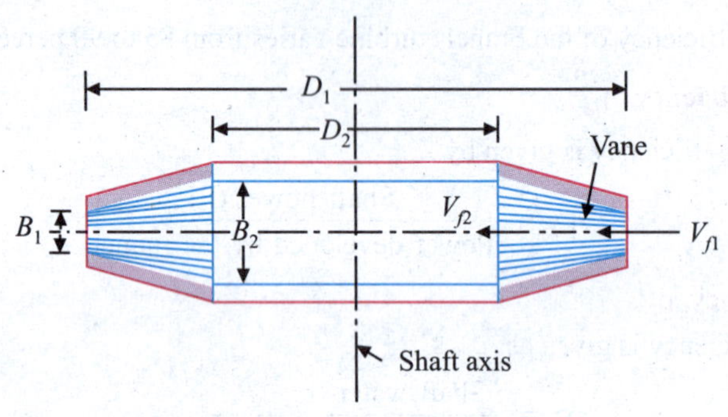

Fig. 2.14. Entry of flow to runner vane.

Then, total area at the outer periphery (*i.e.*, at the runner inlet),

$$A = (\pi D_1 - Z_{t1}) B_1 = K_{t1}\pi D_1 B_1 \quad \text{...(2.25)}$$

where K_{t1} is known as *vane thickness factor/co-efficient*; its value is *always less than unity* (usually of the order of 0.95 or so).

$$\text{Discharge} = \text{Area of flow} \times \text{velocity of flow}$$

i.e.,

$$Q = K_{t1}\pi D_1 B_1 \times V_{f1} \quad \text{...(2.26)}$$

$$\therefore \quad \text{Flow velocity, } V_{f1} = \frac{Q}{K_{t1}\pi D_1 B_1} = \frac{Q}{K_t \pi n D_1^2} \quad (\because B_1 = nD_1) \quad \text{...[2.26(a)]}$$

Also,

$$V_{f1} = K_f\sqrt{2gH}, \quad \therefore \; K_f\sqrt{2gH} = \frac{Q}{K_t \pi n D_1^2}$$

or, $$D_1 = \left[\frac{Q}{(K_f \sqrt{2gH})\, K_t \pi n}\right]^{1/2} \quad ...(2.27)$$

Then, width $B_1 = nD_1$ [Eqn. (2.22)]

4. Find the rim velocity (tangential velocity) u_1 from the relation:

$$u_1 = \frac{\pi D_1 N}{60}$$

5. Find the velocity of whirl at inlet (V_{w1}) from Eq [2.19)

$$\eta_h = \frac{V_{w1}\, u_1}{gH}, \quad \text{or,} \quad V_{w1} = \frac{\eta_h gH}{u_1} \quad ...(2.28)$$

6. Obtain the guide vane angle (α) and the runner vane angle (θ) from the following relations obtained from inlet velocity triangle (Fig. 2.13):

$$\tan\alpha = \frac{V_{f1}}{V_{w1}} \quad ...(2.29)$$

and, $$\tan\theta = \frac{V_{f1}}{V_{w1} - u_1} \quad ...(2.30)$$

7. Assume runner diameter D_2 at the outlet to be approximately one-half the diameter at inlet.

Thus, $$D_2 = \frac{D_1}{2} \text{ and } u_2 = \frac{u_1}{2}.$$

8. The velocity of flow at the exit (V_{f2}) is obtained as follows :

$$Q = K_{t1}\pi D_1 B_1 V_{f1} = K_{t2}\pi D_2 B_2 V_{f2} \quad \textit{...Continuity equation}$$

Thus, $$\frac{V_{f1}}{V_{f2}} = \frac{K_{t2}\pi D_2 B_2}{K_{t1}\pi D_1 B_1}$$

Usually, it is presumed that $V_{f1} = V_{f2}$ and $K_{t1} = k_{t2}$, that gives $B_2 = 2B_1$

9. Find the runner vane angle at exit (ϕ) from the velocity outlet triangle, assuming the discharge at the runner exit to be radial (β = 90°). Thus,

$$\tan\theta = \frac{V_{f2}}{u_2} \quad ...(2.31)$$

10. The number of vanes varies from 16 to 24. In order to *avoid periodic impulse, the number of vanes should be either one more or one less than the number of guide vanes.*

2.4.1.4. Advantages and disadvantages of a Francis turbine over a Pelton wheel

Advantages :

The Francis turbine claims the following *advantages* over Pelton wheel :

1. In Francis turbine the variation in the operating head can be more easily controlled.
2. In Francis turbine the ratio of maximum and minimum operating heads can be even two.
3. The operating head can be utilized even when the variation in the tail water level is relatively large when compared to the total head.
4. The mechanical efficiency of Pelton wheel decreases faster with wear than Francis turbine.

5. The size of the runner, generator and power house required is small and economical if the Francis turbine is used instead of Pelton wheel for same power generation.

Disadvantages/Drawbacks :

As compared with Pelton wheel, the Francis turbine has the following *drawbacks/ shortcomings*:

1. Water which is not clean can cause very rapid wear in high head Francis turbine.
2. The overhaul and inspection is much more difficult comparatively,
3. Cavitation is an ever-present danger.
4. The water hammer effect is more troublesome with Francis turbine.
5. If Francis turbine is run below 50 percent head for a long period it will not only lose its efficiency but also the cavitation danger will become more serious.

Example 2.16. *An inward flow reaction turbine has external and internal diameters as 1·08 m and 0.54 m. The turbine is running at 200 r.p.m. The width of the turbine at inlet is 240 mm and velocity of flow through the runner is constant and is equal to 2.16 m/s. The guide blades make an angle of 10° to the tangent of the wheel and discharge at the outlet of the turbine is radial. Draw the inlet and outlet velocity triangles and determine :*

(*i*) *The absolute velocity of water at inlet of the runner,*
(*ii*) *The velocity of whirl at inlet,*
(*iii*) *The relative velocity at inlet,*
(*iv*) *The runner blade angles,*
(*v*) *Width of runner at outlet,*
(*vi*) *Weight of water flowing through the runner per second,*
(*vii*) *Head at inlet of the turbine,*
(*viii*) *Power developed, and*
(*ix*) *Hydraulic efficiency of the turbine.*

Solution. External diameter, $D_1 = 1.08$ m

Internal diameter, $D_2 = 0.54$ m

Speed, $N = 200$ r.p.m.

Width at inlet, $B_1 = 240$ mm $= 0.24$ m

Velocity of flow, $V_{f1} = V_{f2} = 2.16$ m/s

Guide blade angle, $\alpha = 10°$

Discharge at outlet .... radial

$$\therefore \quad \beta = 90° \text{ and } V_{w2} = 0$$

Tangential velocity of wheel at inlet, $u_1 = \dfrac{\pi D_1 N}{60} = \dfrac{\pi \times 1.08 \times 200}{60} = 11.31$ m/s

Tangential velocity of wheel at outlet, $u_2 = \dfrac{\pi D_2 N}{60} = \dfrac{\pi \times 0.54 \times 200}{60} = 5.65$ m/s

(*i*) Absolute velocity of water at inlet of the runner, V_1 :

From inlet velocity triangle, we have:

$$V_1 \sin \alpha = V_{f1}$$

$$\therefore \quad V_1 = \frac{V_{f1}}{\sin \alpha} = \frac{2.16}{\sin 10°} = \mathbf{12.44\ m/s\ (Ans.)}$$

(ii) Velocity of whirl at inlet, V_{w1} :

$$V_{w1} = V_1 \cos \alpha = 12.44 \times \cos 10° = \mathbf{12.25\ m/s\ (Ans.)}$$

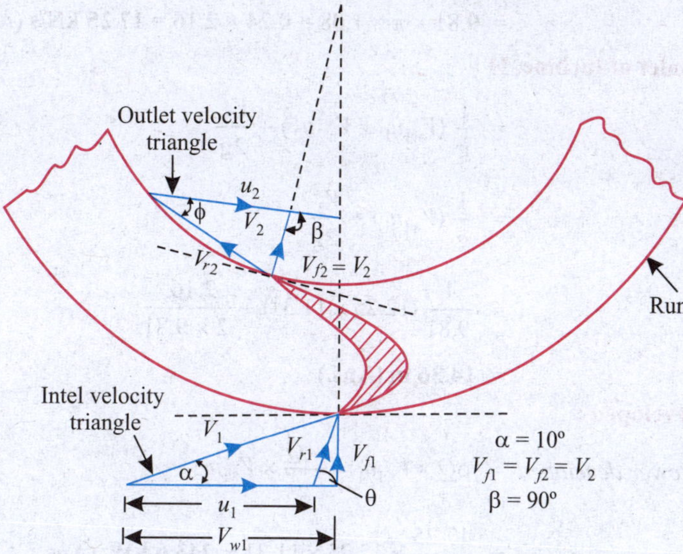

Fig. 2.15

(iii) Relative velocity at inlet, V_{r1} :

$$V_{r1} = \sqrt{(V_{w1} - u_1)^2 + V_{f1}^2}$$

$$= \sqrt{(12{\cdot}25 - 11.31)^2 + (2.16)^2} = \mathbf{2.35\ m/s\ (Ans.)}$$

(iv) Runner blade angles, θ, ϕ :

Again, from inlet velocity triangle, we have:

$$\tan \theta = \frac{V_{f1}}{V_{w1} - u_1} = \frac{2.16}{12.25 - 11.31} = 2.298$$

$$\therefore \quad \theta = \tan^{-1} 2.298 = \mathbf{66.48° \ (Ans.)}$$

From the outlet velocity triangle, we have:

$$\tan \theta = \frac{V_{f2}}{u_2} = \frac{2.16}{5.65} = 0.382$$

$$\therefore \quad \phi = \tan^{-1} 0.382 = \mathbf{20.9° \ (Ans.)}$$

(v) Width of runner at outlet, B_2 :

$$\pi D_1 B_1 V_{f1} = \pi D_2 B_2 V_{f2} (= Q) \qquad \text{...Continuity equation}$$

or, $$D_1 B_1 = D_2 B_2 \qquad (\because V_{f1} = V_{f2})$$

$$\therefore \quad B_2 = \frac{D_1 B_1}{D_2} = \frac{1.08 \times 0.24}{0.54} = 0.48 \text{ m or } \mathbf{480\ mm\ (Ans.)}$$

(vi) Weight of water flowing through the runner per second :

Weight of water flowing per second

$$= wQ = w \times \pi D_1 B_1 V_{f1}$$

$$= 9.81 \times \pi \times 1.08 \times 0.24 \times 2.16 = \textbf{17.25 kN/s (Ans.)}$$

(vii) Head at inlet of turbine, H :

$$= \frac{1}{g}(V_{w1}u_1 \pm V_{w2}u_2) + \frac{V_2^2}{2g}$$

$$= \frac{1}{g}(V_{w1}u_1) + \frac{V_2^2}{2g} \qquad (\because V_{w2} = 0)$$

$$= \frac{1}{9.81}(12.25 \times 11.31) + \frac{2.16^2}{2 \times 9.81} \qquad (\because V_2 = V_{f2})$$

$$= \textbf{14.36 m (Ans.)}$$

(viii) Power developed :

$$\textit{Power developed} = \rho Q \times V_{w1}u_1 = \frac{wQ}{g} \times V_{w1}u_1$$

$$= \frac{17.25}{9.81} \times 12.25 \times 11.31 = \textbf{243.6 kW (Ans.)}$$

(ix) Hydraulic efficiency, η_h :

$$\eta_h = \frac{V_{w1}u_1}{gH} \qquad \text{...[Eqn (2.19)]}$$

$$= \frac{12.25 \times 11.31}{9.81 \times 14.36} = 0.9835 \quad \text{or} \quad \textbf{98.35\% (Ans.)}$$

Example 2.17. *A reaction turbine works at 450 r.p.m. under a head of 120 m. Its diameter at inlet is 1.2 m and the flow area is 0.4 m^2. The angles made by absolute and relative velocities at inlet are 20° and 60° respectively with the tangential velocity. Determine :*

(i) The volume flow rate,

(ii) The power developed, and

(iii) The hydraulic efficiency. **[PTU]**

Solution.

Speed of turbine, N = 450 r.p.m

Head, H = 120 m

Diameter at inlet, D_1 = 1.2 m

Flow area, $\pi D_1 B_1$ = 0.4 m^2

Angle made by absolute velocity, α = 20°

Angle made by the relative velocity at inlet, θ = 60°

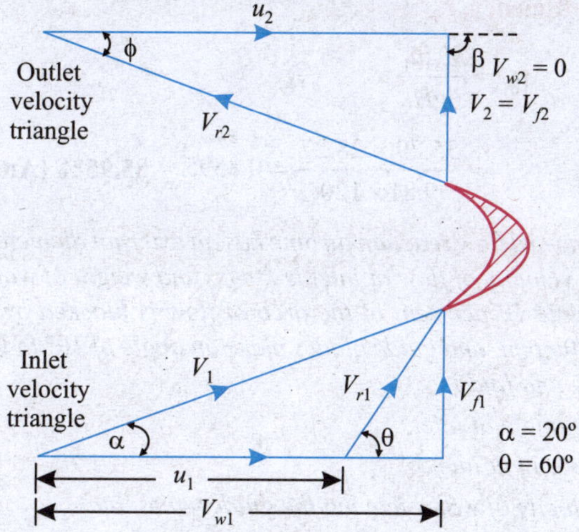

Fig. 2.16

(i) The volume flow rate, Q :

Tangential velocity of the turbine,

$$u_1 = \frac{\pi D_1 N}{60} = \frac{\pi \times 1.2 \times 450}{60} = 28.27 \text{ m/s}$$

From inlet velocity triangle, we have:

$$\tan \alpha = \frac{V_{f1}}{V_{w1}}, \quad \text{or,} \quad \tan 20° = \frac{V_{f1}}{V_{w1}}$$

$$\therefore \quad V_{f1} = V_{w1} \tan 20° = 0.364\, V_{w1} \qquad ...(i)$$

Also, $$\tan \theta = \frac{V_{f1}}{V_{w1} - u_1} = \frac{0.364\, V_{w1}}{V_{w1} - 28.27} \qquad (\because V_{f1} = 0.364\, V_{w1}).$$

or, $$\tan 60° = \frac{0.364\, V_{w1}}{V_{w1} - 28.27}, \text{ or, } 1.732 = \frac{0.364\, V_{w1}}{V_{w1} - 28.27}$$

or, $1.732\,(V_{w1} - 28.27) = 0.364\, V_{w1}$, or , $1.732\, V_{w1} - 48.96 = 0.364\, V_{w1}$

$$\therefore \quad V_{w1} = \frac{48.96}{(1.732 - 0.364)} = 35.79 \text{ m/s}$$

From eqn. (*i*), we have:

$$V_{f1} = 0.364 \times 35.79 = 13.027 \text{ m/s}$$

$\therefore$ Volume flow rate, $Q = \pi D_1 B_1 \times V_{f1}$

But, $\pi D_1 B_1 = 0.4 \text{ m}^2$ ...(*Given*)

$$\therefore \quad Q = 0.4 \times 13.027 = \mathbf{5.211\ m^3/s}$$ **(Ans.)**

(ii) Power developed :

Work done per second $= \rho Q\,(V_{w1} u_1)$ $\quad [\because V_{w2} = 0$ ...Given]

$= 1000 \times 5.211 \times 35.79 \times 28.27 = 5272402$ Nm/s or J/s

$\therefore$ Power developed $= 5272402$ J/s, or, $W =$ **5272.4 kW (Ans.)**

(iii) The hydraulic efficiency, η_h :

$$\eta_h = \frac{V_{w1} u_1}{gH} \quad \text{...[Eqn (2.19)]}$$

$$= \frac{35.79 \times 28.27}{9.81 \times 120} = 0.8595 = \mathbf{85.95\%\ (Ans.)}$$

Example 2.18. *An inward flow reaction turbine has an external diameter of 1 m and its breadth at inlet is 250 mm. If the velocity of flow at inlet is 2 m/s, find weight of water passing through the turbine per second. Assume 10 per cent of the area of flow is blocked by blade thickness. If the speed of the runner is 210 r.p.m. and guide blades make an angle of 10° to the wheel tangent, draw the inlet velocity triangle and find :*

(i) The runner vane angle at inlet,

(ii) The velocity of wheel at inlet,

(iii) The absolute velocity of water leaving the guide vanes, and

(iv) The relative velocity of water entering the runner blade. **[Rookee University]**

Solution.

External diameter, D_1 = 1 m

Breadth of inlet, B_1 = 250 m = 0·25 m

Velocity of flow at inlet, V_{f1} = 2.0 m/s

Area of flow blocked by blade thickness = 10 %

Speed of the runner, N = 210 r.p.m.

Guide blade angle, α = 10°

Weight of water passing through the turbine :

$$\text{Area blocked by vane thickness} = \frac{10}{100} \times \pi D_1 B_1 = 0.1\ \pi D_1 B_1$$

∴ Actual area through which flow takes place,

$$A = \pi D_1 B_1 - 0.1\ \pi D_1 B_1 = 0.9\ \pi D_1 B_1$$
$$= 0.9\ \pi \times 1 \times 0.25 = 0.7068\ \text{m}^2$$

∴ Weight of water passing per second through the turbine

$$= w \times A \times V_{f1} = 9.81 \times 0.7068 \times 2$$
$$= \mathbf{13.68\ kN/s\ (Ans.)}$$

Inlet velocity triangle is shown in Fig 2.17.

(i) The runner vane angle at inlet, θ :

Tangential velocity of wheel at inlet,

$$u_1 = \frac{\pi D_1 N}{60} = \frac{\pi \times 1 \times 10}{60} = 10.99\ \text{m/s}$$

From inlet velocity triangle, we have:

$$\tan \alpha = \frac{V_{f1}}{V_{w1}} = \frac{2.0}{V_{w1}}, \text{ or, } V_{w1} = \frac{2.0}{\tan \alpha} = \frac{2.0}{\tan 10°} = 11.34\ \text{m/s}$$

$$\tan \theta = \frac{V_{f1}}{V_{f1} - u_1} = \frac{2.0}{11.34 - 10.99} = 5.714$$

or,

$$\theta = \tan^{-1} 5.714$$
$$= \mathbf{80.07°\ (Ans.)}$$

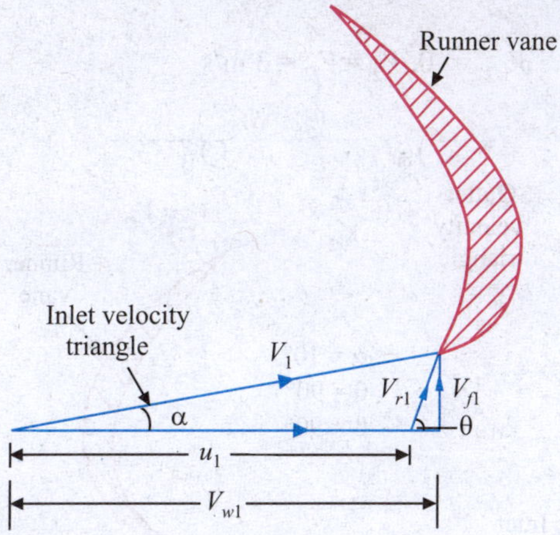

Fig. 2.17

(*ii*) The velocity of wheel at inlet u_1 :

The velocity of wheel at inlet = $\boldsymbol{u_1}$ **= 10.99 m/s (Ans.)**

(*iii*) The absolute velocity of water leaving the guide vanes, V_1 :

Again, from inlet velocity triangle , we have:

$$\sin\alpha = \frac{V_{f1}}{V_1}, \quad \text{or,} \quad \sin 10° = \frac{2{\cdot}0}{V_1}$$

$$\therefore \qquad V_1 = \frac{2.0}{\sin 10°} = \mathbf{11.52\ m/s\ (Ans.)}$$

(*iv*) The relative velocity of water entering the runner blade, V_{r1} :

$$\sin\theta = \frac{V_{f1}}{V_{r1}}, \quad \text{or,} \quad \sin 80.07° = \frac{2.0}{V_{r1}}$$

$$\therefore \qquad V_{r1} = \frac{2.0}{\sin 80.07°} = \mathbf{2.03\ m/s\ (Ans.)}$$

Example 2.19. *In an inward flow reaction turbine the head on the turbine is 32 m. The external and internal diameters are 1.44 m and 0.72 m respectively. The velocity of flow through the runner is constant and equal to 3 m/s. The guide blade angle is 10° and the runner vanes are rigid at inlet. If the discharge at outlet is radial, determine :*

(*i*) *The speed of the turbine,*

(*ii*) *The vane angle at outlet of the runner, and*

(*iii*) *Hydraulic efficiency.*

Solution. Head on the turbine, H = 32 m

External diameter, D_1 = 1.44 m

Internal diameter, D_2 = 0.72 m

Velocity of flow, V_f = constant; $V_{f1} = V_{f2} = 3$ m/s

The guide blade angle, α = 10°

Runner vanes are *radial at inlet,*

$$\therefore \qquad \theta = 90°; \qquad \therefore \qquad V_{w1} = u_1$$

Discharge is radial,

$$\therefore \quad V_{w2} = 0, V_2 = V_{f2} = 3 \text{ m/s}$$

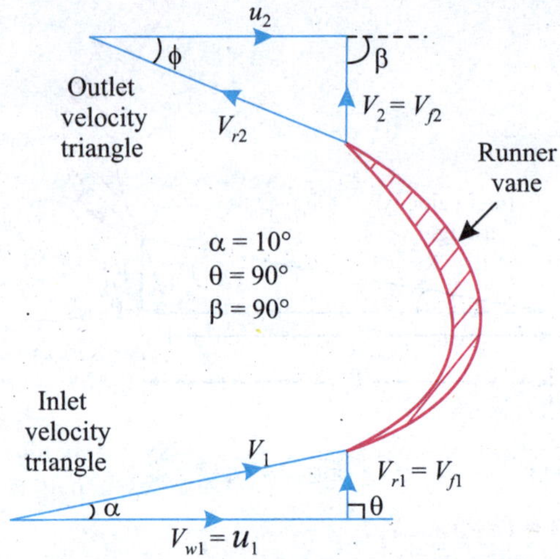

Fig. 2.18

(*i*) The speed of the turbine, *N* :

From inlet velocity triangle, we have:

$$\tan\alpha = \frac{V_{f1}}{u_1}, \quad \text{or,} \quad u_1 = \frac{V_{f1}}{\tan\alpha} = \frac{3}{\tan 10°}$$

$$\text{or,} \quad u_1 = 17.01 \text{ m/s}$$

$$\text{Also,} \quad u_1 = \frac{\pi D_1 N}{60}, \quad \text{or,} \quad N = \frac{60\, u_1}{\pi D_1}$$

$$\text{or,} \quad N = \frac{60 \times 17.01}{\pi \times 1.44} = \mathbf{225.6 \text{ r.p.m.}} \textbf{ (Ans.)}$$

(*ii*) The vane angle at outlet of the runner, ϕ :

$$u_2 = \frac{\pi D_2 N}{60} = \frac{\pi \times 0.72 \times 225.6}{60} = 8.505 \text{ m/s}$$

From the outlet velocity triangle, we have:

$$\tan\phi = \frac{V_{f2}}{u_2} = \frac{3.0}{8.505} = 0.3527$$

$$\therefore \quad \phi = \tan^{-1} 0.3527 = \mathbf{19.43°} \textbf{ (Ans.)}$$

(*iii*) Hydraulic efficiency, η_h :

$$\eta_h = \frac{V_{w1} u_1}{gH} \quad [V_{w2} = 0, \text{ the discharge being } radial \text{ at outlet.}]$$

$$= \frac{17.01 \times 17.01}{9.81 \times 32} = 0.9217 \text{ or } \mathbf{92.17\%} \textbf{ (Ans.)} \quad \left[\begin{array}{l}\because V_{w1} = u_1, \text{ the runner } V \\ \text{at inlet being radial}\end{array}\right]$$

Example 2.20. *An inward flow reaction turbine is supplied 0·233 m³/s of water under a head of 11 m. The wheel vanes are radial at inlet and the inlet diameter is twice the outlet diameter. The velocity of flow is constant and equal to 1.83 m/s. The wheel makes 370 r.p.m. Determine :*

(i) *Guide vane angle,*

(ii) *Inlet and outlet diameters of the wheel, and*

(iii) *The width of the wheel at inlet and exit.*

Assume that the discharge is radial and there are no losses in wheel.

Take speed ratio = 0.7

Neglect the thickness of the vanes. **[Anna University]**

Solution. Flow rate, $Q = 0.233 \text{ m}^3/\text{s}$

Head, $H = 11$ m

Velocity of flow = constant; $V_{f1} = V_{f2} = 1.83$ m/s

Speed of the wheel = 370 r.p.m

Speed ratio, $K = 0.7$

Wheel vanes are radial at inlet, thus $\theta = 90°$, $V_{r1} = V_{f1}$

Inlet diameter = 2 × outlet diameter *i.e.*, $D_1 = 2D_2$

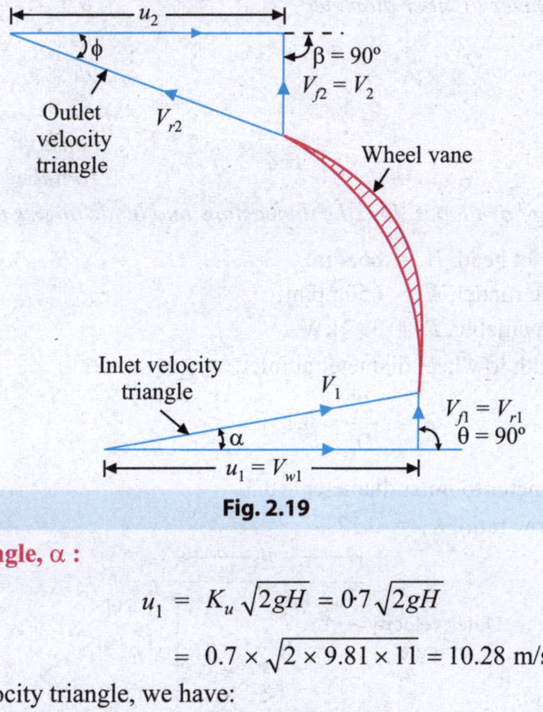

Fig. 2.19

(i) Guide vane angle, α :

$$u_1 = K_u\sqrt{2gH} = 0{\cdot}7\sqrt{2gH}$$

$$= 0.7 \times \sqrt{2 \times 9.81 \times 11} = 10.28 \text{ m/s}$$

From inlet velocity triangle, we have:

$$\tan\alpha = \frac{V_{f1}}{u_1} = \frac{1.83}{10.28} = 0.178$$

$$\therefore \quad \alpha = \tan^{-1} 0.178 = \mathbf{10.09°} \text{ (Ans.)}$$

(ii) Inlet and outlet diameters of the wheel, D_1 and D_2 :

$$u_1 = \frac{\pi D_1 N}{60}, \quad \text{or,} \quad 10.28 = \frac{\pi D_1 \times 370}{60}$$

$$\therefore \quad \text{Inlet diameter, } D_1 = \frac{10.28 \times 60}{\pi \times 370} = \mathbf{0.53 \text{ m}} \text{ (Ans.)}$$

$$\text{Outlet diameter, } D_2 = \frac{D_1}{2} = \frac{0.53}{2} = \mathbf{0.265 \text{ m}} \text{ (Ans.)}$$

(iii) The width of the wheel at inlet and outlet, B_1 and B_2 :

$$Q = \pi D_1 B_1 V_{f1} = \pi D_2 B_2 V_{f2} \quad \text{...Continuity equation}$$

or, $$D_1 B_1 = D_2 B_2 \quad (\because V_{f1} = V_{f2})$$

Now, $$Q = \pi D_1 B_1 V_{f1}, \text{ or, } 0.233 = \pi \times 0.53 \times B_1 \times 1.83$$

or, $$B_1 = \frac{0.233}{\pi \times 0.53 \times 1.83} = 0.0765 \text{ m}, \text{ or, } \mathbf{76.5 \text{ mm (Ans.)}}$$

$$B_2 = 2B_1 = 2 \times 76.5 = \mathbf{153 \text{ mm (Ans.)}}$$

Example 2.21. *The following data pertain to an inward flow reaction turbine :*

Net head	*... 86.4 m*
Speed of the runner	*... 650 r.p.m*
Shaft power available	*.... 397 kW*
Ratio of wheel width to wheel diameter at inlet	*... 0·10*
Ratio of outer diameter to inner diameter	*.... 0.5*
Flow ratio	*... 0.17*
Hydraulic efficiency	*... 95 %*
Overall efficiency	*.... 85 %*
Flow velocity	*... constant.*
Discharge	*... radial*

Neglecting blockage by blades, find the dimensions and blade angles of the turbine.

Solution. Net head, $H = 86.4$ m.

Speed of the runner, $N = 650$ r.p.m.

Shaft power available, $P = 397$ kW

Ratio of wheel width to wheel diameter at inlet,

$$n = \frac{B_1}{D_1} = 0.1$$

Ratio of outer diameter to inner diameter = 0.5

Flow ratio, $K_f = 0.17$

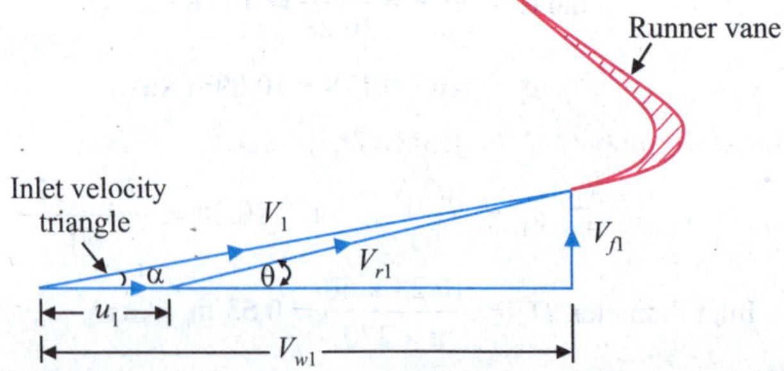

Fig. 2.20

Hydraulic efficiency, η_h = 95%

Overall efficiency, η_0 = 85%

Flow velocity = constant *i.e.*, $V_{f1} = V_{f2}$

Discharge is radial *i.e.*, $\beta = 90°$ or $V_2 = V_{f2}$

Main dimensions of the turbine :

$$\text{Flow velocity, } V_{f1} = K_f\sqrt{2gH}$$

$$= 0.17\sqrt{2 \times 9.81 \times 85.4} = 7.0 \text{ m/s}$$

$$\therefore \quad V_{f1} = V_{f2} = 7.0 \text{ m/s}$$

The shaft power available from the turbine,

$$P = wQH \times \eta_0$$

$$\text{or,} \quad 397 = 9.81 \times Q \times 86.4 \times 0.85$$

$$\therefore \quad \text{Discharge, } Q = \frac{397}{9.81 \times 86.4 \times 0.85} = 0.551 \text{ m}^3/\text{s}$$

Also $Q = \pi D_1 B_1 V_{f1}$ (Neglecting blockage by blades)

[where, D_1 and B_1 are the diameter and width of the wheel at the inlet respectively.]

$$\text{or,} \quad 0.551 = \pi D_1 \times 0.1 D_1 \times 7.0 \qquad \left[\because \frac{B_1}{D_1} = 0.1 \text{ ...Given}\right]$$

$$\therefore \quad D_1 = \left(\frac{0.551}{\pi \times 0.1 \times 7.0}\right)^{1/2} = \mathbf{0.5 \text{ m}} \textbf{ (Ans.)}$$

$$B_1 = 0.1 D_1 = 0.1 \times 0.5 = \mathbf{0.05 \text{ m}} \textbf{ (Ans.)}$$

Diameter of wheel at outlet, $D_2 = 0.5\, D_1 = 0.5 \times 0.5 =$ **0.25 m (Ans.)**

($\because D_2 = 0.5\, D_1$ ...*Given*)

Since the discharge of water at inlet and outlet tips is same, therefore,

$$Q = \pi D_1 B_1 V_{f1} = \pi D_2 B_2 V_{f2} \qquad \text{....Continuity equation}$$

$$\text{or,} \quad D_1 B_1 = D_2 B_2 \qquad (\because V_{f1} = V_{f2})$$

$$\text{or,} \quad B_2 = \frac{D_1 B_1}{D_2} = \frac{D_1 B_1}{0.5 D_1} = 2B_1 = 2 \times 0.05 = \mathbf{0.1 \text{ m}} \textbf{ (Ans.)}$$

Refer to Fig 2.20.

Angles at inlet :

$$\text{Peripheral velocity at inlet, } u_1 = \frac{\pi D_1 N}{60} = \frac{\pi \times 0.5 \times 650}{60} = 17.0 \text{ m/s}$$

$$\text{Hydraulic efficiency, } \eta_h = \frac{V_{w1} u_1}{gH} \qquad \left[\because V_{w2} = 0, \text{ the discharge being radial at outlet.}\right]$$

$$0.95 = \frac{V_{w1} \times 17.0}{9.81 \times 86.4}$$

$$\text{or,} \quad V_{w1} = \frac{0.95 \times 9.81 \times 86.4}{17.0} = 47.36 \text{ m/s}$$

Since $u_1 < V_{w1}$ the inlet triangle will be as shown in Fig 2.20.

From inlet triangle, we have:

$$\tan\alpha = \frac{V_{f1}}{V_{w1}} = \frac{7.0}{47.36} = 0.1478$$

$\therefore$ Guide vane angle, $\alpha = \tan^{-1} 0.1478 = \mathbf{8.4°}$ **(Ans.)**

Again,
$$\tan\theta = \frac{V_{f1}}{V_{w1} - u_1} = \frac{7.0}{47.36 - 17.0} = 0.23$$

$\therefore$ Vane inlet angle, $\theta = \tan^{-1} 0.23 = \mathbf{12.95°}$ **(Ans.)**

Angles at outlet :

From outlet triangle, we have:

$$u_2 = \frac{\pi D_2 N}{60} = \frac{\pi \times 0.25 \times 650}{60} = 8.5 \text{ m/s}$$

$$\tan\phi = \frac{V_{f2}}{u_2} = \frac{7.0}{8.5} = 0.823$$

$\therefore$ Vane angle at outlet, $\phi = \tan^{-1} 0.823 = \mathbf{39.45°}$ **(Ans.)**

$\beta = 90°$ ($\because$ Discharge is *radial*.) **(Ans.)**

Example 2.22. *The following data pertain to an inward flow reaction turbine :*
Diameter of wheel at inner periphery = 540 mm
Width of wheel at inner periphery = 60 mm
Diameter of wheel at outer periphery = 360 mm
Width of wheel at outer periphery = 90 mm
Area occupied by the vanes = 8% of the periphery
Guide vane angle = 25° to the tangent to the runner
Moving vane angle at inlet = 95° (vane inclined forward to the direction of motion)
Exit angle = 30°
Hydraulic losses = 10% of the supply head
Mechanical friction losses = 5% of the supply head
Pressure in the outer casing = 66 m more than that at discharge from the runner.
Determine the following :
(i) Speed of the runner (for no shocks at entry), and
(ii) Power available at the turbine shaft.

Solution. At inner periphery : D_1 = 540 mm or 0.54 m; B_1 = 60 mm or 0.06 m

At outer periphery : D_2 = 360 mm = 0.36 m; B_2 = 90 mm = 0.09 m

Guide vane angle, α = 25° to the tangent to the runner

Moving vane angle at inlet = 95° (vanes inclined forward to the direction of motion)

Exit angle ϕ = 30°

Hydraulic losses = 10% of the supply head

Mechanical losses = 5% of the supply head

Pressure in the outer casing = 66 m more than that at discharge from the runner.

(i) Speed of the runner, N :

Refer to Fig. 2.21.

$$Q = K_{t1}\pi D_1 B_1 \times V_{f1} = K_{t2}\pi D_2 B_2 \times V_{f2} \quad \text{...Continuity equation}$$

Taking, $K_{t1} = K_{t2}$; $V_{f1} = \dfrac{D_2}{D_1} \times \dfrac{B_2}{B_1} \times V_{f2} = \dfrac{0.36}{0.54} \times \dfrac{0.09}{0.06} \times V_{f2} = V_{f2}$

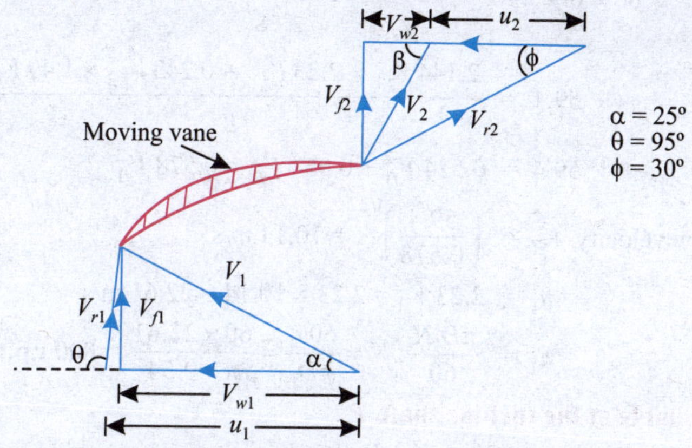

Fig. 2.21

$$\therefore \quad V_{f1} = V_{f2} = V_f$$

From *inlet velocity triangle*, we have:

$$\tan\alpha = \frac{V_{f1}}{V_{w1}}, \quad \text{or,} \quad V_{w1} = \frac{V_{f1}}{\tan\alpha} = \frac{V_{f1}}{\tan 25°}$$

$$= 2.144\ V_{f1} \qquad \text{...(i)}$$

$$\tan(180° - \theta) = \frac{V_{f1}}{u_1 - V_{w1}};$$

$$\tan(180° - 95°) = \frac{V_{f1}}{u_1 - 2.144V_{f1}} \qquad (\because V_{w1} = 2.144\ V_{f1})$$

or, $$\frac{V_{f1}}{u_1 - 2.144\ V_{f1}} = 11.43$$

or, $$V_{f1} = 11.43\ (u_1 - 2.144\ V_{f1}) = 11.43\ u_1 - 24.5\ V_{f1}$$

or, $$25.5\ V_{f1} = 11.43\ u_1, \text{ or, } u_1 = 2.23\ V_{f1}$$

From outlet *velocity triangle*, we have:

Since blade speed is proportional to diameter, therefore,

$$u_2 = u_1 \times \frac{D_2}{D_1} = 2.23V_{f1} \times \frac{0.36}{0.54} = 1.49V_{f1}$$

$$\frac{V_{f2}}{V_{w2} + u_2} = \tan\phi, \quad \text{or,} \quad \frac{V_{f2}}{V_{w2} + 1.49V_{f1}} = \tan 30° = 0.577$$

or, $$\frac{V_{f2}}{V_{w2} + 1.49V_{f2}} = 0.577 \qquad (\because V_{f1} = V_{f2})$$

or, $V_{f2} = 0.577\, V_{w2} + 0.86\, V_{f2}$, or, $V_{w2} = 0.243\, V_{f2}$

Absolute velocity of water at outlet,

$$V_2 = \sqrt{V_{f_2}^2 + V_{w2}^2} = \sqrt{V_{f2}^2 + (0.243V_{f2})^2} = 1.03V_{f2}$$

Now, Head supplied = Head utilised (*i.e.,* work done) + energy or head at outlet (neglecting loss of head in the runner)

$$66 \times 0.9^* = \frac{V_{w1}u_1 + V_{w2}u_2}{g} + \frac{V_2^2}{2g} \qquad \left[\because {}^*\text{Hydraulic losses} = 10\% \ \ldots\textit{Given}\right]$$

$$59.4 = \frac{2.144\,V_{f1} \times 2.23\,V_{f1} + 0.243\,V_{f2} \times 1.49\,V_{f2}}{g} + \frac{(1.03\,V_{f2})^2}{2g}$$

$$59.4 = 0.524\,V_{f1}^2 + 0.054\,V_{f1}^2 = 0.578\,V_{f1}^2 \qquad (\because V_{f1} = V_{f2})$$

$\therefore$ Flow velocity, $V_{f1} = \left(\dfrac{59.4}{0.578}\right)^{1/2} = 10.14$ m/s

Now, $u_1 = 2.23\,V_{f1} = 2.23 \times 10.14 = 22.61$ m/s

Also, $u_1 = \dfrac{\pi D_1 N}{60}$; $N = \dfrac{60u_1}{\pi D_1} = \dfrac{60 \times 22.61}{\pi \times 0.54} =$ **800 r.p.m. (Ans.)**

(ii) Power available at the turbine shaft, *P* :

$$\text{Discharge, } Q = K_{f1}\pi D_1 B_1 \times V_{f1}$$
$$= (1 - 0.08)\,\pi \times 0.54 \times 0.06 \times 10.14 = 0.949 \text{ m}^3/\text{s}$$

$$\text{Power developed by the turbine} = wQ \times \left(\frac{V_{w1}u_1 + V_{w1}u_2}{g}\right)$$

$$= 9.81 \times 0.949 \times 0{\cdot}524\,V_{f1}^2$$

$$\left[\because \frac{V_{w1}u_1 + V_{w2}u_2}{g} = 0.524\,V_{f1}^2, \text{ as calculated above}\right]$$

$$= 9.81 \times 0.949 \times 0.524 \times (10.14)^2 = 501.6 \text{ kW}$$

As the mechanical losses amount to 5 percent,

Shaft/brake power, $P = 501.6 \times 0.95 =$ **476.5 kW (Ans.)**

Example 2.23. *An inward flow reaction turbine (vertical shaft) running at 400 r.p.m requiring a discharge of 15.0 m³/s has an overall efficiency of 90 per cent. The velocity at inlet of the spiral casing is 8·5 m/s and pressure head at this point 230 m. The centre-line of the spiral casing inlet is 2.5 m above the tail water level. The diameter of the runner at inlet is 2.0 m and width at inlet is 0.25 m. If the hydraulic efficiency is 94 per cent and the flow is radial at the outlet from the runner, determine:*

(i) Power developed by the turbine,

(ii) Specific speed,

(iii) Guide vane angle,

(iv) Runner blade angle at inlet, and

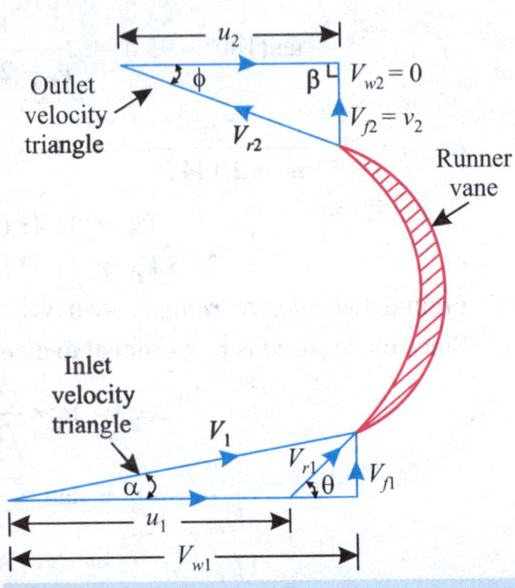

Fig. 2.22

(v) *Percentage of net head which is kinetic at entry to the runner.*
Assume the thickness of the blades to be negligible.

Solution. Speed of the turbine/runner, N = 400 r.p.m.

Flow rate/discharge, Q = 15.0 m^3/s

Overall efficiency, η_0 = 90 %

Velocity at inlet of spiral casing = 8.5 m/s

Pressure head = 230 m

Diameter of the runner at inlet, D_1 = 2.0 m

Width at inlet, B_1 = 0.25 m

Hydraulic efficiency, η_h = 94 %.

(i) Power developed by the turbine :

Head at inlet to the spiral casing,

$$H = 230 + \frac{8.5^2}{2 \times 9.81} + 2.5 = 236.2 \text{ m}$$

$$\text{Overall efficiency, } \eta_0 = \frac{\text{Power developed by the turbine}}{wQH}$$

or,
$$0.9 = \frac{\text{Power developed by the turbine}}{9.81 \times 15.0 \times 236.2}$$

∴ Power developed by the turbine

$$= 0.9 \times 9.81 \times 15.0 \times 236.2 = \mathbf{31281\ kW\ (Ans.)}$$

(ii) Specific speed, N_s :

$$N_s = \frac{N\sqrt{P}}{H^{5/4}} = \frac{400 \times \sqrt{31281}}{(236.2)^{5/4}} = \mathbf{76.4\ (Ans.)}$$

(iii) Guide vane angle, α :

$$\text{Discharge, } Q = \pi D_1 B_1 V_{f1}, \text{ or, } 15.0 = \pi \times 2.0 \times 0.25 \times V_{f1}$$

$$\therefore \text{ Velocity of flow at inlet, } V_{f1} = \frac{15.0}{\pi \times 2.0 \times 0.25} = 9.55 \text{ m/s}$$

Peripheral velocity of blade at inlet,

$$u_1 = \frac{\pi D_1 N}{60} = \frac{\pi \times 2.0 \times 400}{60} = 41.88 \text{ m/s}$$

Also $V_{w2} = 0$, since the discharge is *radial*.

$$\text{Hydraulic efficiency, } \eta_h = \frac{V_{w1} u_1}{gH} \qquad [\because V_{w2} = 0]$$

or,
$$0.94 = \frac{V_{w1} \times 41.88}{9.81 \times 236.2}$$

$$\therefore \text{ Velocity of whirl at inlet, } V_{w1} = \frac{0.94 \times 9.81 \times 236.2}{41.88} = 52.0 \text{ m/s}$$

From *inlet velocity triangle*, we have:

$$\tan\alpha = \frac{V_{f1}}{V_{w1}} = \frac{9.55}{52.0} = 0.1836$$

$$\therefore \quad \alpha = \tan^{-1} 0.1836 = \mathbf{10.4°\ (Ans.)}$$

(iv) Runner blade angle at inlet, θ :

Again, from inlet velocity triangle, we have :

$$\tan\theta = \frac{V_{f1}}{V_{w1} - u_1} = \frac{9.55}{52.0 - 41.88} = 0.9436$$

$$\therefore \quad \theta = \tan^{-1} 0.9436 = \mathbf{43.34°\ (Ans.)}$$

(v) Kinetic head as percentage of net head :

Absolute velocity at inlet, $V_1 = \dfrac{V_{w1}}{\cos\alpha} = \dfrac{52.0}{\cos 10.4°} = 52.87$ m/s

Percentage of net head which is kinetic at entry to the runner

$$= \frac{V_1^2/2g}{H} = \frac{V_1^2}{2gH} = \frac{52.87^2}{2 \times 9.81 \times 236.2} = 0.603 \text{ or } \mathbf{60.3\%\ (Ans.)}$$

Example 2.24. *In an inward flow reaction turbine the diameter of the outer periphery is two times the diameter of inner one, and the turbine operates under a head of 20 m. The turbine has radial tips at the inlet while at the exit the blades make an angle of 30° with the forward tangent. Assuming a constant radial velocity of flow and that the blade friction accounts for a dissipation of energy equivalent to 10 per cent of kinetic energy at the outlet, determine :*

(i) Runner velocity at the rim, and

(ii) Hydraulic efficiency of the turbine.

Assume the turbine discharges radially at the outlet.

Solution. Diameter of the outer periphery, $D_1 = 2D_2$ (diameter of the inner periphery)

Head under which the turbine operates, $H = 20$ m

Tips at inlet = radial *i.e.*, $\theta = 90°$

Blade angle at the outlet, $\phi = 30°$ (with the forward tangent)

Dissipation of energy (due to blade friction) = 10 percent of K.E. at the outlet

Turbine discharge is radial *i.e.*, $\beta = 90°$

Constant radial velocity of flow *i.e.*, $V_{f1} = V_{f2} = V_f$

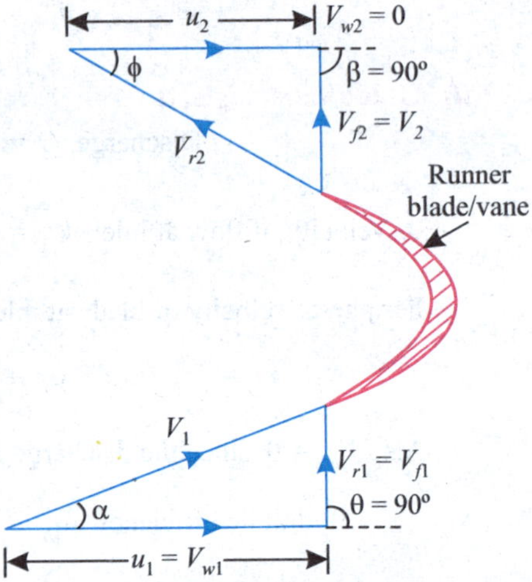

Fig. 2.23

(i) Runner velocity at the rim, u_1 :

From energy considerations, we have:

Energy supplied = Energy transferred to the runner + energy lost in blade friction + K.E. at outlet

$$H = \frac{V_{w1}u_1}{g} + 0.1\frac{V_2^2}{2g} + \frac{V_2^2}{2g} = \frac{u_1^2}{2g} + 1.1\frac{V_2^2}{2g} \qquad (\because V_{w1} = u_1)$$

From outlet velocity triangle (Fig. 2.23),

$$\tan\phi = \frac{V_{f2}}{u_2}, \quad \text{or,} \quad V_{f2} = u_2 \tan\phi = u_2 \tan 30°$$

$$\therefore \quad H = \frac{u_1^2}{2g} + 1.1\frac{(u_2 \tan 30°)^2}{2g} \qquad (\because V_{f2} = V_2)$$

Also, $u_1 = \omega R_1$ and $u_2 = \omega R_2$

$$\therefore \quad \frac{u_1}{u_2} = \frac{\omega R_1}{\omega R_2} = \frac{\omega D_1}{\omega D_2} = 2\text{, or,} \quad u_1 = 2u_2 \qquad [\because D_1 = 2D_2 \text{ ...Given}]$$

$$\therefore \quad H = \frac{(2u_2)^2}{g} + 1.1\frac{(0.577u_2)^2}{2g} = \frac{4u_2^2}{g} + \frac{0.183u_2^2}{g} = \frac{4.183u_2^2}{9.81} = 0.426u_2^2$$

$$\text{or,} \quad 20 = 0.426\,u_2^2\text{, or, } u_2 = \left(\frac{20}{0426}\right)^{1/2} = 6.85 \text{ m/s}$$

$\therefore$ Runner velocity at the rim,

$$u_1 = 2u_2 = 2 \times 6.85 = \mathbf{13.70 \text{ m/s}} \text{ (Ans.)}$$

(*ii*) Hydraulic efficiency, η_h :

$$\eta_h = \frac{V_{w1}u_1}{gH} = \frac{u_1^2}{gH} = \frac{13.70^2}{9.81 \times 20} = 0.9566 \text{ or } \mathbf{95.66\%} \text{ (Ans.)} \qquad \left[\because V_{w2} = 0;\; V_{w1} = u_1\right]$$

Example 2.25. *A Francis turbine with an overall efficiency of 76% is required to produce 150 kW. It is working under a head of 8 m. The peripheral velocity = $0.25\sqrt{2gH}$ and the radial velocity of flow at inlet is $0.95\sqrt{2gH}$. The wheel runs at 150 r.p.m. and the hydraulic losses in the turbine are 20% of the available energy. Assuming radial discharge, determine :*

(*i*) *The guide blade angle,*

(*ii*) *The wheel vane angle at inlet,*

(*iii*) *Diameter of the wheel at inlet, and*

(*iv*) *Width of the wheel at inlet.*

Solution. Overall efficiency, η_0 = 76%

Shaft power produced, P = 150 kW.

Head, H = 8 m

Peripheral velocity, $u = 0.25\sqrt{2gH}$

Radial velocity of flow at inlet, $V_{f1} = 0.95\sqrt{2gH}$

Wheel speed, N = 150 r.p.m.

Since discharge at the outlet is *radial;* $V_{w2} = 0$, $V_{f2} = V_2$

Hydraulic losses in the turbine = 20% of available energy

Now,
$$u_1 = 0.25\sqrt{2 \times 9.81 \times 8} = 3.13 \text{ m/s}$$

$$V_{f1} = 0.95\sqrt{2 \times 9.81 \times 8} = 11.9 \text{ m/s}$$

$$\text{Hydraulic efficiency, } \eta_h = \frac{\text{Total head at inlet} - \text{hydraulic losses}}{\text{Total head at inlet}}$$

$$= \frac{H - 0.2H}{H} = 0.8$$

Also, $$\eta_h = \frac{V_{w1} u_1}{gH} \qquad [\because V_{w2} = 0]$$

$$\therefore \quad 0.8 = \frac{V_{w1} \times 3.13}{9.81 \times 8}, \text{ or, } V_{w1} = \frac{0.8 \times 9.81 \times 8}{3.13} = 20.0 \text{ m/s}$$

(i) The guide blade angle, α :

From inlet velocity triangle (Fig. 2.24),

$$\tan \alpha = \frac{V_{f1}}{V_{w1}} = \frac{11.9}{20.0} = 0.595$$

$$\alpha = \tan^{-1} 0.595 = \mathbf{30.75° \text{ (Ans.)}}$$

(ii) The wheel vane angle at inlet, θ :

$$\tan \theta = \frac{V_{f1}}{(V_{w1} - u_1)} = \frac{11.9}{(20.0 - 3.13)} = 0.705$$

$$\therefore \quad \theta = \tan^{-1} 0.705 = \mathbf{35.18° \text{ (Ans.)}}$$

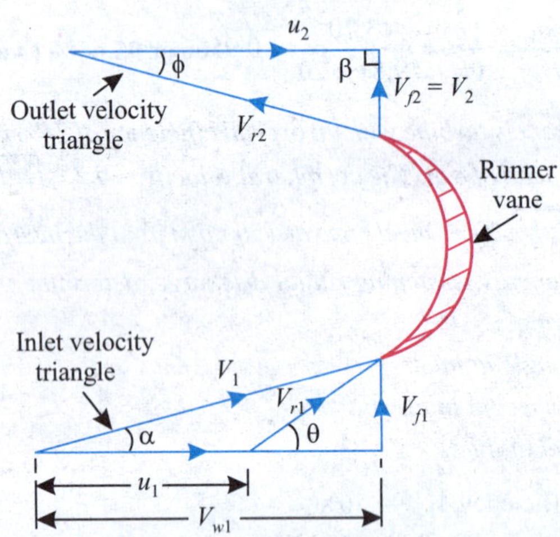

Fig. 2.24

(iii) Diameter of the wheel at inlet, D_1 :

Using the relation, $u_1 = \frac{\pi D_1 N}{60}$, we get:

$$D_1 = \frac{60 u_1}{\pi N} = \frac{60 \times 3.13}{\pi \times 150} = \mathbf{0.398 \text{ m (Ans.)}}$$

(iv) Width of the wheel at inlet, B_1 :

Overall efficiency,

$$\eta_0 = \frac{\text{Shaft power}}{\text{Water power}} = \frac{P}{wQH}$$

or, $$0.76 = \frac{150}{9.81 \times Q \times 8}$$

or, $$Q = \frac{150}{0.76 \times 9.81 \times 8} = 2.515 \text{ m}^3/\text{s}$$

Also, $$Q = \pi D_1 B_1 \times V_{f1}$$

$$2.515 = \pi \times 0.398 \times B_1 \times 11.9$$

$$\therefore \quad B_1 = \frac{2.515}{\pi \times 0398 \times 11.9} = \mathbf{0.169 \text{ m (Ans.)}}$$

Example 2.26. *The following data pertain to a Francis turbine :*

Net head	*... 70 m*
Speed	*... 700 r.p.m.*
Shaft power	*... 330 kW*
Overall efficiency	*... 85 %*
Hydraulic efficiency	*... 92 %*
Flow ratio	*... 0·22*
Breadth ratio	*... 0·1*
Outer diameter of runner	*... 2 × inner diameter of runner*
Velocity of flow	*... constant*
Outlet discharge	*... radial*

The thickness of vanes occupy 6 per cent of circumferential area of the runner.

Determine:

(*i*) *Diameters of runner at inlet and outlet,*

(*ii*) *Width of the wheel at inlet,*

(*iii*) *Guide blade angle, and*

(*iv*) *Runner vane angles at inlet and outlet.*

Solution. Net head, H = 70 m; Speed, N = 700 r.p.m.;

Shaft power = 330 kW; Overall efficiency, η_0 = 85%;

Hydraulic efficiency, η_h = 92%;

Flow ratio = 0.22; Breadth ratio, $\frac{B_1}{D_1} = 0.1$;

D_1 (outer diameter) = $2D_2$ (inner diameter);

Thickness of vanes = 6% of circumferential area of runner;

$V_{f1} = V_{f2}$ ...Velocity of flow is constant (*Given*).

(*i*) Diameters of runner at inlet and outlet D_1, D_2 :

Now, $$\text{flow ratio} = 0.22 = \frac{V_{f1}}{\sqrt{2gH}}$$

$$\therefore \quad V_{f1} = 0.22\sqrt{2gH} = 0.22\sqrt{2 \times 9.81 \times 70} = 8.15 \text{ m/s}$$

$$\text{Actual area of flow} = \left(1 - \frac{6}{100}\right)\pi D_1 B_1 = 0.94\,\pi D_1 B_1$$

Since discharge is *radial* at outlet,

$$\therefore \quad V_{w2} = 0 \text{ and } V_{f2} = V_2$$

Using relation : $\eta_0 = \dfrac{\text{Shaft power}}{\text{Water power}}$, or, $0.85 = \dfrac{330}{wQH} = \dfrac{330}{9.81 \times Q \times 70}$

$$\therefore \quad Q = \frac{330}{0.85 \times 9.81 \times 70} = 0.565 \text{ m}^3/\text{s}$$

But, Q = Actual area of flow × velocity of flow

$$= 0.94\, \pi D_1 B_1 \times V_{f1}$$

$$\therefore \quad 0.565 = 0.94\, \pi D_1 \times 0.1 D_1 \times 8.15 \qquad \left[\because \frac{B_1}{D_1} = 0.1 \quad ...Given\right]$$

$$\therefore \quad D_1 = \left(\frac{0.565}{0.94\pi \times 0.1 \times 8.15}\right)^{1/2} = \mathbf{0.484\ m}\ \textbf{(Ans.)}$$

$$D_2 = \frac{D_1}{2} = \frac{0.484}{2} = \mathbf{0.242\ m}\ \textbf{(Ans.)}$$

(*ii*) Width of the wheel at inlet, B_1 :

$$B_1 = 0.1 D_1 = 0.1 \times 0.484 = 0.0484, \quad \text{or,} \quad \mathbf{48.4\ mm}\ \textbf{(Ans.)}$$

Tangential speed of the runner at inlet,

$$u_1 = \frac{\pi D_1 N}{60} = \frac{\pi \times 0.484 \times 700}{60} = 17.74 \text{ m/s}$$

Using relation for hydraulic efficiency,

$$\eta_h = \frac{V_{w1} u_1}{gH} \qquad [\because V_{w2} = 0]$$

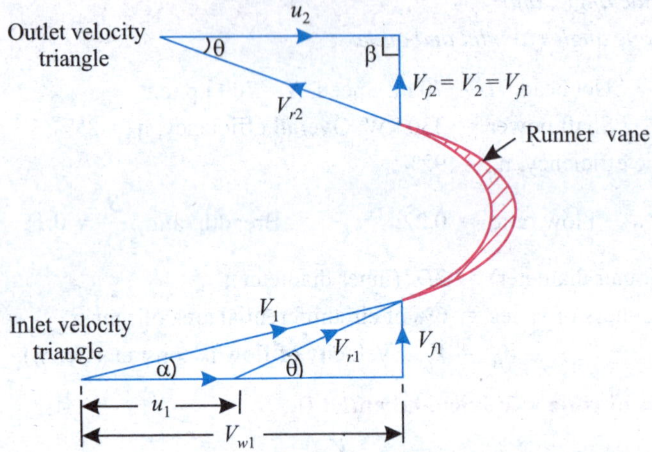

Fig. 2.25

$$\text{or,} \quad 0.92 = \frac{V_{w1} \times 17.74}{9.81 \times 70}$$

$$\text{or,} \quad V_{w1} = \frac{0.92 \times 9.81 \times 70}{17.74} = 35.6 \text{ m/s}$$

(iii) Guide blade angle, α :

From inlet velocity triangle,

$$\tan\alpha = \frac{V_{f1}}{V_{w1}} = \frac{8.15}{35.6} = 0.229$$

$$\therefore \qquad \alpha = \tan^{-1}(0.229) = \mathbf{12.89^\circ\ (Ans.)}$$

(iv) Runner vane angles at inlet and outlet θ, φ :

$$\tan\theta = \frac{V_{f1}}{V_{w1} - u_1} = \frac{8.15}{35.6 - 17.74} = 0.456$$

$$\therefore \qquad \theta = \tan^{-1}(0.456) = \mathbf{24.5^\circ\ (Ans.)}$$

From outlet velocity triangle,

$$\tan\phi = \frac{V_{f2}}{u_2} = \frac{V_{f1}}{u_2} \qquad ...(i)$$

But,
$$u_2 = \frac{\pi D_2 N}{60} = \frac{\pi \times D_1}{2} \times \frac{N}{60} \qquad \left[\because D_2 = \frac{D_1}{2} \quad ...\textit{Given}\right]$$

$$= \frac{\pi \times 0.484 \times 700}{2 \times 60} = 8.87 \text{ m/s}$$

Putting the value of u_2 in eqn. (*i*), we get

$$\tan\phi = \frac{8.15}{8.87} = 0.9188\text{, or,} \quad \phi = \tan^{-1}(0.9188) = \mathbf{42.58^\circ\ (Ans.)}$$

Example 2.27. (*a*) *Show that the hydraulic efficiency for a Francis turbine having velocity of flow through runner as constant, is given by the relation:*

$$\eta_h = \cfrac{1}{1 + \cfrac{\frac{1}{2}\tan^2\alpha}{\left(1 - \cfrac{\tan\alpha}{\tan\theta}\right)}}$$

where, α = guide blade angle, θ = runner vane angle at inlet.

The turbine is having radial discharge at outlet.

(*b*) *If the vanes are radial at inlet, then show that,*

$$\eta_h = \frac{2}{2 + \tan^2\alpha}$$

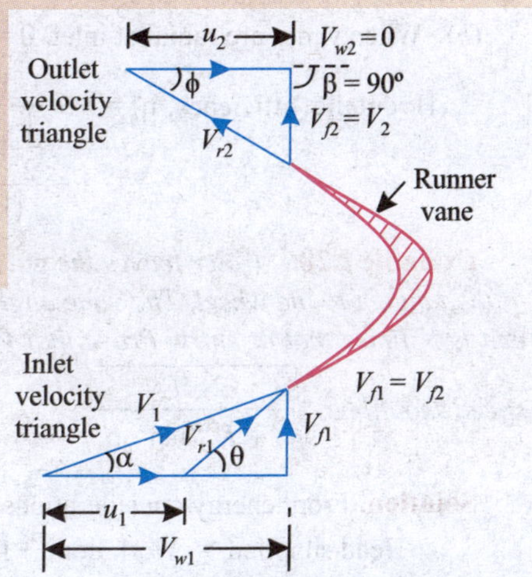

Fig. 2.26

Solution. Velocity of flow is constant *i.e.*, $V_{f1} = V_{f2}$

Discharge is radial at outlet *i.e.*, $V_{w2} = 0$, $V_{f2} = V_2$

From *inlet velocity triangle* : $\tan\alpha = \frac{V_{f1}}{V_{w1}}$

$$\therefore \qquad V_{f1} = V_{w1}\tan\alpha \qquad ...(i)$$

Also, $\tan\theta = \frac{V_{f1}}{V_{w1} - u_1}$, or, $V_{w1} - u_1 = \frac{V_{f1}}{\tan\theta}$

or, $$V_{w1} - u_1 = \frac{V_{w1} \tan \alpha}{\tan \theta}$$

$$\therefore \quad u_1 = V_{w1} - \frac{V_{w1} \tan \alpha}{\tan \theta} = V_{w1}\left(1 - \frac{\tan \alpha}{\tan \theta}\right) \quad ...(ii)$$

Head under which turbine is working,

$$H = \frac{V_{w1} u_1}{g} + \frac{V_2^2}{2g} \quad [\because V_{w2} = 0]$$

or, $$H = \frac{V_{w1} u_1}{g} + \frac{V_{f1}^2}{2g} \quad [\because V_2 = V_{f2} = V_{f1}]$$

Substituting the values of V_{f1} and u_1 from eqns. (*i*) and (*ii*), we get:

$$H = \frac{V_{w1}}{g} \times V_{w1}\left(1 - \frac{\tan \alpha}{\tan \theta}\right) + \frac{(V_{w1} \tan \alpha)^2}{2g}$$

$$= \frac{V_{w1}^2}{g}\left(1 - \frac{\tan \alpha}{\tan \theta}\right) + \frac{V_{w1}^2}{2g} \tan^2 \alpha = \frac{V_{w1}^2}{g}\left(1 - \frac{\tan \alpha}{\tan \theta} + \frac{\tan^2 \alpha}{2}\right)$$

Using relation for hydraulic efficiency, we have:

$$\eta_h = \frac{V_{w1} u_1}{gH} = \frac{V_{w1} \times V_{w1}\left(1 - \frac{\tan \alpha}{\tan \theta}\right)}{g \times \frac{V_{w1}^2}{g}\left(1 - \frac{\tan \alpha}{\tan \theta} + \frac{\tan^2 \alpha}{2}\right)} = \frac{\left(1 - \frac{\tan \alpha}{\tan \theta}\right)}{\left(1 - \frac{\tan \alpha}{\tan \theta} + \frac{\tan^2 \alpha}{2}\right)}$$

or, $$\eta_h = \frac{1}{1 + \frac{\frac{1}{2}\tan^2 \alpha}{\left(1 - \frac{\tan \alpha}{\tan \theta}\right)}} \quad \left[\text{Dividing numerator and denominator by } \left(1 - \frac{\tan \alpha}{\tan \theta}\right)\right] \quad \textbf{...(Proved)}$$

(*b*) When vanes are radial at inlet, $\theta = 90°$

$$\therefore \text{ Hydraulic efficiency, } \eta_h = \frac{1}{1 + \frac{\frac{1}{2}\tan^2 \alpha}{(1-0)}} = \frac{2}{2 + \tan^2 \alpha} \quad \textbf{...(Proved)}$$

Example 2.28. *Water leaves the guide vanes of an inward radial flow turbine at an angle α to the tangent to the wheel. The vane angle at entry to the wheel is 90° and the velocity of flow at exit is K times that at entry. Prove that for maximum efficiency under a head H, the peripheral speed should be* $\sqrt{\dfrac{2gH}{2 + K^2 \tan^2 \alpha}}$.

Solution. From energy considerations,

Head supplied = Work done + kinetic head at exit

$$H = \frac{V_{w1} u_1 \pm V_{w2} u_2}{g} + \frac{V_2^2}{2g}$$

(Assuming the *losses* within the runner to be *negligible*)

From inlet velocity triangle, we have:

$u_1 = V_{w1}$, and , $V_{f1} = u_1 \tan \alpha$

($\because$ Flow at inlet is *radial*.)

For conditions of maximum efficiency, the flow leaves the runner *radially*, *i.e.*, $V_{w2} = 0$

Also, $V_2 = V_{f2} = KV_{f1} = Ku_1 \tan \alpha$

$$\therefore \; H = \frac{V_{w1}u_1}{g} + \frac{V_2^2}{2g}$$

$$\text{or, } H = \frac{u_1^2}{g} + \frac{K^2u_1^2 \tan^2\alpha_1}{2g} = \frac{u_1^2}{2g}(2 + K^2 \tan^2\alpha)$$

$\therefore$ Peripheral speed,

$$u_1 = \sqrt{\frac{2gH}{2 + K^2 \tan \alpha}} \quad \text{...(Proved)}$$

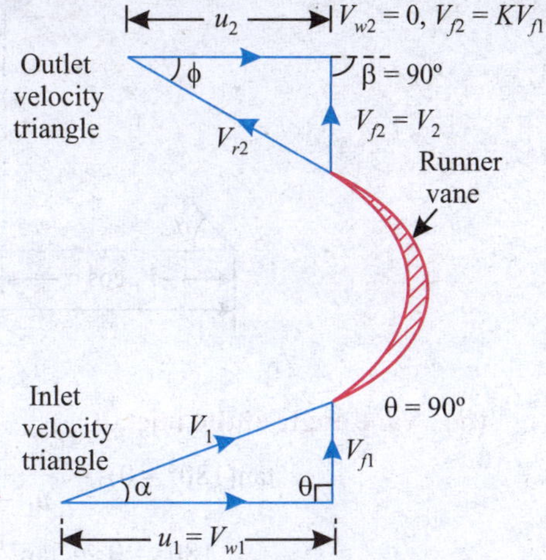

Fig. 2.27

Example 2.29. *A vertical shaft Francis turbine runs at 420 r.p.m. while the discharge is 15 m³/s. The velocity and pressure head at entrance of the runner are 10 m/s and 230 m respectively. The elevation above the tail race is 5 m. The diameter of the runner is 2 m and the width at the inlet is 270 mm. The overall and hydraulic efficiencies are 92% and 98% respectively. Calculate :*

(i) Total head across the turbine;

(ii) Power output;

(iii) The guide vane angle;

(iv) Vane angle at the inlet.

Density of water may be taken as 1000 kg/m³. **[GATE]**

Solution. *Given* : $N = 420$ r.p.m.; $Q = 15$ m³/s; $V_1 = 10$ m/s; Head at the entrance = 230 m; Elevation above the tail race, $z = 5$ m; $D_1 = 2$ m; $B_1 = 270$ mm = 0.27 m; $\eta_0 = 92$ %; $\eta_h = 98$ %.

(*i*) Total head across the turbine :

$$h = 230 + \frac{V_1^2}{2g} + z = 230 + \frac{10^2}{2 \times 9.81} + 5 = \mathbf{240.09\ m\ (Ans.)}$$

(*ii*) Power output $= \eta_0 \times$ power delivered to the fluid

$= 0.92 \times wQH$

$= 0.92 \times (1000 \times 9.81) \times 15 \times 240.09 \times 10^{-6}$ MW = **32.5 MW (Ans.)**

(*iii*) Guide vane angle, α : Refer to Fig. 2.28.

$$u_1 = \frac{\pi D_1 N}{60} = \frac{\pi \times 2 \times 420}{60} = 43.98 \text{ m/s}$$

Now discharge, $Q = \pi D_1 B_1 \times V_{f1}$

or, $15 = \pi \times 2 \times 0.27 \times V_{f1}$

$$\text{or,} \quad V_{f1} = \frac{15}{\pi \times 2 \times 0.27} = 8.84 \text{ m/s}$$

$$\sin \alpha = \frac{V_{f1}}{V_1} = \frac{8.84}{10} = 0.884$$

$$\therefore \qquad \alpha = \sin^{-1}(0.884) = \mathbf{62.13° \ (Ans.)}$$

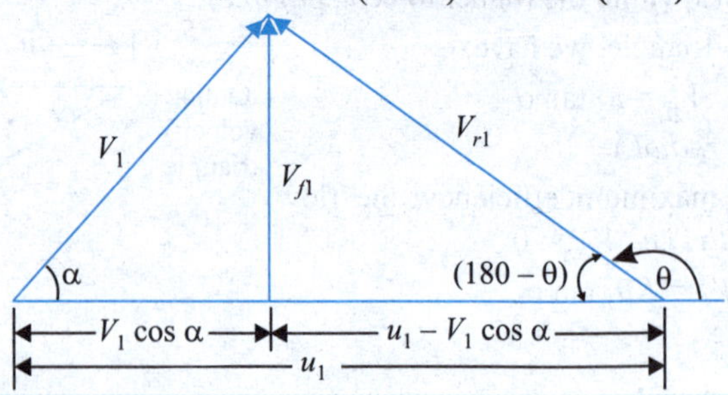

Fig. 2.28

(*iv*) Vane angle at the inlet, θ :

$$\tan(180° - \theta) = \frac{V_{f1}}{u_1 - V_1 \cos \alpha} = \frac{8.84}{43.98 - 10 \cos(62.13°)} = 0.2249$$

$$\therefore \qquad 180° - \theta = \tan^{-1}(0.2249) = 12.67°$$

$$\therefore \qquad \theta = 180° - 12.67° = \mathbf{167.33° \ (Ans.)}$$

Example 2.30. *An inward flow turbine runner has an outer diameter of 0.6 m and an inner diameter of 0.3 m and runs at 750 r.p.m. The radial velocity of flow at inlet and exit is 6 m/s. Water enters the runner making an angle of 12° to the direction of motion of the blades at inlet. It leaves the runner radially. The mass flow rate is 1 kg/s. Calculate :*

(*i*) *Power developed.*

(*ii*) *Angle between the relative velocity of water and tangential velocity of the runner at exit.*

[UPTU]

Solution. *Given* : $D_1 = 0.6$ m; $D_2 = 0.3$ m; $N = 750$ r.p.m.; $V_{f1} = V_{f2} = 6$ m/s; $\alpha = 12°$; $\beta = 90°$; $m = 1$ kg/s.

Refer to Fig. 2.29.

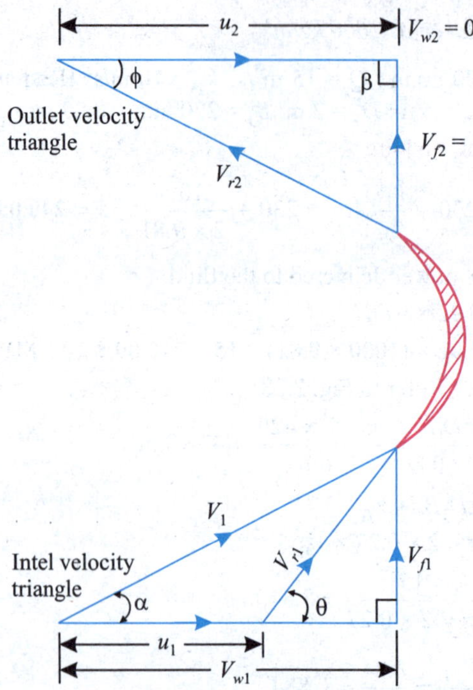

Fig. 2.29

(i) Power developed, P :

Peripheral velocity at inlet,

$$u_1 = \frac{\pi D_1 N}{60} = \frac{\pi \times 0.6 \times 750}{60} = 23.56 \text{ m/s}$$

Peripheral velocity at exit:

$$u_2 = \frac{\pi D_2 N}{60} = \frac{\pi \times 0.3 \times 750}{60} = 11.78 \text{ m/s}$$

From *inlet velocity triangle*, we get

$$\tan \alpha = \frac{V_{f1}}{V_{w1}}$$

or, $$\tan 12° = \frac{6}{V_{w1}}, \quad \text{or,} \quad V_{w1} = \frac{6}{\tan 12°} = 28.23 \text{ m/s}$$

Power developed by the turbine,

$$P = wQ \times \left(\frac{V_{w1}u_1 + V_{w2}u_2}{g}\right)$$

$$= wQ \times \frac{V_{w1}u_1}{g} \qquad (\because V_{w2} = 0)$$

$$\therefore \quad P = mg \times \frac{V_{w1}u_1}{g}$$

$$= 1 \times 9.81 \times \frac{28.23 \times 23.56}{9.81} = \mathbf{666.09\ W} \text{ (Ans.)}$$

(ii) Angle ϕ :

From *outlet velocity triangle*, we have:

$$\tan \phi = \frac{V_{f2}}{u_2} = \frac{6}{11.78} = 0.5093$$

$$\therefore \quad \phi = \tan^{-1}(0.5093) \simeq \mathbf{27°} \text{ (Ans.)}$$

Example 2.31. *An inward flow reaction turbine operating under 30 m head, develops 4000 kW while running at 300 r.p.m. The overall efficiency of the turbine is 0.85; the hydraulic efficiency is 0.9; and the radial velocity of flow at inlet is 7 m/s; the inlet guide vane angle at full gate opening is 30°. Calculate the diameter and width of the runner at inlet. Blade thickness co-efficient is 5%.* **[UPSC]**

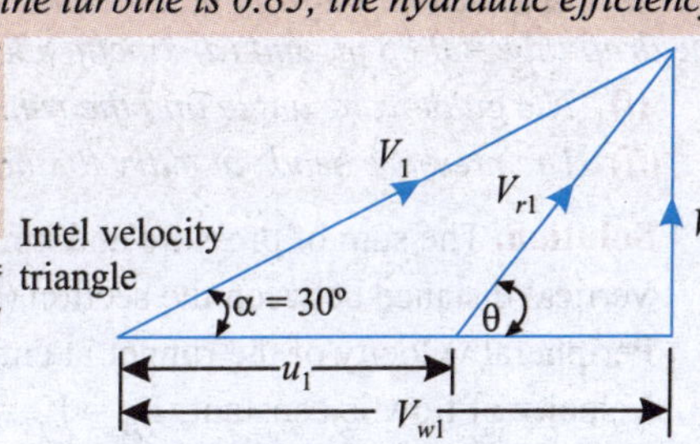

Fig. 2.30

Solution. *Given*: H = 30 m; Power developed = 4000 kW; N = 300 r.p.m.; $\eta_0 = 0.85$; $\eta_h = 0.9$; $V_{f1} = 7$ m/s, $\alpha = 30°$, Vane thickness co-efficient,

$$K_{t_1} = \left(1 - \frac{5}{100}\right) = 0.95$$

Diameter and width of the runner at the inlet (D_1, B_1):

$$\text{Overall efficiency, } \eta_0 = \frac{\text{Output power}}{\text{Power of water supplied}} = \frac{4000 \times 1000}{w \times Q \times H}$$

$$0.85 = \frac{4000 \times 1000}{9810 \times Q \times 30}$$

or,
$$Q = \frac{4000 \times 1000}{0.85 \times 9810 \times 30} = 15.99 \text{ m}^3/\text{s}$$

Hydraulic efficiency,
$$\eta_h = \frac{V_{w1} \times u_1}{gH} \quad \text{...[Eqn. 2.19]}$$

Now, from *inlet velocity triangle* (Refer to Fig. 2.30), we get:

$$\frac{V_{f1}}{V_{w1}} = \tan 30°, \quad \text{or,} \quad V_{w1} = \frac{V_{f1}}{\tan 30°} = \frac{7}{\tan 30°} = 12.12 \text{ m/s}$$

Now, substituting the values in the above eqn., we get:

$$0.9 = \frac{12.12 \times u_1}{9.81 \times 30}$$

or,
$$u_1 = \frac{0.9 \times 9.81 \times 30}{12.12} = 21.85 \text{ m}$$

Also,
$$u_1 = \frac{\pi D_1 N}{60}, \quad \text{or,} \quad 21.85 = \frac{\pi \times D_1 \times 300}{60}$$

or,
$$D_1 = \frac{21.85 \times 60}{\pi \times 300} = \mathbf{1.39 \text{ m (Ans.)}}$$

Further, discharge $Q = K_{t1} \pi D_1 B_1 V_{f1}$

Substituting the values, we get:

$$15.99 = 0.95\, \pi \times 1.39 \times B_1 \times 7$$

$$\therefore \quad B_1 = \frac{15.99}{0.95\, \pi \times 1.37 \times 7} = \mathbf{0.55 \text{ m (Ans.)}}$$

Example 2.32. *In an inward flow reaction turbine (vertical shaft) the sum of the pressure and kinetic heads at entrance to the spiral casing is 132 m and vertical distance between this section and the tail race level is 3·3 m. The peripheral velocity of the runner at entry is 33 m/s, the radial component of velocity of water (velocity of flow) is constant at 11·0 m/s and the discharge from the runner is without whirl, i.e. radial discharge. The hydraulic losses are: (a) losses between turbine entrance and discharge from guide vanes = 4·95 m, (b) losses in the runner = 8·8 m, (c) losses in the draft tube = 0·88 m, and (d) kinetic energy rejected to the tail race = 0·55 m. Determine :*

(i) The guide vane angle and the runner blade angle at inlet;

(ii) The pressure heads at entry to and discharge from the runner.

Solution. The sum of pressure and kinetic heads at entrance to the spiral casing = 132 m

Vertical distance between the section (entrance) and the tail race level = 3.3 m

Peripheral velocity of the runner at entry, $u_1 = 33$ m/s.

Velocity of flow is constant, $V_{f1} = V_{f2} = 11.0$ m/s

Discharge is radial, $V_{w2} = 0$

Losses between turbine entrance and discharge from guide vanes = 4.95 m

Losses in the runner = 8.8 m

Losses in the draft tube = 0.88 m

Kinetic energy rejected to the tail race = 0.55 m.

(i) The guide vane angle (α), the runner blade angle, (θ) :

Head utilized by the runner, $H = 132 + 3.3 - \text{hydraulic losses}$

$$= 132 + 3.3 - (4.95 + 8.8 + 0.88 + 0.55) = 120.12 \text{ m}$$

Head utilized by the runner is given by,

$$H = \frac{V_{w1}u_1}{g} \qquad (\because V_{w2} = 0)$$

$$\therefore \quad V_{w1} = \frac{gH}{u_1} = \frac{9.81 \times 120.12}{33} = 35.71 \text{ m/s}$$

From *inlet velocity triangle*, we have:

$$\tan\alpha = \frac{V_{f1}}{V_{w1}} = \frac{11.0}{35.71} = 0.308$$

$$\therefore \quad \alpha = \tan^{-1} 0.308 = \mathbf{17.12°} \textbf{ (Ans.)}$$

Again, $$\tan\theta = \frac{V_{f1}}{V_{w1} - u_1} = \frac{11.0}{35.71 - 33} = 4.06$$

$$\therefore \quad \theta = \tan^{-1} 4.06 = \mathbf{76.16} \textbf{ (Ans.)}$$

(ii) The pressure heads at entry and discharge from the runner :

Taking tail race as datum, applying Bernoulli's equation to the turbine inlet and the runner, we obtain :

$$132 + 3.3 = \frac{V_1^2}{2g} + \frac{p_1}{w} + z_1 + 4.95$$

Substituting $z_1 = 3{\cdot}3$ m, and,

$$V_1 = \frac{V_{f1}}{\sin\alpha} = \frac{11.0}{\sin 17.12°} = 37.37 \text{ m/s, we have :}$$

The *pressure head at inlet,*

$$\frac{p_1}{w} = (132 + 3.3) - \frac{37.37^2}{2 \times 9.81} - 3.3 - 4.95$$

$$= \mathbf{55.87 \text{ m}} \textbf{ (Ans.)}$$

Now, applying Bernoulli's equation to the turbine entrance and *runner outlet*, we get:

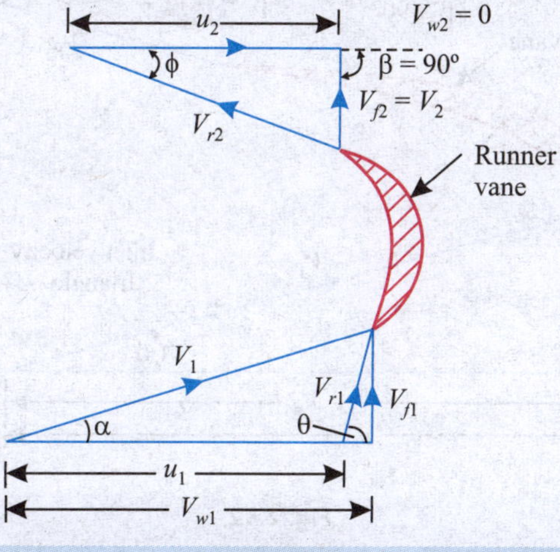

Fig. 2.31

$$132 + 3.3 = \frac{V_2^2}{2g} + \frac{p_2}{w} + z_2 + H + 4{\cdot}95 + 8.8$$

Substituting, $V_2 = 11.0$ m/s, $z_2 = 3.3$ *m*, $H = 120.12$ m, we get:

$$135.3 = \frac{11.0^2}{2 \times 9.81} + \frac{p_2}{w} + 3.3 + 120.12 + 4.95 + 8.8$$

$$= 6.17 + \frac{p_2}{w} + 137.17$$

$$\therefore \quad \frac{p_2}{w} = 135.3 - 61.7 - 137.17 = -\mathbf{8.04\ m}$$ **(Ans.)**

Example 2.33. *The following data pertain to a vertical shaft inward flow reaction turbine :*

Net head under which the turbine operates = 24.5 m

Discharge through the turbine = 10.5 m³/s

Speed of the turbine = 225 r.p.m

Inlet angle of the runner vane = 115° (measured from the direction of runner rotation)

Velocity of flow at inlet to runner = 6.5 m/s

Velocity with which water enters the draft tube without swirl = 6 m/s

Discharge velocity from the exit of draft tube = 2.5 m/s

The mean height of the runner entry surface = 1.5 m } *above tail race level*
The mean height of entrance to the draft tube = 1.2 m }

Hydraulic efficiency = 90 %

Determine the following :

(i) Diameter of the runner at entry surface, and

(ii) Pressure head at entry to the runner and at entrance to the draft tube; friction loss in the runner is 0·9 m and that in the draft tube 0·6 m of water.

Solution. Refer to Fig. 2.32. Inlet angle of runner vane, $\theta = 115°$

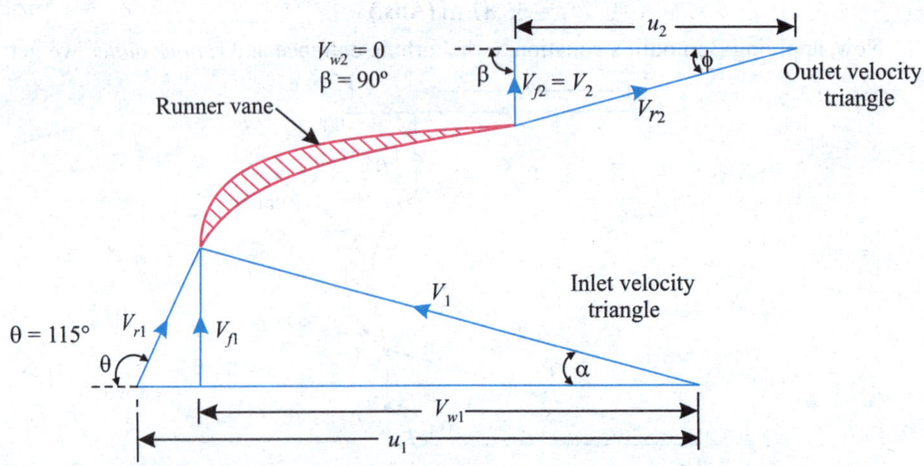

Fig. 2.32

From *inlet velocity triangle*, we have:

$$\frac{V_{f1}}{(u_1 - V_{w1})} \tan(180° - 115°), \quad \text{or,} \quad u_1 - V_{w1} = \frac{V_{f1}}{\tan(180° - 115°)} = \frac{6.5}{0.466} = 3.03$$

$$\therefore \quad u_1 = V_{w1} + 3.03 \qquad ...(i)$$

Hydraulic efficiency, $\eta_h = \dfrac{V_{w1}u_1}{gH}$. $\qquad$ [$\because V_{w2} = 0$ ... *Given*]

$$0.9 = \frac{V_{w1}u_1}{9.81 \times 24.5}, \quad \text{or,} \quad V_{w1}u_1 = 0.9 \times 9.81 \times 24.5 = 216.31$$

i.e., $\qquad V_{w1}u_1 = 216.31 \qquad ...(ii)$

From eqns. (*i*) and (*ii*), we obtain

$$V_{w1}(V_{w1} + 3.03) = 216.31$$

or, $\quad V_{w1}^2 + 3.03\, V_{w1} - 216.31 = 0$

or, $\quad V_{w1} = \dfrac{-3.03 \pm \sqrt{3.03^2 + 4 \times 216.31}}{2} = \dfrac{-3.03 \pm 29.57}{2} = 13.27$ m/s

$\therefore \quad u_1 = 13.27 + 3.03 = 16.3$ m/s $\qquad$ (neglecting –ve sign)

(*i*) Diameter of the runner at entry surface, D_1 :

$$u_1 = \frac{\pi D_1 N}{60} = \frac{\pi D_1 \times 225}{60}$$

$$\therefore \quad D_1 = \frac{60\, u_1}{\pi \times 225} = \frac{60 \times 16.3}{\pi \times 225} = \mathbf{1.38\ m}\ \textbf{(Ans.)}$$

(*ii*) Pressure head at entry to the runner (p_1/w) and at entrance to the draft tube (p_2/w) :

Absolute velocity at entry to runner,

$$V_1 = \sqrt{V_{f1}^2 + V_{w1}^2} = \sqrt{6.5^2 + 13.27^2} = 14.78 \text{ m/s}$$

Taking tail race as datum, the total head across the turbine,

$$H = \frac{p_1}{w} + \frac{V_1^2}{2g} + z_1 \qquad ...(i)$$

(neglecting losses in scroll and guide blades and the velocity head with which the water comes out of the draft tube)

$\therefore$ ***Pressure head at entry to turbine runner,***

$$\frac{p_1}{w} = H - \frac{V_1^2}{2g} - z_1 = 24.5 - \frac{14.78^2}{2 \times 9.81} - 1.5 = \mathbf{11.87\ m\ of\ water}\ \textbf{(Ans.)}$$

From *energy considerations* at the inlet and outlet of the turbine runner, we have:

$$\frac{p_1}{w} + \frac{V_1^2}{2g} + z_1 = \frac{p_2}{w} + \frac{V_2^2}{2g} + z_2 + \text{work done} + \text{losses in runner} \qquad ...(ii)$$

Now, work done (or head utilized) = 0.9 H $\qquad$ (since, $\eta_h = 90\%$ ...*Given*)

Also, $\quad \dfrac{p_1}{w} + \dfrac{V_1^2}{2g} + z_1 = 24.5 \qquad$ ...From eqn. (*i*)

Substituting the above value in eqn. (*ii*), we obtain:

$$24.5 = \frac{p_2}{w} + \frac{V_2^2}{2g} + z_2 + 22.05 + 0.9$$

or, $$\frac{p_2}{w} + \frac{V_2^2}{2g} + z_2 = 24.5 - (22.05 + 0.9) = 1.55 \text{ m}$$

∴ *Pressure head at exit from the turbine runner* (or at inlet to draft tube),

$$\frac{p_2}{w} = 1.55 - \frac{V_2^2}{2g} - z_2 = 1.55 - \frac{6^2}{2 \times 9.81} - 1.2 = -\mathbf{1.485\ m\ (Ans.)}$$

[∵ $V_2 = 6$ m/s ... *Given*]

Example 2.34. *In a vertical shaft inward flow reaction turbine, water enters the runner from the guide blades at an angle of 155° with the runner blade angle at entry being 100°. Both these angle are measured from the tangent at runner periphery drawn in the direction of runner rotation. The flow velocity through the runner is constant, water enters the draft tube from the runner without whirl and the discharge from the draft tube into the tail race takes place with a velocity of 3.0 m/s. The runner has the dimensions of 480 mm external diameter and 45.6 mm inlet width. The turbine works with a net head of 50·4 m and the loss of head in the turbine due to fluid resistance is 5.76 m of water. Determine :*

(i) Speed of the runner;

(ii) Runner blade angle at a point on the outlet edge where the radius of rotation is 108 mm;

(iii) Power generated by the turbine and its specific speed;

(iv) Inlet diameter of the draft tube.

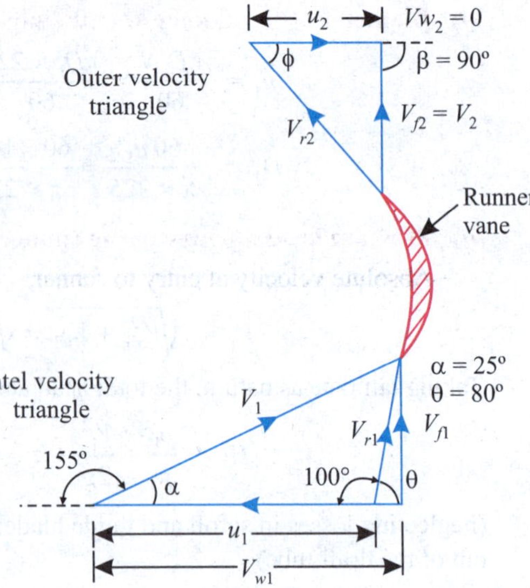

Fig. 2.33

Solution. External diameter of the runner, $D_1 = 480$ mm = 0.48 m

Inlet width of the runner, $B_1 = 45.6$ mm = 0.0456 m

Net head, $H = 50.4$ m

Loss of head in the turbine due to fluid resistance = 5.76 m

Refer to Fig. 2.33. From the *inlet velocity triangle*,

$$\alpha = 180° - 155° = 25° \; ; \; \theta = 180° - 100° = 80°$$

Also, $$\frac{V_{f1}}{V_{w1}} = \tan \alpha, \quad \text{or,} \quad V_{w1} = \frac{V_{f1}}{\tan \alpha}$$

∴ $$V_{w1} = \frac{V_{f1}}{\tan 25°} = 2.144\, V_{f1}$$

Again, $$\tan \theta = \frac{V_{f1}}{V_{w1} - u_1}, \quad \text{or,} \quad V_{w1} - u_1 = \frac{V_{f1}}{\tan \theta}$$

or, $$u_1 = V_{w1} - \frac{V_{f1}}{\tan 80°}$$

$$= 2.144V_{f1} - 0.1763V_{f1} = 1.968V_{f1}$$

Since the discharge is *radial,* therefore,

$$V_{w2} = 0$$

Now, from energy considerations, we have:

Head supplied = Work done + kinetic head at exit + losses in the runner

$$50.4 = \frac{V_{w1}u_1}{g} + \frac{V_2^2}{2g} + 4.8$$

$$50.4 = \frac{2.144\,V_{f1} \times 1.968\,V_{f1}}{9.81} + \frac{3.0^2}{2 \times 9.81} + 5.76 = 0.43\,V_{f1}^2 + 0.459 + 5.76$$

$$\therefore \quad 0.43\,V_{f1}^2 = 44.18, \quad \text{or,} \quad V_{f1} = \left(\frac{44.18}{0.43}\right)^{1/2} = 10.14 \text{ m/s}$$

$$u_1 = 1.968\,V_{f1} = 1.968 \times 10.14 = 19.95 \text{ m/s}$$

(*i*) Speed of the runner, N :

$$u_1 = \frac{\pi D_1 N}{60}, \quad \text{or,} \quad N = \frac{60\,u_1}{\pi D_1} = \frac{60 \times 19.95}{\pi \times 0.48} = \mathbf{794 \text{ r.p.m.}} \textbf{ (Ans.)}$$

(*ii*) Runner blade angle at a point on the outlet edge where the radius of rotation is 108 mm, ϕ :

In the *outlet velocity triangle*, $V_2 = V_{f2} = V_{f1} = 10{\cdot}14$ m/s

Periphery velocity of the outer edge at 108 mm radius,

$$u_2 = u_1 \times \frac{R_2}{R_1} = 19.95 \times \frac{108}{(480/2)} = 8.97 \text{ m/s}$$

$$\tan\phi = \frac{V_{f2}}{u_2} = \frac{10.14}{8.97} = 1.13$$

$$\therefore \quad \phi = \tan^{-1}(1.13) = \mathbf{48.5°} \textbf{ (Ans.)}$$

(*iii*) Power generated by the turbine and its specific speed :

$$\text{Power developed by the turbine} = wQ \times \frac{V_{w1}u_1}{g}$$

where, Q = Discharge through the turbine

$$= \pi D_1 B_1 \times V_{f1} = \pi \times 0.48 \times 0.0456 \times 10.14 = 0.697 \text{ m}^3\text{/s};$$

and

$$\frac{V_{w1}u_1}{g} = \frac{2.144\,V_{f1} \times 1.968\,V_{f1}}{9.81} = 0.43\,V_{f1}^2 = 0.43 \times (10.14)^2 = 44.2$$

$\therefore$ Power developed by the turbine = 9.81 × 6.97 × 44.2 = **302.2 kW (Ans.)**

Assuming a mechanical efficiency of 97%,

Power available at the turbine shaft, P = 302.2 × 0.97 = 293.13 kW

$$\text{Specific speed, } N_s = \frac{N\sqrt{P}}{H^{5/4}} = \frac{794 \times \sqrt{293.13}}{(50.4)^{5/4}} = \mathbf{101.23} \textbf{ (Ans.)}$$

(*iv*) Inlet diameter of the draft tube, d_i :

$$\text{Inlet area of draft tube} = \frac{\text{Discharge}}{\text{Flow velocity}} = \frac{0.697}{10.14} = 0.0687 \text{ m}^2$$

$$\therefore \quad \frac{\pi}{4} d_i^2 = 0.0687, \quad \text{or,} \quad d_i = \left(\frac{0.0687 \times 4}{\pi}\right)^{1/2} = \mathbf{0.295\ m}\ \textbf{(Ans.)}$$

Example 2.35. *A Francis turbine supplied through a 6 m diameter penstock has the following particulars :*

Output of installation = 63500 kW; Flow = 117 m³/s; Speed = 150 r.p.m. Hydraulic efficiency = 92 per cent; Mean diameter of turbine at entry = 4 m; Mean blade height at entry = 1 m; Entry diameter of draft tube = 4·2 m; Velocity in tail race = 2·4 m/s.

The static pressure head in the penstock measured just before entry to the runner is 57·4 m. The point of measurement is 3 m above the level of the tail race. The loss in the draft tube is equivalent to 30 percent of the velocity head at entry to it. The exit plane of the runner is 2 m above the tail race and the flow leaves the runner without swirl. Determine :

(*i*) *Overall efficiency;*

(*ii*) *Direction of flow relative to the runner at inlet;*

(*iii*) *Pressure head at entry to the draft tube.* **[M.U.]**

Solution. Output of installation, $P = 63500$ kW

Diameter of the penstock, $D_p = 6$ m

Discharge through the turbine, $Q = 117\ \text{m}^3/\text{s}$

Speed, $N = 150$ r.p.m.

Hydraulic efficiency, $\eta_h = 92\%$

Mean diameter of turbine at inlet, $D_1 = 4$ m

Mean blade height at entry, $B_1 = 1$m

Entry diameter of draft tube, $d_i = 4{\cdot}2$ m

Velocity in the tail race, $V_{tr} = 2{\cdot}4$ m/s

The static pressure head in the penstock just before entry to the runner = 57·4 m

Loss of head in the draft tube = 30 per cent of the velocity head at entry to it.

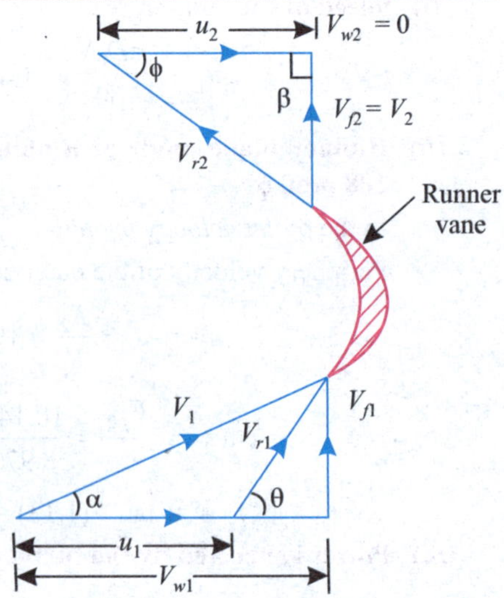

Fig. 2.34

(*i*) **Overall efficiency, η_0 :**

Velocity in the penstock,

$$V_p = \frac{Q}{\frac{\pi}{4} \times D_p^2} = \frac{117}{\frac{\pi}{4} \times 6^2} = 4.138\ \text{m/s}$$

Velocity at entry to the draft tube,

$$V_{ed} = \frac{Q}{\frac{\pi}{4} \times d_i^2} = \frac{117}{\frac{\pi}{4} \times 4.2^2} = 8.44\ \text{m/s}$$

Head just before entry to the runner

$$\frac{p}{w} + \frac{V^2}{2g} + z = 57.4 + \frac{4.138^2}{2 \times 9.81} + 3 = 61.27\ \text{m}$$

(where $z = 3$ m ...*Given*)

Net or effective head, $\quad H$ = Head at entry to runner – kinetic energy in tail race

$$= 61.27 - \frac{V_{tr}^2}{2g}$$

$$\therefore \quad H = 61.27 - \frac{2.4^2}{2 \times 9.81} = 60.97 \text{ m}$$

$$\text{Overall efficiency, } \eta_0 = \frac{\text{Output power}}{\text{Input power}} = \frac{63500}{wQH} = \frac{63500}{9.81 \times 117 \times 60.97}$$

$$= 0.907 \text{ or } \mathbf{90.7\%} \text{ (Ans.)}$$

(*ii*) Direction of flow relative to the runner at inlet, θ :

$$\text{Hydraulic efficiency, } \eta_h = \frac{V_{w1} u_1}{gH}$$

[∵ $V_{w2} = 0$, since the flow leaves the runner without swirl]

$$\text{or,} \quad V_{w1} = \frac{\eta_h \times gH}{u_1}$$

$$\text{where,} \quad u_1 = \frac{\pi D_1 N}{60} = \frac{\pi \times 4 \times 150}{60} = 31.41 \text{ m/s}$$

$$\therefore \quad V_{w1} = \frac{0.92 \times 9.81 \times 60.97}{31.41} = 17.52 \text{ m/s}$$

From *inlet velocity triangle*, we have : $\tan\theta = \dfrac{V_{f1}}{V_{w1} - u_1}$

$$\text{But,} \quad V_{f1} = \frac{Q}{\pi D_1 B_1} = \frac{117}{\pi \times 4 \times 1} = 9.3 \text{ m/s} \qquad (\because Q = \pi D_1 B_1 \times V_{f1})$$

$$\therefore \quad \tan\theta = \frac{9.31}{17.52 - 31.41} = -\,0.67 \text{ m/s}$$

$$\therefore \quad = \tan^{-1}(-0.67) = 180° - 33.82° = \mathbf{146.18°} \text{ (Ans.)}$$

(Fig. 2.34 to be modified accordingly)

(*iii*) Pressure head at entry to the draft tube, (p_2/w) :

Applying Bernoulli's equation between the entrance to the draft tube and the tail race, we obtain:

$$\frac{p_2}{w} + \frac{V_2^2}{2g} + z_2 = 0 + \frac{V_{tr}^2}{2g} + 0.3 \times \frac{V_2^2}{2g}$$

$$\text{or,} \quad \frac{p_2}{w} = -\,z_2 - 0.7\,\frac{V_2^2}{2g} + \frac{V_{tr}^2}{2g} = -\,2 - 07 \times \frac{8.44^2}{2 \times 9.81} + \frac{2.4^2}{2 \times 9.81}$$

[∵ $V_2 = V_{ed} = 8{\cdot}44$ m/s]

$$= -\,2 - 2.54 + 0.29 = \mathbf{-\,4.25 \text{ m}} \text{ (Ans.)}$$

Example 2.36. (Outward flow reaction turbine). **(*a*)** *What are the disadvantages of an onward-flow radial turbine as compared with a radial inward-flow turbine.*

(*b*) *An outward reaction turbine is running at 300 r.p.m. and the rate of flow of water through the turbine is 7.2 m*^{3}*/s. The internal and external diameters of the turbine are 2 m and 2.75 m* *respectively. The width of the runner is constant at inlet and outlet and is equal to 300 mm. The head on the turbine is 216 m. The discharge at the outlet is radial. Neglecting thickness of the vanes, determine :*

(i) Velocities of flow at inlet and outlet, and

(ii) Vane angles at inlet and outlet. **[Anna University]**

Solution. (a) At inlet and exit of the turbine, the relative velocities (V_{r1}, V_{r2}) and blade velocities (u_1, u_2) are related as follows :

$$\frac{V_{r2}^2 - V_{r1}^2}{2g} = \frac{u_2^2 - u_1^2}{2g}, \quad \text{or,} \quad \frac{V_{r2}^2}{2g} = \frac{V_{r1}^2}{2g} + \frac{(u_2 - u_1)^2}{2g}$$

In case of an inward flow turbine the relative velocity *decreases* at exit from the turbine since $u_2 < u_1$. But in an outward flow turbine the relative velocity increases at outlet as $u_2 > u_1$; this aspect makes the task of speed control of an outward flow turbine more difficult than that of an inward flow turbine.

(b) Speed of the turbine, $N = 300$ r.p.m.

Rate of flow of water, $Q = 7{\cdot}2\ \text{m}^3/\text{s}$

Internal diameter, $D_1 = 2$ m

External diameter, $D_2 = 2.75$ m

Width of runner = constant, $B_1 = B_2 = 300$ mm $= 0.3$ m

Head, $H = 216$ m

Discharge at outlet = radial, $V_{w2} = 0$, $V_{f2} = V_2$

Tangential velocity at inlet,

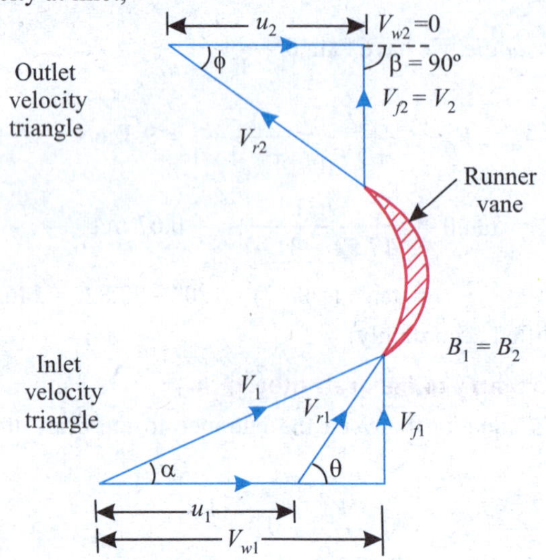

Fig. 2.35

$$u_1 = \frac{\pi D_1 N}{60} = \frac{\pi \times 2 \times 300}{60} = 31.4 \text{ m/s}$$

Tangential velocity at outlet,

$$u_2 = \frac{\pi D_2 N}{60} = \frac{\pi \times 2.75 \times 300}{60} = 43.19 \text{ m/s}$$

(i) Velocities of flow at inlet and outlet V_{f1}, V_{f2} :

The discharge (Q) through the turbine is given by,

$$Q = \pi D_1 B_1 V_{f1} = \pi D_2 B_2 V_{f2}$$

$$\therefore \quad V_{f1} = \frac{Q}{\pi D_1 B_1} = \frac{7.2}{\pi \times 2 \times 0.3} = \mathbf{3.82\ m/s\ (Ans.)}$$

and,
$$V_{f2} = \frac{Q}{\pi D_2 B_2} = \frac{7.2}{\pi \times 2.75 \times 0.3} = \mathbf{2.77\ m/s\ (Ans.)}$$

Now,
$$\text{head, } H = \frac{V_{w1}u_1}{g} + \frac{V_2^2}{2g} = \frac{V_{w1}u_1}{g} + \frac{V_{f2}^2}{2g} \qquad (\because V_{w2} = 0, V_2 = V_{f2})$$

$$216 = \frac{V_{wl} \times 31.4}{981} + \frac{2.77^2}{2 \times 9.81}, \quad \text{or,} \quad 216 = 3.2\ V_{wl} + 0.391$$

$$\therefore \quad V_{wl} = \frac{(216 - 0.391)}{3.2} = 67.38 \text{ m/s}$$

(*ii*) Vane angles at inlet and outlet, θ, φ :

From *inlet velocity triangle*,
$$\tan\theta = \frac{V_{f1}}{V_{wl} - u_1} = \frac{3.82}{67.38 - 31.4} = 0.1062$$

$$\therefore \quad \theta = \tan^{-1}(0.1062) = \mathbf{6.06° \ (Ans.)}$$

From *outlet velocity triangle*,
$$\tan\phi = \frac{V_{f2}}{u_2} = \frac{2.77}{43.9} = 0.0641$$

$$\therefore \quad \phi = \tan^{-1}(0.0641) = \mathbf{3.66° \ (Ans.)}$$

2.4.2. Propeller and Kaplan turbines-Axial Flow Reaction Turbines

It has been observed that with *increasing specific speed* the flow tends to be *axial*. If water flows parallel to the axis of the rotation of the shaft, the turbine is known as *axial flow turbine*; when the head at inlet of the turbine is the sum of pressure energy and kinetic energy and during the flow of water through runner a part of the pressure energy is converted into kinetic energy, the turbine is known as *reaction turbine*. The shaft of an axial flow reaction turbine is *vertical*. The lower end of the shaft is made larger which is known as '*hub*' or '*boss*'. The vanes are fixed on the hub and it acts as runner for axial flow reaction turbine. Two important axial flow reaction turbines are:

(*i*) *Propeller turbine*, and

(*ii*) *Kaplan turbine.*

In these turbines all parts such as *spiral casing*, *stay vanes*, *guide vanes*, *control vanes*, *and draft tube* are similar to mixed-flow turbines in design. But the *water enters the runner in an axial direction* and during the process of energy transfer, it travels across the blade passage in axial direction and *leaves axially*. *The pressure at the inlet of the blades is larger than the pressure at the exit of the blades. The energy transfer is due to the reaction effect, i.e. the change in the magnitude of relative velocity across the blades.*

In an axial flow turbine the number of blades are *fewer* and hence the *loading on the blade is larger. Smaller contact area causes less frictional loss compared to mixed flow turbines*, but the peripheral speed of the turbine is *larger*. Axial flow rotors *do not have a rim at the outer end* like the Francis rotors; but the *blades are enclosed in a cylindrical casing.*

The tip clearance between the blades and the cylindrical casing is small; hence the flow past blades can be considered *two-dimensional*. The water coming out from the guide vanes undergoes a whirl which is assumed to satisfy the law of free vortex ($V_w = C/r$). Accordingly the *whirl is largest near the hub and smallest at the outer end of blade*. Hence the *blade is twisted along its axis.*

2.4.2.1. Propeller turbine

The need to utilize *low heads where large volume of water is available* makes it essential to provide a large flow area and to run the machine at very low speeds. The propeller turbine is a reaction turbine used for heads between 4 m and 80 m. It is purely *axial-flow* device providing the largest possible flow area that will *utilize a large volume of water and still obtain flow velocities which are not too large.*

The propeller turbine (Fig. 2.36) consists of an axial-flow runner with four to six or at the most ten blades of air-foil shape. The runner is generally kept horizontal, *i.e* the shaft is vertical. The blades resemble the propeller of a ship. In the propeller turbine, as in Francis turbine, the runner blades *are fixed* and *non-adjustable*. The *spiral casing* and *guide blades* are similar to those in Francis turbine. The *guide mechanism* is similar to that in a Francis turbine.

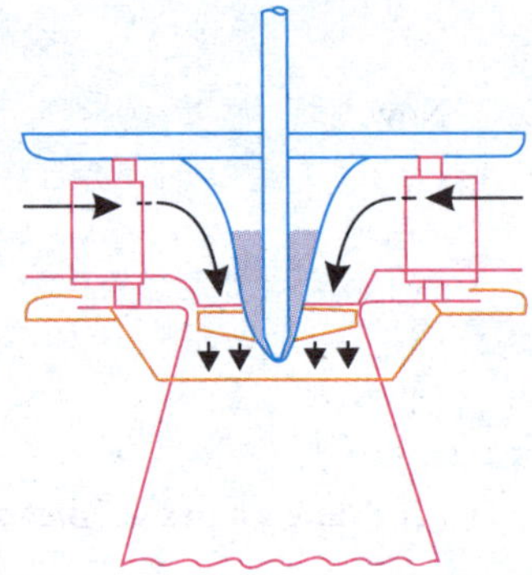

Fig. 2.36. Propeller turbine.

2.4.2.2. Kaplan turbine

A propeller turbine is quite suitable when the *load on the turbine remains constant*. At *part load* its efficiency is very low; since the blades are *fixed*, the water *enters with shock* (at part load) and *eddies are formed which reduce the efficiency*. This defect of the propeller turbine is removed in Kaplan turbine. In a Kaplan turbine the *runner blades are adjustable* and *can be rotated about pivots fixed to the boss of the runner*. The *blades are adjusted automatically by servomechanism so that at all loads the flow enters them without shock*. Thus, *a high efficiency is maintained even at part load*. The servomotor cylinder is usually accommodated in the hub. Figs. 2.37 and 2.38 show the Kaplan turbine runner and Kaplan turbine (schematic diagram) respectively.

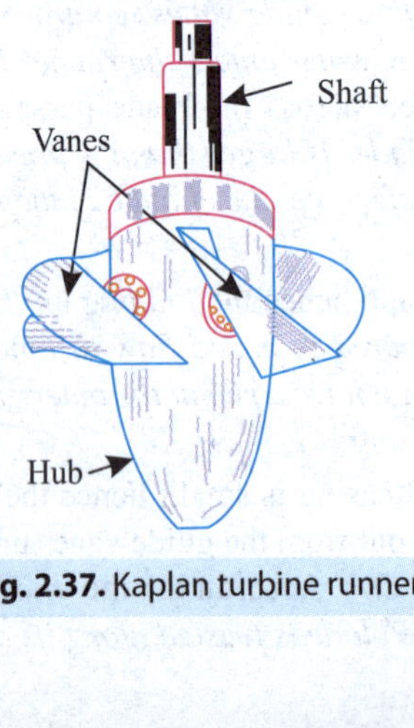

Fig. 2.37. Kaplan turbine runner.

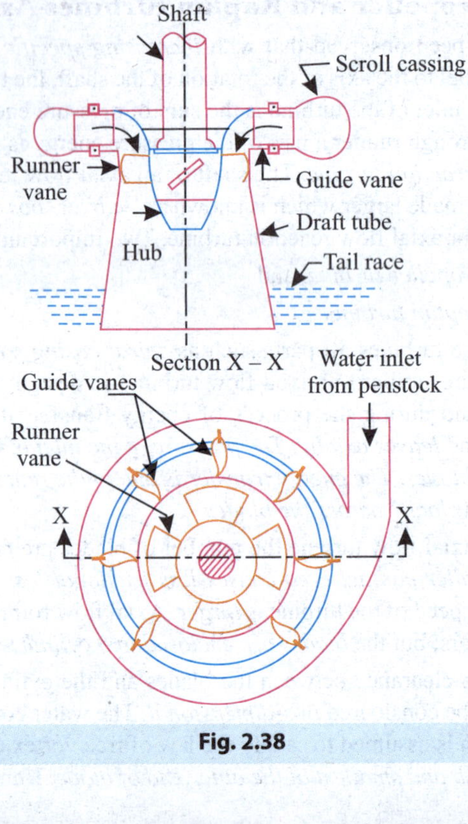

Fig. 2.38

The Kaplan turbine has *purely axial flow*. Usually it has 4 to 6 blades having no outside rim. It is also known as a *variable-pitch propeller turbine* since the pitch of the turbine can be changed because of adjustable vanes. The Kaplan turbine behaves like a propeller turbine at full-load conditions.

The scroll casing, guide mechanism and draft tube are similar to that in the Francis turbine. The shape of runner blades is different from that of Francis turbine. The blades of Kaplan turbine are made of *stainless steel.*

Kaplan turbine, like every propeller turbine, is a *high speed turbine* and is used for smaller heads; as the speed is high, the number of runner-vanes is small.

Kaplan turbines have taken the place of Francis turbines for certain medium head installations. Kaplan turbines with sloping guide vanes to reduce the overall dimensions are being used.

Important Kaplan Turbine Installations in India :

S.No.	Scheme/Project	Location (State)	Source of water
1	Bhakra-Nangal Project	Gangwal & Kota (Punjab)	Nangal hydel
2	Hirakud Dam Project	Hirakud (Orissa)	Mahanadi river
3	Nizam Sagar Project	Nizam Sagar (Andhra Pradesh)	Nanjira river
4	Radhanagri Hydroelectric Scheme	Kolhapur (Maharashtra)	Bhagvati river
5	Tungbhadra Hydroelectric Scheme	Tungbadhra (Karnataka)	Tungbadhra river

Differences between Francis Turbine and Kaplan Turbine :

S.No.	Aspects	Francis turbine	Kaplan turbine
1.	*Type of turbine*	Radially inward or mixed flow.	Partially axial flow.
2.	*Disposition of shaft*	Horizontal or vertical	Only vertical.
3.	*Adjustability of runner vanes*	Runner vanes are *not* adjustable.	Runner vanes are adjustable.
4.	*Number of vanes*	Large, 16 to 24 blades	Small, 3 to 8 blades
5.	*Resistance to be overcome*	Large, (owing to large number of vanes and greater area of contact with water)	Less (owing to fewer number of vanes and less wetted area)
6.	*Head*	Medium (60 m to 250 m)	Low (up to 30 m)
7.	*Flow rate*	Medium	Large
8.	*Specific speed*	50-250	250-850
9.	*Type of governor*	Ordinary	Heavy duty

Working proportions :

The expressions for work done, efficiency and power developed by axial flow propeller and Kaplan turbines are identical to those of a Francis turbine, and the working proportions are obtained in an identical fashion. However, the following *deviations need to be noted carefully*:

1. In case of a propeller/Kaplan turbine, the ratio n is taken as $\frac{D_b}{D_0}\left(\text{and not } \frac{B}{D}\right)$

where, D_0 = Outside diameter of the runner, and

D_b = Diameter of boss (or hub).

Discharge, Q = Area of flow × velocity of flow

$$= \frac{\pi}{4}(D_0^2 - D_b^2) \times V_f$$

$$= \frac{\pi}{4}(D_0^2 - D_b^2) \times K_f\sqrt{2gH} \quad \text{(where, } K_f = \text{flow ratio)}$$

or, $$Q = \frac{\pi}{4}D_0^2\,(1 - n^2) \times K_f\sqrt{2gH} \quad ...(2.31)$$

$$\left(\because n = \frac{D_b}{D_0}, \text{ or, } D_b = nD_0\right)$$

The value of n ranges from 0.35 to 0.60.

The value of $K_f \simeq 0.70$.

2. The peripheral velocity u of the runner vanes depends upon the radius of the point under consideration and thus the blade angles vary from the rim to the boss and the vanes are *warped*; this is necessary to *ensure shock free entry and exit.*

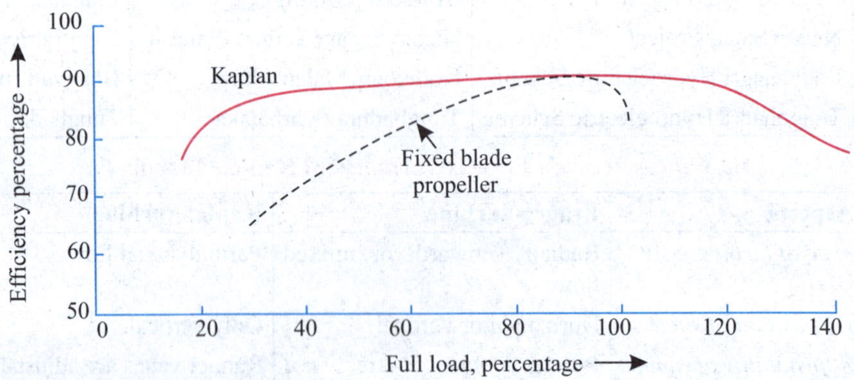

Fig. 2.39. Comparison of efficiencies of propeller (fixed blades) and Kaplan turbines.

3. The *velocity of flow remains constant throughout.*

Fig. 2.39 shows the comparison of efficiencies of propeller (fixed blades) and Kaplan turbines.

2.4.2.3. Kaplan turbine versus Francis turbine :

Kaplan turbine claims the following *advantages* over Francis turbine :

1. For the same power developed Kaplan turbine is more compact in construction and smaller in size.
2. Part-load efficiency is considerably high.
3. Low frictional losses (because of small number of blades used).

Example 2.37. *A Kaplan turbine develops 22000 kW at an average head of 35 m. Assuming a speed ratio of 2, flow ratio of 0.6, diameter of the boss equal to 0·35 times the diameter of the runner and an overall efficiency of 88 per cent, calculate the diameter, speed and specific speed of the turbine.*

Solution. Shaft power, P = 22000 kW; Head, H = 35 m;

Speed ratio, K_u = 2.0; Flow ratio, K_f = 0.6;

Diameter of boss (D_b) = 0.35 × diameter of the runner (D_0), *i.e.* $D_b = 0.35D_0$;

Overall efficiency, η_0 = 88 per cent.

Diameter of the runner, D_0 :

$$K_u = \frac{u_1}{\sqrt{2gH}} = 2.0;\ u_1 = 2.0 \times \sqrt{2gH} = 2.0 \times \sqrt{2 \times 9.81 \times 35} = 52.4 \text{ m/s}$$

$$K_f = \frac{V_{f1}}{\sqrt{2gH}} = 0.6, \quad \text{or,} \quad V_{f1} = 0.6 \times \sqrt{2gH}$$

$$= 0.6 \times \sqrt{2 \times 9.81 \times 35} = 15.7 \text{ m/s}$$

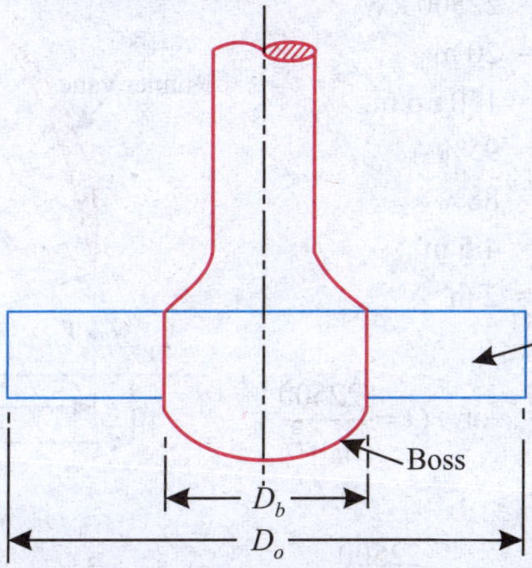

Fig. 2.40

Overall efficiency,

$$\eta_0 = \frac{\text{Shaft power (P)}}{\text{Water power}} = \frac{22000}{wQH}$$

or, $$0.88 = \frac{22000}{9.81 \times Q \times 35}$$

$$Q = \frac{22000}{0.88 \times 9.81 \times 35} = 72.8 \text{ m}^3/\text{s}$$

Also $$Q = \text{Area of flow} \times \text{velocity of flow}$$

$$= \frac{\pi}{4} \times (D_0^2 - D_b^2) \times V_{f1}$$

or, $$72.8 = \frac{\pi}{4}[D_0^2 - (0.35 D_0)^2] \times 15.7 \quad [\because D_b = 0.35\, D_0]$$

$$= \frac{\pi}{4} D_0^2\,[1 - 0.35^2] \times 15.7 = 10.82 D_0^2$$

$\therefore$ $$D_0 = \left(\frac{72.8}{10.82}\right)^{1/2} \simeq \mathbf{2.6\ m}\ \textbf{(Ans.)}$$

Speed of the turbine, N :

$$u_1 = \frac{\pi D_0 N}{60}, \quad \text{or,} \quad N = \frac{60\, u_1}{\pi D_0} = \frac{60 \times 52.4}{\pi \times 2.6} = \mathbf{384.9\ r.p.m.}\ \textbf{(Ans.)}$$

Specific speed of the turbine, N_s :

$$N_s = \frac{N\sqrt{P}}{H^{5/4}} = \frac{384.9 \times \sqrt{22000}}{(35)^{5/4}} = \mathbf{670.6}\ \textbf{(Ans.)}$$

Example 2.38. *The following data pertain to a Kaplan turbine :*

Power available at shaft = 22500 kW; Head = 20 m; Speed = 150 r.p.m. Hydraulic efficiency = 95 %; Overall efficiency = 88 %; Outer diameter of runner = 4·5 m; Diameter of the hub = 2 m.

Assuming that the turbine discharges without whirl at exit, determine the runner vane angles at the hub and at the outer periphery.

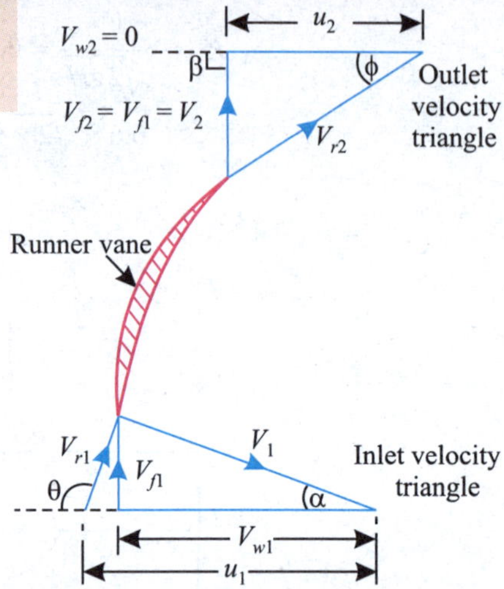

Fig. 2.41

Solution. Shaft power, P = 22500 kW

Head, H = 20 m

Speed, N = 150 r.p.m.

Hydraulic efficiency, η_h = 95%

Overall efficiency, η_0 = 88%

Outer diameter of runner, D_0 = 4·5 m

Diameter of the hub, D_b = 2 m

Overall efficiency,

$$\eta_0 = \frac{\text{Shaft power (P)}}{\text{Water power}} = \frac{22500}{wQH}, \text{ or, } Q = \frac{22500}{\eta_0 wH}$$

∴ Discharge,

$$Q = \frac{22500}{0.88 \times 9.81 \times 20} = 130.32 \text{ m}^3/\text{s}$$

Also,
$$Q = \frac{\pi}{4}(D_0^2 - D_b^2) \times V_{f1}$$

or,
$$130.32 = \frac{\pi}{4}(4.5^2 - 2^2) \times V_{f1}$$

∴
$$V_{f1} = \frac{130.32}{\frac{\pi}{4} \times (4.5 - 2^2)} = 10.21 \text{ m/s}$$

Vane angles at hub :

$$u_1 = \frac{\pi D_b N}{60} = \frac{\pi \times 2 \times 150}{60} = 15.7 \text{ m/s}$$

Hydraulic efficiency,
$$\eta_h = \frac{V_{w1} u_1}{gH}, \text{ or, } V_{wl} = \frac{\eta_h gH}{u_1} = \frac{0.95 \times 9.81 \times 20}{15.7} = 11.87 \text{ m/s}$$

From *inlet velocity triangle*, $\tan(180° - \theta) = \frac{V_{f1}}{u - V_{wl}} = \frac{10.21}{15.7 - 11.87} = 2.26$

∴
$$180° - \theta = \tan^{-1}(2.66) = 69.4°$$

∴ Runner vane angle at inlet, $\theta = 180° - 69.4° =$ **110.6° (Ans.)**

From *outlet velocity triangle*, $\tan\phi = \frac{V_{f2}}{u_2} = \frac{V_{f1}}{u_1} = \frac{10.21}{15.7} = 0.6503$

∴ Runner vane angle at exit, $\phi = \tan^{-1}(0.6503) =$ **33.03° (Ans.)**

Vane angles at extreme edge of the runner (outer periphery) :

$$u_1 = \frac{\pi D_0 N}{60} = \frac{\pi \times 4.5 \times 150}{60} = 35.34 \text{ m/s}$$

$$\text{Hydraulic efficiency, } \eta_h = \frac{V_{w1} u_1}{gH}$$

$$0.95 = \frac{V_{w1} \times 35.34}{9.81 \times 20}, \quad \text{or, } V_{w1} = \frac{0.95 \times 9.81 \times 20}{35.34} = 5.27 \text{ m/s}$$

From *inlet velocity triangle*, $\tan(180° - \theta) = \dfrac{V_{f1}}{u_1 - V_{w1}} = \dfrac{10.21}{35.34 - 5.27} = 0.339$

or, $(180° - \theta) = \tan^{-1}(0.339) = 18.72°$

∴ Runner vane angle at inlet, $\theta = 180° - 18.72° =$ **161.28° (Ans.)**

From *outlet velocity triangle*, $\tan \phi = \dfrac{V_{f2}}{u_2} = \dfrac{V_{f1}}{u_1} = \dfrac{10.21}{35.34} = 0.2889$

∴ Runner vane angle at exit, $\phi = \tan^{-1}(0.2889) =$ **16.11° (Ans.)**

Example 2.39. *Calculate the diameter and speed of the runner of a Kaplan turbine developing 6000 kW under an effective head of 5 m. Overall efficiency of the turbine is 90%. The diameter of the boss is 0·4 times the external diameter of the runner. The turbine speed ratio is 2·0 and flow ratio 0·6. What is the specific speed of the turbine?* **[UPSC]**

Solution. *Given* : Power developed, $P = 6000$ kW; $H = 5$m; $\eta_0 = 90\%$; $D_b = 0.4 D_0$; Speed ratio, $K_u = 2.0$; Flow ratio, $K_f = 0.6$.

Diameter (D_0) and speed (N) of the runner :

$$K_u = 2.0 = \frac{u_1}{\sqrt{2gH}}, \quad \text{or, } u_1 = 2\sqrt{2 \times 9.81 \times 5} = 19.81 \text{ m/s}$$

$$K_f = 0.6 = \frac{V_{f1}}{\sqrt{2gH}}, \quad \text{or, } V_{f1} = 0.6\sqrt{2 \times 9.81 \times 5} = 5.94 \text{ m/s}$$

Also, $$\eta_0 = \frac{P}{wQH}$$

or, $$Q = \frac{P}{\eta_0 w H} = \frac{6000 \times 1000}{0.9 \times 9810 \times 5} = 135.9 \text{ m}^3/\text{s}$$

But, $$Q = \frac{\pi}{4}(D_0^2 - D_b^2) \times V_{f1}$$

or, $$135.9 = \frac{\pi}{4}\left[D_0^2 - (0.4 D_0)^2\right] \times 5.94 = \frac{\pi}{4} D_0^2 (1 - 0.16) \times 5.94$$

∴ $$D_0 = \left[\frac{135.9 \times 4}{\pi (1 - 0.16) \times 5.94}\right]^{1/2} = \mathbf{5.89 \text{ m}} \text{ (Ans.)}$$

Now, $$u_1 = \frac{\pi D_0 N}{60}$$

or, $$N = \frac{u_1 \times 60}{\pi D_0} = \frac{19.81 \times 60}{\pi \times 5.89} = \mathbf{64.23\ r.p.m.\ (Ans.)}$$

Specific speed of the turbine, N_s :

$$N_s = \frac{N\sqrt{P}}{H^{5/4}}$$

$$= \frac{64.23\sqrt{6000}}{(5)^{5/4}} = \mathbf{665.4\ (Ans.)}$$

Example 2.40. *The propeller reaction turbine of runner diameter 4.5 m is running at 48 r.p.m. The guide blade angle at inlet is 145° and the runner blade angle at outlet is 25° to the direction of vane. The axial flow area of water through the runner is 30 m². If the runner blade angle at inlet is radial, determine :*

(*i*) *Hydraulic efficiency of the turbine,*
(*ii*) *Discharge through the turbine, and*
(*iii*) *Power developed by the runner.* **[Roorkee University]**

Solution. Diameter of the runner, $D_0 = 4{\cdot}5$ m

Speed of the runner, $N = 48$ r.p.m.

Guide blade angle, $\alpha = 145°$

Runner blade angle at inlet is radial *i.e.* $\theta = 90°$, $V_{f1} = V_{r1}$

Runner blade angle at outlet, $\phi = 25°$

Area of flow, $A_f = 30\ \text{m}^2$

(*i*) Hydraulic efficiency of the turbine, η_h :

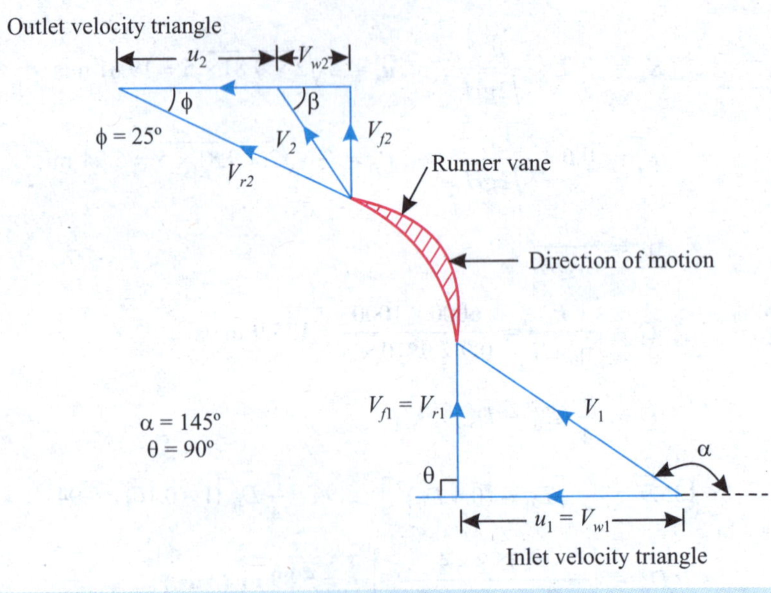

Fig. 2.42

$V_{f1} = V_{f2}$, because area of flow is constant.

The tangential speed of turbine at inlet, $u_1 = \frac{\pi D_0 N}{60} = \frac{\pi \times 4.5 \times 48}{60} = 11.31$ m/s

Also, $u_2 = u_1 = 11.31 \text{ m/s}$

From *inlet velocity triangle*, we have:

$$\tan(180° - \alpha) = \frac{V_{f1}}{u_1}, \text{ or, } \tan(180° - 145°) = \frac{V_{f1}}{u_1}$$

$$\therefore \quad V_{f1} = u_1 \times \tan 35° = 11.31 \times \tan 35° = 7.92 \text{ m/s}$$

Also, $V_{w1} = u_1 = 11.31 \text{ m/s}$

From *outlet velocity triangle*, we have:

$$\tan \phi = \frac{V_{f2}}{u_2 + V_{w2}} = \frac{7.92}{11.31 + V_{w2}} \qquad (V_{f2} = V_{f1};\ u_2 = u_1)$$

$$\therefore \quad \tan 25° = \frac{7.92}{11.31 + V_{w2}}, \text{ or, } 11.31 + V_{w2} = \frac{7.92}{\tan 25°} = 16.98$$

$$\therefore \quad V_{w2} = 16.98 - 11.31 = 5.67 \text{ m/s}$$

$$\therefore \quad V_2 = \sqrt{V_{f2}^2 + V_{w2}^2} = \sqrt{7.92^2 + 5.67^2} = 9.74 \text{ m/s}$$

Also, $$H = \frac{1}{g}(V_{w1}u_1 - V_{w2}u_2) + \frac{V_2^2}{2g}$$

(Negative sign is taken since the absolute velocities at inlet and outlet are in the same direction and so are the velocities of whirl).

$$\therefore \quad H = \frac{1}{9.8}(11.31 \times 11.31 - 5.67 \times 11.31) + \frac{9.74^2}{2 \times 9.81} = 6.5 + 4.83 = 11.33 \text{ m}$$

$$\therefore \text{ Hydraulic efficiency, } \eta_h = \frac{V_{w1}u_1 - V_{w2}u_2}{gH}$$

$$= \frac{(11.31 \times 11.31 - 5.67 \times 11.31)}{9.81 \times 11.33} = 0.575 \text{ or } \mathbf{57.5\%} \text{ (Ans.)}$$

(*ii*) Discharge through the turbine, Q :

$$Q = \text{Area of flow} \times \text{velocity of flow} = A_f \times V_{f1}$$

$$= 30 \times 7.92 = \mathbf{237.6\ m^3/s} \text{ (Ans.)}$$

(*iii*) Power developed by the turbine :

$$\text{Power developed by the turbine} = \frac{wQ}{g}(V_{w1}u_1 - V_{w2}u_2)$$

$$= \frac{9.81 \times 237.6}{9.81}(11.31 \times 11.31 - 5.67 \times 11.31) = \mathbf{15156.12\ kW} \text{ (Ans.)}$$

2.5. DERIAZ TURBINE

Fig. 2.43 shows a schematic view of a Deriaz (or diagonal) turbine which is a reaction turbine. It is named in the honour of its inventor P. Deriaz. This turbine is intermediate between the mixed-flow and the axial-flow turbines, because the flow of water as it passes through the runner is at an angle of 45° to axis and hence it is also known as *Diagonal turbine*. Deriaz turbine has the following *features* :

- It can be employed for the heads varying from 30 m to 150 m.
- The blades of the runner are pivoted to the hub and unlike in the Kaplan turbines, the axes of the blades are inclined to the axis of shaft. The direction of flow of water is as in the Francis runner.
- Guide vanes are provided ahead of the blades to regulate and direct the flow.
- The runner has no outer rim connecting all the blades as these blades are movable.

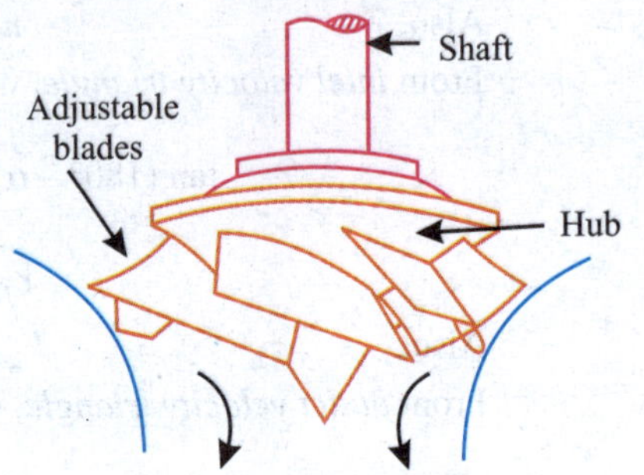

Fig. 2.43. Deriaz turbine.

- The casing of the turbine (not shown in Fig. 2.43) is so shaped that there is only small clearance between the blade tips and the casing to reduce leakage loss.

The runner of the Deriaz turbine is so shaped that it can be used both as a *turbine as well as a pump* and hence it may be classified as a *reversible type turbine*. As such Deriaz turbines are amply suitable for *pumped storage hydropower plants.*

Advantages of Deriaz turbine :

The Deriaz turbine entails the following *advantages* :

1. Improved part load efficiency.
2. Can be conveniently used as a pump-turbine unit.
3. By adjusting the runner to shut position the starting torque under water can be reduced.
4. Unlike axial flow turbines at shut position the flow area is completely closed.
5. Due to oblique location of blades the loading on the outer trunnion journal bearing is reduced.
6. The arrangement for varying the blade angle can be housed with greater convenience as compared to the Kaplan turbine.

2.6. TUBULAR OR BULB TURBINES

Invariably the electric generator coupled to the Kaplan turbine is enclosed and works *inside a straight passage* having the shape of a *bulb*. The *water tight bulb* is submerged directly into a stream of water, and the bends at inlet to casing, draft tube, etc. which are *responsible for the loss of head are dispensed with*. The unit then needs *less installation space* with a consequent *reduction in excavation and other civil engineering works*. These turbines are referred to as **tubular or bulb turbines**. The tubular turbine, a modified axial flow turbine, was developed in Germany by Arno Fischer in 1937. The economical harnessing of fairly low heads on major rivers is now possible with high-output bulb turbines. The following **features** are worth noting :

- A *tubular or bulb turbine* is an axial flow turbine with either adjustable or non-adjustable runner vanes (and hence similar to Kaplan or propeller turbines).
- In such a turbine the scroll casing is *not* provided but the runner is placed in a tube extending from head water to the tail water (and hence it is called tubular turbine).
- It is a low head turbine and is employed for heads varying from 3 m to 15 m.
- The disposition of shaft in a tubular turbine may be vertical, or inclined or horizontal.

The turbo-generator set using tubular turbine has an outer casing having the shape of a bulb. Such a set is now termed as **bulb set** and the turbine used for the set is called a *bulb turbine* (Fig. 2.44). The *advantages* and *disadvantages* of bulb sets *compared to Kaplan turbines* are as follows :

Advantages :

The bulb sets claim the following *advantages* over the Kaplan turbines.

1. Due to absence of spiral casing the plant width is small.
2. Can be used for the sites having very low head.
3. Because of almost straight flow and straight draft tube the maximum turbine efficiency is increased by about 3 per cent.
4. Bulb units can pass higher discharge (than conventional Kaplan turbine) under equivalent conditions.
5. At part loads there is reduced loss of efficiency.
6. Quite suitable for operation on widely varying heads.
7. Because of small dimensions of the power house there is saving in excavation and civil engineering works.

Disadvantages :

1. Leakage of water into generator chamber and condensation are source of trouble (leading to gradual deterioration of electrical insulation).
2. The erection techniques may be time consuming.

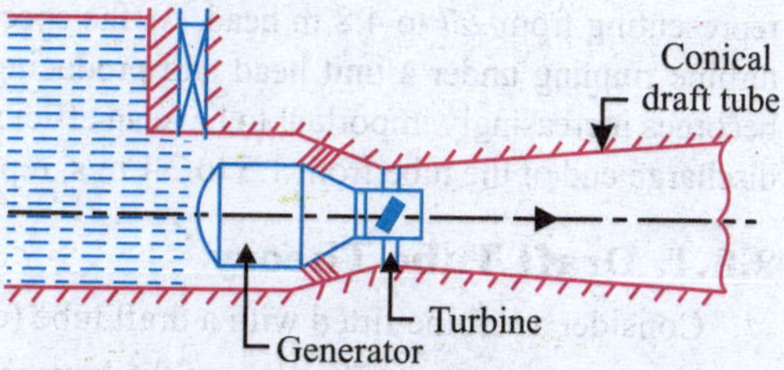

Fig. 2.44. Bulb turbine.

The use of bulb turbines offers the saving in the equipment of low head developments and great flexibility of operation and hence are highly suitable for ***tidal power station.***

2.7. RUNAWAY SPEED

Runaway speed *is the maximum speed, governor being disengaged, at which a turbine would run when there is no external load but operating under design head and discharge.* All the rotating parts including the rotor of alternator should be designed for the centrifugal stresses caused by this maximum speed.

The practical values of run away speeds for various turbines with respect to their rated speed N are as follows :

Pelton wheel	...1.8 to 1.9 N
Francis turbine (mixed flow)	...2.0 to 2.2 N
Kaplan turbine (axial flow)	...2.5 to 3.0 N

2.8. DRAFT TUBE

In the case of mixed and axial flow turbines only a part of available energy is converted into velocity energy at the inlet to the runner; the rest is in the form of pressure energy. This residual pressure is converted into velocity in the runner, as a consequence of which the outlet velocity increases. With the increase in the value of specific speed N_s, the exit velocity energy $\frac{V_2^2}{2g}$ increases compared with H (the available energy).

In the *Pelton wheel* all the available energy is converted into velocity energy before it strikes the wheel. As such it works under *atmospheric conditions* and the wheel *has to be placed above the maximum tail water level.* The loss of energy due to exit velocity varies from 1 to 4% .

In the case of *mixed and axial flow turbines* a large portion of the energy is associated with the water as it leaves the runner. This exit energy varies from 4 to 25% for mixed flow turbines and from 20 to 50% of the total head for axial flow turbines. As this energy cannot be used in the runner, therefore, it becomes necessary to find a way out to extract this energy. *An expanding pressure conduit hermetically fixed at runner outlet and having the other end below the minimum tail water level helps to convert the velocity head into pressure or potential head. This expanding device is called* **draft tube.** Draft tube is an *integral part* of mixed and axial flow turbines. Because of the draft tube *it is possible to have the pressure at runner outlet much below the atmospheric pressure.* The draft tube serves the following two *purposes* :

1. *It allows the turbine to be set above tail-water level, without loss of head, to facilitate inspection and maintenance.*
2. *It regains, by diffuse action, the major portion of the kinetic energy delivered to it from the runner.*

At rated load, the velocity at the upstream end of the tube for modern units ranges from 7 to 9 m/s, representing from 2.7 to 4.8 m head. As the specific speed (it is the speed of a geometrically similar turbine running under a unit head and producing unit power) is increased and the head reduced, it becomes increasingly important to have an efficient draft tube. Good practice limits the velocity at the discharge end of the tube from 1.5 to 2.1 m/s, representing less than 0·3 m velocity head loss.

2.8.1. Draft Tube Theory

Consider a turbine fitted with a draft tube (conical) as shown in Fig. 2.45.

Let, y = Distance of the bottom of draft tube from tail race, and

p_a = Atmospheric pressure at the surface of tail race.

Applying Bernoulli's equation to the section 2-2 (representing the runner exit or inlet of the draft tube) and the section 3-3 (representing the draft tube exit); assuming section 3-3 as the datum line, we have:

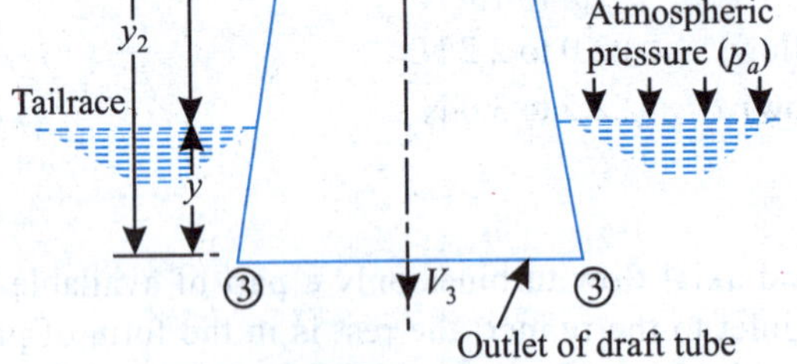

Fig. 2.45. Draft tube theory.

$$\frac{p_2}{w} + \frac{V_2^2}{2g} + y_2 = \frac{p_3}{w} + \frac{V_3^2}{2g} + 0 + h_f \qquad ...(i)$$

where, h_f = Loss of energy between sections 2-2 and 3-3.

Rewriting the above expression (*i*) for $\frac{p_2}{w}$, we obtain:

$$\frac{p_2}{w} = \frac{p_3}{w} - y_2 - \left(\frac{V_2^2 - V_3^2}{2g} - h_f\right) \quad ...(ii)$$

Substituting $\frac{p_3}{w} = \frac{p_a}{w} + y$ in expression (*ii*), we get:

$$\frac{p_2}{w} = \frac{p_a}{w} + (y - y_2) - \left(\frac{V_2^2 - V_3^2}{2g} - h_f\right)$$

The term $(y_2 - y)$ which represents the vertical distance between the runner exit and the tail water level is called the **suction head of draft tube** and is denoted by H_s. Correspondingly the factor $\frac{V_2^2 - V_3^2}{2g}$ is called the **dynamic head**.

$$\therefore \quad \frac{p_2}{w} = \frac{p_a}{w} - H_s - \left(\frac{V_2^2 - V_3^2}{2g} - h_f\right) \quad ...(2.32)$$

In eqn. (2.32), $\frac{p_2}{w}$ is *less than atmospheric pressure.*

The **efficiency of a draft tube** (η_d) is defined as the *ratio of net gain in pressure head to the velocity head at entrance of draft tube*. Thus,

$$\eta_d = \frac{\text{Net gain in pressure head}}{\text{Velocity head at entrance of draft tube}}$$

$$= \frac{\left(\frac{V_2^2 - V_3^2}{2g} - h_f\right)}{\frac{V_2^2}{2g}} \quad ...(2.33)$$

where, V_2 = Velocity of water at section 2-2 (inlet of draft tube), and

V_3 = Velocity of water at section 3-3 (outlet of draft tube).

$$\left[h_f = \frac{V_2^2 - V_3^2}{2g} - \eta_d \times \frac{V_2^2}{2g}\right] \quad ...2.33\ (a)$$

2.8.2. Types of Draft Tubes

The following two types of draft tubes are commonly used :

1. The straight conical or concentric tube.
2. The elbow type.

Properly designed, the two types are about equally efficient, *over 85 per cent.*

1. Conical type. The conical type draft tube is *generally* used on *low-powered units for all specific speeds, frequently, on large-head units*. The side angle of flare ranges from 4 to 6°, the length from 3 to 4 times the diameter and the discharge area from four to five times the throat area. Fig. 2.46 shows a straight *conical* draft tube.

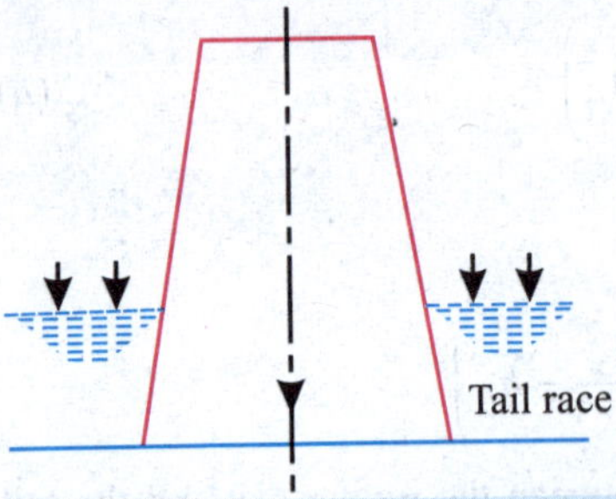

Fig. 2.46. Straight conical draft tube.

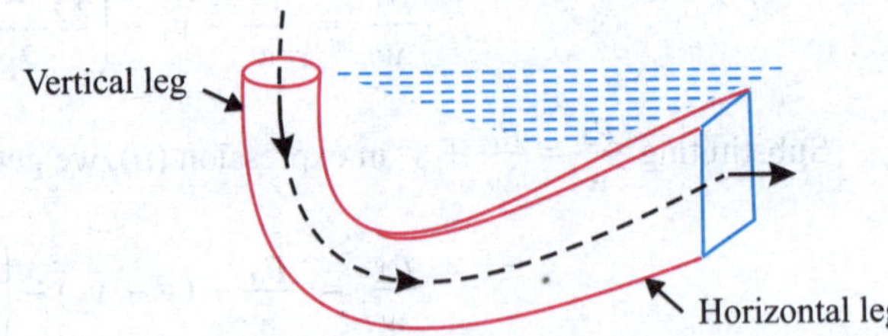

Fig. 2.47. Elbow type draft tube.

2. Elbow type. The elbow type of tube is used with *most* turbine installations. This type of draft tube is designed to turn the water from the *vertical to the horizontal direction with a minimum depth of excavation and at the same time having a high efficiency*. The transition from a circular section in the vertical leg to a rectangular section in the horizontal leg takes place in the bend. The horizontal portion of the draft tube is generally *inclined upwards to lead the water gradually to the level of the tail race and to prevent entry of air from the exit end*. The *exit* end of the draft tube *must be totally immersed in water*. Fig. 2.47 shows an elbow type draft tube. One or two vertical piers are placed in the horizontal portion of the tube, for structural and hydraulic reasons.

Moody's spreading draft tube. Fig. 2.48 shows a *Moody's spreading draft tube*. It is provided with a *solid central core* of conical shape which *reduces whirling action of discharged water*. The efficiency of such a draft tube is about 85%. It is suited particularly for helical flows which occur when the water leaves the runner with a whirl component.

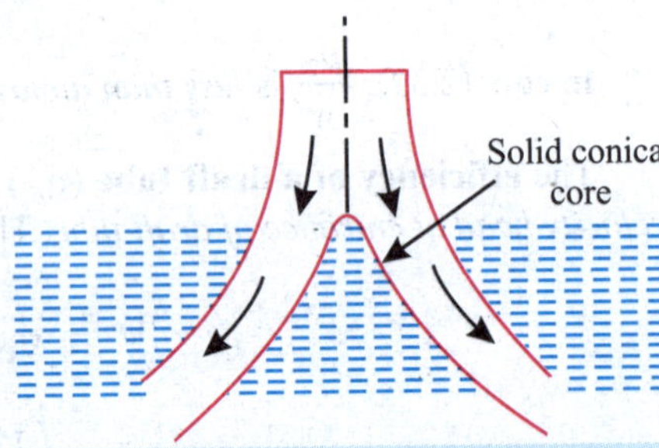

Fig. 2. 48. Moody's spreading draft tube or 'Hydrocone'.

Example 2.41. *A Kaplan turbine develops 1500 kW under a head of 6 m. The turbine is set 2.5 m above the tail race level. A vacuum gauge inserted at the turbine outlet records a suction head of 3·1 m. If the hydraulic efficiency is 82 per cent, what would be the efficiency of draft tube having inlet diameter of 3 m ?*

What will be the reading of suction gauge if power developed is reduced to 750 kW, the head and speed remaining constant.

Solution. Power developed = 1500 kW; Head, $H = 6$ m

Height of turbine above tail race level = 2·5 m; Hydraulic efficiency, $\eta_h = 82\%$

Draft tube inlet diameter, $d_i = 3$ m.

Efficiency of draft tube, η_d :

$$\text{Hydraulic efficiency, } \eta_h = \frac{\text{Power developed}}{\text{Water power}} = \frac{\text{Power developed}}{wQH}$$

$$\therefore \quad \text{Power developed} = wQH \times \eta_h$$

$$1500 = 9.81 \times Q \times 6 \times 0.82, \quad \text{or,} \quad Q = \frac{1500}{9.81 \times 6 \times 0.82} = 31.08 \text{ m}^3/\text{s}$$

Velocity of water at inlet of draft tube,

$$V_2 = \frac{Q}{\frac{\pi}{4} d_i^2} = \frac{31.08}{\frac{\pi}{4} \times 3^2} = 4.397 \text{ m/s}$$

$$\text{Pressure head required} = 3.1 - 2.5 = 0.6 \text{ m}$$

$\therefore$ Efficiency of draft tube, $\eta_d = \dfrac{0.6}{\dfrac{V_2^2}{2g}} = \dfrac{0.6}{\dfrac{4.397^2}{2 \times 9.81}} = 0.6088$ or **60.88% (Ans.)**

Reading of suction gauge :

For reduced output of 750 kW assuming constant efficiency, we have:

$$\text{Discharge, } Q_1 = \frac{Q}{2} = \frac{31.08}{2} = 15.54 \text{ m}^3/\text{s}$$

Also,
$$V_2 = \frac{15.54}{\frac{\pi}{4} \times 3^2} = 2.198 \text{ m/s}$$

$$\text{Head gained in draft tube} = \eta_d \times \frac{2.198^2}{2g}$$

$$= 0.6088 \times \frac{2.198^2}{2 \times 9.81} \simeq 0.15 \text{ m}$$

$\therefore$ Reading of gauge $= 2.5 + 0.15 =$ **2.65 m (Ans.)**

Example 2.42. *Determine the overall efficiency of a Kaplan turbine developing 2850 kW under a head of 5·2 m. It is provided with a draft tube with its inlet (diameter 3 m) set 1·8 m above the tail race level. A vacuum gauge connected to the draft tube indicates a reading of 5·2 m of water. Assume draft tube efficiency as 75 per cent.*

Solution. Power developed = 2850 kW; Head, $H = 5.2$ m

Height of draft inlet tube above tail race level, $H_s = 1.8$ m

Reading of the gauge = – 5.2 m

Draft tube efficiency, $\eta_d = 75\%$

Overall efficiency of the turbine, η_0 :

$$\frac{p_2}{w} = \frac{p_a}{w} - H_s - \left(\frac{V_2^2 - V_3^2}{2g} - h_f\right) \qquad \text{...[Eqn. (2.32)]}$$

$$-5.2 = 0 - 1.8 - \left(\frac{V_2^2 - V_3^2}{2g}\right), \text{ neglecting } h_f \text{ (head loss in draft tube)}$$

or,
$$\frac{V_2^2 - V_3^2}{2g} = 3.4$$

Also,
$$\eta_d = \frac{(V_2^2 - V_3^2)/2g}{(V_2^2/2g)} \qquad \text{...[Eqn. (2.33)]}$$

or,
$$0.75 = \frac{3.4}{(V_2^2/2g)}, \text{ or, } \frac{V_2^2}{2g} = \frac{3.4}{0.75} = 4.533$$

$\therefore$
$$V_2 = \sqrt{4.533 \times 2g} = \sqrt{4.533 \times 2 \times 9.81} = 9.43 \text{ m/s}$$

$$\text{Discharge, } Q = \frac{\pi}{4} \times 3^2 \times 9.43 = 66.65 \text{ m}^3/\text{s}$$

$$\therefore \text{ Overall efficiency, } \eta_0 = \frac{\text{Power developed}}{\text{Water power}} = \frac{2850}{wQH}$$

$$= \frac{2850}{9.81 \times 66.65 \times 5.2} = 0.8382 \quad \text{or} \quad \mathbf{83.82\%\ (Ans.)}$$

Example 2.43. *A conical draft tube having inlet and outlet diameters 1.2 m and 1.8 m discharges water at outlet with a velocity of 3 m/s. The total length of the draft tube is 7.2 m and 1·44 m of the length of draft tube is immersed in water. If the atmospheric pressure head is 10.3 m of water and loss of head due to friction in the draft tube is equal to 0.2 × velocity head at outlet of the tube, determine :*

(i) Pressure head at inlet, and

(ii) Efficiency of the draft tube.

Solution. Inlet diameter of the draft tube, $d_i = 1.2$ m

Outlet diameter, $d_0 = 1.8$ m

Velocity at outlet, $V_3 = 3$ m/s

Total length of draft tube, $H_s + y = 7.2$ m

Length of draft tube in water, $y = 1.44$ m

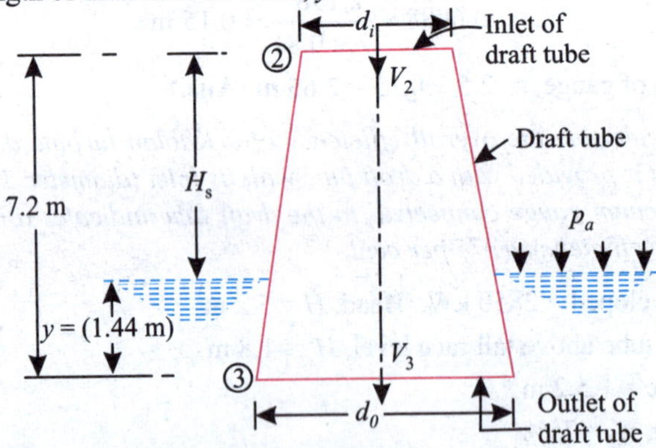

Fig. 2.49

$$\therefore \quad H_s = 7.2 - 1.44 = 5.76 \text{ m}$$

Atmospheric pressure head, $\frac{p_a}{w} = 10.3$ m

Loss of head due to friction,

$$h_f = 0.2 \times \text{velocity head at outlet}$$

$$= 0.2\,\frac{V_3^2}{2g}$$

(*i*) Pressure head at inlet, $\frac{p_2}{w}$:

Discharge through the draft tube,

$$Q = A_3V_3 = \frac{\pi}{4} \times d_0^2 \times V_3 = \frac{\pi}{4} \times 1.8^2 \times 3 = 7.634 \text{ m/s}$$

$$\text{Velocity of inlet, } V_2 = \frac{Q}{A_2} = \frac{7.634}{\frac{\pi}{4}\,d_i^2} = \frac{7.634}{\frac{\pi}{4} \times 1.2^2} = 6.75 \text{ m/s}$$

Using eqn. (2.32),

$$\frac{p_2}{w} = \frac{p_a}{w} - H_s - \left(\frac{V_2^2 - V_3^2}{2g} - h_f\right) = \frac{p_a}{w} - H_s - \left(\frac{V_2^2 - V_3^2}{2g} - 0.2\frac{V_3^2}{2g}\right)$$

$$= 10.3 - 5.76 - \left(\frac{6.75^2 - 3^2}{2 \times 9.81} - 0.2 \times \frac{3^2}{2 \times 9.81}\right)$$

or, $$\frac{p_2}{w} = 4.54 - (1.863 - 0.092) = \mathbf{2.769\ m\ (abs)}\ \textbf{(Ans.)}$$

***(ii)* Efficiency of the draft tube, η_d :**

$$\eta_d = \frac{\left(\frac{V_2^2 - V_3^2}{2g} - h_f\right)}{\frac{V_2^2}{2g}} = \frac{\frac{V_2^2 - V_3^2}{2g} - 0.2\frac{V_3^2}{2g}}{\frac{V_2^2}{2g}} = \frac{\frac{V_2^2}{2g} - \left(\frac{V_3^2}{2g} + 0.2\frac{V_3^2}{2g}\right)}{\frac{V_2^2}{2g}}$$

$$= 1 - 1.2\left(\frac{V_3}{V_2}\right)^2 = 1 - 1.2\left(\frac{3}{6.75}\right)^2 = 0.763, \text{ or, } \mathbf{76.3\ \%}\ \textbf{(Ans.)}$$

Example 2.44. *A reaction turbine and its draft tube have a vertical axis. The pressure head in the spiral casing at inlet is 48 m above atmospheric pressure and the velocity of water is 6 m/s. The water flow through the tube is 2.1 m^3/s, and the hydraulic and overall efficiencies are 83 per cent and 80 per cent respectively. The top of the draft tube is 1.2 m below the centre line of the spiral casing while the tail race is 3.9 m below the top of the draft tube. The diameter of the draft tube at inlet is 0·75 m and that at the tail race level 1.05 m. Determine :*

(i) *Total head across the turbine,* *(ii)* *Shaft power,*

(iii) *Head lost in friction in turbine and draft tube, and*

(iv) *Power lost in mechanical friction.*

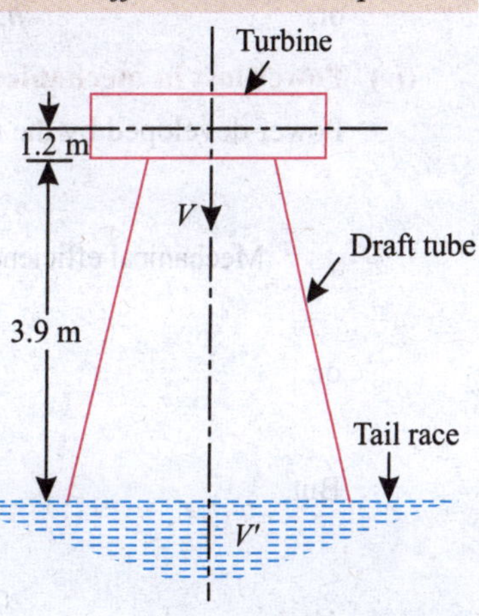

Fig. 2.50

Solution. Pressure head in the spiral casing = 48 m (above atmospheric pressure)

Velocity of water, V = 6 m/s

Discharge through the draft tube, Q = 2.1 m^3/s

Hydraulic efficiency, η_h = 83%

Overall efficiency, η_0 = 80%

Inlet diameter of draft tube, d_i = 0.75 m

Outlet Diameter, d_0 = 1.05 m

***(i)* Total head across the turbine :**

Total head across the turbine above the tail race level = Total head in the spiral casing measured above the tail race.

or, $$H = \frac{p}{w} + \frac{V^2}{2g} + (1.2 + 3.9)$$

$$= 48 + \frac{6^2}{2 \times 9.81} + (1.2 + 3.9) = \mathbf{54.93\ m}\ \textbf{(Ans.)}$$

(ii) Shaft power, P :

$$P = \eta_0 \times wQH = 0.8 \times 9.81 \times 2.1 \times 54.93$$
$$= \mathbf{905.3\ kW\ (Ans.)}$$

(iii) Head lost in friction in turbine and draft tube, $(h_{ft} + h_{fd})$:

$$V' = \frac{Q}{\frac{\pi}{4} d_0^2} = \frac{4Q}{\pi d_0^2} = \frac{4 \times 2.1}{\pi \times 1.05^2} = 2.425 \text{ m/s}$$

Head utilized by the turbine $H - h_{ft} - h_{fd} - \dfrac{V'^2}{2g}$

$$= 54.93 - (h_{ft} + h_{fd}) - \frac{2.425^2}{2 \times 9.81} = 54.63 - (h_{ft} + h_{fd})$$

(where, h_{ft} and h_{fd} are the heads lost due to friction in the turbine and draft tube respectively.)

$$\text{Hydraulic efficiency, } \eta_h = \frac{\text{Head utilized by the turbine}}{\text{Head supplied to the turbine}}$$

$$0.83 = \frac{54.63 - (h_{ft} + h_{fd})}{54.93}$$

or, $$h_{ft} + h_{fd} = 54.63 - 0.8 \times 54.93 = \mathbf{10.68\ m\ (Ans.)}$$

(iv) Power lost in mechanical friction, P_f :

Power developed by the runner = Shaft power (P) + power lost in mechanical friction (P_f)
$$= P + P_f$$

$$\text{Mechanical efficiency, } \eta_m = \frac{\text{Shaft power}}{\text{Power developed by the runner}}$$

or, $$\eta_m = \frac{P}{P + P_f}$$

But, $$\eta_0 = \eta_h \times \eta_m, \text{ or, } \eta_m = \frac{\eta_0}{\eta_h} = \frac{0.8}{0.83} = 0.9638$$

$\therefore$ $$0.9638 = \frac{905.3}{905.3 + P_f}, \text{ or, } 0.9638\,(905.3 + P_f) = 905.3$$

or, $$P_f = \frac{905.3}{0.9638} - 905.3 = \mathbf{34\ kW\ (Ans.)}$$

2.9. SPECIFIC SPEED

The **specific speed** *of a turbine is defined as the speed of a turbine which is identical in shape, geometrical dimensions, blade angles, gate opening, etc. which would develop unit power when working under a unit head.*

The specific speed may be *derived* as follows :

The overall efficiency (η_0) of any turbine is given by,

$$\eta_0 = \frac{\text{Power available at the shaft of the turbine (shaft power)}}{\text{Power supplied at the inlet of the turbine (water power)}}$$

$$= \frac{P}{wQH} \quad ...(i)$$

where, P = Shaft power,

Q = Discharge through turbine,

H = Head under which turbine is working, and

w = Weight density of water.

From eqn. (*i*), $P = \eta_0 \times wQH$

or, $P \propto Q \times H$ (as η_0 and w are constant.) ...(*ii*)

Now, let D = Diameter of actual turbine,

N = Speed of actual turbine,

u = Tangential velocity of the turbine,

N_s = Specific speed of the turbine, and

V = Absolute velocity of water.

Then relation between V, u and H is as given below :

$$u \propto V \text{ where, } V \propto \sqrt{H}$$

or, $$u \propto \sqrt{H} \quad ...(iii)$$

But the tangential velocity u is given by:

$$u = \frac{\pi DN}{60}$$

or, $$u \propto DN \quad ...(iv)$$

∴ From eqns. (*iii*) and (*iv*), we have:

$$\sqrt{H} \propto DN$$

or, $$D = \frac{\sqrt{H}}{N} \quad ...(v)$$

The discharge (Q) through the turbine is given by:

$$Q = \text{Area} \times \text{velocity}$$

But, Area $\propto B \times D$ (where, B = width)

∴ $\propto D^2$ $(\because B \propto D)$

∴ $$Q \propto D^2 \sqrt{H}$$

$$\propto \left(\frac{\sqrt{H}}{N}\right)^2 \sqrt{H} \quad \left[\because \text{From eqn. } (v),\ D \propto \frac{\sqrt{H}}{N}\right]$$

$$\propto \frac{H}{N^2}\sqrt{H} \propto \frac{H^{3/2}}{N^2}$$

Substituting the value of Q in eqn. (*i*), we get:

$$P \propto \frac{H^{3/2}}{N^2} \times H \propto \frac{H^{5/2}}{N^2}$$

∴ $$P = K\frac{H^{5/2}}{N^2}$$ where, K = constant of proportionality.

If, $P = 1$ kW and $H = 1$ m, the speed N = specific speed N_s, then by substituting these values in the above equation, we get:

$$1 = \frac{K \times (1)^{5/2}}{N_s^2}, \text{ or, } N_s^2 = K$$

$$\therefore \quad P = N_s^2 \frac{H^{5/2}}{N^2}, \text{ or, } N_s^2 = \frac{N^2 P}{H^{5/2}}$$

$$\therefore \quad N_s = \frac{N\sqrt{P}}{H^{5/4}} \qquad ...(2.33)$$

where, P is in kW and H in metres.

[N_s (S.I units) = 0.86 N_s (metric)]

- Specific speed plays an important role in the *selection of the type of turbine*. By knowing the specific speed of turbine the *performance of the turbine can also be predicted*.
- If a runner of *high specific speed* is used for a given head and power output, the *overall cost of installation is lower*. The selection of too high specific speed reaction runner would reduce the size of the runner to such an extent that the discharge velocity of water into the throat of draft tube would be excessive. This is objectionable because a *vacuum* may be created in the extreme case.
- The runner of *too high specific speed* with high available head *increases the cost of turbine* on account of high mechanical strength required. The runner of *too low specific speed* with low available head increases the cost of generator due to the low turbine speed.
- *An increase in specific speed of turbine is accompanied by lower maximum efficiency and greater depth of excavation of the draft tube*. In choosing a high specific speed turbine, an increase in cost of excavation of foundation and draft tube should be considered in addition to the efficiency. The weighted *efficiency over the operating range of turbine is more important in the selection of a turbine instead of maximum efficiency.*

Note : For N_s–range refer to Table 2.2. (P-163).

Example 2.45. *A turbine is to operate under a head of 25 m at 200 r.p.m . The discharge is 9 m³/s. If the overall efficiency is 90 per cent, determine :*

(i) Power generated; *(ii) Specific speed of the turbine;*

(iii) Type of turbine. **[N.U.]**

Solution. Head, $H = 25$ m; Speed, $N = 200$ r.p.m.;

Discharge, $Q = 9$ m³/s; Overall efficiency, $\eta_0 = 90\%$.

(i) Power generated, P :

$$P = \eta_0 \times wQH = 0.9 \times 9.81 \times 9 \times 25 = \mathbf{1986.5 \text{ kW}} \textbf{ (Ans.)}$$

(ii) Specific speed of the turbine, N_s :

$$N_s = \frac{N\sqrt{P}}{H^{5/4}} = \frac{200 \times \sqrt{1986.5}}{(25)^{5/4}} = \mathbf{159.4 \text{ r.p.m.}} \textbf{ (Ans.)}$$

(iii) Type of Turbine :

As the specific speed lies between 80 and 400 (Refer to table 2.2), the turbine is a **Francis turbine. (Ans.)**

Example 2.46. *A Pelton wheel generates 8000 kW under a net head of 130 m at a speed of 200 r.p.m. Assuming the co-efficient of velocity for the nozzle 0.98, hydraulic efficiency 87 percent, speed ratio 0.46, and jet diameter to wheel diameter ratio* $\frac{1}{9}$, *determine :*

(i) *Discharge required,* **(ii)** *Diameter of the wheel,*

(iii) *Diameter and number of jets required, and* **(iv)** *Specific speed.*

Mechanical efficiency is 75 per cent. **[GATE]**

Solution. Power generated (shaft power), $P = 8000$ kW

Net head, $H = 130$ m

Speed, $N = 200$ r.p.m.

Co-efficient of velocity, $C_v = 0.98$

Hydraulic efficiency, $\eta_h = 87\%$

Speed ratio, $K_u = 0.46$

Jet diameter to wheel diameter, $\frac{d}{D} = \frac{1}{9}$

Mechanical efficiency, $\eta_m = 75\ \%$

(i) Discharge required, Q :

Overall efficiency, $\eta_0 = \eta_h \times \eta_m = 0.87 \times 0.75 = 0.6525$

Also, $$\eta_0 = \frac{\text{Shaft power}}{\text{Water power}} = \frac{P}{wQH}$$

$$\therefore \quad Q = \frac{P}{\eta_0 wH} = \frac{8000}{0.6525 \times 9.81 \times 130} = \mathbf{9.614\ m^3/s}\ \textbf{(Ans.)}$$

(ii) Diameter of the wheel, D :

$$\text{Speed ratio, } K_u = \frac{u_1}{\sqrt{2gH}}, \quad \text{or,} \quad 0.46 = \frac{u_1}{\sqrt{2 \times 9.81 \times 130}}$$

$$\therefore \quad u_1 = 0.46\sqrt{2 \times 9.81 \times 130} = 23.23 \text{ m/s}$$

Also, $$u_1 = \frac{\pi DN}{60}, \quad \text{or,} \quad D = \frac{60\, u_1}{\pi N}$$

or, $$D = \frac{60 \times 23.23}{\pi \times 200} = \mathbf{2.218\ m}\ \textbf{(Ans.)}$$

(iii) Diameter and number of jets required :

$$\frac{d}{D} = \frac{1}{9};\ d = \frac{D}{9} = \frac{2.218}{9} = 0.2464 \text{ m or } \mathbf{246.4\ mm}\ \textbf{(Ans.)}$$

$$\therefore \quad \text{Area of jet, } a = \frac{\pi}{4} \times d^2 = \frac{\pi}{4} \times (0.2464)^2 = 0.04768 \text{ m}^2$$

$$\text{Velocity of the jet, } V = C_v\sqrt{2gH} = 0.98\sqrt{2 \times 9.81 \times 130} \simeq 49.5 \text{ m/s}$$

$\therefore$ Discharge through one jet, $q = a \times V = 0.04768 \times 49.5 = 2.36 \text{ m}^3/\text{s}$

$$\therefore \quad \text{Number of jets} = \frac{Q}{q} = \frac{9.614}{2.36} = 4.07 \text{ say } \mathbf{4}\ \textbf{(Ans.)}$$

***(iv)* Specific speed, N_s :**

$$N_s = \frac{N\sqrt{P}}{H^{5/4}} = \frac{200 \times \sqrt{8000}}{(130)^{5/4}} = \mathbf{40.75\ r.p.m.\ (Ans.)}$$

Example 2.47. *Give the range of specific speed values of the Kaplan, Francis turbines and Pelton wheels. What factors decide whether Kaplan, Francis or a Pelton wheel type turbine would be used in a hydroelectric project.* **[UPSC]**

Solution. • The **specific speed** *of a turbine is defined* as *the speed of a turbine which is identical in shape, geometrical dimensions, blade angles, gate opening, etc. which would develop unit power when working under a unit head.*

- Based on specific speed, the turbines for the project are selected as shown in the Fig. 2.51.
- In general, the selection of a turbine for hydroelectric project is based on the following considerations :

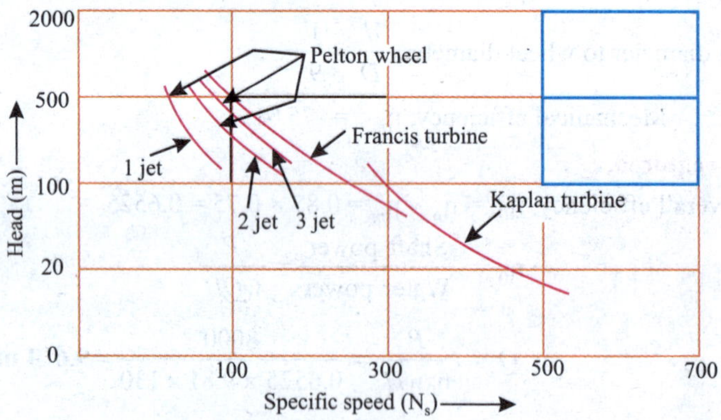

Fig. 2.51

1. For *high heads*, *Pelton wheels* are invariably selected.
2. For *intermediate heads*, *Francis turbines* are selected.
3. For *low head* and *high discharge*, Kaplan turbines are selected.

Example 2.48. *In a hydroelectric station, water is available at the rate of 175 m³/s under a head of 18 m. The turbines run at a speed of 150 r.p.m. with overall efficiency of 82%. Find the number of turbines required if they have the maximum specific speed of 460.* **[GATE]**

Solution. *Given* : $Q = 175\ m^3/s$; $H = 18$ m; $N = 150$ r.p.m.; $\eta_0 = 82\%$; $N_s = 460$.

Number of turbines required :

$$\text{Specific speed of the turbine,} = \frac{N\sqrt{P}}{H^{5/4}} \quad \text{...[Eqn. (2.33)]}$$

$$460 = \frac{150\sqrt{P}}{(18)^{5/4}} \quad \text{(where, P is in kW and H is in metres.)}$$

$$\text{or, Power available at turbine shaft, } P = \left[\frac{460 \times (18)^{5/4}}{150}\right]^2 = 12927.5 \text{ kW}$$

$$\text{Power available from turbines} = wQH \times \eta_0 = 9.81 \times 175 \times 18 \times 0.82 = 25339.23 \text{ kW}$$

$$No.\ of\ turbines\ required = \frac{25339.23}{12927.5} = 1.96 \text{ say } \mathbf{2\ (Ans.)}$$

2.10. UNIT QUANTITIES

Let us consider a *single unit*. When the head on the unit is changed/varied then the speed of an ungoverned turbine changes. The velocities at various points do not change direction but their magnitudes vary in proportion to the *square root of the head.*

At a given point in the turbine under a head H, let

V = Absolute velocity,

V_r = Relative velocity,

u = Peripheral velocity, and

V', V_r', u' = Corresponding values at a different head H', then as velocity is proportional to $\sqrt{H}$, we have

$$\frac{u}{u'} = \frac{V_r}{V_r'} = \frac{V}{V'} = \frac{\sqrt{H}}{\sqrt{H'}} \qquad ...(2.34)$$

If the discharges are Q and Q' then,

$$\frac{Q}{Q'} = \frac{V}{V'} = \frac{N}{N'} = \frac{\sqrt{H}}{\sqrt{H'}} \qquad ...(2.35)$$

If the power outputs are P and P' then,

$$\frac{P}{P'} = \frac{QH}{Q'H'} = \frac{\sqrt{H}}{\sqrt{H'}} \times \frac{H}{H'} = \left(\frac{H}{H'}\right)^{3/2} \qquad ...(2.36)$$

$$\left(\because \frac{Q}{Q'} = \frac{\sqrt{H}}{\sqrt{H'}}\right)$$

The hydraulic efficiency of the turbine under these two heads may be considered to be nearly same, as the velocity triangles at these heads are similar at a point.

If the various quantities are *reduced to a theoretical one metre head* the comparison of performance data and computations of experimental values on a single unit are *considerably simplified.*

Then,

$$N_u = \frac{N}{\sqrt{H}} \qquad ...(2.37)$$

$$Q_u = \frac{Q}{\sqrt{H}} \qquad ...(2.38)$$

$$P_u = \frac{P}{H^{3/2}} \qquad ...(2.39)$$

The above quantities are called **unit quantities** of a turbine. *Unit speed is the hypothetical speed of the turbine operating under one metre head.* Similarly, other proportionality constants in Eqns. 2.38 and 2.39 are defined.

For presenting the performance of geometrically similar turbines independent of the actual head, discharge and power output the ***unit characteristics*** prove quite helpful. **Geometrically similar turbines will have the same unit characteristics under similar operating conditions.** Thus with the help of a model the performance of a prototype can be predicted within certain limits.

If a turbine is working under different heads the behaviour of the turbine can be easily known from the values of the *unit quantities* as follows :

Let,

H_1, H_2 = Heads under which a turbine works,
N_1, N_2 = Corresponding speeds,
Q_1, Q_2 = Corresponding discharges, and
P_1, P_2 = Corresponding powers developed.

Then using eqns. (2.37), (2.38), (2.39), respectively, we obtain

$$N_u = \frac{N_1}{\sqrt{H_1}} = \frac{N_2}{\sqrt{H_2}} \quad ...(2.40)$$

$$Q_u = \frac{Q_1}{\sqrt{H_1}} = \frac{Q_2}{\sqrt{H_2}} \quad ...(2.41)$$

$$P_u = \frac{P_1}{H_1^{3/2}} = \frac{P_2}{H_2^{3/2}} \quad ...(2.42)$$

Example 2.49. *A turbine is to operate under a head of 25 m at 200 r.p.m. The discharge is 9 m³/s. If the efficiency is 90 per cent determine the performance of the turbine under a head of 20 m.* **[M.U]**

Solution. Head under which turbine works, H_1 = 25 m

Speed of the turbine, N_1 = 200 r.p.m.

Discharge through the turbine, Q_1 = 9 m³/s

Efficiency (overall), η_0 = 90%

Performance of turbine under a head of 20 m; N_2, Q_2, P_2 :

Performance of the turbine under a head, H_2 = 20 m means to find speed (N_2), discharge (Q_2), and power generated (P_2) by the turbine when working under a head of 20 m.

Overall efficiency,

$$\eta_0 = \frac{\text{Shaft power}}{\text{Water power}} = \frac{P}{wQH} = \frac{P_1}{wQ_1H_1}$$

$$\therefore \quad P_1 = \eta_0 \times wQ_1H_1 = 0.9 \times 9.81 \times 9 \times 25 = 1986.5 \text{ kW}$$

Now,

$$\frac{N_1}{\sqrt{H_1}} = \frac{N_2}{\sqrt{H_2}} \quad ...[\text{Eqn. (2.40)}]$$

$$\therefore \quad N_2 = \frac{N_1\sqrt{H_2}}{\sqrt{H_1}} = \frac{200 \times \sqrt{20}}{\sqrt{25}} = \mathbf{178.88 \text{ r.p.m.}} \text{ (Ans.)}$$

and,

$$\frac{Q_1}{\sqrt{H_1}} = \frac{Q_2}{\sqrt{H_2}} \quad ...[\text{Eqn. (2.41)}]$$

$$\therefore \quad Q_2 = \frac{Q_1\sqrt{H_2}}{\sqrt{H_1}} = \frac{9 \times \sqrt{20}}{\sqrt{25}} = \mathbf{8.05 \text{ m}^3/\text{s}} \text{ (Ans.)}$$

and,

$$\frac{P_1}{H_1^{3/2}} = \frac{P_2}{(H_2)^{3/2}} \quad ...[\text{Eqn (2.42)}]$$

$$\therefore \quad P_2 = \frac{P_1 \times (H_2)^{3/2}}{(H_1)^{3/2}} = \frac{1986.5 \times (20)^{3/2}}{(25)^{3/2}} = \mathbf{1421.4 \text{ kW}} \text{ (Ans.)}$$

2.11. MODEL RELATIONSHIP

(i) Head co-efficient, C_H :

The tangential velocity of the runner, $u = K_u\sqrt{2gH} = \dfrac{\pi DN}{60}$

or, $$N = \frac{60K_u\sqrt{2gH}}{\pi D}, \text{ or, } N \propto \frac{\sqrt{H}}{D}$$

$$\therefore \quad ND = \sqrt{H}, \text{ or, } \frac{H}{N^2D^2} = \text{constant} \qquad ...(2.43)$$

The parameter $\dfrac{H}{N^2D^2}$ is called **head co-efficient**, C_H.

(ii) Capacity or flow co-efficient, C_Q :

Discharge through the turbine, Q = Area × velocity = $A \times V_f$

But, $$A \propto D^2, \text{ and, } V_f = K_f\sqrt{2gH} \propto \sqrt{H}$$

$$\therefore \quad Q \propto D^2\sqrt{H}$$

Substituting the value of Q in eqn. (2.43), we obtain:

$$Q \propto D_2 \times ND \propto ND^3$$

or, $$\frac{Q}{ND^3} = \text{constant} \qquad ...(2.44)$$

The parameter, $\dfrac{Q}{ND^3}$ is called the **capacity or flow co-efficient, C_Q**.

(iii) Power co-efficient C_P :

The shaft power available from a turbine,

$$P = \eta_0 \times wQH \propto QH$$

But, $Q \propto ND^3$ and $H \propto N^2D^2$ $\therefore$ $P \propto ND^3 \times N^2D^2$, or, $\propto N^3D^5$

or, $$\frac{P}{N^3D^5} = \text{constant} \qquad ...(2.45)$$

The parameter $\dfrac{P}{N^3D^5}$ is called the **power co-efficient, C_P.**

With the use of above relations it is possible to present the behaviour of a prototype from the test runs made on a geometrically similar model; the model is presumed to have the same values of speed ratio K_u, flow ratio K_f and specific speed N_s. A group of geometrically similar machines are said to belong to a homologous series. All machines of such a series have the same values of C_H, C_Q or C_P or their combinations.

Example 2.50. *A hydro-turbine is required to give 25 MW at 50 m head and 90 r.p.m. runner speed. The laboratory facilities available permit testing of 20 kW model at 5 m head. What should be the model runner speed and model to prototype scale ratio?* **[UPTU]**

Solution. *Given* : $P_p = 25$ MW; $H_p = 50$ m; $N_p = 90$ r.p.m.; $P_m = 20$ kW; $H_m = 5$ m

$$\mathbf{N_m;\ \frac{D_p}{D_m}\ (= L_r):}$$

Prototype specific speed, $(N_s)_p = \dfrac{N_p\sqrt{P_p}}{(H_p)^{5/4}}$ (where, P is in kW)

$$= \frac{90 \times \sqrt{25 \times 10^3}}{(50)^{5/4}} = 107$$

For model, $107 = \dfrac{N_m\sqrt{P_m}}{(H_m)^{5/4}}$ $[\because (N_s)_p = (N_s)_m]$

or, $N_m = \dfrac{107 \times (H_m)^{5/4}}{\sqrt{P_m}} = \dfrac{107 \times (5)^{5/4}}{\sqrt{20}} = \mathbf{178.89\ r.p.m.}$ **(Ans.)**

For similar turbines $\dfrac{P}{H^{3/2}D^2}$ should be equal.

$$\therefore \quad \frac{P_p}{H_p^{3/2}D_p^2} = \frac{P_m}{H_m^{3/2}D_m^2}$$

or, $\dfrac{D_p}{D_m}(= L_r) = \sqrt{\dfrac{P_p}{P_m} \times \left(\dfrac{H_m}{H_p}\right)^{3/2}} = \sqrt{\dfrac{25 \times 10^3}{20} \times \left(\dfrac{5}{50}\right)^{3/2}} = \mathbf{6.287}$ **(Ans.)**

Example 2.51. *A water turbine delivering 10 MW power is to be tested with the help of a geometrically similar 1 : 8 model, which runs at the same speed as the prototype.*

(i) Find the power developed by the model assuming the efficiencies of the model and the prototype are equal.

(ii) Find the ratio of the heads and the ratio of mass flow rates between the prototype and the model. **[PTU]**

Solution. *Given* : $P_p = 10$ MW; $N_p = N_m$; $\dfrac{L_m}{L_p} = \dfrac{D_m}{D_p} = \dfrac{1}{8}$; $\eta_p = \eta_m$.

(i) Power developed by the model, $\mathbf{P_m}$:

We know that, $P \propto N^3 \times D^5$ ... [Eqn. (2.45)]

(where, N is the speed and D is the diameter.)

$$\therefore \quad P_p \propto N_p^3\, D_p^5, \text{ and, } P_m \propto N_m^3\, D_m^5$$

or, $\dfrac{P_p}{P_m} = \left(\dfrac{N_p}{N_m}\right)^3 \times \left(\dfrac{D_p}{D_m}\right)^5 = (1)^3 \times \left(\dfrac{8}{1}\right)^5 = 8^5$ $(\because N_p = N_m)$

$\therefore$ $P_m = \dfrac{P_p}{(8)^5} = \dfrac{10 \times 10^6}{(8)^5} = \mathbf{305.2\ W}$ **(Ans.)**

(ii) Ratio of heads $\left(\mathbf{\dfrac{H_p}{H_m}}\right)$ and ratio of mass flow rates $\left(\mathbf{\dfrac{m_p}{m_m}}\right)$:

We know that, $H \propto N^2D^2$ ... [Eq. (2.43)]

$\therefore$ $\dfrac{H_p}{H_m} = \left(\dfrac{N_p}{N_m}\right)^2 \times \left(\dfrac{D_p}{D_m}\right)^2 = (1)^2 \times (8)^2 = \mathbf{64}$ **(Ans.)**

Also, discharge, $$Q \propto ND^3 \quad \text{... [Eqn. (2.47)]}$$

$\therefore$ Ratio of mass flow rates, $$\frac{Q_p}{Q_m} = \frac{m_p}{m_m} = \left(\frac{N_p}{N_m}\right)\left(\frac{D_p}{D_m}\right)^3 = 1 \times (8)^3 = \mathbf{512} \text{ (Ans.)}$$

Example 2.52. *Obtain an expression for the specific speed of a hydraulic tubine and explain its significance. Give the range of speed values of the Kaplan, Francis tubines and Pelton wheels.* **[UPSC]**

Solution. Refer to Article 2.9 and Table 2.2.

Example 2.53. *A 1/5 scale model of a centrifugal pump absorbs 20 kW when pumping against a test head of 8 m at its best speed of 400 r.p.m. If the actual pump works against 32 m head, find the speed and power required for the actual pump. Determine also the quantities of water discharged by the two pumps.* **[UPSC]**

Solution. *Given* : $L_r = \dfrac{D_p}{D_m} = 5$; $P_m = 20$ kW; $H_m = 8$ m, $N_m = 4000$ r.p.m.;

$$H_p = 32 \text{ m.}$$

N_p, P_p; Q_p, Q_m :

Now, $$\frac{\sqrt{H_p}}{D_p N_p} = \frac{\sqrt{H_m}}{D_m N_m}$$ (where, suffix p stands for prototype and suffix m for model)

$$\therefore \quad N_p = N_m \times \frac{D_m}{D_p}\sqrt{\frac{H_p}{H_m}} = 400 \times \frac{1}{5} \times \sqrt{\frac{32}{8}} = \mathbf{160 \text{ r.p.m.}} \text{ (Ans.)}$$

Also, $$\frac{P_p}{D_p^5 N_p^3} = \frac{P_m}{D_m^5 N_m^3}$$

$$\therefore \quad P_p = \left(\frac{D_p}{D_m}\right)^5 \times \left(\frac{N_p}{N_m}\right)^3 \times P_m = (5)^5 \times \left(\frac{160}{400}\right)^3 \times 20 = \mathbf{4000 \text{ kW}} \text{ (Ans.)}$$

We know that $$P_m = w_m Q_m H_m$$

or, $$20 \times 20^3 = 9810 \times Q_m \times 8$$

$$\therefore \quad Q_m = \frac{20 \times 10^3}{9810 \times 8} = \mathbf{0.255 \text{ m}^3/\text{s}} \text{ (Ans.)}$$

Also, $$\frac{Q_p}{D_p^3 N_p} = \frac{Q_m}{D_m^3 N_m}$$

or, $$Q_p = \left(\frac{D_p}{D_m}\right)^3 \times \frac{N_p}{N_m} \times Q_m = (5)^3 \times \frac{160}{400} \times 0.255 = \mathbf{12.75 \text{ m}^3/\text{s}} \text{ (Ans.)}$$

Example 2.54. *A hydraulic turbine is to develop 1015 kW when running at 120 r.p.m. under a net head of 12 m. Work out the maximum flow rate and specific speed for the turbine if the overall efficiency at the best operating point is 92 per cent. In order to predict its performance, a 1 : 10 scale model is tested under a head of 7.2 m. What would be the speed, power output and water consumption of the model if it runs under the conditions similar to the prototype ?*

Solution. Shaft power, $P = 1015$ kW; Speed, $N = 120$ r.p.m.

Overall efficiency, $\eta_0 = 92$ %; Head, $H = 12$ m

Flow rate (Q), Specific speed (N_s) :

$$\eta_0 = \frac{\text{Shaft power}}{\text{Water power}} = \frac{P}{wQH}; \; Q = \frac{P}{\eta_0 wH}$$

or, $$\text{Flow rate, } Q = \frac{1015}{0.92 \times 9.81 \times 12} = \mathbf{9.372 \ m^3/s} \ \textbf{(Ans.)}$$

$$\text{Specific speed, } N_s = \frac{N\sqrt{P}}{H^{5/4}} = \frac{120\sqrt{1015}}{(12)^{5/4}} = \mathbf{171.2 \ r.p.m.} \ \textbf{(Ans.)}$$

Model scale = 1 : 10 (*Given*)

Head under which under model is tested, $H_m = 7.2$ m (*Given*)

N_m, P_m, Q_m :

For similar turbines each of the following parameters must be same for both model and prototype:

(*i*) Head co-efficient, $C_H = \dfrac{H}{N^2D^2}$;

(*ii*) Flow co-efficient, $C_Q = \dfrac{Q}{ND^3}$

(*iii*) Power co-efficient, $C_p = \dfrac{P}{N^3D^5}$

(*i*) $$\left(\frac{H}{N^2D^2}\right)_m = \left(\frac{H}{N^2D^2}\right)_p, \text{ or, } \frac{H_m}{N_m^2 D_m^2} = \frac{H_p}{N_p^2 D_p^2}, \text{ or, } N_m^2 = N_p^2 \frac{D_p^2}{D_m^2} \times \frac{H_m}{H_p}$$

$$\therefore \textit{Model speed, } N_m = N_p \times \frac{D_p}{D_m} \times \left(\frac{H_m}{H_p}\right)^{1/2} = 120 \times 10 \times \left(\frac{7.2}{12}\right)^{1/2} = \mathbf{929.5 \ r.p.m.} \ \textbf{(Ans.)}$$

(*ii*) $$\left(\frac{Q}{ND^3}\right)_m = \left(\frac{Q}{ND_3}\right)_p, \text{ or, } \frac{Q_m}{N_m D_m^3} = \frac{Q_p}{N_p D_p^3}$$

$\therefore$ *Discharge in the model,*

$$Q_m = Q_p \times \frac{N_m}{N_p} \times \left(\frac{D_m}{D_p}\right)^3 = 9.372 \times \frac{929.5}{120} \times \left(\frac{1}{10}\right)^3 = \mathbf{0.0726 \ m^3/s} \ \textbf{(Ans.)}$$

(*iii*) $$\left(\frac{P}{N^3D^5}\right)_m = \left(\frac{P}{N^3D^5}\right)_p, \text{ or, } \frac{P_m}{N_m^3 D_m^5} = \frac{P_p}{N_p^3 D_p^5}$$

$\therefore$ *Power produced by the model,*

$$P_m = P_p \times \left(\frac{N_m}{N_p}\right)^3 \times \left(\frac{D_m}{D_p}\right)^5 = 1015 \times \left(\frac{929.5}{120}\right)^3 \times \left(\frac{1}{10}\right)^5 = \mathbf{4.72 \ kW} \ \textbf{(Ans.)}$$

Example 2.55. *In a hydroelectric generating plant there are four similar turbines of total output 220000 kW. Each turbine is 90 per cent efficient and runs at 100 r.p.m. under a head of 65 m. It is proposed to test the model of the above turbine in a flume where discharge is 0.4 m³/s under a head of 4 m. Determine the size (scale ratio) of the model. Also calculate the model speed and power results expected from the model.* **[P.E.C.]**

Solution. Power available from each turbine $= \dfrac{220000}{4} = 55000$ kW

Efficiency (overall) of each turbine, $\eta_0 = 90\%$

Speed of the turbine, $N = 100$ r.p.m.

Head, $H = 65$ m

Discharge through the model, $Q_m = 0.4\ \text{m}^3/\text{s}$

Head under which model is tested, $H_m = 4$ m

Scale ratio :

Using the relation, $P = \eta_0 \times wQH$, or, $Q = \dfrac{P}{\eta_0 wH} = \dfrac{55000}{0.9 \times 9.81 \times 65} = 95.84\ \text{m}^3/\text{s}$

For similar turbines the following dimensionless parameters must be same for model and prototype :

(*i*) Head co-efficient, $C_H = \dfrac{H}{N^2D^2}$; (*ii*) Flow co-efficient, $C_Q = \dfrac{Q}{ND^3}$;

(*iii*) Power co-efficient, $C_P = \dfrac{P}{N^3D^5}$

From the head co-efficient and flow co-efficient it follows that:

$$\frac{Q}{D^2\sqrt{H}} = \text{Constant}; \left(\frac{Q}{D^2\sqrt{H}}\right)_m = \left(\frac{Q}{D^2\sqrt{H}}\right)_p, \text{ or, } \frac{Q_m}{D_m^2\sqrt{H_m}} = \frac{Q_p}{D_p^2\sqrt{H_p}}$$

∴ **Scale ratio :**

$$\frac{D_m}{D_p} = \left[\left(\frac{Q_m}{Q_p}\right)\left(\frac{H_p}{H_m}\right)^{1/2}\right]^{1/2} = \left[\left(\frac{0.4}{95.84}\right)\left(\frac{65}{4}\right)^{1/2}\right]^{1/2} = 0.1297 \text{ or } \mathbf{1 : 7{\cdot}71} \textbf{ (Ans.)}$$

Speed of the model, N_m :

$$\left(\frac{H}{N^2D^2}\right)_m = \left(\frac{H}{N^2D^2}\right)_p, \text{ or, } \frac{H_m}{N_m^2D_m^2} = \frac{H_p}{N_p^2D_p^2}, \text{ or, } \frac{N_m^2}{N_p^2} = \frac{H_m}{H_p} \times \frac{D_p^2}{D_m^2}$$

$$\therefore \quad N_m = N_p\sqrt{\frac{H_m}{H_p} \times \frac{D_p}{D_m}} = 100\sqrt{\frac{4}{65}} \times \left(\frac{1}{0.1297}\right) = \mathbf{191.26 \text{ r.p.m.}}$$

Power available from the model P_m :

$$\left(\frac{P}{N^3D^5}\right)_m = \left(\frac{P}{N^3D^5}\right)_p, \text{ or, } \frac{P_m}{N_m^3D_m^5} = \frac{P_p}{N_p^3D_p^5}$$

$$P_m = P_p \times \left(\frac{N_m}{N_p}\right)^3 \times \left(\frac{D_m}{D_p}\right)^5 = 55000\left(\frac{191.26}{100}\right)^3 \times (0.1297)^5 = \mathbf{14.12 \text{ kW}}$$

Example 2.56. *(a) Prove that specific speed can be expressed as* $N_s = 3.13 N_u \sqrt{Q_u \eta_0}$, *where,* N_u = *unit speed,* Q_u = *unit discharge, and* η_0 = *overall efficiency.*

(b) A Kaplan turbine working under a head of 10 m and at a design speed of 250 r.p.m. has a flow rate of 24 m^3*/s. The diameters of the runner and boss/hub are 2 m and 1 m respectively. The inlet and outlet diameters of the draft tube are 2 m and 3 m respectively. The pressure recorded at inlet to the draft tube is 3 m vacuum. The vapour and barometric pressures are 1·6 m and 10 m respectively. The efficiency of the draft tube is 80 per cent. The Thoma's cavitation factor for the turbine is given by the relation,* $\sigma = \eta_d K_f^2 + \lambda K_u^2$, *in which* η_d *is the efficiency of the draft tube,* K_f *is the flow ratio,* K_u *is the speed ratio and* λ *is a dimensionless factor defined by* $\lambda = (p_2/w - p_{min}/w)/(u^2/2g)$ *in which u is the tangential velocity,* (p_2/w) *is the pressure head at inlet to draft tube, and* (p_{min}/w) *is the minimum pressure head at a point on the blade. If overall efficiency is 90 per cent determine :*

(i) The minimum pressure on the blade;

(ii) The value of N_s.

Solution. *(a)* Specific speed (N_s) is given by : $N_s = \dfrac{N\sqrt{P}}{H^{5/4}}$ ...(*i*)

Also power output (shaft power), $P = wQH \times \eta_0 = 9.81\, QH\eta_0$ kW

Substituting the value of P in eqn. (*i*), we get:

$$N_s = \frac{N}{H^{5/4}}\sqrt{9.81\, QH\eta_0} = 3.13 \frac{N}{H^{5/4}}\sqrt{QH\eta_0}$$

$$= 3.13 \times \frac{N}{\sqrt{H}} \times \sqrt{\frac{Q}{\sqrt{H}} \times \eta_0} = 3.13\, N_u \sqrt{Q_u \eta_0} \quad \textbf{...(Proved)}$$

$$\left(\because N_u = \frac{N}{\sqrt{H}},\ Q_u = \frac{Q}{\sqrt{H}}\right)$$

(b) Head under which the turbine is working, H = 10 m

Speed of the turbine runner, N = 250 r.p.m.

Flow rate, Q = 24 m^3/s

Runner Diameter, D_0 = 2 m

Diameter of the boss/hub, D_b = 1 m

Inlet diameter of the draft tube, d_i = 2 m

Outlet diameter of the tube, d_0 = 3 m

Vapour pressure, H_v = 1.6 m

Barometric pressure, $\dfrac{p_a}{w}$ = 10 m

The pressure head inlet to the draft tube, $\dfrac{p_2}{w}$ = 3 m vacuum

Efficiency of draft tube, η_d = 80%

Overall efficiency, η_0 = 90%

(*i*) The minimum pressure on the blade, p_{min} :

Discharge, $$Q = \frac{\pi}{4}(D_0^2 - D_b^2) \times V_f = \frac{\pi}{4}(D_0^2 - D_b^2)\, K_f \sqrt{2gH}$$

or, $$= \frac{\pi}{4}(2^2 - 1^2) \times K_f \times \sqrt{2 \times 9.81 \times 10} = 33.0 K_f$$

$$\therefore \quad K_f = \frac{24}{33.0} = 0.727$$

Also, $$u = \frac{\pi D_0 N}{60} = K_u \sqrt{2gH}, \text{ or, } \frac{\pi \times 2 \times 250}{60} = K_u \times \sqrt{2 \times 9.81 \times 10}$$

$$\therefore \quad K_u = \frac{\pi \times 2 \times 250}{60\sqrt{2 \times 9.81 \times 10}} = 1.87$$

Also, $$u = \frac{\pi D_0 N}{60} = \frac{\pi \times 2 \times 250}{60} = 26.18 \text{ m/s}$$

Using the Eqn. (2.32), we have:

$$\frac{p_2}{w} = \frac{p_a}{w} - H_s - \left(\frac{V_2^2 - V_3^2}{2g} - h_f\right)$$

But h_f (head lost in draft tube) $= \dfrac{V_2^2 - V_3^2}{2g} - \eta_d \dfrac{V_2^2}{2g}$ ...[Eqn. (2.33 (*a*)]

$$\therefore \quad \frac{p_2}{w} = \frac{p_a}{w} - H_s - \left[\frac{V_2^2 - V_3^2}{2g} - \left(\frac{V_2^2 - V_3^2}{2g} - \eta_d \frac{V_2^2}{2g}\right)\right] \quad \text{(By substitution)}$$

or, $$\frac{p_2}{w} = \frac{p_a}{w} - H_s - \eta_d \frac{V_2^2}{2g}$$

where, $$\frac{p_2}{w} = -3 + 10 = 7 \text{ m absolute, and}$$

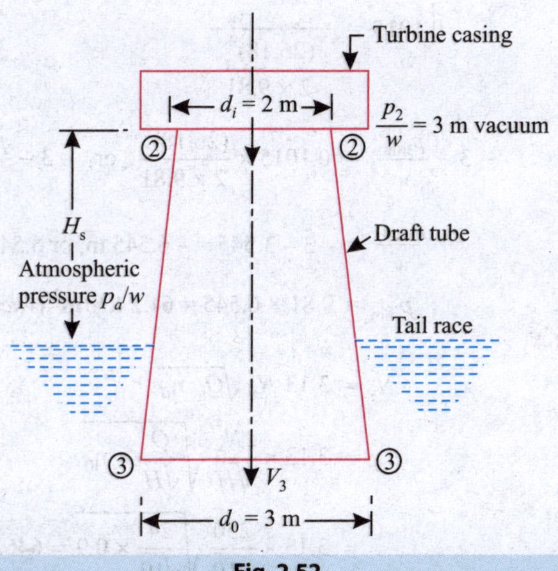

Fig. 2.52

$$V_2 = \frac{Q}{\frac{\pi}{4} d_i^2} = \frac{24}{\frac{\pi}{4} \times 2^2} = 7.64 \text{ m/s}$$

By substituting these values in the above equation, we get:

$$7 = 10 - H_s - 0.8 \times \frac{7.64^2}{2 \times 9.81}$$

$$\therefore \quad H_s = 10 - 7 - 0.8 \times \frac{7.64^2}{2 \times 9.81} = 0.62 \text{ m}$$

The cavitation factor (σ) is given by:

$$\sigma = \frac{H_a - H_v - H_s}{H} \quad \text{...[Eqn. (2.48)]}$$

By substituting the values, we get:

$$\sigma = \frac{10 - 1.6 - 0.62}{10} = 0.778$$

Also, $\quad \sigma = \eta_d K_f^2 + \lambda K_u^2 \quad$ ...(*Given*)

Thus, by substituting the values, we obtain:

$$0.778 = 0.8 \times 0.727^2 + \lambda \times 1.87^2$$

or, $\quad 0.778 = 0.423 + 3.497\lambda$

$$\therefore \quad \lambda = \frac{0.778 - 0.423}{3.497} = 0.1015$$

But, $$\lambda = \frac{\frac{p_2}{w} - \frac{p_{min}}{w}}{\left(\frac{u^2}{2g}\right)} \quad \text{...(Given)}$$

or, $$0.1015 = \frac{-3 - \frac{p_{min}}{w}}{\frac{(26.18)^2}{2 \times 9.81}}$$

or, $$-3 - \frac{p_{min}}{w} = 0.1015 \times \frac{(26.18)^2}{2 \times 9.81}, \text{ or, } -3 - \frac{p_{min}}{w} = 3.545$$

or, $$\frac{p_{min}}{w} = -3 - 3.545 = -6.545 \text{ m, or } 6.545 \text{ m (vacuum)}$$

or, $\quad p_{min} = 9.81 \times 6.545 =$ **64.2 kN/m²** (*vacuum*) **(Ans.)**

(*ii*) The value of N_s :

$$N_s = 3.13\, N_u \sqrt{Q_u}\, \eta_0 \quad \text{...[Proved at (a)]}$$

$$= 3.13 \times \frac{N}{\sqrt{H}} \sqrt{\frac{Q}{\sqrt{H}}} \times \eta_0$$

$$= 3.13 \times \frac{250}{\sqrt{10}} \sqrt{\frac{24}{\sqrt{10}}} \times 0.9 = \mathbf{646.7} \text{ (Ans.)}$$

2.12. SCALE EFFECT

However smooth a model is made, the geometric similarity between the prototype and model cannot be extended to surface roughness. This variation of surfaceness with respect to the size of turbine will cause a *small but appreciable variation in the proportion of the effective head lost due to hydraulic friction*. Thus, the *efficiency of prototype will be different from the corresponding model efficiency*. This aspect is referred to as **scale effect**. It has been observed that with increase in size a geometrically similar mixed or axial flow turbine has greater efficiency than that of the model operating under hydraulically similar conditions.

In order to express the difference of efficiencies as found in tests of model and prototypes, various laws have been proposed. One of the earliest and most generally accepted is the semi-empirical formula suggested by Moody;

$$\frac{1-\eta_p}{1-\eta_m} = \left(\frac{D_m}{D_p}\right)^{0.2} \quad ...(2.46)$$

Ackert suggested the following formula, *considering the frictional loss as a function of Reynolds number*,

$$\frac{1-\eta_p}{1-\eta_m} = \frac{1}{2}\left[1+\left(\frac{D_m}{D_p}\right)^{0.2}\left(\frac{H_m}{H_p}\right)^{01}\right] \quad ...(2.47)$$

where, η_p = Overall efficiency of the prototype,

D_p = Linear dimension of the prototype,

H_p = Head of the prototype,

and, η_m, D_m, H_m = Corresponding values of overall efficiency, linear dimension and head of the model.

These formulae are applicable for the *point of best efficiency*.

Note : (*i*) Moody's formula is based on the assumptions that all losses are due to fluid friction only and the flow is completely turbulent (and that eliminates any effect of Reynolds number).

(*ii*) No scale effect has been observed in the case of Pelton wheel; this may be due to the deterioration in the smoothness of the jet with increasing size, which offsets the benefits due to reduced frictional losses.

Example 2.57. *A model turbine constructed to a scale of 1:10 when tested under a head of 8 m at 400 r.p.m. gave an efficiency of 77 per cent. Determine the r.p.m. of the prototype and the ratio of powers developed by the model and prototype if the prototype works under a head of 100 m. What will be the efficiency of the prototype if scale effect is considered ?*

[Anna University]

Solution. *Given* : Scale ratio = 1 : 10; $\frac{D_p}{D_m} = 10$; $H_m = 8$ m; $N_m = 400$ r.p.m.; $\eta_m = 77\%$; $H_p = 100$ m.

Speed of the prototype, N_p:

For similarity between model and prototype :

$$\left(\frac{H}{N^2D^2}\right)_m = \left(\frac{H}{N^2D^2}\right)_p, \text{ or, } \frac{H_m}{N_m^2D_m^2} = \frac{H_p}{N_p^2D_p^2}, \text{ or, } N_p^2 = N_m^2 \times \frac{D_m^2}{D_p^2} \times \frac{H_p}{H_m}$$

$$\therefore \quad N_p = N_m \times \frac{D_m}{D_p} \times \left(\frac{H_p}{H_m}\right)^{1/2} = 400 \times \frac{1}{10} \times \left(\frac{100}{8}\right)^{1/2} = \textbf{141.42 r.p.m. (Ans.)}$$

Ratio of powers developed, $\frac{P_p}{P_m}$:

$$\therefore \quad \left(\frac{P}{N^3 D^5}\right)_m = \left(\frac{P}{N^3 D^5}\right)_p, \text{ or, } \frac{P_m}{N_m^3 D_m^5} = \frac{P_p}{N_p^3 D_p^5}$$

$$\therefore \quad \frac{P_p}{P_m} = \left(\frac{N_p}{N_m}\right)^3 \times \left(\frac{D_p}{D_m}\right)^5 = \left(\frac{141.42}{400}\right)^3 \times (10)^5 = \textbf{4419.3 (Ans.)}$$

Efficiency of the prototype when scale effect is considered :

The efficiencies are related by:

$$\frac{N_p}{N_m} = \frac{D_m}{D_p}\sqrt{\frac{H_p}{H_m}}\sqrt{\frac{\eta_p}{\eta_m}}$$

$$\frac{\eta_p}{\eta_m} = \left(\frac{N_p}{N_m}\right)^2 \left(\frac{D_p}{D_m}\right)^2 \frac{H_m}{H_p} = \left(\frac{141.42}{400}\right)^2 (10)^2 \times \frac{8}{100} = 0.9999$$

$$\therefore \quad \eta_p = \eta_m \times 0.9999 = 0.77 \times 0.9999 = 0.7699 \text{ or } \textbf{76.99\% (Ans.)}$$

2.13. PERFORMANCE CHARACTERISTICS OF HYDRAULIC TURBINES

The turbines are normally designed for specific values of head, speed, discharge, power and efficiency (known as the *designed conditions*). But oftenly turbines may be required to operate under conditions different from those for which these have been designed. Thus, to know about their exact behaviour under varying conditions it becomes necessary to conduct tests either on the actual turbines at the site or on their small scale models in a research laboratory. The results so obtained are usually represented graphically and the curves obtained are known as "*Characteristic curves*". These curves are usually plotted in terms of unit quantities (for sake of convenience). The characteristic curves are of the following types :

1. Main or constant head characteristic curves.
2. Operating or constant speed characteristic curves.
3. Constant efficiency or iso-efficiency or Muschel curves.

2.13.1. Main or Constant Head Characteristic Curves

- *Head and gate opening are maintained constant.*
- *Speed is varied* by allowing a variable quantity of water to flow through the inlet opening.
- The brake power (P) is then measured mechanically by means of a dynamometer.
- The overall efficiency and unit quantities are then calculated by using the basic data; these are then plotted *against unit speed as abscissa.*

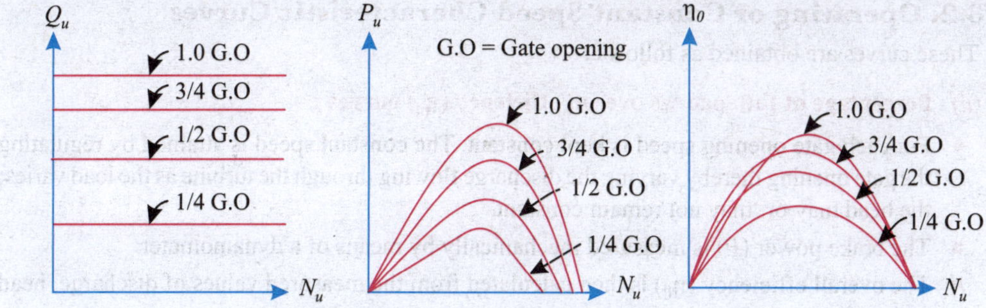

Fig. 2.53. Main characteristic curves of Pelton wheel.

Figs. 2.53, 2.54 and 2.55 show the main characteristic curves of Pelton wheel, Francis turbine and Kaplan turbine respectively.

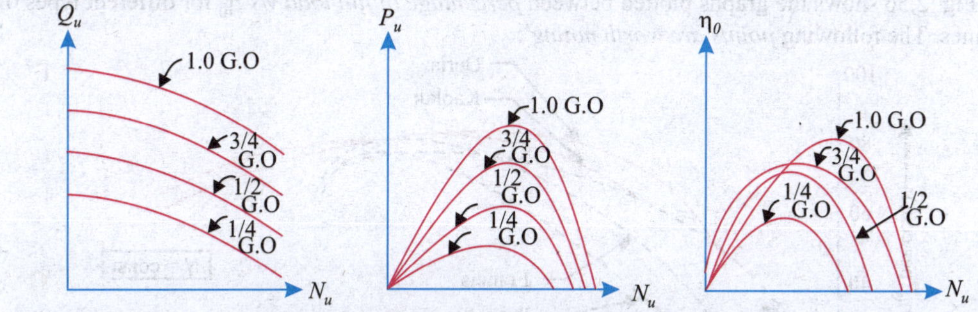

Fig. 2.54. Main characteristic curves of Francis turbine.

The main characteristic curves yield the following information :

- The discharge Q_u for a *Pelton wheel* depends only upon the gate opening and is independent of N_u; the curves for Q_u are horizontal.
- The curves between Q_u and N_u for a *Francis turbine are falling curves*. This is due to the fact that a *centrifugal head develops which* acts outwards and *opposes* the external head causing flow, eventually decreasing the discharge as the speed increases.
- The curves between Q_u and N_u for a *Kaplan turbine* are *rising curves*; the discharge increases with the increase in speed.

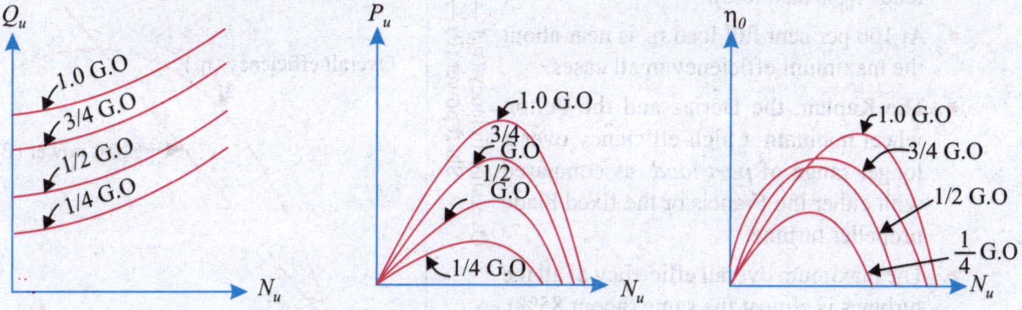

Fig. 2.55. Main characteristic curves of Kaplan turbine.

- The curves between P_u and N_u and those between η_0 and N_u indicate that at a particular speed the efficiency is maximum.

The maximum efficiency for a *Pelton wheel* usually occurs at the *same speed for all gate openings; this speed usually corresponds to a speed ratio of 0.45 . However, the maximum efficiency for a reaction turbine usually occurs at different speeds for different gate openings.*

2.13.2. Operating or Constant Speed Characteristic Curves

These curves are obtained as follows :

(a) Percentage of full load v/s overall efficiency (η_0) curves :

- For each gate opening speed is kept constant. The constant speed is attained by regulating the gate opening thereby varying the discharge flowing through the turbine as the load varies; the head may or may not remain constant.
- The brake power (P) is measured mechanically by means of a dynamometer.
- The overall efficiency (η_0) is then calculated from the measured values of discharge, head and power.
- Further knowing the total load capacity of the turbine the percentage of full load is computed from the measured power and a plot of η_0 *v/s* percentage of full load is prepared.

Fig. 2.56 shows the graphs plotted between *percentage of full load v/s* η_0 for different types of turbines. The following *points are worth noting* :

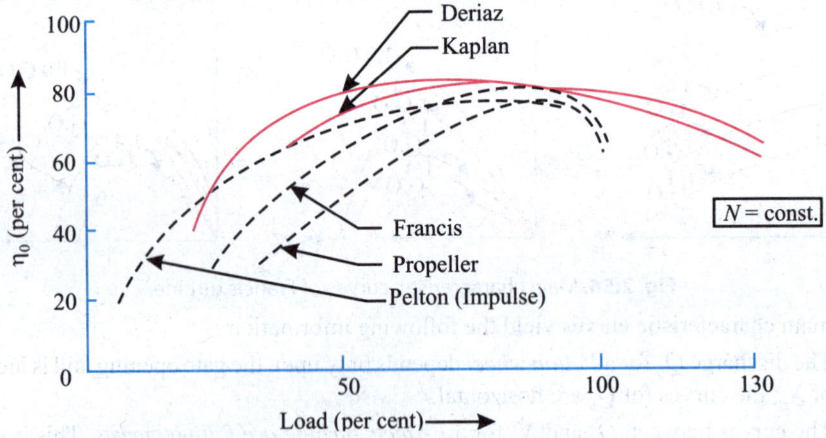

Fig. 2.56. Percentage of full load v/s η_0 curves for hydraulic turbines.

- As the percentage full load increases η_0 also increases (In other words, at reduced loads η_0 is also less).
- At 100 per cent full load η_0 is near about the maximum efficiency in all cases.
- The Kaplan, the Deriaz and the Pelton wheel maintain a high efficiency over a longer range of *part load* as compared with either the Francis or the fixed blade propeller turbine.
- The maximum overall efficiency of all the turbines is almost the same (about 85%).

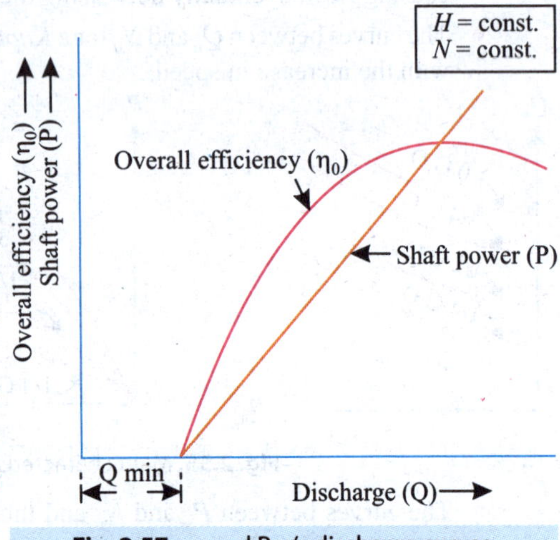

Fig. 2.57. η_0 and P v/s discharge curves.

(b) Overall efficiency (η_0) and output (shaft) power (P) v/s discharge (Q) curves:

Fig. 2.57 shows overall efficiency (η_0) and shaft power (P) *v/s* discharge curves. Q_{min} is the minimum discharge required to set the turbine runner into motion from its state of rest. These curves yield the following information :

- Shaft power or output power (P) is a straight line, since $P \propto Q$ if H (head) is constant.
- η_0 *v/s* discharge (Q) graph is curvilinear and η_0 increases with Q and remains *nearly* constant beyond a particular value of discharge.

2.13.3. Constant efficiency or iso-efficiency or Muschel curves

Refer to Fig. 2.58. As η–N curve is of parabolic nature, there exits two speeds for one value of efficiency except for maximum efficiency which occurs at one speed only. Corresponding to these values of speeds there are also two values of discharge for each value of efficiency (Q-N curve). Hence on Q-N curve we can plot two points for each value of efficiency and one point for maximum efficiency. By adopting this procedure for different gate openings or heads we can get number of Q-N curves and we can plot on them efficiency points (as described above). The points denoting the same efficiency can now be joined to get constant iso-efficiency curves or Muschel curves (The German word 'Muschel' means shell, indicating shape of curve). The diagram showing these curves is also called Hill diagram (since it looks like top view of a hill). In actual practice unit speed and unit discharge are taken along the co-ordinate axes.

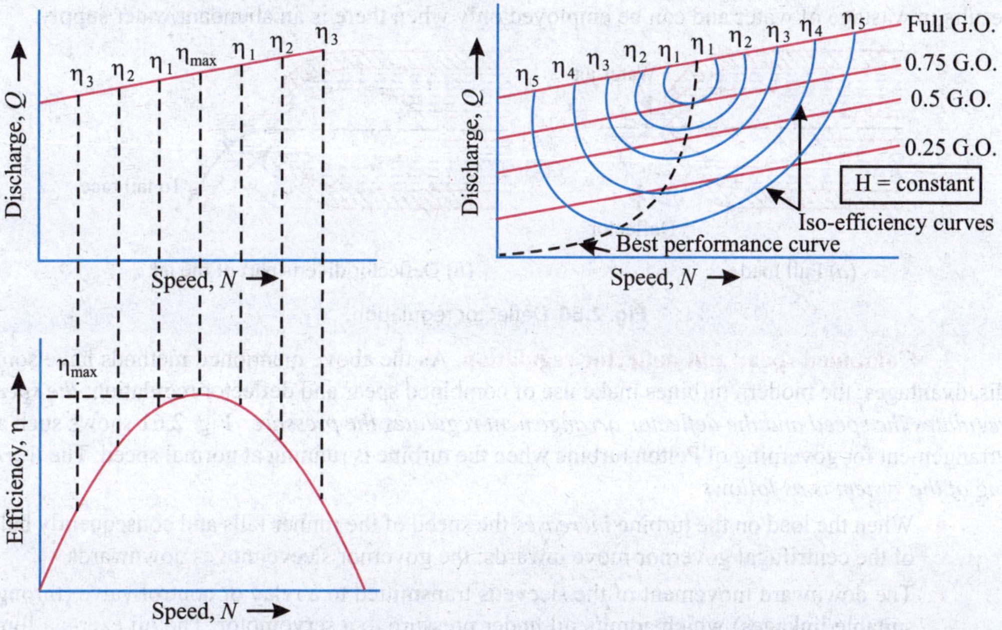

Fig. 2.58. Constant efficiency curves for turbines.

The *curve for the best performance is obtained by joining the peak points of the various efficiency curves.*

The constant efficiency curves are helpful for determining the zone of constant efficiency and for predicting the performance of the turbine at various efficiencies.

2.14. GOVERNING OF HYDRAULIC TURBINES

Governing of hydraulic turbine means *speed regulation*. Governing of a turbine is necessary as a turbine is directly coupled to an electric generator, which is required to run at constant speed under all fluctuating load conditions. This is achieved by means of a *governor* called oil pressure governor.

2.14.1. Governing of Impulse Turbines

In order to regulate the quantity of water rejected from the turbine nozzle and from striking the buckets one of the following methods of regulation may be adapted :

1. Spear regulation. 2. Deflector regulation. 3. Combined spear and deflector regulation.

1. Spear regulation. Refer to Fig. 2.59. In this method the rate of flow is regulated by altering the cross-sectional area of stream by moving the spear to and from inside nozzle. This method of speed regulation is suitable when the *fluctuation of load is small and a relatively large penstock feeds a small turbine.* The disadvantages of this method is that when the load falls all of sudden, the turbine nozzle has to close suddenly which may cause water *hammer* in the *penstock.*

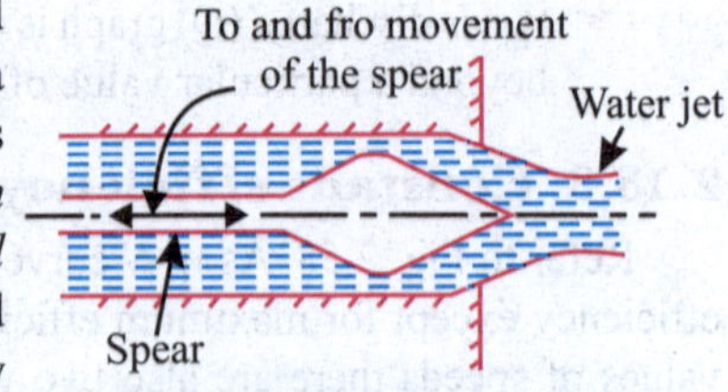

Fig. 2.59. Spear regulation in Pelton wheel.

2. Deflector regulation. Refer to Fig. 2.60. The deflector is generally a plate connected to the oil pressure governor by means of levers. When necessity arises to deflect the jet, the plate can be brought in between the nozzle and buckets, thereby diverting the water away from the runner and directing into the tail race. The use of deflector regulation is restored to *when the supply of water is constant* but the *load fluctuates*. The position of spear can be adjusted by hand. As the nozzle has always a constant opening, it results in wastage of water and can be employed only when there is an abundant water supply.

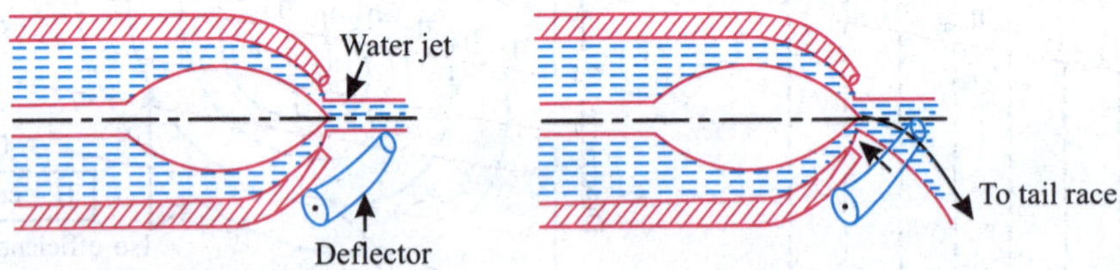

(*a*) Full load (*b*) Deflector diverts part of the jet.

Fig. 2.60. Deflector regulation.

3. Combined spear and deflector regulation. As the above mentioned methods have some disadvantages, the modern turbines make use of combined spear and deflector regulation; *the spear regulates the speed and the deflector arrangement regulates the pressure*. Fig. 2.61 shows such an arrangement for governing of Pelton turbine when the turbine is running at normal speed. The *working of the system is as follows* :

- When the load on the turbine *increases* the speed of the runner falls and consequently balls of the centrifugal governor move inwards; the governor sleeve moves downwards.
- The downward movement of the sleeve is transmitted to a relay or control valve (through suitable linkages) which admits oil under pressure to a servomotor. The oil exerts a force on the piston of the servomotor, and that pushes the spear to a position which *increases* the annular area of the nozzle flow passage; the quantity of water striking the buckets is then increased and the turbine regains its normal speed.
- When the load on the turbine *decreases* the direction of movement of the servomotor is such the nozzle area *decreases* and that allows a smaller quantity of water to strike the runner of the turbine.
- A deflector arrangement safeguards against excessive water hammer pressure.

2.14.2. Governing of Reaction Turbines

In a reaction turbine the discharge is controlled *by varying the area of flow between adjacent guide vanes*. The guide vanes are connected to the regulating ring through links. The regulating ring is connected to the regulating lever through two regulating rods. The regulating ring is thus connected to the regulating shaft which is operated by a servomotor (Fig. 2.62). The servomotor, oil sump, control valve and system of pipes, etc. are similar to that in the governing arrangement of an impulse turbine. The component parts are, however, stronger as the greater energy is required to move the gates as compared to the spear in the nozzle of a Pelton turbine.

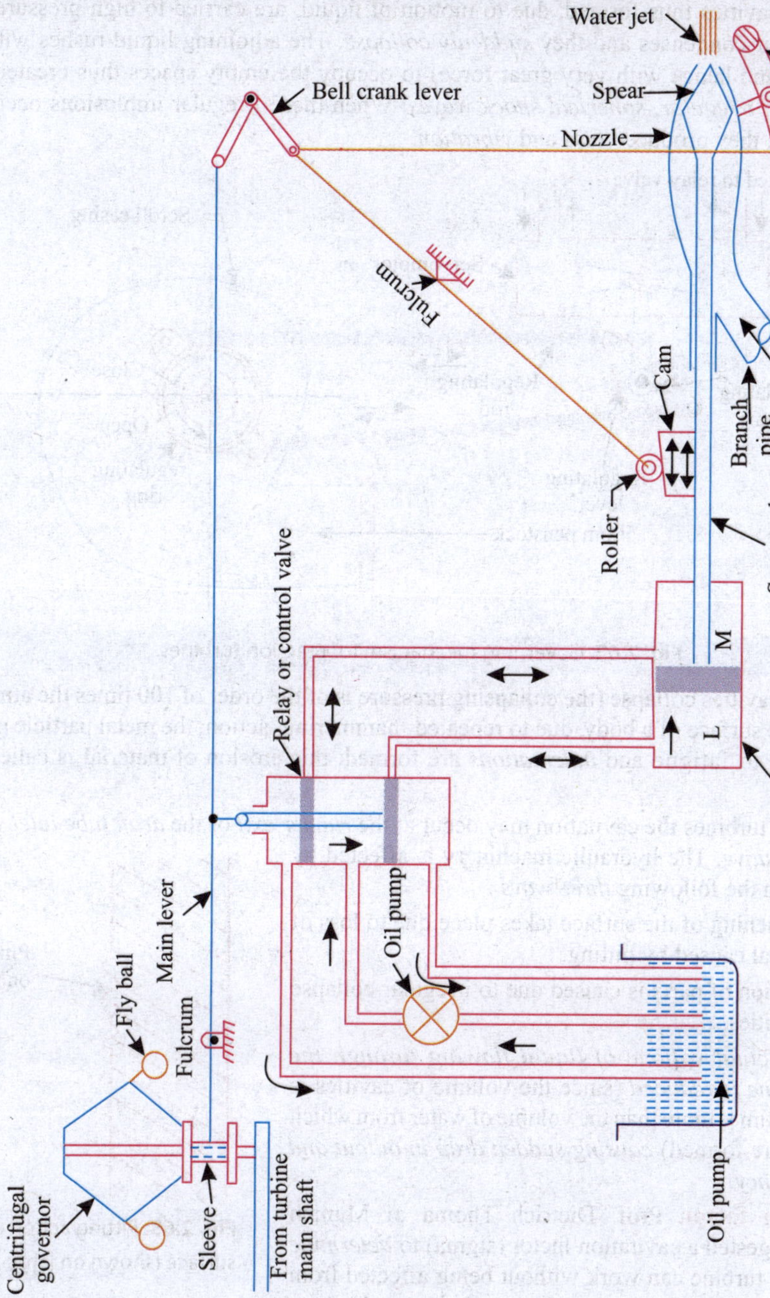

Fig. 2.61

2.15. CAVITATION

The formation, growth, and collapse of vapour filled cavities or bubbles in a flowing liquid due to local fall in fluid pressure is called **cavitation**. When the pressure at any point in a flow field equals the vapour pressure of the liquid at that temperature vapour cavities (bubbles of vapour) begin to appear. It is presumed that a vapour cavity is formed around a dust nuclei which is in the liquid (The vapour pressure values of water at 15° and 20° C are 1.74 m and 2.38 m of water column absolute). The cavities thus formed, due to motion of liquid, are carried to high pressure regions where the vapour condenses and they *suddenly collapse*. The adjoining liquid rushes with a very great velocity (and hence with very great force) to occupy the empty spaces thus created, *causes series of violent, irregular, spherical shock waves*. When these irregular implosions occur on the metallic surface, they produce *noise* and *vibration*.

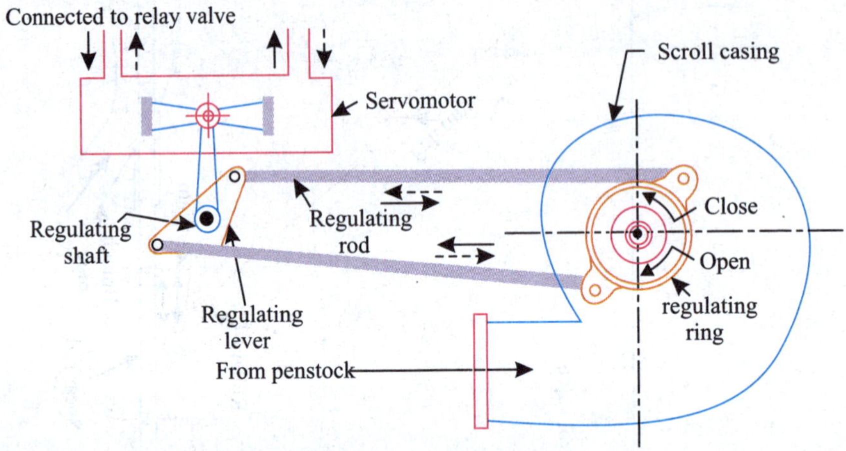

Fig. 2.62. Governing mechanism for reaction turbines.

When the cavities collapse (the collapsing pressure is of the order of 100 times the atmospheric pressure) on the surface of a body, due to repeated 'hammering' action, the metal particle gives way ultimately due to fatigue and *indentations* are formed; this erosion of material is called **pitting** (Fig. 2.63).

In reaction turbines the cavitation may occur at the *runner exit* or the *draft tube inlet* where the *pressure is negative*. The hydraulic machinery is affected by the cavitation in the following *three ways* :

1. Roughening of the surface takes place due to loss of material caused by pitting.
2. Vibration of parts is caused due to irregular collapse of cavities.
3. The *actual volume of liquid flowing through the machine is reduced* (since the volume of cavities is many times more than the volume of water from which they are formed) *causing sudden drop in output and efficiency*.

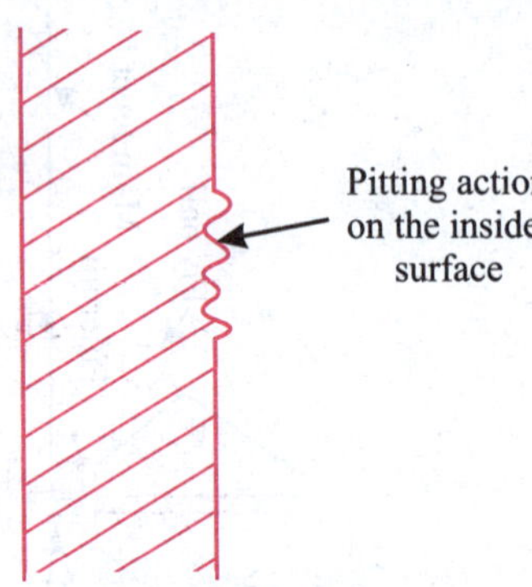

Fig. 2.63. Pitting action on the surface (shown on large scale).

Cavitation factor. Prof. Dietrich Thoma of Munich (Germany) suggested a cavitation factor (sigma) to *determine the zone* where turbine can work without being affected from cavitation. The *critical* value of cavitation factor (σ_c) is given by,

$$\sigma_c = \frac{(H_a - H_v) - H_s}{H} \quad ...(2.48)$$

where, H_a = Atmospheric pressure head in metres of water,

H_v = Vapour pressure in metres of water corresponding to the water temperature,

H = Working head of turbine (difference between head race and tail race level in metres), and

H_s = Suction pressure head (or height of turbine outlet above tail race level in metres).

The values of critical factor *depends upon the specific speed of the turbine.*

The value for σ_c for different materials may be determined with the help of the following empirical relations :

For Francis turbine : $$\sigma_c = 0.625\left(\frac{N_s}{380.78}\right)^2 \quad ...(2.49)$$

For propeller turbine : $$\sigma_c = 0.28 + \left[\frac{1}{7.5}\left(\frac{N_s}{380.78}\right)^3\right] \quad ...(2.50)$$

For Kaplan turbines, values of σ_c obtained by eqn. (2.50) should be *increased by* 10%.

(In the above expressions N_s is in (r.p.m., kW, m) units.

Suction specific speed $(N_s)_{suc.}$: In addition to Thoma's criterion the consideration of suction specific speed provides very useful criterion for *establishing similarity in respect of cavitation in the turbines. The* **suction speed** *may be defined as the speed of a geometrically similar turbine such that when it is developing a power equal to 1 kW, the total suction head H_{sv}* is equal to 1 m (absolute units). It can be proved that specific speed is given by:

$$(N_s)_{suc.} = \frac{N\sqrt{P}}{(\sigma H)^{4/5}} \quad ...(2.51(a))$$

$$\sigma = \left[\frac{N_s}{(N_s)_{suc.}}\right]^{4/5} \quad ...(2.51(b))$$

The eqns. 2.51(a) and (b) give the relation between the two parameters σ and $(N_s)_{suc.}$, both of which are useful for establishing a similarity in respect of cavitation in the model and prototype turbines. The *concept of suction speed, however, is more commonly used in pumps.*

Methods to avoid cavitation :

The following methods may be used to *avoid cavitation* :

1. Runner/turbine may be *kept under water*. But it is not advisable as the inspection and repair of the turbine is difficult. The other method to avoid cavitation zone without keeping the runner under water is *to use the runner of low specific speed.*
2. The *cavitation free runner* may be designed to fulfil the given conditions with extensive research.
3. It is possible to reduce the cavitation effect by *selecting materials which resist better the cavitation effect*. The cast steel is better than cast iron and stainless steel or alloy steel is still better than cast steel.
4. The cavitation effect can be reduced by *polishing* the surface. That is why the cast steel runners and blades are coated with stainless steel.
5. The cavitation may be avoided by selecting a runner of proper specific speed for given head.

Example 2.58. *A Francis turbine works under a head of 25 m and produces 11800 kW while running at 120 r.p.m. The turbine has been installed at a station where atmospheric pressure is 10 m of water and vapour pressure is 0·2 m of water. Calculate the maximum height of the straight draft tube for the turbine.*

Solution. Head under which the turbine works, $H = 25$ m

Power output, $P = 11800$ kW

Speed of the turbine, $N = 120$ r.p.m.

Atmospheric pressure, $p_a = 10$ m of water

Vapour pressure, $H_v = 0.2$ m.

Maximum height of the draft tube, H_s :

$$\text{Specific speed, } N_s = \frac{N\sqrt{P}}{H^{5/4}} = \frac{120\sqrt{11800}}{(25)^{5/4}} = 233.2 \text{ r.p.m.}$$

Critical value of Thoma's cavitation factor for a Francis turbine

$$\sigma_c = 0.625\left(\frac{N_s}{380.78}\right)^2 \qquad \text{...[Eqn. (2.49)]}$$

$$= 0.625\left(\frac{233.2}{380.78}\right)^2 = 0.2344$$

Also,
$$\sigma_c = \frac{H_c - H_v - H_s}{H} \qquad \text{...By definition}$$

or,
$$0.2344 = \frac{10 - 0.2 - H_s}{25}, \text{ or, } 0.2344 \times 25 = 10 - 0.2 - H_s$$

$$\therefore \quad H_s = 10 - 0.2 - 0.2344 \times 25 = 3.94 \text{ m}$$

Hence, *maximum permissible height of the draft tube* = **3.94 m (Ans.)**

2.16. SELECTION OF HYDRAULIC TURBINES

The following points should be considered while selecting right type of hydraulic turbines for hydroelectric power plant :

1. **Specific speed.** High specific speed is essential where head is low and output is large, because otherwise the rotational speed will be low which means cost of turbo-generator and power- house will be high. On the other hand, there is practically no need of chosing a high value of specific speed for high installations, because even with low specific speed high rotational speed can be attained with medium capacity plants. Refer to Table 2.2.

2. **Rotational speed.** Rotational speed depends on specific speed. Also the rotational speed of an electrical generator with which the turbine is to be directly coupled, depends on the frequency and number of pair of poles. The *value of specific speed adopted should be such that it will give the synchronous speed of the generator.*

3. **Efficiency.** The turbine selected should be such that it gives the *highest overall efficiency for various operating conditions.*

4. **Partload operation.** In general the efficiency at partloads and overloads is less than normal. For the sake of economy the turbine should always run with maximum possible efficiency to get more revenue.

When the turbine has to run at part or overload conditions *Deriaz turbine* is employed. Similarly, for low heads, Kaplan turbine will be useful for such purposes in place of propeller turbine.

5. **Cavitation.** The installation of water turbines of reaction type over the tail race is affected by *cavitation*. The critical value of cavitation factor must be obtained to see that the turbine works in *safe zone*. Such a value of cavitation factor also affects the design of turbine, especially of Kaplan, propeller and bulb types.

6. Disposition of turbine shaft. Experience has shown that the *vertical shaft* arrangement is better for large-sized reaction turbines, therefore, it is *almost universally adopted.* In case of *large size impulse turbines*, *horizontal shaft arrangement* is mostly employed.

7. Head. (*i*) *Very high heads* (*350 m and above*). For heads greater than 350 m, Pelton turbine is generally employed and there is practically no choice except in very special cases.

(*ii*) *High heads* (*150 m to 350 m*). In this range either Pelton or Francis turbine may be employed. *For higher specific speeds Francis turbine is more compact and economical than the Pelton turbine* which for the same working conditions would have to be much bigger and rather cumbersome.

(*iii*) *Medium heads* (*60 m to 150 m*). A Francis turbine is usually employed in this range. Whether a high or low specific speed unit would be used depends on the selection of the speed.

(*iv*) *Low heads* (*below 60 m*). Between 30 and 60 m heads both Francis and Kaplan turbines may be used. The latter is more expensive but yields a higher efficiency at partloads and overloads. It is therefore preferable for *variable loads*. *Kaplan turbine is generally employed for heads under 30 m. Propeller turbines are however, commonly used for heads up to 15 m. They are adopted only when there is practically no load variations.*

(*v*) *Very low heads.* For very low heads bulb turbines are employed these days. Although Kaplan turbines can also be used for heads from 2 m to 15 m but they are *not economical.*

Table 2.2. Criteria for Selection of Turbines

S. No.	Type of turbine	Head H(m)	Specific speed (N_s)	Speed ratio (K_u)	Maximum hydraulic efficiency (%)	Remarks
1.	*Pelton*: 1 jet	up to 2000	12 to 30	0.43 to 0.48	89	Employed for very high head.
	2 jets	up to 1500	17 to 50			
	4 jets	up to 500	24 to 70			
2.	*Francis*:	up to 300	80 to 150	0.6 0.9 to	93	Full load efficiency high; partload efficiency lower than Pelton wheel.
	High-head	50 to 150	150 to 250			
	Medium head	30 to 60	250 to 400			
	Low head					
3.	*Propeller and Kaplan*	4 to 60	300 to 1000	1.4 to 2	93	High part load efficiency; high discharge with low head.
4.	*Bulb or tubular turbines*	3 to 10	1000 to 1200	6 to 8	91	Employed for very low head–*tidal power plants.*

Overall efficiency (η_0) of all turbines ≃ 85 per cent.

2.17. SURGE TANKS

A **surge tank** *is a small reservoir or tank in which the water level rises or falls to reduce the pressure swings so that they are not transmitted in full to a closed circuit.* In general a surge tank is employed to serve the following *purposes* :

1. To reduce the distance between the free water surface and turbine thereby reducing the water hammer effect (the *water hammer* is defined as the change in pressure rapidly above or below normal pressure caused by sudden changes in rate of flow through the pipe according to the demand of the prime mover) on penstock and also protect upstream tunnel from high pressure rises.
2. To serve as *supply tank* to the turbine when water in the pipe is accelerating during increased load conditions and *storage tank* when the water is decelerating during reduced load conditions.

Types of surge tanks :

The different types of surge tanks in use are :

1. Simple surge tank
2. Inclined surge tank
3. The expansion chamber and gallery type surge tank
4. Restricted orifice surge tank
5. Differential surge tank.

1. Simple surge tank. A simple surge tank is a vertical standpipe connected to the penstock as shown in Fig. 2.64. In the surge tank if the overflow is allowed, the rise in pressure can be eliminated but *overflow surge tank* is seldom satisfactory and usually uneconomical. Surge tanks are built high enough so that water cannot overflow even with a full load change on the turbine. It is always desirable to place the surge tank on ground surface, above the penstock line, at the point where the latter drops rapidly to the powerhouse as shown in Fig. 2.64. Under the circumstances when site for its location is not available the height of the tank should be increased with the help of a support.

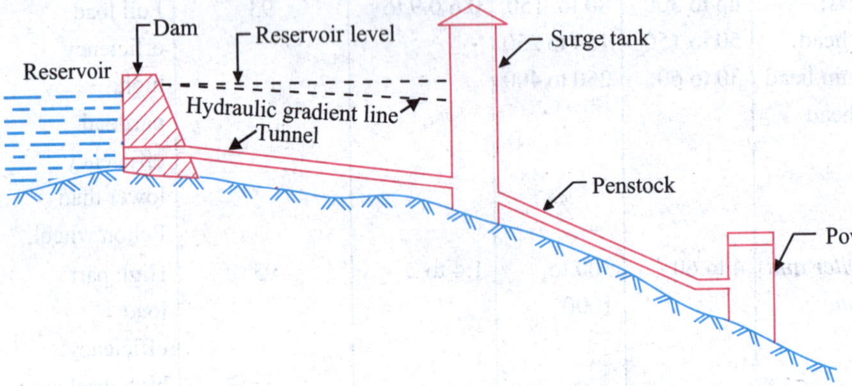

Fig. 2.64. Surge tank on ground level.

2. Inclined surge tank. When a surge tank is inclined (Fig.2.65) to the horizontal its effective water surface increases and therefore, lesser height surge tank is required of the same diameter if it is inclined or lesser diameter tank is required for the same height. But this type of surge tank is more costlier than ordinary type as construction is difficult and is rarely used unless the topographical conditions are in favour.

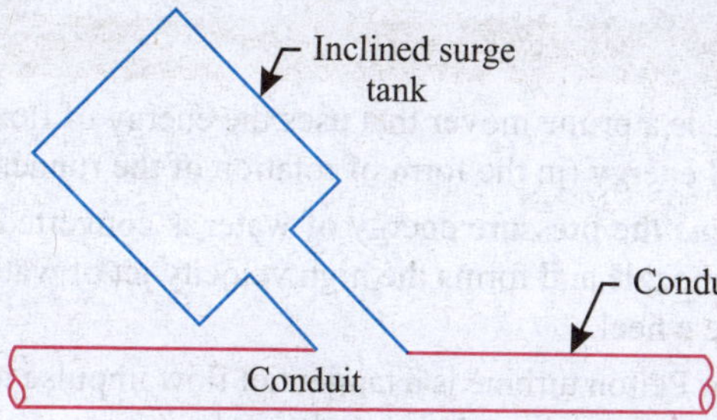

Fig. 2.65. Inclined surge tank.

3. **Expansion chamber surge tank.** Refer to Fig. 2.66. This type of a surge tank has an expansion tank at top and expansion gallery at the bottom; these expansions *limit the extreme* surges. The 'upper expansion chamber' must be above the maximum reservoir level and 'bottom gallery' must be *below the lowest steady running level in the surge tank.* Besides this the intermediate shaft should have stable minimum diameter.

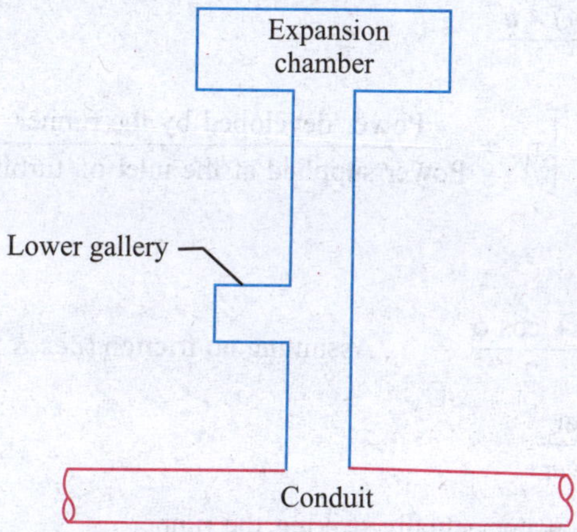

Fig. 2.66. Expansion chamber surge tank.

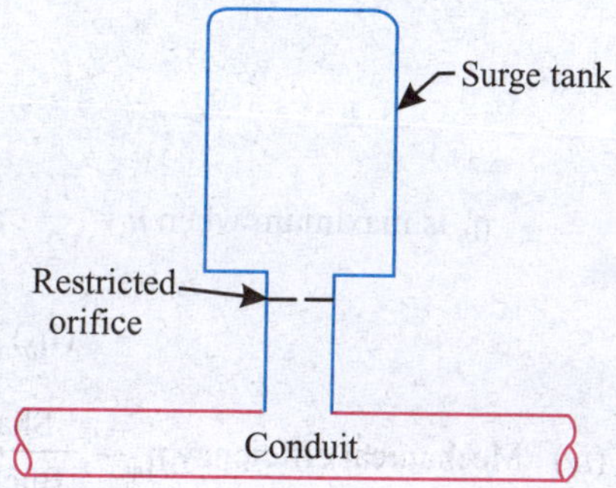

Fig. 2.67. Restricted orifice surge tank.

4. **Restricted orifice surge tank.** Refer to Fig. 2.67. It is also called *throttled surge tank.* The main object of providing a throttle or restricted orifice is to *create an appreciable friction loss when the water is flowing to or from the tank.* When the load on the turbine is reduced, the surplus water passes through the throttle and a retarding head equal to the loss due to throttle is built up in the conduit. The size of the throttle can be designed for any desired retarding head. *The size of the throttle adopted is usually such as the initial retarding head is equal to the rise of water surface in the tank when the full load is rejected by the turbine* (a case when there is closure of the gate valve).

Advantage. Storage function of the tank can be separated from accelerating and retarding functions.

Disadvantage. Considerable portion of water hammer pressure is transmitted directly into the low pressure conduit.

In comparison to other types of surge tanks these are *less popular*.

5. **Differential surge tank.** Refer to Fig. 2.68. A differential surge tank has a riser with a small hole at its lower end through which water enters in it. The function of the surge tank depends upon the area of hole.

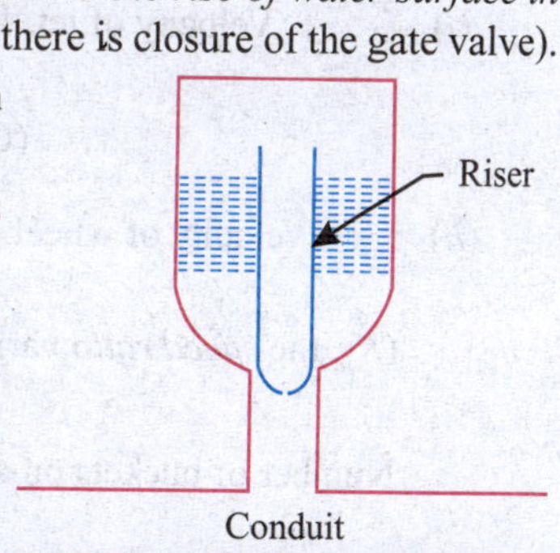

Fig. 2.68. Differential surge tank.

HIGHLIGHTS

1. A *hydraulic turbine* is a prime mover that uses the energy of flowing water and converts it into the mechanical energy (in the form of rotation of the runner).
2. In an *impulse turbine* the pressure energy of water is converted into kinetic energy when passed through the nozzle and forms the high velocity jet of water. The formed water jet is used for driving the wheel.

   The Pelton wheel or Pelton turbine is a tangential flow impulse turbine and is used for high head. Some *important formulae* relating *Pelton wheel* are :

Work done and efficiencies :

(*i*) The work done by the jet on runner per second = $\rho a V_1 (V_{w1} \pm V_{w2})$

(*ii*) The work done per second per unit weight of water striking

$$= \frac{1}{g} (V_{w1} \pm V_{w2}) \times u$$

(*iii*) Hydraulic efficiency, $\eta_h = \dfrac{2 (V_{w1} \pm V_{w2}) \times u}{V_1^2}$

$$\left[\eta_h = \frac{\text{Power developed by the runner}}{\text{Power supplied at the inlet of turbine}}\right]$$

η_h is maximum when $u = \dfrac{V_1}{2}$, and

$$(\eta_h)_{max} = \frac{1 + \cos\phi}{2} \quad \text{...Assuming no friction (i.e., } K = 1)$$

(*iv*) Mechanical efficiency, $\eta_m = \dfrac{\text{Shaft power}}{\text{Bucket power}}$

(*v*) Volumetric efficiency, $\eta_v = \dfrac{\text{Volume of water actually striking the runner}}{\text{Total water supplied by the jet to the turbine}}$

(*vi*) Overall efficiency, $\eta_0 = \dfrac{\text{Shaft power}}{\text{Water power}} = \dfrac{P}{wQH}$

Also, $\eta_0 = \eta_h \times \eta_m \times \eta_v$

Design aspects :

(*i*) Velocity of jet, $V_1 = C_v \sqrt{2gH}$ $\left(\text{or, } C_v = \dfrac{V_1}{\sqrt{2gH}}\right)$

$(C_v = 0.98 \text{ or } 0.99)$

(*ii*) Velocity of wheel, $u = (u_1 = u_2) = K_u \sqrt{2gH}$ $\left[K_u = \dfrac{u}{\sqrt{2gH}}\right]$

(K_u, the *speed ratio* varies from 0.43 to 0.48).

Number of buckets on a runner $Z = 15 + \dfrac{D}{2d} = 15 + 0.5\, m$

[Where m (jet ratio) $= \frac{D}{d}$; D and d being the pitch diameters of Pelton wheel and the jet diameter respectively. $\frac{D}{d}$ lies between 11 and 16 for maximum hydraulic efficiency; normally jet ratio is adopted as 12 in practice.]

3. In a reaction turbine the runner utilizes both potential and kinetic energies.

Formulae for various reaction turbines are as follows :

(*a*) Francis turbine :

(*i*) Francis turbine is an inward radial flow reaction turbine having discharge *radial* at outlet which means the angle made by absolute velocity at outlet is 90°, *i.e.* $\beta = 90°$ Then $V_{w2} = 0$ and work done by water on the runner per second per unit weight of water is

$$= \frac{1}{g} V_{w1} u_1$$

(*ii*) Flow ratio, $K_f = \frac{V_{f1}}{\sqrt{2gH}}$; K_f varies from 0.15 to 0.30.

(*iii*) Speed ratio, $K_u = \frac{u}{\sqrt{2gH}}$; K_u ranges from 0.6 to 0.9.

(*iv*) The ratio of width (B_1) to the diameter of the wheel (D_1), $n = \frac{B_1}{D_1}$; n varies from 0.10 to 0.45.

(*v*) Discharge, $Q = K_{t1}\pi D_1 B_1 V_{f1} = K_{t2}\pi D_2 B_2 V_{f2}$

[where K_t is known as vane thickness factor/co-efficient; its value is usually of the order of 0·95 or so (always less than unity)]

(*b*) Kaplan turbine :

It is an axial flow turbine in which the vanes on the hub are *adjustable*. It is used for low heads where large volumes of water are available. In this turbine a high efficiency is maintained even at partload. The peripheral velocities at inlet and outlet are equal, *i.e.* $u_1 = u_2$

$$\text{Discharge, } Q = \frac{\pi}{4} \times (D_0^2 - D_b^2) \times V_f$$

where, D_0 = Outside diameter of the runner, and

D_b = Diameter of boss (or hub).

V_f = Velocity of flow; $(V_{f1} = V_{f2} = V_f)$

4. *Deriaz turbine*. It is also known as diagonal turbine. Its runner is so shaped that it can be used both as a turbine as well as a pump and hence it may be classified as a reversible type turbine. As such these turbines are amply suitable for pumped storage hydropower plants.

5. *Tubular or bulb turbine*. It is an axial flow turbine with either adjustable or non-adjustable runner vanes. It is employed for low heads, varying from 3 m to 15 m.

6. *Runaway speed* is the maximum speed, governor being disengaged, at which a turbine would run when there is no external load but operating under design head and discharge.

7. A *draft tube* is a pipe of gradually increasing area used for discharging water from the exit of a reaction turbine. It is an integral part of mixed and axial flow turbines. The efficiency of a draft tube (η_d) is given by :

$$\eta_d = \frac{\text{Net gain in pressure head}}{\text{Velocity head at entrance of draft tube}} = \frac{\left(\frac{V_2^2 - V_3^2}{2g} - h_f\right)}{\frac{V_2^2}{2g}}$$

where, V_2 = Velocity of water at inlet of the draft tube, and

V_3 = Velocity of water at outlet of the draft tube.

$$\left[\text{or,} \quad h_f = \frac{V_2^2 - V_3^2}{2g} - \eta_d \times \frac{V_2^2}{2g}\right]$$

8. *Specific speed* (N_s) of a turbine is defined as the speed of a geometrically similar turbine which would develop unit power when working under a unit head. It is given by the relation:

$$N_s = \frac{N\sqrt{P}}{H^{5/4}}$$

where, P = Shaft power, and

H = Net head on the turbine.

Specific speed plays an important role in the *selection of the type of turbine.*

9. *Unit quantities* are the quantities which are obtained when the head on the turbine is unity. They are given as:

$$\text{Unit speed, } N_u = \frac{N}{\sqrt{H}}$$

$$\text{Unit discharge, } Q_u = \frac{Q}{\sqrt{H}}$$

$$\text{Unit power, } P_u = \frac{P}{H^{3/2}}$$

10. The important characteristic curves of a turbine are:
    (*i*) Main or constant head characteristic curves.
    (*ii*) Operating or constant speed characteristic curves.
    (*iii*) Constant efficiency or iso-efficiency or Muschel curves.

11. The formation, growth and collapse of vapour filled cavities or bubbles in a flowing liquid due to local fall in fluid pressure is called **cavitation.** The critical value of cavitation factor (σ_c) is given by

$$\sigma_c = \frac{(H_a - H_v) - H_s}{H}$$

where, H_a = Atmospheric pressure head in metres of water,

H_v = Vapour pressure in metres of water corresponding to the water temperature,

H = Working head of turbine (difference between head race and tail race levels in metres), and

H_s = Suction pressure head (or height of turbine inlet above tail race level) in metres.

The value of critical factor depends upon specific speed of the turbine.

12. A '*surge tank*' is a small reservoir or tank in which the water level rises or falls to reduce the pressure swings so that they are not transmitted in full to a closed circuit.

OBJECTIVE TYPE QUESTIONS

Choose the Correct Answer

1. For an impulse turbine which of the following statements is *correct* :
   (*a*) It makes use of a draft tube
   (*b*) It is not exposed to atmosphere
   (*c*) It is most suited for low head installations
   (*d*) It operates with initial complete conversion of pressure head to velocity head.

2. Which of the following statements is *correct* in case of a Pelton wheel :
   (*a*) It can operate at optimum efficiency at all high speeds
   (*b*) It is kept entirely submerged in water below the tail race.
   (*c*) It gives optimum efficiency at runaway speed
   (*d*) It operates by converting the available energy fully into kinetic energy before entering the rotor.

3. The effective (or net) head at the turbine is
   (*a*) the sum of gross head plus head loss in penstock and the velocity head at the turbine exit.
   (*b*) the difference between gross head minus the head loss in penstock
   (*c*) the difference between the gross head minus head loss in penstock and the velocity head at the turbine exit
   (*d*) the sum of gross head plus the head loss in the penstock.

4. The difference between the power obtained from the turbine shaft and power supplied by water at its entry to the turbine is equal to
   (*a*) sum of hydraulic and mechanical losses
   (*b*) sum of mechanical and volumetric losses
   (*c*) mechanical losses
   (*d*) hydraulic losses.

5. Which of the following statements is a definition of the hydraulic efficiency of a turbine?
   (*a*) The ratio of power available at the turbine shaft to that supplied to it by runner.
   (*b*) The ratio of the power supplied by the runner to the power available at the shaft.
   (*c*) The ratio of power utilized by runner to that supplied by the water at entry to the turbine.
   (*d*) The ratio of power supplied by water at entry to the power utilized by runner.

6. The power which appears in the expression for the specific speed is the:
   (*a*) shaft power
   (*b*) water power
   (*c*) power into the turbine
   (*d*) none of the above.

7. Which of the following statements is *correct*: Runaway speed of a hydraulic turbine is the speed
   (*a*) at which there would be no damage to the turbine runner
   (*b*) at which the turbine runner can be allowed to run freely without load and with wicket gates wide open
   (*c*) corresponding to maximum overload permissible
   (*d*) at full load.

8. The specific speed of a turbine is expressed as
   (*a*) $\dfrac{N\sqrt{P}}{H}$ (*b*) $\dfrac{N\sqrt{P}}{H^2}$
   (*c*) $\dfrac{N\sqrt{P}}{H^{3/4}}$ (*d*) $\dfrac{N\sqrt{P}}{H^{5/4}}$

9. Which of the following statements is *correct* for a reaction turbine?
   (*a*) The outlet must be above the tail race.
   (*b*) Water may be allowed to enter a part or whole of wheel circumference.
   (*c*) Flow can be regulated without loss.
   (*d*) There is only partial conversion of available head to velocity head before entry to rotor.

10. In a reaction turbine the function of a draft tube is to
   (*a*) provide safety to turbine
   (*b*) prevent air from entering
   (*c*) reconvert the kinetic energy to flow energy
   (*d*) increase the rate of flow.

11. Which of the following statements with respect to a reaction water turbine is *incorrect*:

(*a*) The spiral casing serves to uniformly distribute water into guide blades

(*b*) The water leaves the turbine at atmospheric pressure

(*c*) The draft tube allows setting of the turbine above the tail race with minimum reduction of available energy

(*d*) The guide vanes direct the flow at proper angle.

12. Which of the following turbines is suitable for specific speed ranging from 300 to 1000 and heads below 30 m:

(*a*) Francis (*b*) Kaplan

(*c*) Propeller (*d*) Pelton.

13. Specific speed of a turbo-machine

(*a*) relates the shape rather than the size of the machine

(*b*) remains unchanged under different conditions of operation

(*c*) has the dimensions of rotational speed

(*d*) is the speed of a machine having unit dimensions.

14. In an outward radial flow turbine energy conversion process is

(*a*) purely by reaction only

(*b*) purely by impulse only

(*c*) partly by impulse and partly by reaction

(*d*) none of the above.

15. Which of the following turbines is least efficient under part load conditions:

(*a*) Propeller (*b*) Kaplan

(*c*) Francis (*d*) Pelton.

16. A surge tank is used to

(*a*) prevent occurrence of hydraulic jump

(*b*) smoothen the flow

(*c*) relieve the pipeline of excessive pressure transients

(*d*) avoid reversal of flow.

17. Which of the following turbines is most efficient at partload operation?

(*a*) Kaplan (*b*) Propeller

(*c*) Francis (*d*) Pelton wheel.

18. With respect to a Kaplan turbine which of the following statements is *incorrect*?

(*a*) It employs large guide vane angles than is the case for a Francis turbine.

(*b*) It is designed for flow velocity of mixed flow type.

(*c*) It has blades of small camber to prevent separation.

(*d*) It can adjust both guide vane and blade angles according to rate of discharge.

19. Specific speed of an impulse turbine (Pelton wheel) ranges from

(*a*) 12 to 70 (*b*) 80 to 400

(*c*) 300 to 1000 (*d*) 1000 to 1200.

20. A turbo-machine becomes more susceptible to cavitation if

(*a*) velocity attains a high value

(*b*) pressure become very high

(*c*) temperature rises above the critical value

(*d*) Thoma's cavitation parameter exceeds a certain limit

(*e*) pressure falls below the vapour pressure.

21. Cavitation damage in turbine runner occurs near the

(*a*) inlet on the convex side of blades

(*b*) outlet on the convex side of blades

(*c*) inlet on the concave side of blades

(*d*) outlet on the concave side of blades.

22. Which of the following serious problems arise from cavitation?

(*a*) Noise and vibration.

(*b*) Damage to blade surface.

(*c*) Fall in efficiency.

(*d*) All of the above.

23. Which of the following statements is *correct*?

(*a*) Muschel curves are the performance plots pertaining to constant efficiency.

(*b*) Operating characteristics curves of a turbine refer to the performance curves drawn at constant speed.

(*c*) Main characteristic curves of a turbine are the performance curves obtained under condition of constant head.

(*d*) All of the above.

(*e*) None of the above.

24. A Kaplan turbine is

(*a*) an inward flow impulse turbine

(*b*) low head axial flow turbine

(*c*) high head axial flow turbine

(*d*) high head mixed flow turbine.

25. An impulse turbine requires

(*a*) high head and small quantity of flow

(*b*) low head and small quantity of flow

(*c*) low head and high rate of flow

(*d*) none of the above.

26. ......of a turbine is defined as the ratio of power available at the turbine shaft to the power supplied by the water jet.
(*a*) Mechanical efficiency
(*b*) Hydraulic efficiency
(*c*) Overall efficiency
(*d*) Volumetric efficiency.

27. The ratio of power developed by the runner to the power supplied by the jet at entrance to the turbine is known as
(*a*) hydraulic efficiency
(*b*) mechanical efficiency
(*c*) volumetric efficiency
(*d*) overall efficiency.

28. The water which acts on the runner blades of a reaction turbine is under a pressure
(*a*) equal to atmospheric
(*b*) below atmospheric
(*c*) above atmospheric
(*d*) none of the above.

29. The runner passages of a reaction turbine are
(*a*) partially filled with water
(*b*) always completely filled with water
(*c*) never filled with water
(*d*) none of the above.

30. The value of speed ratio (K_u) in case of a Francis turbine ranges from
(*a*) 0.2 to 0.3 (*b*) 0.4 to 0.5
(*c*) 0.6 to 0.9 (*d*) none of the above.

31. The value of flow ratio (K_f) in case of a Francis turbine varies from:
(*a*) 0.1 to 0.14 (*b*) 0.15 to 0.30
(*c*) 0.35 to 0.5 (*d*) 0.6 to 0.9.

32. A Kaplan turbine claims which of the following advantages over a Francis turbine?
(*a*) More compact in construction and smaller in size.
(*b*) Partload efficiency is considerably high.
(*c*) Low frictional losses.
(*d*) All of the above.

33. Which of the following draft tubes is suited particularly for helical flow?
(*a*) Conical type draft tube.
(*b*) Elbow type draft tube.
(*c*) Moody's spreading draft tube.
(*d*) None of the above.

34. Critical value of cavitation factor (σ_c) is given by:
(*a*) $\dfrac{(H_a + H_v) - H_s}{H}$
(*b*) $\dfrac{(H_a - H_v) + H_s}{H}$
(*c*) $\dfrac{(H_a - H_v) - H_s}{H}$
(*d*) $\dfrac{(H_a + H_v) + H_s}{H}$.

35. Which of the following surge tank is also called a throttled surge tank?
(*a*) Inclined surge tank.
(*b*) Expansion chamber surge tank.
(*c*) Restricted orifice surge tank.
(*d*) None of the above.

ANSWERS

1. (*d*)	**2.** (*d*)	**3.** (*c*)	**4.** (*a*)	**5.** (*c*)	**6.** (*a*)
7. (*b*)	**8.** (*d*)	**9.** (*d*)	**10.** (*c*)	**11.** (*b*)	**12.** (*b*)
13. (*a*)	**14.** (*d*)	**15.** (*a*)	**16.** (*c*)	**17.** (*a*)	**18.** (*b*)
19. (*a*)	**20.** (*e*)	**21.** (*b*)	**22.** (*d*)	**23.** (*d*)	**24.** (*b*)
25. (*a*)	**26.** (*c*)	**27.** (*a*)	**28.** (*c*)	**29.** (*b*)	**30.** (*c*)
31. (*b*)	**32.** (*d*)	**33.** (*c*)	**34.** (*c*)	**35.** (*c*).	

THEORETICAL QUESTIONS

1. What is a hydraulic turbine ?
2. How are hydraulic turbines classified ?
3. Give the comparison between impulse and reaction turbines.
4. With the help of neat diagram explain the construction and working of a Pelton wheel turbine.
5. Derive an expression for hydraulic efficiency of a Pelton wheel.
6. What is the condition for hydraulic efficiency of a Pelton wheel to be maximum ?
7. Derive an expression for maximum hydraulic efficiency of a Pelton wheel.

8. Draw a general layout of a hydroelectric power plant using an impulse turbine and define the following:
   (*i*) Gross head, (*ii*) Net head,
   (*iii*) Hydraulic efficiency, and
   (*iv*) Overall efficiency of the impulse turbine.
9. Draw a schematic diagram of a Francis turbine and explain briefly its construction and working.
10. Draw a general layout of a hydroelectric power plant using a reaction turbine.
11. State the advantages and disadvantages of a Francis turbine over a Pelton wheel.
12. What are the functions of a draft tube ?
13. Why does a Pelton wheel not possess any draft tube ?
14. How do the losses in the draft tube effect the pressure at runner exit ?
15. What is the difference between a propeller turbine and a Kaplan turbine ?
16. Where is Kaplan turbine used ?
17. State the advantages of a Kaplan turbine over Francis turbine.
18. Write a short note on Deriaz turbine.
19. What are tubular or bulb turbines ?
20. What are the advantages and disadvantages of bulb sets compared to Kaplan turbines ?
21. How is specific speed of a turbine defined ?
22. Write a short note on 'scale effect.'
23. What is cavitation? How can it be avoided in reaction turbines ?
24. On what factors does the cavitation in water turbines depend ?
25. Enumerate some methods to avoid cavitation in water turbines.
26. What is governing and how it is accomplished for different types of water turbines ?
27. Sketch and describe a modern method of regulation to maintain a constant speed for either (*a*) Pelton wheel or (*b*) Francis turbine.
28. Show with the help of a line sketch as to how the speed of a reaction water turbine is governed by servomotor ?
29. What is a surge tank ?
30. Which points should be considered while selecting right type of hydraulic turbines for hydroelectric power plant ?

UNSOLVED EXAMPLES

1. A Pelton wheel has a mean bucket speed of 10 m/s with a jet of water flowing at the rate of 0.7 m^3/s under a head of 30 m. The buckets deflect the jet through an angle of 160°. Calculate the power and the efficiency of the turbine. Assume co-efficient of velocity as 0.98.
   [Ans. 186.9 kW; 94.54%**]**
2. A Pelton wheel having a mean bucket diameter of 1·0 m is running at 1000 r.p.m. The net head on the Pelton wheel is 700 m. If the side clearance angle is 15° and discharge through the nozzle is 0·1 m^3/s, determine power available at the nozzle and hydraulic efficiency of the turbine.
   [Ans. 686.5 kW; 97.18%**]**
3. The shaft power of a Pelton wheel, the buckets of which are struck by two jets, is 15445 kW. The diameter of each jet is 200 mm. If the net head on the turbine is 400 m, find the overall efficiency of the turbine. Take $C_v = 1.0$.
   [Ans. 70.8%**]**
4. The jet of water coming out of nozzle strikes the buckets of a Pelton wheel which when stationary would deflect the jet through 165°. The velocity of water at exit is 0·9 times at the inlet and the bucket speed is 0·45 times the jet speed. If the speed of the Pelton wheel is 300 r.p.m. and the effective head is 150 m, determine:
   (*i*) Hydraulic efficiency, and
   (*ii*) Diameter of the Pelton wheel.
   Take co-efficient of velocity, $C_v = 0.98$
   [Ans. (*i*) 92.5%; (*ii*) 1.55 m**]**
5. A Pelton wheel is required to develop 9193.7 kW at the shaft when working under a head of 300 m. Assuming the values of C_v, K_u (speed ratio) and m (= D/d) as 0.98, 0.45 and 12 respectively, determine:
   (*i*) The number of jets,
   (*ii*) The diameter of the wheel,
   (*iii*) The quantity of water required, and
   (*iv*) The diameter of the jet.
   Take the speed of the wheel as 550 r.p.m. and overall efficiency as 85%.
   [Ans. (*i*) 7; (*ii*) 1.2 m; (*iii*) 3.68 m^3/s; (*iv*) 0.1 m**]**
6. A Pelton wheel is to be designed to develop 735.5 kW at 400 r.p.m. It is to be supplied with water from a reservoir whose level is 250 m above the wheel through a pipe 900 m long. The pipeline losses are to be 5 per cent of gross head. The co-efficient of friction is 0.005. The bucket speed is to be 0.46 of the jet speed and efficiency of wheel is 85%. Calculate :
   (*i*) Pipeline diameter,
   (*ii*) Jet diameter, and

(*iii*) Wheel diameter. **[Panjab University]**

[Ans. (*i*) 440 mm; (*ii*) 84 mm; (*iii*) 1.48 m**]**

7. The following data is related to a Pelton turbine:

Brake/shaft power = 126.5 kW
Head = 300 m
Speed = 600 r.p.m.
Co-efficient of velocity, $C_v = 0.98$
Speed ratio, $K_u = 0.45$
Overall efficiency, $\eta_0 = 75\%$

Determine the following :

(*i*) The discharge,
(*ii*) The least jet diameter,
(*iii*) The mean runner diameter jet ratio, and
(*iv*) The number of buckets.

[Panjab University]

[Ans. (*i*) 0.0573 m^3/s; (*ii*) 31 mm; (*iii*) 35.6; (*iv*) 33**]**

8. A Pelton wheel has a mean bucket speed of 12 m/s and is supplied with water at the rate of 0.7 m^3/s under a head of 30 m. If the buckets deflect the jet through an angle of 160°, find the power and the efficiency of the turbine.

[Madras University and UPSC]

[Ans. 194.12 kW; 93.4%**]**

9. A jet of water impinges on a series of curved vanes at an angle of 30° to the direction of motion of the vanes while entering and leaves the vanes horizontally. The head under which the jet issues from the nozzle is 30 m, the co-efficient of velocity for the nozzle is 0·9 and the diameter of the jet after leaving the nozzle is 50 mm. The speed of the vanes is 10 m/s and the relative velocity of the water at outlet is 0·8 times the relative velocity at inlet. Calculate :

(*i*) The angle of vane tips at inlet;
(*ii*) The power developed by the jet, and
(*iii*) The efficiency of the system.

[Rajputana University]

[Ans. 50.8°; 8.6 kW; 69.2 %**]**

10. A Pelton wheel is to be designed to the following specifications :

Power	... 11948 kW
Head	... 381 m
Speed	... 750 r.p.m.
Overall efficiency	... 86%

Jet diameter not to exceed $\frac{1}{16}$ times the wheel diameter. Determine :

(*i*) The wheel diameter.
(*ii*) The number of jets required.
(*iii*) The diameter of the jet.

[Rajasthan University]

[Ans. (*i*) 1 m; (*ii*) 4; (*iii*) 118 mm**]**

11. The buckets of a Pelton impulse turbine deflect the jet through a total angle of 165° and owing to surface friction the relative velocity of water leaving the bucket is 0·85 times that at entry. Draw the velocity vector diagram at entry and exit and find the ratio of bucket velocity to jet velocity in order that the water shall leave the buckets without whirl. In such a turbine the available head at the nozzle is 650 m, the co-efficient of velocity for the nozzle is 0·97, the jet diameter 100 mm and mean bucket diameter 1·2 m. Using the conditions referred to above determine :

(*i*) Best running speed in r.p.m,
(*ii*) Impulsive force of the buckets at this speed,
(*iii*) Power developed by the buckets, and
(*iv*) Efficiency of buckets.

[UPSC, Ravi Shanker University]

[Ans. (*i*) 786 r.p.m.; (*ii*) 94470 N; (*iii*) 4660 kW; (*iv*) 90.5%**]**

12. A Pelton wheel, 2.45 m in diameter, operates under the following conditions :

Net head = 370 m; co-efficient of velocity = 0.98; speed ratio = 0.47; relative velocity of water at outlet = 0.90 times that at inlet; deflection of jet = 160°; diameter of the jet = 0·88 m.

Determine the following :

(*i*) The input power to the shaft, and
(*ii*) The r.p.m. of the wheel.

[UPSC Exams.]

[Ans. (*i*) 7549.8 kW; (*ii*) 312.15 r.p.m.**]**

13. A Pelton wheel has to develop 13230 kW under a net head of 800 m while running at a speed of 600 r.p.m. If the co-efficient of the jet $C_v = 0.97$, speed ratio $K_u = 0{\cdot}46$ and the ratio of jet diameter is $\frac{1}{16}$ of wheel diameter, determine the following :

(*i*) The diameter of the pitch circle,
(*ii*) The diameter of each jet,
(*iii*) The quantity of water supplied to the wheel, and
(*iv*) The number of jets required.

Assume overall efficiency as 85 percent.

[UPSC; AMIE]

[Ans. 1.834 m; 114.6 mm; 1.254 m^3/s; 2**]**

14. In a Pelton wheel the buckets deflect the jet by 170° and the relative velocity is reduced by 12% due to bucket friction. For a speed ratio of 0.47, calculate from first principles the hydraulic efficiency of the wheel.

The bucket circle diameter of the wheel is 0.9 m and there is one jet for which $C_v = 0.98$. The actual efficiency of the wheel is 0.9 times its theoretical efficiency. The wheel develops 1700 kW under a head of 550 m. Calculate :

(*i*) The speed of wheel in r.p.m. and

(*ii*) The diameter of the nozzle. **[UPSC]**

[Ans. (*i*) 1035.6 r.p.m; (*ii*) 66 mm]

15. An inward flow reaction turbine has external and internal diameters as 0.9 m and 0.45m respectively. The turbine is running at 200 r.p.m. and width of turbine at inlet is 0.2 m. The velocity of flow through the runner is constant and is equal to 1.8 m/s. The guide blades make an angle of 10° to the tangent of the wheel and discharge at the outlet of turbine is radial. Draw the inlet and outlet velocity triangles and determine: (*i*) Relative velocity at inlet, (*ii*) The runner blade angles, (*iii*) Width of the runner at outlet, (*iv*) Head at the inlet of the turbine, (*v*) Power developed; and (*vi*) Hydraulic efficiency of the turbine.

[Ans. (*i*) 1.963 m/s; (*ii*) $\theta = 66.48°$; $\phi = 20.9°$; (*iii*) 0.4 m; (*iv*) 9.97 m; (*v*) 97.8 kW; (*vi*) 98.34%]

16. An inward flow reaction turbine of inlet diameter 1·2 m operates under a head of 150 m and requires a discharge of 6 m^3/s at a rotational speed of 400 r.p.m. The guide vane angle is 20° and the water leaves the runner blade axially. If the runner is 0.1 m wide at the inlet, calculate :

(*i*) The torque and power supplied to the shaft, and

(*ii*) The efficiency of the turbine.

[Ravi Shanker University]

[Ans. (*i*) 156 kNm; 6533 kW; (*ii*) 74%]

17. The inward flow reaction turbine develops 735 kW at 750 r.p.m. under a net head of 100 m. The guide vanes makes an angle of 15° with the tangent at inlet. The axial length of the blade at inlet is 0.1 times the outer diameter. The radial velocity of flow through the wheel is constant and the discharge from the wheel is radial. The blade thickness blocks 5 per cent of the area of flow at inlet. The hydraulic efficiency of the wheel is 88% and overall efficiency is 84%. Determine :

(*i*) The wheel diameter,

(*ii*) The wheel width, and

(*iii*) The blade angle at inlet.

[Ans. (*i*) 0.574 m; (*ii*) 0.0574 m; (*iii*) $\theta = 38.7°$]

18. A Francis turbine has to be designed to develop 367.5 kW under a head of H = 70 m while running at N = 750 r.p.m. Ratio of width of runner to diameter of runner, $n = 0.1$, inner diameter is half the outer diameter. Flow ratio = 0.15, hydraulic efficiency = 95%, mechanical efficiency = 84%. Four percent of the circumferential area of runner to be occupied by the thickness of vanes, velocity of flow is constant and the discharge is radial at exit. Calculate: (*i*) The diameter of the wheel, (*ii*) The quantity of water supplied, and (*iii*) The guide vane angle at inlet and runner vane angles at inlet and exit. **[UPSC Exams.]**

[Ans. (*i*) 0.633 m; (*ii*) 0.67 m^3/s; (*iii*) $\alpha = 11°\ 58'$, $\theta = 76°\ 50'$ $\phi = 24°\ 5'$]

19. A Francis turbine with an overall efficiency of 75 percent is required to produce 149.26 kW. It is working under a head of 7.62 m. The peripheral velocity = $0.26\sqrt{2gH}$ and the radial velocity of flow at inlet is $0.96\sqrt{2gH}$. The wheel runs at 150 r.p.m. and the hydraulic losses in the turbine are 22 percent of the available energy. Assuming radial discharge, determine :

(*i*) The guide blade angle,

(*ii*) The wheel vane angle at inlet,

(*iii*) Diameter of the wheel at inlet, and

(*iv*) Width of the wheel at inlet.

[AMIE, Fluid Power Engg.]

[Ans. (*i*) $\alpha = 32.6°$; (*ii*) $\theta = 37.7°$; (*iii*) 0.404 m; (*iv*) 0.17 m]

20. Design a Francis turbine runner with the following data :

Net head	... 68 m
Speed of the runner	... 750 r.p.m.
Output	... 330·9 kW.
Hydraulic efficiency	... 94%
Overall efficiency	... 85%
Flow ratio	... 0.15
Breadth ratio	... 0·1

Inner diameter of runner = $\frac{1}{2}$ × outer diameter

Also assume 5 per cent of circumferential area of the runner to be occupied by the thickness of the vanes.

Velocity of flow remains constant throughout and flow is radial at exit.

[Ans. $D_1 = 0.6$ m; $D_2 = 0.3$ m, $B_1 = 0.06$ m; $\alpha = 11°\ 38'$, $\theta = 60°\ 40'$; $\phi = 24°.57'$

21. A Francis turbine has a wheel diameter of 1 m at the entrance and 0.5 m at the exit. The vane angle at the entrance is 90° and the guide vane angle is 15°. The water at the exit leaves the vanes without any tangential velocity. The head is 30 m and the radial component of flow is constant. What would be the speed of the wheel in r.p.m. and vane angle at exit ? State whether the speed calculated is synchronous one or not. If not, what speed would you recommend to couple the turbine with an alternator of 50 Hz ?

[UPSC Exams.]

[**Ans.** 324 r.p.m; 28° 12′; 300 r.p.m.]

22. The following data pertain to an inward flow reaction turbine :

Overall efficiency ... 75%

Power given by the turbine ... 128·7 kW.

Head ... 6 m

The velocity of the periphery of wheel $= 0.5\sqrt{2gH}$

Radial velocity of flow $= 0.5\sqrt{2gH}$

Speed of the wheel = 250 r.p.m.

Hydraulic losses in the turbine = 22 per cent of the available energy.

Determine the following :

(*i*) Guide blade angle at inlet,

(*ii*) The wheel vane angle at inlet,

(*iii*) The diameter of the wheel, and

(*iv*) The width of wheel at inlet.

Assume the discharge to be radial.

[Panjab University]

[**Ans.** (*i*) 34° 48′, (*ii*) 49° 36′; (*iii*) 662 mm; (*iv*) 371 mm]

23. The following data pertain to an inward flow reaction turbine:

Power to be developed ... 625 kW

Speed of the runner ... 1000 r.p.m.

Head ... 100 m

Guide vane angle ... 16°

Internal diameter = 0·6 times the external diameter

Hydraulic efficiency ... 88%

Overall efficiency ... 86%

Allowance of blade thickness ... 5%

Axial length of blade inlet = 0·1 times the outer diameter

Radial velocity of flow ... constant

Find the leading dimensions of the runner.

[Jadavpur University]

[**Ans.** $D_1 = 0.527$ m; $D_2 = 0.316$ m; $B_1 = 52.7$ mm; $\theta = 67.5°$; $\phi = 31° 37'$]

24. The following data pertain to a Kaplan turbine :

Power available at shaft = 8850 kW; net available head = 5.5 m;

speed ratio = 2.1; flow ratio = 0.67, overall efficiency = 85%.

Assuming that hub diameter of the wheel is 0·35 times the outside diameter, determine :

(*i*) Runner diameter;

(*ii*) Runner speed.

[**Ans.** (*i*) 6.34 m; (*ii*) 65.7 r.p.m.]

25. A Kaplan turbine produces 44000 kW under a head of 24.7 m, with an overall efficiency of 90 per cent. Taking the value of speed ratio as 1.6, flow ratio as 0.5 and the hub diameter as 0.35 times the outside diameter, find the runner diameter and speed of the turbine.

[**Ans.** 5.16 m, 130.7 r.p.m.]

26. The following data pertain to a Kaplan turbine :

Shaft power = 13230 kW; Speed = 75 r.p.m; Head = 8 m; Diameter of boss of runner = 0.35 times the external diameter; speed ratio = 2; Flow ratio = 0.6.

Find the efficiency of the turbine.

[**Ans.** 80.7%]

27. A Kaplan turbine develops 1471 kW under a head of 6 m. The turbine is set 2.5 m above the tail race level. A vacuum gaunge inserted at the turbine outlet records a suction head of 3.1 m. If the hydraulic efficiency is 85 per cent, what would be the efficiency of draft tube having inlet diameter of 3 m ? What will be the reading of suction gauge if power developed is reduced to half (735.5 kW), the head and speed remaining constant.

[**Ans.** 68%; 2.6496 m]

28. Calculate the efficiency of a Kaplan turbine developing 2900 kW under a net head of 5 m. It is provided with a draft tube with its inlet (diameter 3 m) set 1.6 m above the tail race level. A vacuum gauge connected to the draft tube indicates a reading of 5 m of water. Assume draft tube efficiency as 78 per cent.

[**Ans.** 90.4%]

29. A conical draft tube is discharging water at outlet with a velocity of 2.5 m/s. Its inlet and outlet diameters are 1 m and 1.5 m respectively. The total length of the draft tube is 6 m and 1.2 m of the length of draft tube is immersed in water. If the atmospheric pressure head is 10.3 m of water and loss of head due to friction in the draft tube is equal to 0.2 × velocity head at outlet of the draft tube, find: (*i*) Pressure head at inlet, (*ii*) Efficiency of the draft tube.

[**Ans.** (*i*) 4·27 m (abs.); 76·3%]

30. A hydraulic turbine is to develop 845. kW when running at 100 r.p.m. under a head of 10 m. Work out the maximum flow rate and specific speed for the turbine if the overall efficiency at the best operating point is 92 per cent. In order to predict its performance, a 1:10 scale model is tested under a head of 6 m. What would be the speed, power output and water consumption of the model if it runs under the conditions similar to the prototype ?

[**Ans.** $Q = 9.37$ m^3/s; $N_s = 163.5$; $N_m = 774.6$ r.p.m.; $Q_m = 0.0726$ m^3/s; $P_m = 3.93$ kW]

31. Determine the maximum height of straight conical draft tube of 13240 kW Francis turbine running at 150 r.p.m., under a net head of 27 m. The turbine is installed at station where the effective atmospheric pressure is 10·6 m of water. The draft tube must sink at least 0·77 m below the tail race. [**Ans.** 2.22 m]

[**Hint:** Max. height of draft tube = $H_s + 0.77$]

Chapter

3

CENTRIFUGAL PUMPS

3.1. INTRODUCTION

A **pump** *is a contrivance which provides energy to a fluid in a fluid system; it assists to increase the pressure energy or kinetic energy, or both of the fluid by converting the mechanical energy.* The *basic difference* between a *turbine* and *the pump*, from hydrodynamic point of view, is that in the former flow takes place from the *high pressure side to the low pressure side,* whereas in *pump* flow takes place from the *low pressure towards the higher pressure.* Thus in a *turbine* there is *accelerated flow,* while in a *pump*, the flow is *decelerated.*

3.2. CLASSIFICATION OF PUMPS

On the basis of transfer of mechanical energy the pumps can be broadly *classified* as follows:

1. *Rotodynamic pumps:*

(*i*) Radial flow pumps

(*ii*) Axial flow pumps

(*iii*) Mixed flow pumps .

2. *Positive displacement pumps.*

In rotodynamic pumps, *increase in energy level is due to a combination of centrifugal energy, pressure energy, and kinetic energy.*

- The energy transfer, *in a radial flow pump*, occurs mainly when the flow is in its *radial path.*
- In an *axial flow pump*, the energy transfer occurs when the flow is in its *axial direction.*
- The energy transfer in a *mixed flow pump* takes place when the flow comprises radial as well as axial components.

The radial flow type pumps are commonly called *centrifugal pumps* (only these pumps will be discussed in this chapter).

Classification of centrifugal pumps:

On the *basis of characteristic features*, the centrifugal pumps are *classified* as follows:

1. *Type of casing*:
   (*i*) Volute pumps (*ii*) Turbine pump or diffusion pump.

2. *Working head:*
   (*i*) Low lift centrifugal pumps .... they work against heads upto 15 m
   (*ii*) Medium lift centrifugal pumps .... used to build up heads as high as 40 m
   (*iii*) High lift centrifugal pumps .... employed to deliver liquids at heads above 40 m.

3. *Liquid handled:*
   (*i*) Closed impeller pump (*ii*) Semi-open impeller pump (or Non-clog pump)
   (*iii*) Open impeller pump.

4. *Number of impellers per shaft:*
   (*i*) Single stage centrifugal pump ... has *one* impeller, usually a low lift pump.
   (*ii*) Multi- stage centrifugal pump ... has *two or more* impellers and pressure is built in steps; used usually for high working heads and the number of stages depends on the head required.

5. *Number of entrances to the impeller:*
   (*i*) Single entry or single suction pump ... water is admitted on one side of the impeller.
   (*ii*) Double entry or double suction pump ... water is admitted from both sides of the impeller; axial thrust is neutralised.
   ... employed for pumping large quantities of fluid.

6. *Relative direction of flow through impeller:*
   (*i*) Radial flow pump ... normally radial flow impellers are used in *all centrifugal pumps*
   (*ii*) Axial flow pump ... designed to deliver huge quantities of water at comparatively low heads; *ideally suited for irrigation purposes.*
   (*iii*) Mixed flow pump ... mostly employed for *irrigation purposes.*

3.3. ADVANTAGES OF CENTRIFUGAL PUMP OVER DISPLACEMENT (RECIPROCATING) PUMP

The centrifugal pump claims the following *advantages* with reference to a positive displacement (reciprocating) pump.

1. The cost of a centrifugal pump is less as it has fewer parts.
2. Installation and maintenance are easier and cheaper.
3. Its discharging capacity is much greater than that of a reciprocating pump.
4. It is compact and has smaller size and weight for the same capacity and energy transfer.
5. Its performance characteristics are superior.
6. It can be employed for lifting highly viscous liquid such as paper pulp, muddy and sewage water, oil, sugar molasses etc.

7. It can be operated at very high speeds without any danger of separation and cavitation.
8. It can be directly coupled to an electric motor or an oil engine.
9. The torque on the power source is uniform, the output fron the pump is also uniform.

However, because of higher efficiency the reciprocating pumps are still employed for high heads and small discharges. A reciprocating pump can build up very high pressures (as high as 700 bar or even more) and as such these pumps are used for lifting oils from very deep oil wells.

3.4. COMPONENT PARTS OF A CENTRIFUGAL PUMP

Refer to Fig. 3.1. A centrifugal pump consists of the following *main components*:

1. Impeller 2. Casing 3. Suction pipe 4. Delivery pipe.

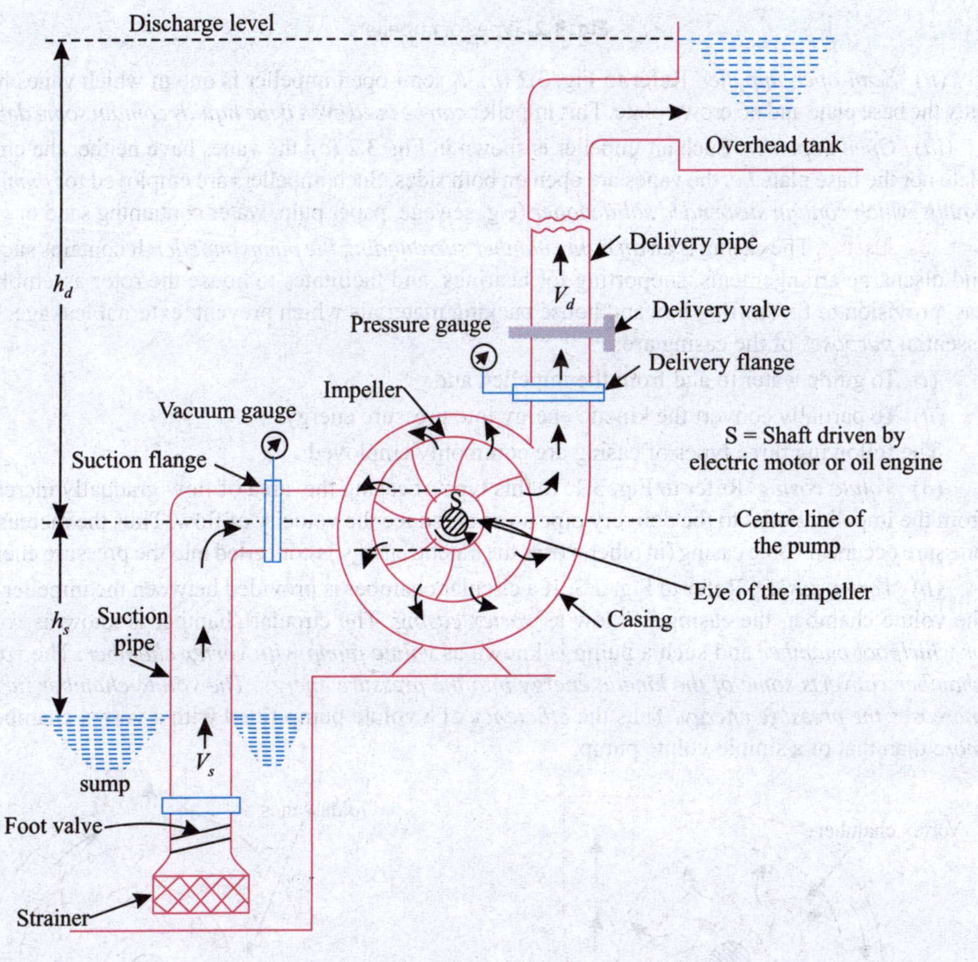

Fig. 3.1. Volute type centrifugal pump–component parts.

1. Impeller. An *impeller* is a wheel (or rotor) with a series of backward curved vanes (or blades). It is mounted on a shaft which is usually coupled to an electric motor.

The impellers are of following *three types*:

(*i*) *Shrouded or closed impeller.* Refer to Fig. 3.2 (*a*). In this type of impeller vanes are provided with metal cover plates or shrouds on both the sides. It provides better guidance for the liquid and has a high efficiency. It is employed when the *liquid to be pumped is pure and relatively free from debris.*

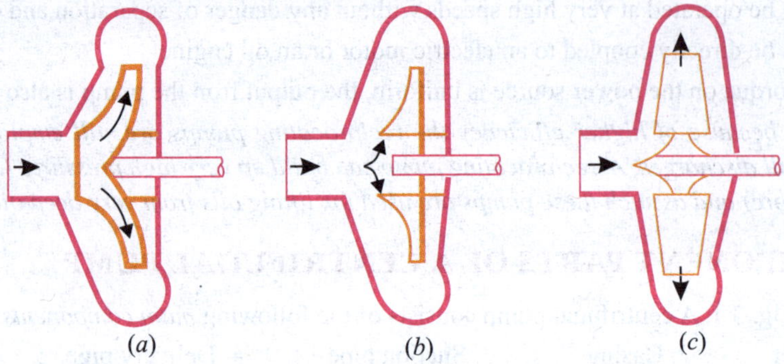

Fig. 3.2. Types of impellers.

(*ii*) *Semi-open impeller.* Refer to Fig. 3.2 (*b*). A semi-open impeller is one in which vanes have only the base plate and no crown plate. This impeller *can be used even if the liquids contain some debris.*

(*iii*) *Open impelller.* Such an impeller is shown in Fig. 3.2 (*c*); the vanes have neither the crown plate nor the base plate *i.e.* the vanes are open on both sides. Such impellers are employed for *pumping liquids which contain suspended solid matter* (e.g. sewage, paper pulp, water containing sand or grit)

2. Casing. The casing is an *airtight chamber surrounding the pump impeller.* It contains suction and discharge arrangements, supporting for bearings, and facilitates to house the rotor assembly. It has provision to fix stuffing box and house packing materials which prevent external leakage. The essential *purposes* of the casing are:

(*i*) To guide water to and from the impeller, and

(*ii*) To partially convert the kinetic energy into pressure energy.

The following three types of casing are commonly employed:

(*a*) *Volute casing.* Refer to Fig. 3.1. In this type of casing the area of flow gradually increases from the impeller outlet to the delivery pipe so as to reduce the velocity of flow. Thus the increase in pressure occurs in voute casing (in other words the kinetic energy is concerted into the pressure energy)

(*b*) *Vortex casing.* Refer to Fig. 3.3. If a circular chamber is provided between the impeller and the volute chamber, the casing is know as *vortex casing.* The circular chamber is know as *vortex or whirlpool chamber* and such a pump is known as *volute pump with vortex chamber*. The *vortex chamber converts some of the kinetic energy into the pressure energy. The volute chamber further increases the pressure energy.* Thus the *efficiency* of a volute pump fitted with a vortex chamber is *more* than that of a simple volute pump.

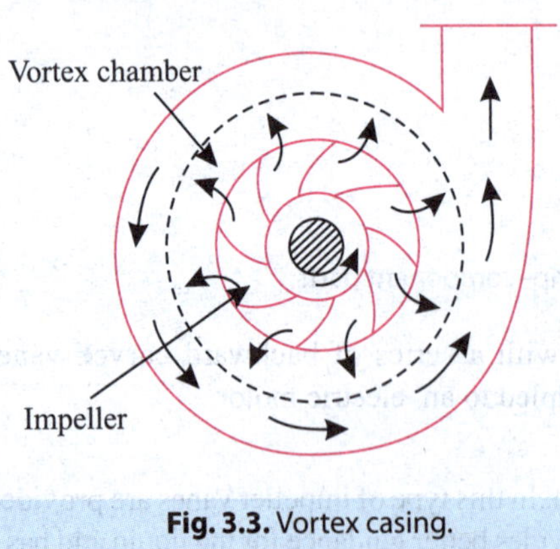

Fig. 3.3. Vortex casing.

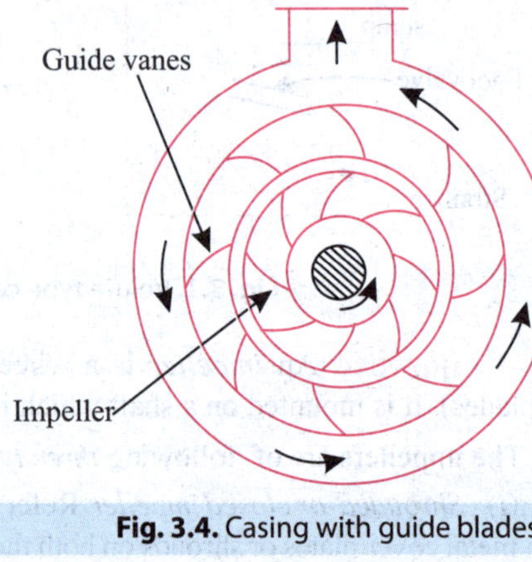

Fig. 3.4. Casing with guide blades.

(*c*) *Casing with guide blades*. Refer to Fig. 3.4. In this type of casing impeller is surrounded by a series of guide blades (or vanes) mounted on a ring which is know as a *diffuser*. The liquid leaving the impeller passes through the passage (having a gradually increasing area) between guide vanes/blades; the velocity of flow *decreases* and the kinetic energy, is converted into pressure energy. *Machines with diffuser blades have rather maximum efficiency, but are less satisfactory when a wide range of operating conditions is required. These pumps are costlier than volute pumps.*

3. Suction pipe. The pipe which connects the centre/eye of the impeller to sump from which liquid is to be lifted is known as *suction pipe.* In order to check the formation of air pockets the pipe is laid *air tight.* To prevent the entry of soid particles, debris etc. into the pump the suction pipe is provided with a strainer at its lower end. The lower end of the pipe is also fitted with a *non-return foot valve* which does not permit the liquid to drain out of the suction pipe when pump is *not working*; this also helps in *priming.*

4. Delivery pipe. The pipe which is connected at its lower end to the outlet of the pump and it delivers the liquid to the required height is known as *delivery pipe*. A *regulating valve* is provided on the delivery pipe to regulate the supply of water.

Following points, regarding *impellers, are* worth noting:

(*i*) Where it is required to pump clear and fresh water, the impeller is *cast as a single piece* and is made of *castiron.* The cast–iron impellers are *cheaper.*

(*ii*) Where *corrosion* due to salt water or chemicals is expected the impellers are made of materials such as *gunmetal, stainless steel etc.*

(*iii*) Machines (pumps) that handle *hot water,* having temperatures above 150°C have to be made of *cast steel impellers with special types of packings.*

(*iv*) Where acids are to pumped, the impeller and all inside surfaces in contact with liquid should be coated with a suitable material to withstand corrosion.

(*v*) Machines (pumps) employed in milk industry are made of *stainless steel* to prevent contamination of the liquids handled.

3.5. WORKING OF A CENTRIFUGAL PUMP

A centrifugal pump works on the *principle that when a certain mass of fluid is rotated by an external source, it is thrown away from the central axis of rotation and a centrifugal head is impressed which enables it to rise to a higher level.*

The *working /operation* of a centrifugal pump is explained *step-wise* below:

1. The delivery valve is closed and the pump is *primed* that is, *suction pipe, casing and portion of the delivery pipe upto the delivery valve are completely filled with the liquid (to be pumped) so that no air pocket is left.*

2. Keeping the delivery valve still closed the electric motor is started to rotate the impeller. The rotation of the impeller causes strong suction or vaccum just at the eye of the casing.

3. The speed of the impeller is gradually increased till the impeller rotates at its normal speed and develops normal energy required for pumping the liquid.

4. After the impeller attains the normal speed the delivery valve is opened when the liquid is continuously sucked (from sump well) up the suction pipe, it passes through the eye of casing and enters the impeller at its centre or it enters the impeller vanes at their inlet tips. This liquid is impelled out by the rotating vanes and it comes out at the outlet tips of the vanes into the casing. Due to impeller action the *pressure head as well as velocity heads of the liquid are increased* (some of this velocity heads is converted into pressure head in the casing and in the diffuser blades/vanes if they are also provided).

5. From casing, the liquid passes into pipe and is lifted to the required height (and discharged from the outlet or upper end of the delivery pipe).

6. So long as motion is given to the impeller and there is supply of liquid to be lifted the process of lifting the liquid to the required height remains continuous.

7. When pump is to be stopped the delivery valve should be first closed, otherwise there may be some backflow from the reservoir.

3.6. WORK DONE BY THE IMPELLER (OR CENTRIFUGAL PUMP) ON LIQUID

The expression for *work done* or energy supplied by the impeller of a centrifugal pump on the liquid flowing through may be derived in the same way as for turbine. Fig . 3.5 shows one vane of the impeller.

The liquid enters the impeller at its centre and leaves at its outer periphery.

Assumptions:

(*i*) Liquid enters the impeller eye in *radial direction,* the whirl component V_{w1} (of the inlet absolute velocity V_1) is *zero* and the flow component V_{f1} equals the absolute velocity itself (*i.e.* $V_{f1} = V_1$); $\alpha = 90°$.

(*ii*) No energy loss in the impeller due to friction and eddy formation.

(*iii*) No loss due to shock at entry.

(*iv*) There is uniform velocity distribution in the narrow passages formed between two adjacent vanes.

Fig. 3.5 shows a portion of the impeller of a centrifugal pump with the one vane and the velocity triangles at the inlet and the outlet tips of the vane.

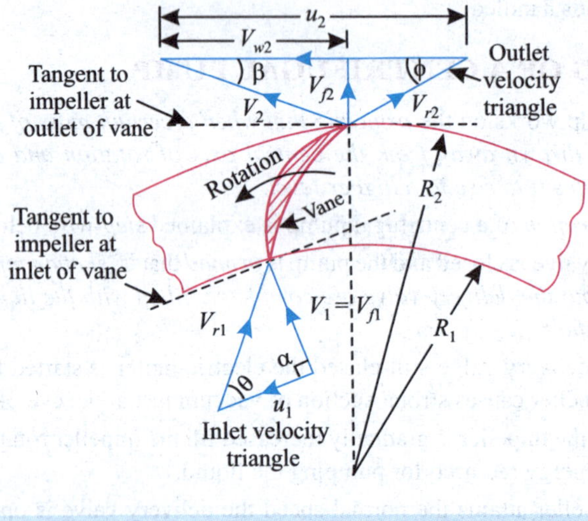

Fig. 3.5. Velocity triangles for an impeller vane.

Let, D_1 = Diameter of the impeller at inlet ($R_1 = D_1/2$).

N = Speed of the impeller in r.p.m.,

ω = Angular velocity $\left(= \frac{2\pi N}{60} \text{ rad/s}\right)$,

u_1 = Tangential velocity of the impeller at inlet

$$= \frac{\pi D_1 N}{60} \left(= \frac{2\pi R_1 N}{60} \right) = \omega R_1,$$

D_2 = Diameter of the impeller at outlet ($R_2 = D_2/2$),

u_2 = Tangential velocity of impeller at outlet

$$= \frac{\pi D_2 N}{60} \left(= \frac{2\pi R_2 N}{60} \right) = \omega R_2,$$

V_1 = Absolute velocity of water at *inlet*,

V_{w1} = Velocity of whirl at inlet,

V_{r1} = Relative velocity of liquid at inlet,

V_{f1} = Velocity of flow at inlet,

α = Angle made by absolute velocity (V_1) at inlet with the direction of motion of vane,

θ = Angle made by the relative velocity (V_{r1}) at inlet with the direction of motion of vane, and

$V_2, V_{w2}, V_{r2}, V_{f2}, \beta$ and ϕ are the corresponding values at *outlet*.

While passing through the impeller, the velocity of whirl changes and there is a change of moment of momentum.

Torque on the impeller = Rate of change of moment of momentum

Moment of momentum at inlet = 0 $\quad (\because V_{w1} = 0)$

$$\text{Moment of momentum at outlet} = \frac{W}{g}(V_{w2}R_2)$$

$$\therefore \quad \text{Torque} = \frac{W}{g}(V_{w2}R_2)$$

Work done per second = Torque × angular velocity

$$= \frac{W}{g}(V_{w2}R_2) \times \omega = \frac{W}{g}(V_{w2}u_2) \qquad (\because u_2 = \omega R_2) \quad ...(3.1)$$

Work done per second per unit weight of liquid

$$= \frac{V_{w2}u_2}{g} \quad ...[3.1\ (a)]$$

Eqn. (3.1) has been developed *assuming* flow at inlet to be *radial* (*i.e.* $V_{w1} = 0$). If the flow is *not radial*, the expression for work done may be written as :

$$\text{Work done per second} = \frac{W}{g}(V_{w2}u_2 - V_{w1}u_1)$$

or, Work done per second per unit weight of liquid

$$= \frac{1}{g}(V_{w2}u_2 - V_{w1}u_1) \quad ...(3.2)$$

Eqn 3.2 is known as the **Euler momentum equation for centrifugal pumps.**

The term $\frac{1}{g}(V_{w2}u_2 - V_{w1}u_1)$ is referred to as ***Euler head*** $\boldsymbol{H_e}$ [*Theoretical head*] where, W = weight of liquid = $w \times Q$, and

Q = volume of liquid

$= \pi D_1 B_1 \times V_{f1} = \pi D_2 B_2 \times V_{f2}$

where, B_1 and B_2 are the widths of impeller at inlet and outlet and V_{f1} and V_{f2} are the velocities of flow at inlet and outlet respectively.

Eqn. 3.1 stipulates that for delivering liquid at high heads the peripheral velocity u_2 must be high and vector V_{w2} must be large (so as to provide adequate whirl to the liquid). The increase in u_2 can be obtained by increasing the impeller diameter and speed of rotation. The whirl component V_{w2} however can be augmented by the providing adequate number of vanes of suitable size and shape.

Further, from *outlet triangle*, we have:

$$V_{r2}^2 = V_{f2}^2 + (u_2 - V_{w2})^2, \quad \text{or,} \quad V_{f2}^2 = V_{r2}^2 - (u_2 - V_{w2})^2 \qquad ...(i)$$

Also,
$$V_{f2}^2 = V_2^2 - V_{w2}^2 \qquad ...(ii)$$

From expressions (*i*) and (*ii*), we have:

$$V_2^2 - V_{w2}^2 = V_{r2}^2 - (u_2 - V_{w2})^2 = V_{r2}^2 - (u_2^2 + V_{w2}^2 - 2u_2 V_{w2})$$

or,
$$V_2^2 - V_{w2}^2 = V_{r2}^2 - u_2^2 - V_{w2}^2 + 2u_2 V_{w2}$$

or,
$$u_2 V_{w2} = \frac{1}{2}(V_2^2 + u_2^2 - V_{r2}^2)$$

Similarly from *inlet triangle*, we can obtain:

$$u_1 V_{w1} = \frac{1}{2}(V_1^2 + u_1^2 - V_{r1}^2)$$

Substituting in eqn. (3.2), we get:

Work done per second per unit weight of liquid (or H_e)

$$= \frac{V_2^2 - V_1^2}{2g} + \frac{u_2^2 - u_1^2}{2g} + \frac{V_{r1}^2 - V_{r2}^2}{2g} \qquad ..(3.3)$$

Eqn. (3.3) indicates that work done on the liquid consists of *three* terms:

– The *first* term $\left(\frac{V_2^2 - V_1^2}{2g}\right)$ represents the increase in kinetic energy or dynamic head.

– The *second* term $\left(\frac{u_2^2 - u_1^2}{2g}\right)$ represents an increase in static pressure.

– The *third* term $\left(\frac{V_{r1}^2 - V_{r2}^2}{2g}\right)$ indicates the change in kinetic energy due to retardation of flow relative to the impeller (this term, therefore, represents conversion of kinetic energy within the impeller itself).

Eqn. (3.3) is sometimes known as the *fundamental equation* of centrifugal pump.

3.7. HEAD OF A PUMP

The head of a centrifugal pump may be expressed in the following ways:

(*i*) Static head; (*ii*) Manometric head; (*iii*) Total, gross or effective head.

(i) Static head. The *sum of suction head and delivery head is known as* ***static head.*** This is represented by H_{stat} and is written as:

$$H_{stat} = h_s + h_d \quad ...(3.4)$$

where, h_s = Suction head (it is the vertical height of the centre line of the pump shaft above the liquid surface in the sump from which the liquid is being raised), and

h_d = Delivery head (it is vertical height of the liquid surface in the tank / reservoir to which the liquid is delivered above the centre line of the pump shaft).

The terms h_s and h_d are known as *static suction lift* and *static delivery lift* respectively.

(ii) Manometric head. *The head against which a centrifugal pump has to work is known* as the ***manometric head.*** It is the head measured across the pump inlet and outlet flanges. It is denoted by H_{mano} and is given by the following expressions:

(i) H_{mano} = Head imparted by the impeller to liquid – loss of head in the pump (*i.e.* impeller, and casing)

$$= \frac{V_{w2}u_2}{g} - (h_{Li} + h_{Lc}) \quad ...(3.5)$$

(where, h_{Li} and h_{Lc} are the losses of head in the *impeller* and *casing* respectively.)

$$= \frac{V_{w2}u_2}{g} \quad \text{...if loss of head in the pump is } \textit{zero}. \quad ...(3.6)$$

(ii)

$$H_{mano} = H_{static} + \text{losses in pipes} + \frac{V_d^2}{2g}$$

$$= (h_s + h_d) + (h_{fs} + h_{fd}) + \frac{V_d^2}{2g} \quad ...(3.7)$$

where, h_s = Suction head,

h_d = delivery head,

h_{fs} = Frictional head loss in the suction pipe,

h_{fd} = Frictional head loss in the delivery pipe, and

V_d = Velocity of liquid in delivery pipe.

(iii) H_{mano} = Total head at outlet of the pump – total head at inlet of the pump

$$= \left(\frac{p_2}{w} + \frac{V_2^2}{2g} + z_2\right) - \left(\frac{p_1}{w} + \frac{V_1^2}{2g} + z_1\right) \quad ...(3.8)$$

where, $\frac{p_2}{w}$ = Pressure head at outlet of pump = h_d,

$\frac{V_2^2}{2g}$ = Velocity head at outlet of the pump

= Velocity head in the delivery pipe = $\frac{V_d^2}{2g}$,

z_2 = Vertical height of the pump outlet from the datum line, and

$\frac{p_1}{w}, \frac{V_1^2}{2g}, z_1$ = corresponding values of pressure head, velocity head and datum head at inlet of the pump $\left(i.e.,\ h_s, \frac{V_s^2}{2g} \text{ and } z_s \text{ respectively}\right)$.

(iii) Total, gross or effective head. *It is equal to the static head plus all the head losses occurring in flow before, through and after the impeller.*

3.8. LOSSES AND EFFICIENCIES OF A CENTRIFUGAL PUMP

3.8.1. Losses in Centrifugal Pump

When a centrifugal pump operates, the various losses which occur are as follows:

1. **Hydraulic losses:**

   (*i*) *Hydraulic losses in the pump:*

   (*a*) Shock or eddy losses at the entrance to and exit from the impeller.

   (*b*) Losses due to friction in the impeller.

   (*c*) Friction and eddy losses in the guide vanes/diffuser and casing.

   (*ii*) *Other hydraulic losses:*

   (*a*) Friction and other minor losses in the suction pipe.

   (*b*) Friction and other minor losses in the delivery pipe.

2. **Mechanical losses:**

   (*i*) Losses due to disc friction between the impeller and the liquid which fills the clearance spaces between the impeller and casing.

   (*ii*) Losses pertaining to friction of the main bearing and glands.

3. **Leakage loss:**

The loss of energy due to leakage of liquid is known *as leakage loss*. The various losses in a centrifugal pump are shown diagramatically in Fig. 3.6.

3.8.2. Efficiencies of a Centrifugal Pump

The various *efficiencies* of a centrifugal pump are:

(*i*) Manometric efficiency (η_{mano}), (*ii*) Volumetric efficiency (η_v),

(*iii*) Mechanical efficiency (η_m), and (*iv*) Overall efficiency (η_0).

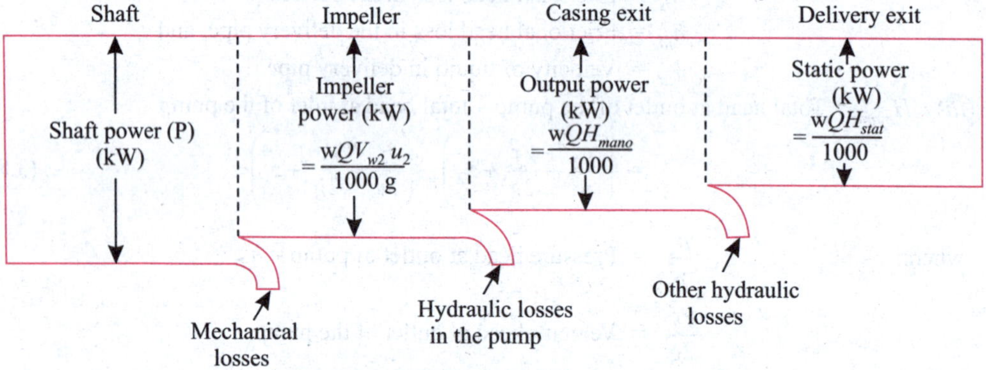

Fig. 3.6. Losses in centrifugal pump.

(*i*) Manometric efficiency (η_{mano}). *The ratio of the manometric head developed by the pump to the head imparted by the impeller to the liquid is known as* ***manometric efficiency.*** Thus,

$$\eta_{mano} = \frac{\text{Manometric head}}{\text{Head imparted by impeller to liquid}}$$

or,

$$\eta_{mano} = \frac{H_{mano}}{\left(\frac{V_{w2}u_2}{g}\right)} = \frac{gH_{mano}}{V_{w2}u_2} \quad \text{...(3.9)}$$

(ii)* Volumetric efficiency (η_v).** *The ratio of quantity of liquid discharged per second from the pump to quantity passing per second through the impeller is known as* ***volumetric efficiency. Thus,

$$\eta_v = \frac{\text{Liquid discharged per second from the pump}}{\text{Quantity of liquid passing per second through the impeller}}$$

or, $$\eta_v = \frac{Q}{Q+q}$$

where, Q = Actual liquid discharged at the pump outlet per second, and

q = Leakage of liquid per second from the impeller (through the clearances between the impeller and casing).

(iii)* Mechanical efficiency (η_m).** *The ratio of the power delivered by the impeller to the liquid to the power input to the pump shaft is known as* ***mechanical efficiency. Thus,

$$\eta_m = \frac{\text{Power delivered by the impeller to the liquid}}{\text{Power input to the pump shaft (P)}}$$

or, $$\eta_m = \frac{w\,(Q+q)\,(V_{w2}u_2/g)}{P} \qquad ...(3.11)$$

$$= \frac{P - P_{mech.loss}}{P} \qquad ...[3.11(a)]$$

(iv)* Overall efficiency (η_0).** *The ratio of power output of the pump to the power input to the pump is known as* ***overall efficiency. Thus,

$$\eta_0 = \frac{\text{Power output of the pump}}{\text{Power input to the pump / shaft}} = \frac{wQH_{mano}}{P} \qquad ...(3.12)$$

Also, $$\eta_0 = \eta_{mano} \times \eta_v \times \eta_m$$

$$= \frac{H_{mano}}{(V_{w2}u_2/g)} \times \frac{Q}{(Q+q)} \times \frac{w(Q+q)\,(V_{w2}u_2/g)}{P}$$

$$= \frac{wQH_{mano}}{P}, \text{ which is the same as eqn. (3.12)}$$

3.8.3. Effect of Outlet Vane Angle (ϕ) on Manometric Efficiency

The total energy of liquid before entering the impeller, with reference to the centre of pump, is taken as zero. After leaving the impeller, the liquid has a pressure energy (H_{mano}) and kinetic energy $\left(\frac{V_2^2}{2g}\right)$. The energy supplied to the impeller is $\frac{V_{w2}u_2}{g}$. Neglecting the losses in the pump and equating the energy given to the impeller to the increase in total energy, we have:

$$\frac{V_{w2}u_2}{g} = H_{mano} + \frac{V_2^2}{2g}$$

or, $$H_{mano} = \frac{V_{w2}u_2}{g} - \frac{V_2^2}{2g} \qquad ...(3.13)$$

From velocity triangle at outlet shown in Fig. 3.5, we have:

$$V_2^2 = V_{w2}^2 + V_{f2}^2, \text{ and, } V_{w2} = u_2 - \frac{V_{f2}}{\tan\phi} = u_2 - V_{f2}\cot\phi$$

Then,
$$H_{mano} = \frac{(u_2 - V_{f2}\cot\phi)\times u_2}{g} - \frac{(u_2 - V_{f2}\cot\phi)^2 + V_{f2}^2}{2g}$$

$$= \frac{2\,(u_2^2 - u_2V_{f2}\cot\phi) - [(u_2^2 + V_{f2}^2\cot^2\phi - 2u_2V_{f2}\cot\phi) + V_{f2}^2]}{2g}$$

$$= \frac{2u_2^2 - 2u_2V_{f2}\cot\phi - u_2^2 - V_{f2}^2\cot^2\phi + 2u_2V_{f2}\cot\phi - V_{f2}^2}{2g}$$

$$= \frac{u_2^2 - V_{f2}^2(1+\cot^2\phi)}{2g} = \frac{u_2^2 - V_{f2}^2\,\text{cosec}^2\theta}{2g} \quad \text{...(3.14)}$$

The manometric efficiency of the pump, under the ideal condition assumed above, will become:

$$\eta_{mano} = \frac{H_{mano}}{(V_{w2}u_2/g)} = \frac{gH_{mano}}{V_{w2}u_2} = \frac{u_2^2 - V_{f2}^2\text{cosec}^2\theta}{2u_2\,(u_2 - V_{f2}\cot\phi)}$$

Let us assume the flow ratio $K_f\left[= \frac{V_{f2}}{\sqrt{2gH_{mano}}}\right] = 0.25$; computing the value of u_2 in terms of H_{mano} from eqn. 3.14 for different values of ϕ, it will be observed that as the value of ϕ varies from 90° to 20° the value of η_{mano} increases from 0.47 to 0.73. A further decrease in the angle ϕ will increase the efficiency, but it is impracticable to have the angle ϕ less than 20°, as it would *result in long and narrow blades, with very high frictional losses.* As such the *minimum value of* ϕ is 20°.

Example 3.1. *The impeller of a centrifugal pump has an external diameter of 450 mm and internal diameter of 200 mm and it runs at 1440 r.p.m. Assuming a constant radial flow through the impeller at 2.5 m/s and that the vanes at exit are set back at an angle 25°, determine:*

(i) Inlet vane angle,

(ii) The angle, absolute velocity of water at exit makes with the tengent, and

(iii) The work done per N of water.

Solution. Internal diameter of the impeller, $D_1 = 200$ mm = 0.2 m

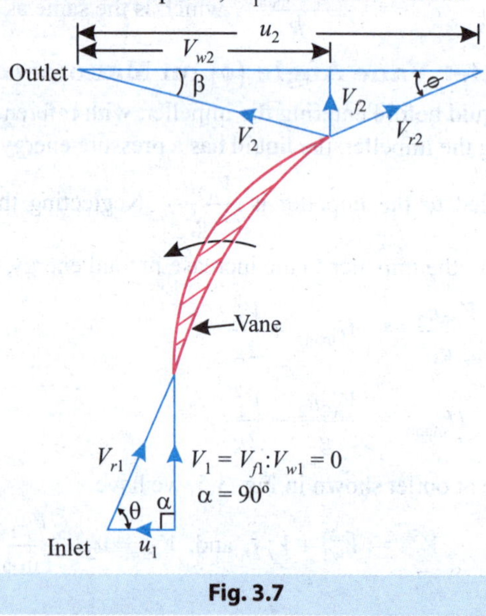

Fig. 3.7

External diameter of the impeller, D_2 = 450 mm = 0.45 m

Speed of impeller, N = 1440 r.p.m.

Velocity of flow, $V_{f1} = V_{f2} = 2.5$ m/s

Vane angle at outlet, $\phi = 25°$

(i) Inlet vane angle, θ:

Tangential velocity of impeller at inlet,

$$u_1 = \frac{\pi D_1 N}{60} = \frac{\pi \times 0.2 \times 1440}{60} = 15.08 \text{ m/s}$$

From velocity triangle at *inlet,* we have:

$$\tan\theta = \frac{V_{f1}}{u_1}, \text{ or, } \tan\theta = \frac{2.5}{15.08} = 0.1658$$

$$\therefore \quad \theta = \tan^{-1} 0.1658 = \mathbf{9.4° \text{ (Ans.)}}$$

(ii) The angle, absolute velocity of water at exit makes with the tangent, β:

Tangential velocity of impeller at outlet,

$$u_2 = \frac{\pi D_2 N}{60} = \frac{\pi \times 0{\cdot}45 \times 1440}{60} = 33.93 \text{ m/s}$$

From velocity triangle at *outlet,* we have:

$$V_{w2} = u_2 - \frac{V_{f2}}{\tan\phi}, \text{ or, } V_{w2} = 33.93 - \frac{2.5}{\tan 25°} = 28.57 \text{ m/s}$$

Now,
$$\tan\beta = \frac{V_{f2}}{V_{w2}} = \frac{2.5}{28.57} = 0.0875$$

$$\therefore \quad \beta = \tan^{-1} 0.0875 = \mathbf{5° \text{ (Ans.)}}$$

(iii) Work done per N of water:

$$\textit{Work done per N of water} = \frac{V_{w2} u_2}{g} = \frac{28.57 \times 33.93}{9.81} = \mathbf{98.81 \text{ Nm}} \text{ (Ans.)}$$

Example 3.2. *A centrifugal pump is to discharge 0.118 m³/s at a speed of 1450 r.p.m against a head of 25 m. The impeller diameter is 250 mm, its width at outlet is 50 mm and manometric efficiency is 75 percent. Determine the vane angle at the outer periphery of the impeller.* **[PTU]**

Solution. Discharge, Q = 0.118 m³/s

Speed N = 1450 r.p.m.

Speed, H_{mano} = 25 m

Diameter of impeller at outlet,

D_2 = 250 mm = 0.25 m

Width at outlet, B_2 = 50 mm = 0.05 m

Manometric efficiency, η_{mano} = 75%

Vane angle at outlet, φ:

Tangential velocity of impeller at outlet,

$$u_2 = \frac{\pi D_2 N}{60} = \frac{\pi \times 0.25 \times 1450}{60} = 18.98 \text{ m/s}$$

Discharge, $Q = \pi D_2 B_2 \times V_{f2}$

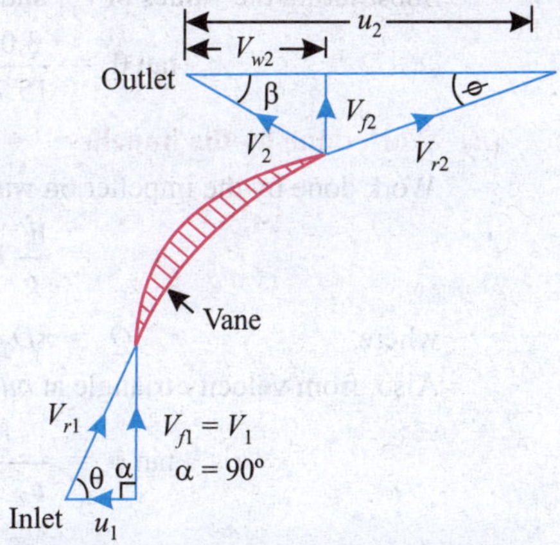

Fig. 3.8

$$\therefore \quad V_{f2} = \frac{Q}{\pi D_2 B_2} = \frac{0.118}{\pi \times 0.25 \times 0.05} = 3.0 \text{ m/s}$$

$$\text{Manometric efficiency, } \eta_{mano} = \frac{gH_{mano}}{V_{w2}u_2}$$

$$\text{or,} \quad 0.75 = \frac{9.81 \times 25}{V_{w2} \times 18.98}, \quad \text{or,} \quad V_{w2} = \frac{9.81 \times 25}{0.75 \times 18.98} = 17.23 \text{ m/s}$$

From velocity triangle at *outlet,* we have:

$$\tan \phi = \frac{V_{f2}}{u_2 - V_{w2}} = \frac{3.0}{18.98 - 17.23} = 1.7143$$

$$\therefore \quad \phi = \tan^{-1} 1.7143 = \mathbf{59.74^\circ} \text{ (Ans.)}$$

Example 3.3. *The impeller of a centrifugal pump having external and internal diameters 500 mm and 250 mm respectively, width at outlet 50 mm and running at 1200 r.p.m. works against a head of 48 m. The velocity of flow through the impeller is constant and equal to 3.0 m/s. The vanes are set back at an angle of 40° at outlet. Determine:*

(*i*) *Inlet vane angle,*

(*ii*) *Work done by the impeller on water per second, and*

(*iii*) *Manometric efficiency.*

Solution. External diameter of impeller, D_2 = 500 mm = 0.05 m

Internal diameter, D_1 = 250 mm = 0.25 m

Width at outlet, B_2 = 50 mm = 0.05 m

Speed, N = 1200 r.p.m.

Head, H_{mano} = 48 m

Velocity of flow, $V_{f1} = V_{f2}$ = 3.0 m/s

Vane angle at outlet, ϕ = 40°

(*i*) Inlet vane angle, θ:

Refer to Fig. 3.8. From velocity triangle at *inlet*, we have:

$$\tan \theta = \frac{V_{f1}}{u_1} \quad \text{where, } u_1 = \frac{\pi D_1 N}{60} = \frac{\pi \times 0.25 \times 1200}{60} = 15.7 \text{ m/s}$$

Substituting the values of V_{f1} and u_1, we get:

$$\tan \theta = \frac{3.0}{15.7} = 0.191 \qquad \therefore \theta = \tan^{-1} 0.191 = \mathbf{10.81^\circ} \text{ (Ans.)}$$

(*ii*) Work done by the impeller:

Work done by the impeller on water per second is given by eqn. (3.1) as

$$= \frac{W}{g} V_{w2} u_2 = \frac{wQ}{g} \times V_{w2} u_2 \quad \text{...(i)}$$

$$\text{where,} \quad Q = \pi D_2 B_2 \times V_{f2} = \pi \times 0.5 \times 0.05 \times 3.0 = 0.2356 \text{ m}^3\text{/s}$$

Also, from velocity triangle at *outlet*, we have:

$$\tan \phi = \frac{V_{f2}}{u_2 - V_{w2}} = \frac{3.0}{31.41 - V_{w2}}, \quad \text{or,} \quad \tan 40^\circ = \frac{3.0}{31.41 - V_{w_2}}$$

$$\left(\text{where, } u_2 = \frac{\pi D_2 N}{60} = \frac{\pi \times 0.5 \times 1200}{60} = 31.41 \text{ m/s} \right)$$

or, $\quad 31.41 - V_{w2} = \dfrac{3.0}{\tan 40°}$, or, $V_{w_2} = 31.41 - \dfrac{3.0}{\tan 40°} = 27.83$ m/s

Substituting the values in eqn (*i*), we get the work done by the impeller

$$= \frac{9.81 \times 0.2356}{9.81} \times 27.83 \times 31.41 = \mathbf{205.95\ kNm\ (Ans.)}$$

$$(\because w = 9.81 \text{ kN/m}^3)$$

(*iii*) Manometric efficiency, (η_{mano}):

Manometric efficiency is given by eqn. (3.9) as:

$$\eta_{mano} = \frac{gH_{mano}}{V_{w2}u_2} = \frac{9.81 \times 48}{27.83 \times 31.41} = 0.5386 \text{ or } \mathbf{53.86\%\ (Ans.)}$$

Example 3.4. *A centrifugal pump running at 800 r.p.m. is working against a total head of 20.2 m. The external diameter of the impeller is 480 mm and outlet width 60 mm. If the vanes angle at outlet is 40° and manometric efficiency is 70 percent, determine:*

(*i*) *Flow velocity at outlet,*

(*ii*) *Absolute velocity of water leaving the vane,*

(*iii*) *Angle made by the absolute velocity at outlet with the dircetion of motion at outlet, and*

(*iv*) *Rate of flow through the pump.*

Solution. Speed, N = 800 r.p.m.; Head, H_{mano} = 20.2 m;

External diameter, D_2 = 480 mm = 0.48 m; Width at outlet, B_2 = 60 mm = 0.06 m;

Outlet vangle angle, ϕ = 40°; Manometric efficiency, η_{mano} = 70 %.

Refer to Fig. 3.9.

(*i*) Flow velocity at outlet, V_{f2}:

Tangential velocity of impeller at outlet,

$$u_2 = \frac{\pi D_2 N}{60} = \frac{\pi \times 0.48 \times 800}{60}$$

$$= 20.1 \text{ m/s}$$

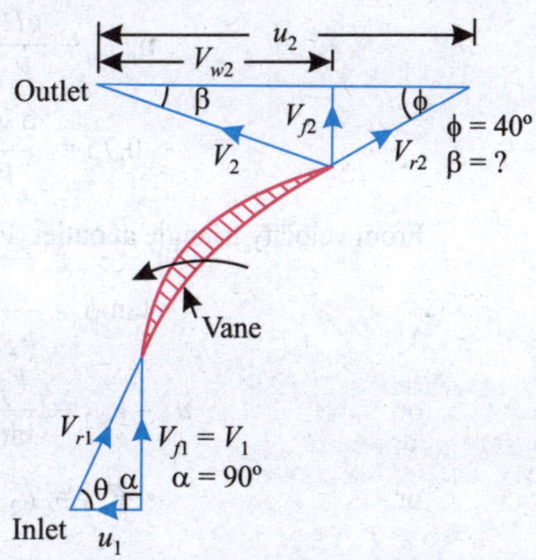

Fig. 3.9

Also, manometric efficiency is given by,

$$\eta_{mano} = \frac{gH_m}{V_{w2}u_2} \quad \text{...(Eqn 3.9)}$$

or, $$0.70 = \frac{9.81 \times 20.2}{V_{w2} \times 20.1}$$

or, $$V_{w2} = \frac{9.81 \times 20.2}{0.70 \times 20.1} = 14.08 \text{ m/s}$$

From velocity triangle at outlet, we have:

$$\tan\phi = \frac{V_{f2}}{u_2 - V_{w2}} = \frac{V_{f2}}{(20.1 - 14.08)}$$

or, $$V_{f_2} = \tan\phi\,(20.1 - 14.08)$$

$$= \tan 40°\,(20.1 - 14.08) = \mathbf{5.05\ m/s\ (Ans.)}$$

(*ii*) Absolute velocity of water leaving the vane, V_2:

$$V_2 = \sqrt{V_{f2}^2 + V_{w2}^2} = \sqrt{5.05^2 + 14.08^2} = \mathbf{14.96\ m/s\ (Ans.)}$$

(iii) Angle made by the absolute velocity at outlet with the direction of motion, β:

From velocity triangle at *outlet*, we have:

$$\tan\beta = \frac{V_{f2}}{V_{w2}} = \frac{5.05}{14.08} = 0.3586 \quad \therefore \ \beta = \tan^{-1}(0.3586) = \mathbf{19.7°\ (Ans.)}$$

(iv) Rate of flow through the pump, Q:

$$Q = \pi D_2 B_2 \times V_{f2} = \pi \times 0.48 \times 0.06 \times 5.05 = \mathbf{0.457\ m^3/s\ (Ans.)}$$

Example 3.5. *A centrifugal pump impeller runs at 80 r.p.m. and has outlet vane angle of 60°. The velocity of flow is 2.5 m/s throughout and diameter of the impeller at exit is twice that at inlet. If the manometric head is 20 m and the manometric efficiency is 75 percent, determine:*

(i) The diameter of the impeller at the exit, and

(ii) Inlet vane angle.

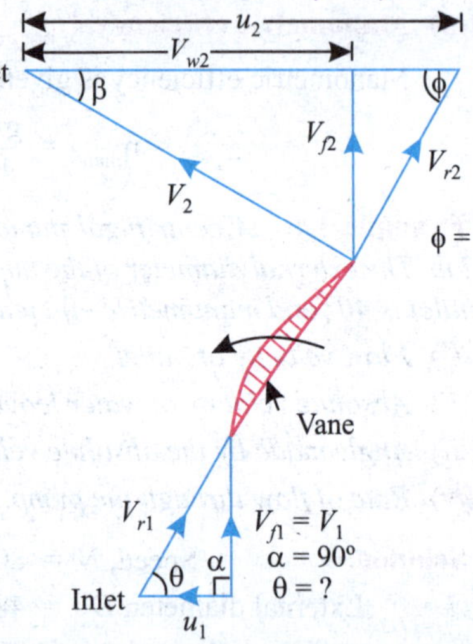

Fig. 3.10

Solution. Speed, N = 80 r.p.m.; Outlet vane angle, ϕ = 60°; Velocity of flow, $V_{f1} = V_{f2}$ = 2.5 m/s; manometric head, H_{mano} = 20 m; Manometric efficiency, η_{mano} = 75%;

Diameter of the impeller at outlet,

$$D_2 = 2D_1 \text{(diameter at inlet)}$$

(i) The diameter of the impeller at the exit, D_2:

$$\eta_{mano} = \frac{gH_{mano}}{V_{w2}u_2} \quad \text{.... Eqn. (3.9)}$$

$$0.75 = \frac{9.81 \times 20}{V_{w2}u_2}, \text{ or, } V_{w2}u_2 = \frac{9.81\times 20}{0.75} = 261.6 \quad \text{...(i)}$$

From velocity triangle at outlet (Fig. 3.10), we have

$$\tan\phi = \frac{V_{f2}}{u_2 - V_{w2}}$$

or, $$u_2 - V_{w2} = \frac{V_{f2}}{\tan\phi}, \text{ or, } V_{w2} = u_2 - \frac{V_{f2}}{\tan\phi}$$

or, $$V_{w2} = u_2 - \frac{2.5}{\tan 60°} = u_2 - 1.44$$

Substituting this value of V_{w2} in (*i*), we get:

$$(u_2 - 1.44)\,u_2 = 261.6, \quad \text{or} \quad u_2^2 - 1.44u_2 - 261.6 = 0$$

or, $$u_2 = \frac{1.44 \pm \sqrt{1.44^2 + 4\times 261.6}}{2} = \frac{1.44 \pm 32.38}{2} = 16.91 \text{ m/s (ignoring –ve sign)}$$

Also, tangential velocity of impeller at outlet, $u_2 = \dfrac{\pi D_2 N}{60}$, or, $D_2 = \dfrac{60u_2}{\pi N}$

$$\therefore \quad D_2 = \frac{60\times 16.91}{\pi \times 80} = 4.037 \text{ m} \simeq \mathbf{4\ m\ (Ans.)}$$

(ii) Inlet vane angle, θ:

Tangential velocity of the impeller at inlet, $u_1 = \dfrac{u_2}{2} = \dfrac{16.91}{2} = 8.455$ m/s $\quad \left(\because D_1 = \dfrac{D_2}{2}\right)$

From velocity triangle at *inlet*, we have:

$$\tan\theta = \frac{V_{f1}}{u_1} = \frac{2.5}{8.455} = 0.2957$$

$$\therefore \qquad \theta = \tan^{-1}(0.2957) = \mathbf{16.47° \ (Ans.)}$$

Example 3.6. *A centrifugal pump impeller having external and intenal diameters 480 mm and 240 mm respectively is running at 100 r.p.m. The rate of flow through the pump is 0.0576 m³/s and velocity of flow is constant and equal to 2.4 m/s. The diameters of the suction and delivery pipes are 180 mm and 120 mm respectively and suction and delivery heads are 6.2 m (abs.) and 30.2 m of water respectively. If the power required to drive the pump is 23.3 kW and the outlet vane angle is 45°, determine:*

(i) Inlet vane angle,

(ii) The overall efficiency of the pump, and

(iii) The manometric efficiency of the pump.

Solution. External diameter of the impeller,

$$D_2 = 480 \text{ mm} = 0.48 \text{ m},$$

$$\text{Internal diameter, } D_1 = 240 \text{ mm} = 0.24 \text{ m},$$

$$\text{Speed, } N = 1000 \text{ r.p.m.},$$

$$\text{Discharge, } Q = 0.0567 \text{ m}^3/\text{s};$$

$$\text{Velocity of flow, } V_{f1} = V_{f2} = 2.4 \text{ m/s}$$

The diameter of suction pipe,

$$D_s = 180 \text{ mm} = 0.18 \text{ m}$$

The diameter of delivery pipe,

$$D_d = 120 \text{ mm} = 0.12 \text{ m}$$

$$\text{Suction head, } h_s = 6.2 \text{ m (abs.)}$$

$$\text{Delivery head, } h_d = 30.2 \text{ m (abs.)}$$

$$\text{Shaft power, } P = 23.3 \text{ kW}$$

$$\text{Outlet vane angle, } \phi = 45°$$

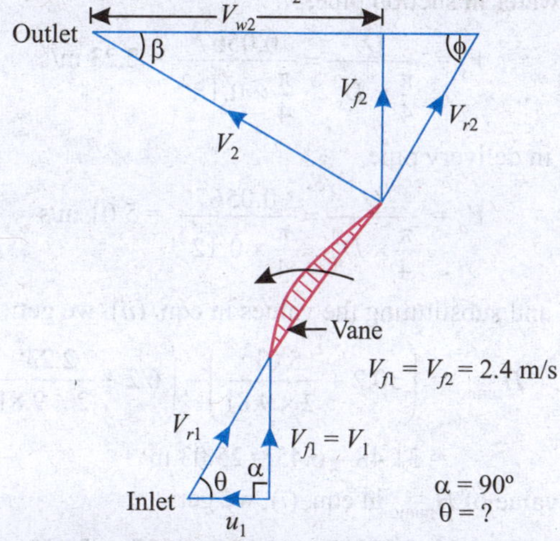

Fig. 3.11

(*i*) Inlet vane angle, θ:

Tangential velocity of impeller at *inlet*,

$$u_1 = \frac{\pi D_1 N}{60} = \frac{\pi \times 0.24 \times 1000}{60} = 12.56 \text{ m/s}$$

From velocity triangle at *inlet*, we have:

$$\tan\theta = \frac{V_{f1}}{u_1} = \frac{2.4}{12.56} = 0.191$$

$$\therefore \quad \theta = \tan^{-1}(0.191) = \mathbf{10.8°} \text{ (Ans.)}$$

(*ii*) The overall efficiency of the pump, η_0:

The overall efficiency of a centrifugal pump is given by,

$$\eta_0 = \frac{wQH_{mano}}{P} = \frac{9.81 \times 0.0567 \times H_{mano}}{23.3} = 0.02387\, H_{mano}$$

(where, $w = 9.81$ kN/m^3)

Also H_{mano} is given by eqn. (3.8) as:

$$H_{mano} = \left(\frac{p_2}{w} + \frac{V_2^2}{2g} + z_2\right) - \left(\frac{p_1}{w} + \frac{V_1^2}{2g} + z_1\right) \quad \text{...(ii)}$$

where, $\frac{p_2}{w}$ = Pressure head at pump outlet, $h_d = 30.2$ m,

$\frac{V_2^2}{2g}$ = Velocity head at pump outlet = $\frac{V_d^2}{2g}$,

z_2 = Vertical height of pump outlet from datum line,

$\frac{p_1}{w}$ = Pressure heat at pump inlet, $h_s = 6.2$ m (abs.),

$\frac{V_1^2}{2g}$ = Velocity head at pump inlet, $\frac{V_s^2}{2g}$, and

z_1 = Vertical height of pump inlet from datum line.

Now, velocity of water in suction pipe,

$$V_s = \frac{Q}{\frac{\pi}{4} \times D_s^2} = \frac{0.0567}{\frac{\pi}{4} \times 0.18^2} = 2.23 \text{ m/s}$$

Velocity of water in delivery pipe,

$$V_d = \frac{Q}{\frac{\pi}{4} \times D_d^2} = \frac{0.0567}{\frac{\pi}{4} \times 0.12^2} = 5.01 \text{ m/s}$$

Assuming $z_2 = z_1$ and substituting the values in eqn. (*ii*), we get:

$$H_{mano} = \left(30.2 + \frac{5.01^2}{2 \times 9.81}\right) - \left(6.2 + \frac{2.23^2}{2 \times 9.81}\right)$$

$$= 31.48 - 6.45 = 25.03 \text{ m}$$

Substituting this value of H_{mano} in eqn. (*i*), we get:

$$\eta_0 = 0.02387 \times 25.03 = 0.597 \text{ or } \mathbf{59.7\%} \text{ (Ans.)}$$

(iii) The manometric efficiency of the pump, η_{mano}:

$$\eta_{mano} = \frac{gH_{mano}}{V_{w2}u_2} \quad \text{...Eqn. (3.9)}$$

$$u_2 = \frac{\pi D_2 N}{60} = \frac{\pi \times 0.48 \times 1000}{60} = 25.13 \text{ m/s}$$

From velocity triangle at outlet (Fig. 3.11), we have:

$$\tan\phi = \frac{V_{f2}}{u_2 - V_{w2}}, \text{ or, } \tan 45° = \frac{2.4}{25.13 - V_{w2}}, \text{ or, } 25.13 - V_{w2} = \frac{2.4}{\tan 45°} = 2.4$$

$$\therefore \quad V_{w2} = 25.13 - 2.4 = 22.73 \text{ m/s}$$

Substituting the values in the above equation, we get:

$$\eta_{mano} = \frac{9.81 \times 25.03}{22.73 \times 25.13} = 0.43 \text{ or } \mathbf{43\%} \text{ (Ans.)}$$

Example 3.7. *It is required to deliver 0.048 m^3/s of water to a height of 24 m through a 150 mm diameter pipe and 120 m long, by a centrifugal pump. If the overall efficiency of the pump is 75 percent and co-efficient of friction, f = 0.01 for thr pipe line, find the power required to drive the pump.*

Solution. Rate of flow, $Q = 0.048 \text{ m}^3/\text{s}$; Height, $H_{stat} = h_s + h_d = 24$ m

Diameter of pipe, $D_s = D_d = D = 150$ mm, or, 0.15 m,

Length, $L_s + L_d = L = 120$ m

Overall efficiency, $\eta_o = 75\%$

Co-efficient of friction, $f = 0.01$

Power required to drive the pump, P:

Velocity of water in pipe, $V_s = V_d = V = \dfrac{Q}{\text{Area of pipe}} = \dfrac{0.048}{\frac{\pi}{4} \times 0.15^2} = 2.7 \text{ m/s}$

Loss of head due to friction in pipe,

$$(h_{fs} = h_{fd}) = \frac{4fLV^2}{D \times 2g} = \frac{4 \times 0.01 \times 120 \times 2.7^2}{0.15 \times 2 \times 9.81} = 11.89 \text{ m}$$

The manometric head (H_{mano}) is given by eqn. (3.7) as:

$$H_{mano} = (h_s + h_d) + (h_{fs} + h_{fd}) + \frac{V_d^2}{2g}$$

$$= 24 + 11.89 + \frac{2.7^2}{2 \times 9.81} = 36.26 \text{ m}$$

Using the relation : $\eta_0 = \dfrac{wQH_{mano}}{P}$, we get:

$$0.75 = \frac{9.81 \times 0.048 \times 36.26}{P} \quad (\because w = 9.81 \text{ kN/m}^3)$$

or,

$$P = \frac{9.81 \times 0.048 \times 36.26}{0.75} = \mathbf{22.76 \text{ kW}} \text{ (Ans.)}$$

Example 3.8. *The impeller of a centrifugal pump is of 300 mm diameter and 50 mm width at the periphery, and has blades whose tip angle incline backwards 60° from the radius. The pump delivers 17 m³/min. of water and the impeller rotates at 1000 r.p.m. Assuming that the pump is designed to admit radially, calculate:*

(i) Speed and direction of water as it leaves the impeller,

(ii) Torque exerted by the impeller on water,

(iii) Shaft power required, and

(iv) Lift of the pump.

Take, mechanical efficiency = 95% and hydraulic efficiency = 75%.

[Rajasthan University]

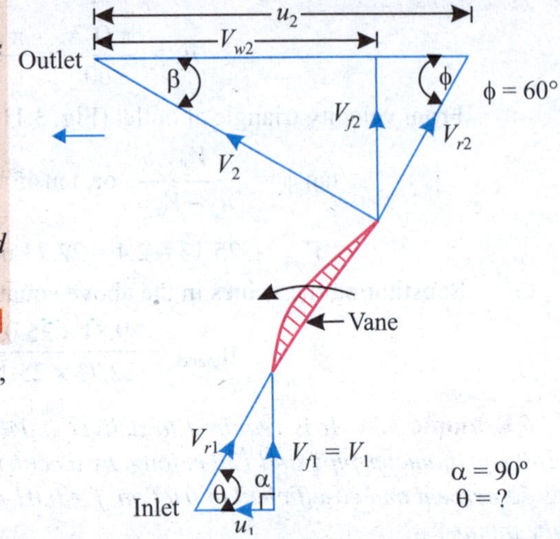

Fig. 3.12

Solution. External diameter of impeller, $D_2 = 300$ mm = 0.3 m

Width at periphery, B_2 = 50 mm = 0.05 m

Outlet vane angle, $\phi = 60°$

Discharge, $Q = 17$ m³/min., or, 0.2833 m³/s

Speed of the impeller, $N = 1000$ r.p.m.

$\eta_m = 95\%$; $\eta_h = 70\%$.

(i) Speed and direction of water as it leaves the impeller, V_2, β:

Tangential velocity of impeller at *outlet*,

$$u_2 = \frac{\pi D_2 N}{60} = \frac{\pi \times 0.3 \times 1000}{60} = 15.71 \text{ m/s}$$

Also,
$$Q = \pi D_2 B_2 \times V_{f2}, \text{ or, } V_{f2} = \frac{Q}{\pi D_2 B_2}$$

$$\therefore \quad V_{f2} = \frac{0.2833}{\pi \times 0.3 \times 0.05} = 6.01 \text{ m/s}$$

From velocity triangle at *outlet*, (Fig. 3.12), we have:

$$\tan\phi = \frac{V_{f2}}{u_2 - V_{w2}}, \text{ or, } u_2 - V_{w2} = \frac{V_{f2}}{\tan\phi}$$

or,
$$V_{w2} = u_2 - \frac{V_{f2}}{\tan\phi} = 15.71 - \frac{6.01}{\tan 60°} = 12.24 \text{ m/s}$$

∴ Absolute velocity of water at outlet tip of impeller,

$$V_2 = \sqrt{V_{w2}^2 + V_{f2}^2} = \sqrt{12.24^2 + 6.01^2} = \mathbf{13.63 \text{ m/s (Ans.)}}$$

$$\tan\beta = \frac{V_{f2}}{V_{w2}} = \frac{6.01}{12.24} = 0.491$$

∴ The direction of outgoing velocity, $\beta = \tan^{-1}(0.491) =$ **26.15° (Ans.)**

(ii) Torque exerted by the impeller on water, T:

$$T = \frac{wQ}{g}(V_{w2} R_2)$$

$$= \frac{9.81 \times 0.2833}{9.81} \times 12.24 \times \left(\frac{0.3}{2}\right) = \mathbf{0.52\ kNm\ (Ans.)}$$

$(\because w = 9.81 \text{ kN/m}^3)$

(iii) Shaft power required, P:

$$\text{Impeller or rotor power} = \frac{2\pi NT}{60} = \frac{2\pi \times 1000 \times 0.52}{60} = 54.45 \text{ kW}$$

$$\eta_m = \frac{\text{Impeller power}}{\text{Shaft power}}, \text{ or, } 0.95 = \frac{54.45}{P}$$

$$P = \frac{54.45}{0.95} = \mathbf{57.31\ kW\ (Ans.)}$$

(iv) Lift of the pump:

$$\text{Impeller power} = w\,(Q + q)\,H_i$$

where, w = Weight density of water, kN/ m^3,

H_i = Ideal head, m (= theoretical head – hydraulic losses), and

q = Leakage of water, m^3/s.

Neglecting leakage effects, we obtain:

$$54.45 = 9.81 \times 0.2833 \times H_i, \text{ or, } H_i = \frac{54.45}{9.81 \times 0.2833} = 19.59 \text{ m}$$

$$\text{Now, hydraulic efficiency, } \eta_h = \frac{\text{Actual head or lift}}{\text{Ideal head}}$$

$\therefore$ Lift of the pump $= \eta_h \times$ ideal head $(H_i) = 0.70 \times 19.59 =$ **13.71 m of water (Ans.)**

Example 3.9. *A centrifugal pump (diffusion type) has suction lift of 1.8 m and the delivery tank is 14.2 m above the pump. The velocity of water in the delivery pipe is 1.6 m/s. The radial velocity of flow through the wheel is 2.8 m/s and the tangent to the vane at the exit from the wheel makes an angle of 120° with the dirction of motion. Assuming that the water enters radially and neglecting friction and other losses, determine:*

(i) Velocity of wheel at the exit,

(ii) Velocity and pressure head at exit from the wheel, and

(iii) Direction of the fixed guide vane.

Solution. Suction lift, h_s = 1.8 m

Delivery head, h_d = 14.2 m

The velocity of water in delivery pipe, V_d = 1.6 m/s

The radial velocity of flow, $V_{f1} = V_{f2} = 2.8$ m/s

Outlet vane angle, $\phi = 180° - 120° = 60°$.

(i) Velocity of wheel at the exit, u_2:

Head against which pump has to work,

$$H = h_s + h_d \quad \text{(neglecting friction and other losses)}$$

$$= 1.8 + 14.2 = 16 \text{ m of water}$$

Also, $$H = \frac{V_{w2} u_2}{g} \quad \text{...Euler equation}$$

$$V_{w2} u_2 = gH = 9.81 \times 16 = 156.96 \quad ...(i)$$

From velocity triangle at outlet (Fig. 3.13), we have:

$$\tan\phi = \frac{V_{f2}}{u_2 - V_{w2}}, \text{ or } u_2 - V_{w2} = \frac{V_{f2}}{\tan\phi}$$

or,
$$V_{w2} = u_2 - \frac{V_{f2}}{\tan\phi}$$

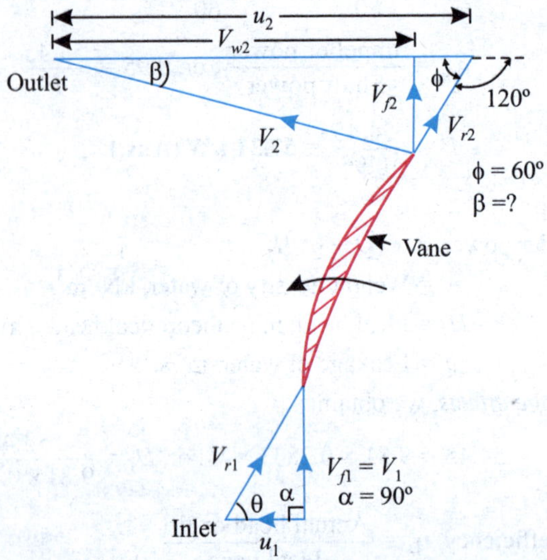

Fig. 3.13

$$\therefore \quad V_{w2} = u_2 - \frac{2.8}{\tan 60^\circ} = u_2 - 1.616 \qquad ...(ii)$$

Substituting this value of V_{w2} in (*i*), we get:

$$(u_2 - 1.616)\, u_2 = 156.96, \text{ or, } u_2^2 - 1.616\, u_2 - 156.96 = 0$$

$$\therefore \quad u_2 = \frac{1.616 \pm \sqrt{1.616^2 + 4 \times 156.96}}{2} = \frac{1.616 \pm 25.108}{2} = 13.36 \text{ m/s (ignoring –ve sign)}$$

Hence, *the velocity of wheel at exit* = **13.36 m/s (Ans.)**

(*ii*) Velocity and pressure head at exit from the wheel:

From eqn. (*i*), we have: $V_{w2} = \frac{156.96}{u_2} = \frac{156.96}{13.36} = 11.75$ m/s

Absolute velocity, $V_2 = \sqrt{V_{w2}^2 + V_{f2}^2} = \sqrt{11.75^2 + 2.8^2} = 12.08$ m/s

$\therefore$ Velocity head at exit $= \frac{V_2^2}{2g} = \frac{12.08^2}{2 \times 9.81} =$ **7.44 m of water (Ans.)**

Pressure head at exit $= \frac{p_2}{w} = h_d - \frac{V_2^2}{2g}$

$= 14.2 - 7.44 =$ **6.76 m of water (Ans.)**

(*iii*) Direction of fixed guide vane, β:

$$\tan\beta = \frac{V_{f2}}{V_{w2}} = \frac{2.8}{11.75} = 0.2383$$

$\beta = \tan^{-1}(0.2383) =$ **13.4° (Ans.)**

Example 3.10. *Show that the rise of pressure in the impeller of a centrifugal pump when frictional and other losses in the impeller are neglected, is given by*

$$\frac{1}{2g}[V_{f1}^2 + u_2^2 - V_{f2}^2 \operatorname{cosec}^2 \phi]$$

where, V_{f1}, V_{f2} = *Velocities of flow at inlet and outlet respectively,*

u_2 = *Tangential velocity of impeller at outlet, and*

ϕ = *Vane angle at outlet.*

Solution. Let the values at inlet and outlet of impeller are represented by the suffices 1 and 2 respectively.

Invoking Bernoulli's equation at the inlet and outlet of the impeller, neglecting losses from inlet to outlet, we have:

$$(\text{Total energy})_{\text{inlet}} = (\text{Total energy})_{\text{outlet}} - \text{work done by impeller on water}$$

$$\frac{p_1}{w} + \frac{V_1^2}{2g} + z_1 = \left(\frac{p_2}{w} + \frac{V_2^2}{2g} + z_2\right)$$

$$- \text{work done by impeller on water per unit weight of water}$$

$$= \frac{p_2}{w} + \frac{V_2^2}{2g} + z_2 - \frac{V_{w2} u_2}{g} \quad \text{(assuming flow to be \textit{radial at inlet})}$$

If the inlet and outlet of the impeller are at the same height (*i.e.* $z_1 = z_2$), then

$$\frac{p_1}{w} + \frac{V_1^2}{2g} = \frac{p_2}{w} + \frac{V_2^2}{2g} - \frac{V_{w2} u_2}{g}$$

$$\therefore \quad \left(\frac{p_2}{w} - \frac{p_1}{w}\right) = \frac{V_1^2}{2g} - \frac{V_2^2}{2g} + \frac{V_{w2} u_2}{g} \quad \text{...(i)}$$

where, $\frac{p_2}{w} - \frac{p_1}{w}$ = Pressure rise in impeller.

Refer to Fig. 3.8. From velocity triangle at *inlet*, we have $V_1 = V_{f1}$ ... (*ii*)

From velocity triangle at outlet, we have:

$$\tan \phi = \frac{V_{f2}}{u_2 - V_{w2}}, \text{ or, } u_2 - V_{w2} = \frac{V_{f2}}{\tan \phi}$$

or, $$V_{w2} = u_2 - \frac{V_{f2}}{\tan \phi} = u_2 - V_{f2} \cot \phi \quad \text{...(iii)}$$

Also, $$V_2^2 = V_{f2}^2 + V_{w2}^2 = V_{f2}^2 + (u_2 - V_{f2} \cot \phi)^2$$

$$= V_{f2}^2 + (u_2^2 + V_{f2}^2 \cot^2 \phi - 2u_2 V_{f2} \cot \phi)$$

$$= V_{f2}^2 + V_{f2}^2 \cot^2 \phi + u_2^2 - 2u_2 V_{f2} \cot \phi$$

$$= V_{f2}^2 (1 + \cot^2 \phi) + u_2^2 - 2u_2 V_{f2} \cot \phi$$

$$= V_{f2}^2 \operatorname{cosec}^2 \phi + u_2^2 - 2u_2 V_{f2} \cot \phi \quad \text{...(iv)}$$

$$(\because 1 + \cot^2 \phi = \operatorname{cosec}^2 \phi)$$

Substituting the values of V_1, V_{w2} and V_2^2 from eqns. (*ii*), (*iii*), and (*iv*), respectively in eqn. (*i*), we get:

$$\textit{Pressure rise in impeller} = \frac{V_{f1}^2}{2g} - \frac{1}{2g}(V_{f2}^2 \text{cosec}^2\phi + u_2^2 - 2u_2V_{f2}\cot\phi) + \frac{(u_2 - V_{f2}\cot\phi)\times u_2}{g}$$

$$= \frac{1}{2g}[V_{f1}^2 - V_{f2}^2 \text{cosec}^2\phi - u_2^2 + 2u_2V_{f2}\cot\phi + 2u_2^2 - 2u_2V_{f2}\cot\phi]$$

$$= \frac{1}{2g}[V_{f1}^2 + u_2^2 - V_{f2}^2 \text{cosec}^2\phi] \quad ...(3.16)$$

Example 3.11. *The following data relate to a centrifugal pump:*
The diameters of the impeller at inlet and outlet = 180 mm and 360 mm respectively
The widths of the impeller at inlet and outlet = 14.4 mm and 7.2 mm respectively
The rate of flow through the pump = 17.28 litres/s
Speed of the impeller = 1500 r.p.m.
Vane angle at the outlet = 45°
The water enters the impeller radially at inlet.
Neglecting losses through the impeller, find the pressure rise in the impeller

Solution. *Given :* D_1 = 180 mm = 0.18 m; D_2 = 360 mm = 0.36 m; B_1 = 14.4 mm = 0.0144 m; B_2 = 7.2 mm = 0.0072 m; Q = 17.26 litres/s = 0.01728 m^3/s ; N=1500 r.p.m.; ϕ = 45°

Pressure rise in the impeller:

Velocity of flow at inlet, $V_{f1} = \dfrac{Q}{\pi D_1 B_1} = \dfrac{0.01728}{\pi \times 0.18 \times 0.0144} = 2.12$ m/s

Velocity of flow at outlet, $V_{f2} = \dfrac{Q}{\pi D_2 B_2} = \dfrac{0.01728}{\pi \times 0.36 \times 0.0072} = 2.12$ m/s

Tangential velocity of impeller at outlet,

$$u_2 = \frac{\pi D_2 N}{60} = \frac{\pi \times 0.36 \times 1500}{60} = 28.27 \text{ m/s}$$

$$\textit{Pressure rise in impeller} = \frac{1}{2g}[V_{f1}^2 + u_2^2 - V_{f2}^2 \text{cosec}^2\phi] \quad ...[\text{Eqn. } 3.16]$$

$$= \frac{1}{2\times 9.81}[2.12^2 + 28.27^2 - 2.12^2 \times \text{cosec}^2\ 45°]$$

$$= \frac{1}{2\times 9.81}(4.494 + 799.193 - 8.988) = \mathbf{40.5\ m} \ \textbf{(Ans.)}$$

Example 3.12. *A centrifugal pump impeller has at outlet a diameter of 360 mm and width 60 mm. The vanes are curved backwards at 35° to the tangent at outer periphery and thickness of vanes occupies 20 percent of the peripheral area and the velocity of flow is constant from inlet to outlet. The impeller rotates at 800 r.p.m. If the rate of flow through the pump is 0.13 m^3/s, determine:*
(*i*) *The pressure rise in the impeller, and*
(*ii*) *The percentage of total work converted to kinetic energy.*

Solution. Diameter of the impeller at outlet, D_2 = 360 mm = 0.36 m
Width of impeller at outlet, B_2 = 60 mm = 0.06 m
Outlet vane angle, ϕ = 35°
Area occupied by thickness of vanes = 20% of the peripheral area

Thickness co-efficient, $K_t = 1 - 0.2 = 0.8$

Speed of impeller, $N = 800$ r.p.m.

Rate of flow through the pump,

$$Q = 0.13 \text{ m}^3/\text{s}$$

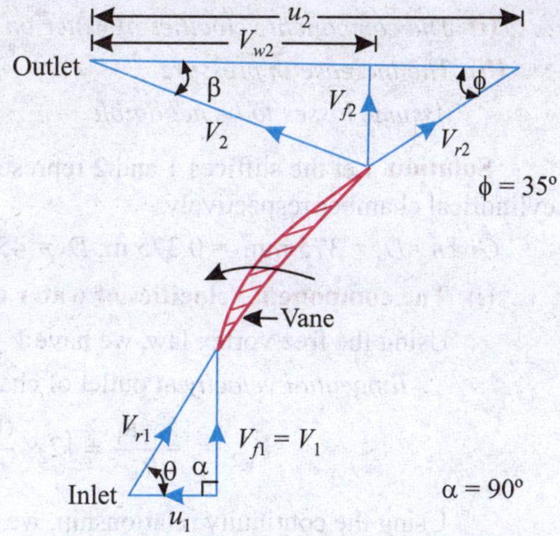

Fig. 3.14

***(i)* Pressure rise in the impeller:**

Tangential velocity of impeller at outlet,

$$u_2 = \frac{\pi D_2 N}{60} = \frac{\pi \times 0.36 \times 800}{60} = 15.08 \text{ m/s}$$

Rate of flow, $Q = K_t \times \pi D_2 B_2 \times V_{f2}$

or, $\quad 0.13 = 0.8 \times \pi \times 0.36 \times 0.06 \times V_{f2}$

$$\therefore V_{f2} = \frac{0.13}{0.8 \times \pi \times 0.36 \times 0.06} = 2.39 \text{ m/s}$$

Since the velocity of flow is constant from inlet to outlet, $V_{f1} = V_{f2} = 2.39$ m/s

From velocity triangle at *outlet* (Fig, 3.14), we have:

$$\tan \phi = \frac{V_{f2}}{u_2 - V_{w2}}, \text{ or, } u_2 - V_{w2} = \frac{V_{f2}}{\tan \phi}, \text{ or, } V_{w2} = u_2 - \frac{V_{f2}}{\tan \phi}$$

$$\therefore \quad V_{w2} = 15.08 - \frac{2.39}{\tan 35°} = 11.66 \text{ m/s}$$

Now, $\quad$ Pressure rise $= \dfrac{1}{2g}[V_{f1}^2 + u_2^2 - V_{f2}^2 \operatorname{cosec}^2\phi]$ ...[Eqn. 3.16]

$$= \frac{1}{2 \times 9.81}[2.39^2 + 15.08^2 - 2.39^2 \times \operatorname{cosec}^2 35°]$$

$$= \frac{1}{2 \times 9{\cdot}81}[5.712 + 227.406 - 17.362] \simeq \textbf{11 m of water} \text{ (Ans.)}$$

***(ii)* The percentage of work converted to kinetic energy:**

Absolute velocity of water leaving the vane,

$$V_2 = \sqrt{V_{f2}^2 + V_{w2}^2} = \sqrt{2.39^2 + 11.66^2} = 11.9 \text{ m/s}$$

Kinetic energy per unit weight of water $= \dfrac{V_2^2}{2g} = \dfrac{11.9^2}{2 \times 9.81} = 7.217$ Nm

Work done per unit weight of water $= \dfrac{V_{w2}u_2}{g} = \dfrac{11.66 \times 15.08}{9.81} = 17.92$ Nm

$\therefore$ Percentage of work converted to kinetic energy

$$= \frac{7.217}{17.92} \times 100 = \textbf{40.27\%} \text{ (Ans.)}$$

Example 3.13. *A pump impeller is 375 mm in diameter and it discharges water with velocity components of 2 m/s and 12 m/s in the radial and tangential directions respectively. The impeller is surrounded by a concentrical cylindrical chamber with parallel sides; the outer diameter being 450 mm. If the flow in the chamber is a free spiral vortex, find:*

(i) The component velocities of water on leaving, and

(ii) The increase in pressure.

Assume losses to be negligible.

Solution. Let the suffices 1 and 2 represent the conditions at inlet and outlet of the concentric cylindrical chamber respectively.

Given : $D_1 = 375$ mm $= 0.375$ m; $D_2 = 450$ mm $= 0.45$ m; $V_{f1} = 2$ m/s; $V_{w1} = 12$ m/s.

(i) The component velocities of water at outlet of the chamber:

Using the free vortex law, we have $V_{w1} R_1 = V_{w2} R_2$

∴ *Tangential velocity* at outlet of chamber,

$$V_{w2} = \frac{V_{w1} R_1}{R_2} = 12 \times \frac{(0.375 / 2)}{(0.45 / 2)} = \mathbf{10\ m/s}\ \textbf{(Ans.)}$$

Using the continuity relationship, we have:

$$\pi D_1 B_1 \times V_{f1} = D_2 B_2 \times Vf_2$$

(where, V_{f2} = *radial velocity* at outlet of chamber)

But, $B_1 = B_2$ ... for a parallel sided chamber

$$\therefore \quad D_1 \times V_{f1} = D_2 \times V_{f2}, \text{ or, } V_{f2} = \frac{D_1 V_{f1}}{D_2} = \frac{0.375 \times 2}{0.45} = \mathbf{1.667\ m/s}\ \textbf{(Ans.)}$$

(ii) The increase in pressure:

Invoking Bernoulli's equation assuming no loss of head and no change in datum, we have:

$$\frac{p_1}{w} + \frac{V_1^2}{2g} = \frac{p_2}{w} + \frac{V_2^2}{2g}$$

$$\therefore \quad \textit{The increase in pressure,} = \frac{p_2 - p_1}{w} = \frac{V_1^2 - V_2^2}{2g}$$

$$= \frac{(12^2 + 2^2) - (10^2 + 1.667^2)}{2 \times 9.81} = \mathbf{2.3\ m\ of\ water}\ \textbf{(Ans.)}$$

Example 3.14. *A centrifugal pump, in which water enters radially, delivers water to a height of 165 mm. The impeller has a diameter of 360 mm and width 180 mm at inlet and the corresponding dimensions at the outlet are 720 mm and 90 mm respectively; its rotational speed is 1200 r.p.m. The blades are curved backward at 30° to the tangent at exit and the discharge is 0.389 m³/s. Determine:*

(i) Theoretical head developed,

(ii) Manometric efficiency

(iii) Pressure rise across the impeller assuming losses equal to 12 percent of velocity head at exit,

(iv) Pressure rise and the loss of head in the volute casing,

(v) The vane angle at inlet, and

(vi) Power required to drive the pump assuming an overall efficiency of 70%. What would be corresponding mechanical efficiency ?

Solution. Manometric head, $H_{mano} = 165$ m

Diameter of impeller at inlet, $D_1 = 360$ mm $= 0.36$ m

Width at inlet, $B_1 = 180$ mm $= 0.18$ m

Diameter of impeller at outlet,

$$D_2 = 720 \text{ mm} = 0.72 \text{ m}$$

Width at outlet, $B_2 = 90$ mm $= 0.09$ m

Speed, $N = 1200$ r.p.m.

Outlet vane angle, $\phi = 30°$

Discharge, $Q = 0.389 \text{ m}^3/\text{s}$

$\eta_0 = 70\%$.

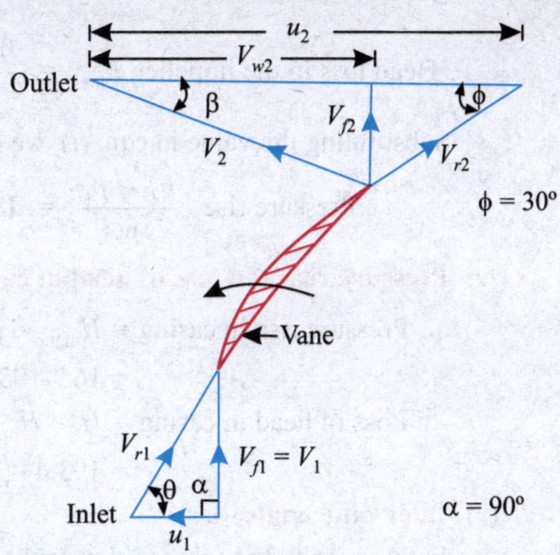

Fig. 3. 15

(i) Theoretical head developed:

In case of a centrifugal pump the Euler or outlet theoretical head is given by:

$$H_e = \frac{V_{w2}u_2 - V_{w1}u_1}{g}$$

As the flow is radial, $\alpha = 90°$, and $V_{w1} = 0$

$$\therefore \quad H_e = \frac{V_{w2}u_2}{g}$$

Now,

$$u_2 = \frac{\pi D_2 N}{60} = \frac{\pi \times 0.72 \times 1200}{60} = 45.24 \text{ m/ s}$$

$$V_{f2} = \frac{Q}{\pi D_2 B_2} = \frac{0.389}{\pi \times 0.72 \times 0.09} = 1.91 \text{ m/s}$$

From velocity triangle at *outlet* (Fig. 3.15), we have:

$$\tan\phi = \frac{V_{f2}}{u_2 - V_{w2}}, \text{ or, } u_2 - V_{w2} = \frac{V_{f2}}{\tan\phi}, \text{ or, } V_{w2} = u_2 - \frac{V_{f2}}{\tan\phi}$$

$$\therefore \quad V_{w2} = 45.24 - \frac{1.91}{\tan 30°} = 41.93 \text{ m/s}$$

$$\therefore \quad H_e \text{ (Euler head)} = \frac{V_{w2}u_2}{g} = \frac{41.93 \times 45.24}{9.81} = \mathbf{193.4 \text{ m}} \text{ (Ans.)}$$

(ii) Manometric efficiency, η_{mano}:

$$\eta_{mano} = \frac{\text{Manometric head}}{\text{Theoretical head}} = \frac{165}{193.4} = 0.853 \text{ or } \mathbf{85.3\%} \text{ (Ans.)}$$

(iii) Pressure rise across the impeller:

Applying energy equation between the inlet (suffix 1) and outlet (suffix 2), we have

$$\frac{p_1}{w} + \frac{V_1^2}{2g} + z_1 + H_e = \frac{p_2}{w} + \frac{V_2^2}{2g} + z_2 + h_{loss}$$

(where, h_{loss} = head loss in impeller)

Pressure rise,

$$\frac{p_2 - p_1}{w} = H_e + \frac{V_1^2 - V_2^2}{2g} - h_{loss} \quad ...(i) \text{ (Assuming } z_1 = z_2)$$

where,

$$V_1^2 = V_{f1}^2 = 1.912 = 3.648$$

$$V_2^2 = V_{f1}^2 + V_{w2}^2 = 1.91^2 + 41.93^2 = 1761.77$$

$$(\because V_{f1} = V_{f2} = 1.91 \text{ m/s})$$

$$\text{Head loss in the impeller, } h_{loss} = \frac{0.12V_2^2}{2g} = \frac{0.12 \times 1761.77}{2 \times 9.81} = 10.77 \text{ m}$$

Substituting the value in eqn. (*i*), we have:

$$\text{Pressure rise, } \frac{p_2 - p_1}{w} = 193.4 + \frac{3.648 - 1761.77}{2 \times 9.81} - 10.77 = \mathbf{93.02\ m}\ \textbf{(Ans.)}$$

(*iv*) Pressure rise and loss of head in casing:

Pressure rise in casing = H_{mano} – pressure rise in impeller

= 165 – 93.02 = **71.98 m (Ans.)**

Loss of head in casing = $H_e - H_{mano} - h_{loss}$

= 193.4 – 165 – 10.77 = **17.63 m (Ans.)**

(*v*) Inlet vane angle, θ:

From velocity triangle at *inlet*, we have:

$$\tan\theta = \frac{V_{f1}}{u_1} = \frac{1.91}{(\pi D_1 N/60)} = \frac{1.91}{(\pi \times 0.36 \times 1200/60)} = 0.0844$$

$$\therefore \quad \theta = \tan^{-1}(0.0844) = \mathbf{4.8°}\ \textbf{(Ans.)}$$

(*vi*) Power required to drive the pump, P:

$$\text{Overall efficiency,} \quad \eta_0 = \frac{wQH_{mano}}{P}$$

$$\text{or} \quad P = \frac{wQH_{mano}}{\eta_0} = \frac{9.81 \times 0.389 \times 165}{0.7} = 899.5 \text{ kW}$$

Mechanical efficiency, η_m:

We know that, $\eta_0 = \eta_{mano} \times \eta_v \times \eta_m$

Assuming volumetric efficiency (η_v) to be 100%, we have:

$$\eta_m = \frac{\eta_0}{\eta_{mano}} = \frac{0.7}{0.853} = 0.82 \text{ or } \mathbf{82\ \%}\ \textbf{(Ans.)}$$

Example 3.15. *A centrifugal pump lifts water under a static head of 36 m of water of which 4 m is suction lift. Suction and delivery pipes are both 150 mm in diameter. The head loss in suction pipe is 1.8 m and in delivery pipe 7 m. The impeller is 380 mm in diameter and 25 mm wide at mouth and revolves at 1200 r.p.m. Its exit blade angle is 35°. If the manometric efficiency of the pump is 82 percent determine:*

(*i*) *The discharge through the pump, and*

(*ii*) *The pressure at the suction and delivery branches of the pump,* **[M.U]**

Solution. Static head, H_{stat} = 36 m; Suction lift, h_s = 4 m;

The diameter of each of the suction and delivery pipes,

$D_s = D_d$ = 150 mm or 0.15 m;

Head lost in suction pipe, h_{fs} = 1.8 m;

Head lost in delivery pipe, h_{fd} = 7 m;

Diameter of impeller at the outlet, D_2 = 380 mm or 0.38 m

Width of impeller at the outlet, B_2 = 25 mm or 0.025 m

Speed of impeller, $N = 1200$ r.p.m.

Outlet blade/ vane angle, $\phi = 35°$

Manometric efficiency, $\eta_{mano} = 82\%$

***(i)* The discharge through the pump, Q :**

Total head to be supplied by the pump,

$$H_{mano} = (h_s + h_d) + h_{fs} + h_{fd} = 36 + 1.8 + 7 = 44.8 \text{ m}$$

[where $(h_s + h_d) = H_{stat} = 36$ m ...(*Given*)]

Tangential/periphery velocity of impeller/ wheel at outlet,

$$u_2 = \frac{\pi D_2 N}{60} = \frac{\pi \times 0.38 \times 1200}{60} = 23.87 \text{ m/s}$$

Assuming flow at the inlet to be *radial*, $\alpha = 90°$

Work done per unit weight of water $= \frac{V_{w2} u_2}{g}$

$$\text{Manometric efficiency, } \eta_{mano} = \frac{H_{mano}}{\frac{V_{w2} u_2}{g}} = \frac{g H_{mano}}{V_{w2} u_2}$$

$$\therefore \quad V_{w2} = \frac{g H_{mano}}{\eta_{mano} \times u_2} = \frac{9.81 \times 44.8}{0.82 \times 2387} = 22.45 \text{ m/s}$$

Now, $\tan \phi = \frac{V_{f2}}{u_2 - V_{w2}}$ (Refer to Fig. 3.8)

or, $\tan 35° = \frac{V_{f2}}{23.87 - 22.45}$, or, $V_{f2} = \tan 35° (23.87 - 22.45) = 0.99$ m/s

$\therefore$ Discharge, $Q = \pi D_2 B_2 \times V_{f2} = \pi \times 0.38 \times 0.025 \times 0.99 =$ **0.0295 m³/s (Ans.)**

***(ii)* The pressure at the suction and delivery branches of the pump :**

$$\text{Velocity in suction or delivery pipe, } V_s \text{ or } V_d = \frac{Q}{\text{Area of pipe}} = \frac{0.0295}{\frac{\pi}{4} \times 0.15^2} = 1.67 \text{ m/s}$$

$$\text{Velocity head, } \frac{V_s^2}{2g} = \frac{V_d^2}{2g} = \frac{1.67^2}{2 \times 9.81} = 0.142 \text{ m}$$

Total effective pressure on the delivery side

$$= h_d + h_{fd} + \frac{V_d^2}{2g} = 32 + 7 + 0.142 = 39.142 \text{ m of water}$$

$= 9.81 \times 39.142$ kN/m² = **383.98 kN/m² (Ans.)**

($\because p = wH; w = 9.81$ kN/m³)

(where $h_d = H_{stat} - h_s = 36 - 4 = 32$ m)

Pressure on the suction side $= h_s + h_{fs} + \frac{V_s^2}{2g} = 4 + 1.8 + 0.142 = 5.942$ m of water *vacuum*

or, $10 - 5.942 = 4.058$ m of water absolute

$= 9.81 \times 4.058$ kN/m² = **39.8 kN/m² absolute (Ans.)**

Example 3.16. *A centrifugal impeller has dimensions and blade angles as given in Fig. 3.16. Water at the rate of 60 litres per second enters the impeller radially and the radial velocity remains constant in the impeller. Determine the impeller speed and the torque produced by it.* **[GATE]**

data: $R_1 = 7.5$ cm

$R_2 = 15$ cm

$\beta_1 = \beta_2 = 30°$

Impeller inlet area, $A_1 = 250\text{ cm}^2$.

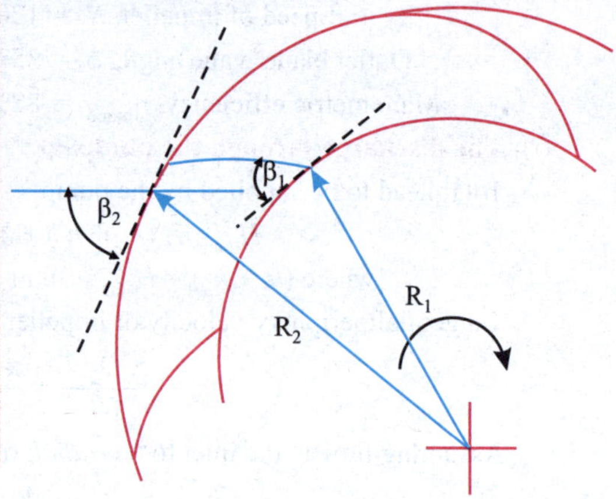

Fig. 3.16

Solution. *Given:* $Q = 60\ l/s = 0.06\text{ m}^3/\text{s}$;

$R_1 = 7.5\text{ cm} = 0.075\text{ m}$;

$R_2 = 15\text{ cm} = 0.15\text{ m}\ \beta_1 = \beta_2 = 30°$

Impeller inlet area, $A_1 = 250\text{ cm}^2$

$= 250 \times 10^{-4}\text{ m}^2$

$$\therefore \quad V_1 = \frac{Q}{A} = \frac{0.06}{250 \times 10^{-4}} = 2.4\text{ m/s}$$

If speed of the impeller is N r.p.m. then,

$$u_1 = \frac{2\pi R_1 N}{60} = \frac{2\pi \times 0.075N}{60} = 0.00785\text{ N}$$

$$u_2 = \frac{2\pi R_2 N}{60} = \frac{2\pi \times 0.15 \times N}{60} = 0.0157\text{ N}$$

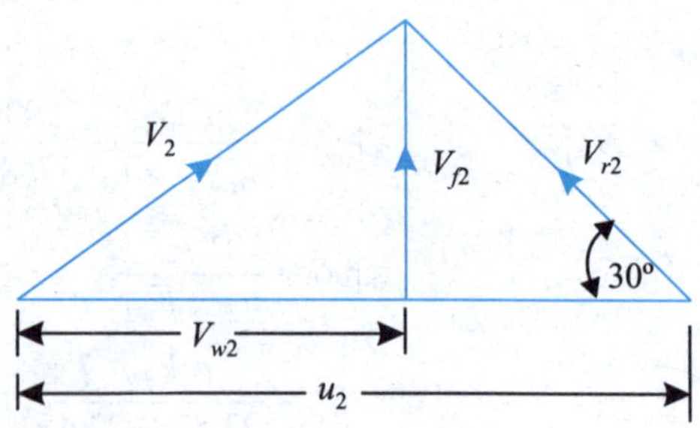

Outlet

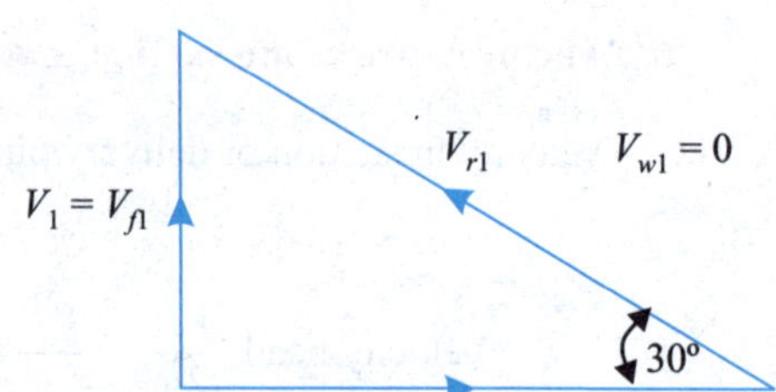

Inlet

Fig. 3.17

From *inlet velocity triangle*, we have:

$$\tan 30° = \frac{V_{f1}}{u_1}, \text{ or, } u_1 = \frac{V_{f1}}{\tan 30°}$$

$$= \frac{2.4}{\tan 30°} = 4.16\text{ m/s}$$

$$\therefore \quad u_1 = 4.16 = 0.00785\text{ N}$$

or, Impeller speed,

$$N = \frac{4.16}{0.00785} = \mathbf{530\ r.p.m.}\ \textbf{(Ans.)}$$

$$V_{r1} = \sqrt{V_{f1}^2 + u_1^2} = \sqrt{(2.4)^2 + (4.16)^2} = 4.8\text{ m/s} = V_{r2}$$

$$u_2 = 0.0157\text{ N} = 0.0157 \times 530 = 8.32\text{ m/s}$$

From *outlet velocity triangle*, we have:

$$V_{w2} = u_2 - V_{r2}\cos 30°$$

$$= 8.32 - 4.8 \times \cos 30° = 4.16\text{ m/s}$$

$$\text{Torque} = \frac{W}{g} \times V_{w2}R_2 = \frac{wQ}{g} \times V_{w2}R_2 = \frac{(1000 \times g) \times 0.06}{g} \times 4.16 \times 0.15 = \mathbf{37.44\ Nm}\ \textbf{(Ans.)}$$

Example 3.17. *A centrifugal pump is required to discharge 0.2 m^3 of water per second against a head of 22 m when the impeller rotates at a speed of 1500 r.p.m. The manometric efficiency is 75 percent. The loss of head in pump in metres due to fluid resistance is 0.03 V_2^2 where V_2 m/s is the velocity of water leaving the impeller. The area of the impeller outlet surface is 1.2 D_2^2 m^2, where D is the impeller diameter in m. Determine:*

(i) The impeller diameter, and (ii) The outlet vane angle.

Assume that the water enters the impeller without whirl.

Solution. Discharge through the centrifugal pump, $Q = 0.2 \text{ m}^3/\text{s}$

Manometric head, $H_{mano} = 22$ m

Speed of the impeller, $N = 1500$ r.p.m.

Manometric efficiency, $\eta_{mano} = 75\%$

Loss of head due to fluid resistance $= 0.03 V_2^2$

Area of impeller outlet surface $= 1.2 D_2^2$

(i) The impeller diameter, D_2 :

$$\eta_{mano} = \frac{gH_{mano}}{V_{w2}u_2} \quad \text{...[Eqn. 3.9]}$$

$$\therefore \quad = \frac{V_{w2}u_2}{g} = \frac{H_{mano}}{\eta_{mano}} = \frac{22}{0.75} = 29.33 \text{ m} \quad \text{...}(i)$$

Also, $$H_{mano} = \frac{V_{w2}u_2}{g} - \text{loss of head in the pump} \quad \text{[Eqn. 3.5]}$$

$\therefore$ Losses in the pump $= 29.33 - 22 = 7.33$ m

Thus, $$0.03 V_2^2 = 7.33, \text{ or, } V_2 = \left(\frac{7.33}{0.03}\right)^{1/2} = 15.63 \text{ m/s}$$

Velocity of flow at outlet, $$V_{f2} = \frac{Q}{\text{Area of flow}} = \frac{0.2}{1.2 D_2^2} = \frac{0.167}{D_2^2} \text{ m/s}$$

Peripheral or tangential velocity of impeller at outlet,

$$u_2 = \frac{\pi D_2 N}{60} = \frac{\pi \times D_2 \times 1500}{60} = 78.54 \text{ D}_2$$

Substituting the value of u_2 in eqn (*i*), we have

$$V_{w2} = \frac{g \times 29.33}{u_2} = \frac{9.81 \times 29.33}{78.54 D_2} = \frac{3.66}{D_2} \text{ m/s}$$

Refer to Fig. 3.8. From velocity triangle at *outlet*, we have:

$$V_{f2} = (V_2^2 - V_{w2}^2)^{1/2}$$

or, $$\frac{0.167}{D_2^2} = \left[(15.63)^2 - \left(\frac{3.66}{D_2}\right)^2\right]^{1/2} = \left[244.3 - \frac{13.4}{D_2^2}\right]^{1/2}$$

Squaring both sides, we have:

or, $$\frac{0.0279}{D_2^4} = 244.3 - \frac{13.4}{D_2^2}, \text{ or, } 0.0279 = 244.3 D_2^4 - 13.4 D_2^2$$

or, $$244.3D_2^4 - 13.4D_2^2 - 0.0279 = 0$$

$$D_2^2 = \frac{13.4 \pm \sqrt{13.4^2 + 4 \times 244.3 \times 0.0279}}{2 \times 244.3} = \frac{13.4 \pm 14.38}{2 \times 244.3} = 0.0568 \text{ m}^2 \text{ (ignoring –ve sign)}$$

$$\therefore \quad D_2 = \sqrt{0.0568} = 0.238 \text{ or } \mathbf{238\ mm}\ \textbf{(Ans.)}$$

(ii) Outlet vane angle, ϕ:

From velocity triangle at *outlet*, we have:

$$\tan\phi = \frac{V_{f2}}{u_2 - V_{w2}} = \frac{(0.167/0.238^2)}{(78.54 \times 0.238) - 3.66/0.238)} = \frac{2.95}{18.69 - 15.38} = 0.8912$$

$$\therefore \quad \phi = \tan^{-1}(0.8912) = \mathbf{41.7°}\ \textbf{(Ans.)}$$

Example 3.18. *The following data refer to a radial, single stage, double suction, centrifugal pump:*

Discharge at the pump outlet = 90 litres / sec; Diameter at inlet = 100 mm;

Diameter at outlet = 290 mm; Head = 36 m;

Speed of impeller = 1750 r.p.m; Width at inlet = 25 mm per side; Width at outlet = 23 mm in total;

Overall efficiency = 60 percent; Leakage losses = 2.7 litres / sec;

Mechanical losses = 1.5 kW; Contraction factor due to vane thickness = 0.87;

Outlet vane angle = 27°.

Assuming that water enters the impeller at inlet radially, determine;

(i) *The inlet vane angle,*
(ii) *The angle at which water leaves the wheel,*
(iii) *The speed ratio,*
(iv) *The absolute velocity of water leaving the impeller,*
(v) *The manometric efficiency,*
(vi) *The volumetric efficiency, and*
(vii) *The mechanical efficiency.*

Solution. *Given* : Q_{po} = 90 litres/sec. = 0.09 m^3/s; D_1 = 100 mm or 0.1 m; D_2 = 290 mm or 0.29 m; H_{mano} = 36 m; N = 1750 r.p.m; B_1 = 0.025 m per side;

B_2 = 0.023 m in total; η_0 = 60%;

Leakage losses, q = 2.7 litres/sec. = 0.0027 m^3/s;

Mechanical losses = 1.5 kW;

Contraction factor due to vane thickness, K_t = 0.87; ϕ = 27°; α = 90°.

(i) The quantity of water handled:

Total quantity of water handled by pump,

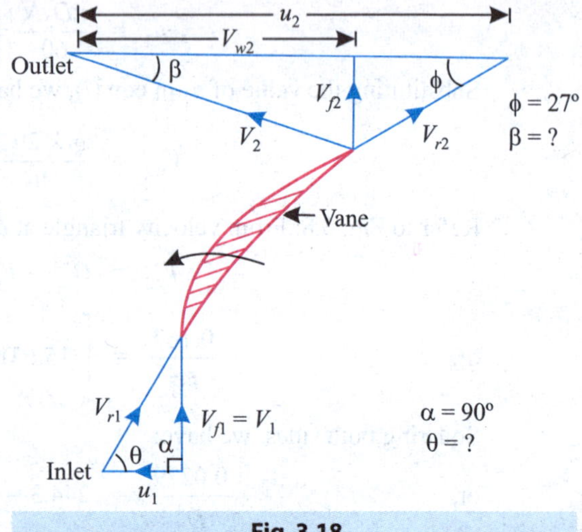

Fig. 3.18

$$Q = Q_{po} + q = 0.09 + 0.0027 = 0.0927 \text{ m}^3/\text{s}$$

(where, Q_{po} = discharge at pump outlet)

$$\therefore \quad Q \text{ per side} = \frac{0.0927}{2} = 0.04635 \text{ m}^3/\text{s}$$

Peripheral speed at inlet,

$$u_1 = \frac{\pi D_1 N}{60} = \frac{\pi \times 0.1 \times 1750}{60} = 9.16 \text{ m/s}$$

Also, $$Q = K_t \times \pi D_1 B_1 \times V_{f1}$$

or, $$V_{f1} = \frac{Q}{K_t \times \pi D_1 B_1}$$

or, $$V_{f1} = \frac{0.04635}{0.87 \times \pi \times 0.1 \times 0.025} = 6.78 \text{ m/s}$$

From velocity triangle at *inlet* (Fig. 3.18), we have:

$$\tan\theta = \frac{V_{f1}}{u_1} = \frac{6.78}{9.16} = 0.74 \quad \therefore \theta = \tan^{-1}(0.74) = \mathbf{36.5°} \quad \textbf{(Ans.)}$$

(*ii*) The angle at which water leaves the wheel, β:

Again, $$Q = K_t \times \pi D_2 B_2 \times V_{f2}$$

$$\left(\text{where, } B_2 = \frac{0.023}{2} = 0.0115 \text{ m from one side}\right)$$

$$\therefore \text{ Velocity of flow at outlet, } V_{f2} = \frac{Q}{K_t \pi D_2 B_2} = \frac{0.04635}{0.87 \times \pi \times 0.29 \times 0.0115} = 5.08 \text{ m/s}$$

$$\text{Peripheral speed at outlet, } u_2 = \frac{\pi D_2 N}{60} = \frac{\pi \times 0.29 \times 1750}{60} = 26.57 \text{ m/s}$$

From velocity triangle at *outlet* (fig. 3.18), we have:

$$\tan\phi = \frac{V_{f2}}{u_2 - V_{w2}}, \quad \text{or,} \quad u_2 - V_{w2} = \frac{V_{f2}}{\tan\phi}$$

$$\therefore \quad V_{w2} = u_2 - \frac{V_{f2}}{\tan\phi} = 26.57 - \frac{5.08}{\tan 27°} = 16.6 \text{ m/s}$$

Further, $$\tan\beta = \frac{V_{f2}}{V_{w2}} = \frac{5.08}{16.6} = 0.306$$

$$\therefore \quad \beta = \tan^{-1}(0.306) = \mathbf{17°} \textbf{ (Ans.)}$$

(*iii*) The speed ratio, K_{u2}:

$$K_{u2} = \frac{u_2}{\sqrt{2gH_{mano}}} = \frac{26.57}{\sqrt{2 \times 9.81 \times 36}} \simeq \mathbf{1} \textbf{ (Ans.)}$$

(*iv*) The absolute velocity of water leaving the impeller, V_2:

Refer to Fig. 3.18: $V_2 \cos\beta = V_{w2}$, or, $V_2 = \dfrac{V_{w2}}{\cos\beta} = \dfrac{16.6}{\cos 17°} = \mathbf{17.36 \text{ m/s}}$ **(Ans.)**

(v) The manometric efficiency, η_{mano}:

$$\eta_{mano} = \frac{gH_{mano}}{V_{w2}u_2} \quad \text{...Eqn. (3.9)}$$

$$= \frac{9.81 \times 36}{16.6 \times 26.57} = 0.8 \text{ or } \mathbf{80\ \%} \text{ (Ans.)}$$

(vi) The volumetric efficiency, η_v:

$$\eta_v = \frac{Q_{po}}{Q_{po} + q} = \frac{0.09}{0.09 + 0.0027} = 0.97 \text{ or } \mathbf{97\ \%} \text{ (Ans.)}$$

(vii) The mechanical efficiency, η_m:

$$\eta_m = \frac{\text{Shaft power} - \text{mechanical losses}}{\text{Shaft power}}$$

But, shaft power $= \dfrac{wQ_{po}H_{mano}}{\eta_0} = \dfrac{9.81 \times (0.09/2) \times 36}{0.6} = 26.49 \text{ kW}$

$$\therefore \quad = \frac{26.49 - 1.5}{26.49} = 0.9433 \text{ or } \mathbf{94.33\ \%} \text{ (Ans.)}$$

Example 3.19. *A centrifugal pump impeller, having outlet diameter 0.35 m is running at 960 r.p.m. The velocity of flow (assumed constant throughout the system) is equal to 2.4 m/s. The vane angle at outlet is 28°. The static suction lift is 4.03 m. The energy losses in suction pipe, impeller and volute casing are 0.88 m, 0.70 m and 1.26 m of water respectively. Determine the readings of vacuum or pressure gauges placed at:*

(i) Inlet to the pump,

(ii) Impeller outlet (in clearance between impeller and outlet), and

(iii) Pump outlet or delivery flange, 0.24 m above the centreline of the pump.

Solution. Diameter of impeller at outlet, $D_2 = 0.35$ m

Speed of impeller, $N = 960$ r.p.m.

Velocity of flow, $V_{f1} = V_{f2} = 2.4$ m/s

The vane angle at outlet, $\phi = 28°$

The static suction lift, $h_s = 4.03$ m

The energy losses in suction pipe, $h_{fs} = 0.88$ m

The energy losses in impeller, $h_{Li} = 0.70$ m

The energy losses in volute casing, $h_{Lc} = 1.26$ m

(i) Reading of the vacuum gauge at inlet to the pump, p_s:

The pressure head at inlet to the pump is given by,

$$\frac{p_s}{w} = -\left[h_s + h_{fs} + \frac{V_s^2}{2g}\right] = -\left(4.03 + 0.88 + \frac{2.4^2}{2 \times 9.81}\right) = \mathbf{-\ 5.2\ m} \text{ (Ans.)}$$

(ii) Reading of the gauge at impeller outlet, $\frac{p_2}{w}$:

$$\frac{p_s}{w} + \frac{V_s^2}{2g} + \frac{V_{w2}u_2}{g} = \frac{p_2}{w} + \frac{V_2^2}{2g} + h_{Li} \quad \text{...(i)}$$

Now, $u_2 = \dfrac{\pi D_2 N}{60} = \dfrac{\pi \times 0.35 \times 960}{60} = 17.59 \text{ m/s}$

Refer to Fig. 3.8.

$$V_{w2} = u_2 - \frac{V_{f2}}{\tan\phi} = 17.59 - \frac{2.4}{\tan 28°} = 13.07 \text{ m/s}$$

$$V_2 = \sqrt{V_{w2}^2 + V_{f2}^2} = \sqrt{13.07^2 + 2.4^2} = 13.29 \text{ m/s}$$

Substituting the above values in eqn. (*i*), we have:

$$-5{\cdot}2 + \frac{2.4^2}{2\times 9.81} + \frac{13.07\times 17.59}{9.81} = \frac{p_2}{w} + \frac{13.29^2}{2\times 9.81} + 0.7$$

or, $$-5.2 + 0.293 + 23.43 = \frac{p_2}{w} + 9.0 + 0.7$$

or, $$\frac{p_2}{w} = -5.2 + 0.293 + 23.43 - 9.0 - 0.7 = \mathbf{8.82\ m}$$ **(Ans.)**

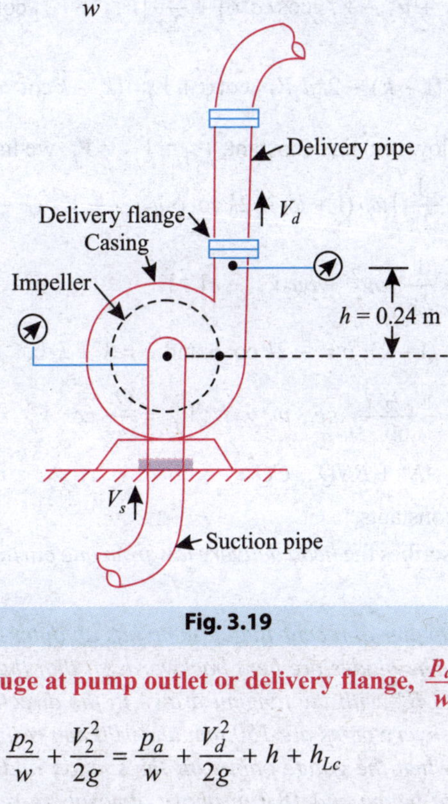

Fig. 3.19

(*iii*) **Reading of the gauge at pump outlet or delivery flange, $\frac{p_d}{w}$:**

Refer to Fig. 3.19: $$\frac{p_2}{w} + \frac{V_2^2}{2g} = \frac{p_a}{w} + \frac{V_d^2}{2g} + h + h_{Lc}$$

or, $$8.82 + \frac{13.29^2}{2\times 9.81} = \frac{p_d}{w} + \frac{2.4^2}{2\times 9.81} + 0.24 + 1.26$$

or, $$8.82 + 9.0 = \frac{p_d}{w} + 0.293 + 0.24 + 1.26$$

∴ $$\frac{p_d}{w} = 8.82 + 9.0 - 0.293 - 0.24 - 1.26 = \mathbf{16.03\ m}$$ **(Ans.)**

Example 3.20. *Show that, in general, for a centrifugal pump running at speed N and giving a discharge Q, the manometric head is expressible in the form:*

$$H_{mano} = AN^2 + BNQ + CQ^2$$

where A, B and C are constants.

Solution. Pressure rise, $\frac{p_2 - p_1}{w} = \frac{1}{2g}[V_{f1}^2 + u_2^2 - V_{f2}^2 \text{cosec}^2\phi]$ ...[Refer to example 3.10.]

The above relationship has been derived by neglecting gravitational effects, any friction losses, and assuming radial entry of water. Further, neglecting any loss of head in the pump, the manometric head is given by *pressure rise through the impeller together with a certain percentage of kinetic head at the impeller exit which is recovered in the volute chamber or the diffuser ring.* Thus,

$$H_{mano} = \frac{p_2 - p_1}{w} + \frac{kV_2^2}{2g}$$

where, $V_2^2 = V_{f2}^2 + V_{f2}^2 \cot^2\phi + u_2^2 - 2u_2 V_{f2} \cot\phi$ ...Refer to example 3.10

$$\therefore \quad H_{mano} = \frac{1}{2g}[V_{f1}^2 + u_2^2 - V_{f2}^2 \text{cosec}^2\phi] + \frac{k}{2g}[V_{f2}^2 + V_{f2}^2\cot^2\phi + u_2^2 - 2u_2V_{f2}\cot\phi]$$

$$= \frac{1}{2g}[u_2^2(1+k) - 2ku_2V_{f2}\cot\phi + V_{f2}^2(k + k\cot^2\phi - \text{cosec}^2\phi) + V_{f1}^2]$$

Assuming, velocity of flow remains constant, $V_{f1} = V_{f2} = V_f$, we have:

$$H_{mano} = \frac{1}{2g}[u_2^2(1+k) - 2k\cot\phi\, u_2V_f + V_f^2(k + k\cot^2\phi - \text{cosec}^2\phi)]$$

$$= \frac{1}{2g}[au_2^2 + bu_2V_f + cV_f^2]$$

where, $a = (1+k);\ b = -2k\cot\phi$ and $c = k + k\cot^2\phi - \text{cosec}^2\phi$

Also, $u_2 = \frac{\pi D_2 N}{60}$ *i.e.,* $u_2 \propto N;\ V_f = \frac{Q}{A}$ *i.e.* $V_f \propto Q$

$$H_{mano} = AN^2 + BNQ + CQ^2$$ **...Proved**

(where A, B and C are constants.)

The above equation prescribes the *head delivery law from one particular pump at one particular speed.*

Example 3.21. *The impeller of a centrifugal pump has an outer diameter of 250 mm and an effective area of 0.017 m². The blades are bent backwards so that the direction of outlet relative velocity makes an angle of 148° with the tangent drawn in the direction of impeller rotaton, the diameters of suction and delivery pipes are 150 mm and 100 mm respectively. The pump delivers 0.031 m³/s at 1450 r.p.m. when the gauge points on the suction and delivery pipes close to the pumps show heads of 4.6 m below and 18.0 m above atmosphere respectively. The head losses in the suction and delivery pipes are 2.0 m and 2.9 m respectively. The motor driving the pump delivers 8.67 kW. Assumimg that water enters the pump without shock and whirl, determine:*

(i) The manometric efficiency, and

(ii) The overall efficiency of the pump. **[UPSC Exams.]**

Solution. Outer diameter of impeller, D_2 = 250 mm = 0.25 m

Effiective area of flow = 0.017 m²

Outlet vane angle, ϕ = 180° – 148° = 32°

Diameter of the suction pipe, D_s = 150 mm = 0.15 m

Diameter of the delivery pipc, D_d = 100 mm = 0.1 m

Discharge through the pump, $Q = 0.031\ m^3/s$

Speed of the pump, $N = 1450$ r.p.m.

Head lost in the suction pipe, $h_{fs} = 2.0$ m

Head lost in the delivery pipe, $h_{fd} = 2.9$ m

Power delivered to the pump, $P = 8.67$ kW

(i) The manometric efficiency, η_{mano}:

Velocity in the suction pipe,

$$V_s = \frac{Q}{\frac{\pi}{4} \times D_s^2} = \frac{0.031}{\frac{\pi}{4} \times 0.15^2}$$

$$= 1.754 \text{ m/s}$$

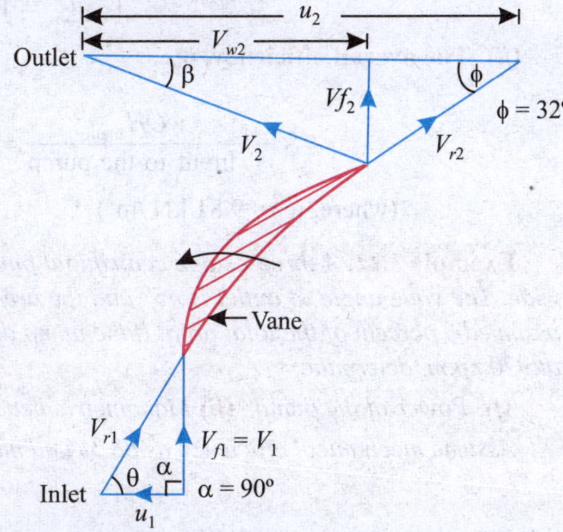

Fig. 3.20

Invoking Bernoulli's equation between the water surface in the sump and the pump inlet where pressure gauge is fitted, we have:

$$0 = h_s + h_{fs} + \frac{p_s}{w} + \frac{V_s^2}{2g}$$

or, $$0 = h_s + 2.0 - 4.6 + \frac{(1.754)^2}{2 \times 9.81}$$

or, $$h_s = -2.0 + 4.6 - \frac{(1.754)^2}{2 \times 9.81} = 2.44 \text{ m}$$

Again, applying Bernoulli's equation between the impeller outlet where the delivery pressure gauge is fitted and outlet of the delivery pipe, we have:

$$\frac{p_d}{w} + \frac{V_d^2}{2g} = h_d + h_{fd} + \frac{V_d^2}{2g}$$

or, $$h_d = \frac{p_d}{w} - h_{fd} = 18 - 2.9 = 15.1 \text{ m}$$

From velocity triangle at *outlet* (Fig. 3.20), we have:

$$\tan\phi = \frac{V_{f2}}{u_2 - V_{w2}}, \text{ or, } u_2 - V_{w2} = \frac{V_{f2}}{\tan\phi}, \text{ or, } V_{w2} = u_2 - \frac{V_{f2}}{\tan 32°}$$

or, $$V_{w2} = u_2 - \frac{V_{f2}}{\tan 32°} = u_2 - 1.6\, V_{f2}$$

where, $$u_2 = \frac{\pi D_2 N}{60} = \frac{\pi \times 0.25 \times 1450}{60} = 18.98 \text{ m/s}$$

and, $$V_{f2} = \frac{Q}{\text{Effective area of flow}} = \frac{0.031}{0.017} = 1.82 \text{ m/s}$$

$\therefore$ $$V_{w2} = 18.98 - 1.6 \times 1.82 = 16.07 \text{ m/s}$$

Velocity in the delivery pipe, $$V_d = \frac{Q}{\frac{\pi}{4} \times D_d^2} = \frac{0.031}{\frac{\pi}{4} \times 0.1^2} = 3.95 \text{ m/s}$$

Manometric head, $H_{mano} = (h_s + h_d) + (h_{fs} + h_{fd}) + \frac{V_d^2}{2g}$

$$= (2.44 + 15.1) + (2.0 + 2.9) + \frac{3.95^2}{2 \times 9.81} = 23.23 \text{ m}$$

Manometric efficiency, $\eta_{mano} = \frac{gH_{mano}}{V_{w2}u_2} = \frac{9.81 \times 23.23}{16.07 \times 18.98} = 0.747$ or **74.7% (Ans.)**

(*ii*) The overall efficiency, η_0:

$$= \frac{wQH_{mano}}{\text{Input to the pump}} = \frac{9.81 \times 0.031 \times 23.23}{8.67} = 0.8148, \text{ or, } \mathbf{81.48\%}$$

(where, w = 9.81 kN/m^3)

Example 3.22. *A three - stage centrifugal pump has impeller 400 mm in diameter and 20 mm wide. The vane angle at outlet is 45° and the area occupied by the thickness of the vanes may be assumed 8 percent of the total area. If the pump delivers 3.6 m^3 of water per minute when running at 920 r.p.m. determine:*

(i) Power of the pump, (ii) Manometric head, and (iii) Specific speed.

Assume mechanical efficiency as 88 % and manometric efficiency as 77 percent.

[Delhi University]

Solution. Number of stage, n = 3

Diameter of impeller at outlet, D_2 = 400 mm = 0.4 m

Width of impeller at outlet, B_2 = 20 mm = 0.02 m

Outlet vane angle, ϕ = 45°

Area occupied by thickness of vanes = 8% of the total area

Discharge through the pump, Q = 3.6 m^3/min. = 0.06 m^3/s

Speed of the impeller, N = 920 r.p.m.

Mechanical efficiency = 88%.

Manometric efficiency = 77%.

(*i*) Manometric head $(H_{mano})_{total}$:

Peripheral or tangential velocity of impeller at outlet,

$$u_2 = \frac{\pi D_2 N}{60} = \frac{\pi \times 0.4 \times 920}{60} = 19.27 \text{ m/s}$$

$$\text{Net outlet area} = \left(1 - \frac{8}{100}\right) \pi D_2 B_2$$

$$= 0.92\pi \times 0.4 \times 0.02 = 0.02312 \text{ m}^2$$

Velocity of flow at outlet,

$$= \frac{Q}{\text{Net outlet area}} = \frac{0.06}{0.02312} = 2.59 \text{ m/s}$$

From velocity triangle at *outlet* (Fig. 3.21), we have:

$$\tan\phi = \frac{V_{f2}}{u_2 - V_{w2}}, \text{ or, } u_2 - V_{w2} = \frac{V_{f2}}{\tan\phi}$$

or, $$V_{w2} = u_2 - \frac{V_{f2}}{\tan \phi} = 19.27 - \frac{2.59}{\tan 45°} = 16.68 \text{ m/s}$$

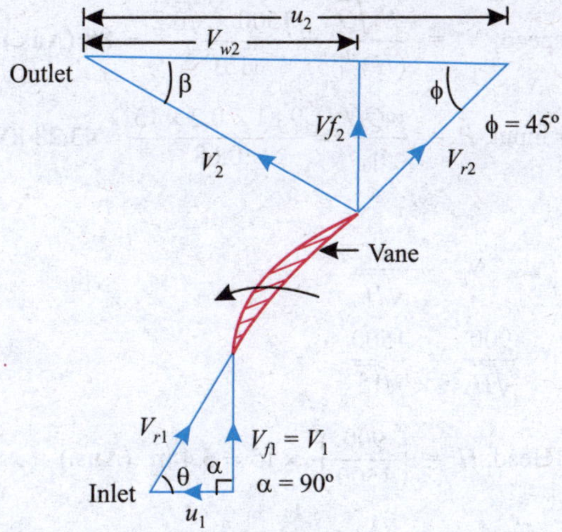

Fig. 3.21

Assuming radial entry, the head developed by each impeller,

$$H_{mano} = \eta_{mano} \times \frac{V_{w2}u_2}{g} = 0.77 \times \frac{16.68 \times 19.27}{9.81} = 25.23 \text{ m}$$

∴ Manometric head $(H_{mano})_{total}$ developed by three impeller,

$$3 \times 25.23 = \mathbf{75.69 \text{ m}} \text{ (Ans.)}$$

(*ii*) Power of the pump, P:

$$\eta_0 = \eta_m \times \eta_{mano} = 0.88 \times 0.77 = 0.6776 \text{ (Assuming } \eta_v \text{ to be unity)}$$

Also, $$\eta_0 = \frac{wQH_{mano}}{\text{Power of pump}}$$

∴ Power of pump, $$P = \frac{wQ(H_{mano})_{total}}{\eta_0} = \frac{9810 \times 0.06 \times 75.69}{1000 \times 0.6776} = \mathbf{65.75 \text{ kW}} \text{ (Ans.)}$$

(*iii*) Specific speed, N_s:

$$N_s = \frac{N\sqrt{Q}}{(H_{mano})^{3/4}},$$ where H_{mano} is the manometric head developed per impeller

...(Eqn. 3.25)

$$= \frac{920 \times \sqrt{0.06}}{(25.23)^{3/4}} = \mathbf{20} \text{ (Ans.)}$$

Example 3.23. *A centrifugal pump rotating at 1500 r.p.m. delivers 0.2 m³/s at a head of 15 m. Calculate the specific speed of the pump and the power input. Assume overall efficiency of the pump as 0.68.*

If this pump were to operate at 900 r.p.m. what would be the head, discharge and power required for homologous conditions? Assume overall efficiency remains unchanged at new r.p.m.

[UPTU]

Solution. *Given:* $N = 1500$ r.p.m.; $Q = 0.2\ m^3/s$; $H = 15$ m; $\eta_0 = 0.68$; N = 900 r.p.m.

At 1500 r.p.m.:

$$\text{Specific speed, } N_s = \frac{N\sqrt{Q}}{(H)^{3/4}} = \frac{1500 \times \sqrt{0.2}}{(15)^{3/4}} = \mathbf{88}\ \textbf{(Ans.)}$$

$$\text{Power input, } P = \frac{wQH}{\eta_0} = \frac{9.81 \times 0.2 \times 15}{0.68} = \mathbf{43.28\ kW}\ \textbf{(Ans.)}$$

At 900 r.p.m.:

$$N_u = \frac{N}{\sqrt{H}}$$

i.e.

$$\frac{900}{\sqrt{H}} = \frac{1500}{\sqrt{15}}$$

or,

$$\text{Head, } H = \left(\frac{900}{1500}\right)^2 \times 15 = \mathbf{5.4\ m}\ \textbf{(Ans.)}$$

$$Q_u = \frac{Q}{\sqrt{H}}$$

i.e.

$$\frac{Q}{\sqrt{5{\cdot}4}} = \frac{0.2}{\sqrt{15}}$$

$\therefore$

$$Q = \sqrt{\frac{5.4}{15}} \times 0.2 = \mathbf{0.12\ m^3/s}\ \textbf{(Ans.)}$$

$$P_u = \frac{P}{(H)^{3/2}}$$

i.e.

$$\frac{P}{(5{\cdot}4)^{3/2}} = \frac{43.28}{(15)^{3/2}}$$

$$P = \left(\frac{5.4}{15}\right)^{3/2} \times 43.28 = \mathbf{9.35\ kW}\ \textbf{(Ans.)}$$

Example 3.24. *Two geometrically similar pumps are running at the same speed of 1000 r.p.m. One has an impeller diameter of 0.4 m and discharge of 30 l/s against a head of 20 m. If the other pump gives half of this discharge rate, determine the head and diameter of the second pump.* **[GATE]**

Solution. Given: $N_1 = N_2 = 1000$ r.p.m.; $D_1 = 0.4$ m; $Q_1 = 30\ l/s = 0.03\ m^3/s$;

$$H_1 = 20 \text{ m}; Q_2 = \frac{30}{2} = 15\, l/s = 0.015\ m^3/s.$$

***H_2, D_2*:**

All geometrically similar pumps will have the same specific speed.

i.e.

$$N_s = \frac{N_1\sqrt{Q_1}}{(H_1)^{3/4}} = \frac{N_2\sqrt{Q_2}}{(H_2)^{3/4}}$$

or,

$$H_2 = \left[\frac{N_2}{N_1} \times \sqrt{\frac{Q_2}{Q_1}} \times (H_1)^{3/4}\right]^{4/3}$$

$$= \left[\frac{1000}{1000} \times \sqrt{\frac{0.015}{0.030}} \times (20)^{3/4}\right]^{4/3} = \mathbf{12.6\ m}\ \textbf{(Ans.)}$$

Also, $$D \propto \frac{\sqrt{H}}{N}$$

or, $$D_1 \propto \frac{\sqrt{H_1}}{N_1}, \text{ and } D_2 \propto \frac{\sqrt{H_2}}{N_2} \qquad (\because N_1 = N_2)$$

$$\therefore \quad \frac{D_1}{D_2} = \sqrt{\frac{H_1}{H_2}} = \sqrt{\frac{20}{12.6}} = 1.256$$

or, $$D_2 = \frac{D_1}{1.256} = \frac{0.4}{1.256} = \mathbf{0.317\ m}\ \textbf{(Ans.)}$$

Example 3.25. *A centrifugal pump running at 750 r.p.m. discharges water at 0.1 m^3/s against a head of 10 m at its best efficiency. A second pump of the same homologous series, when working at 500 r.p.m., is to deliver water at 0.05 m^3/s at its best efficiency. What will be the design head of the second pump and what is the scale ratio between the first and the second?* **[GATE]**

Solution. *Given*: $N_1 = 750$ r.p.m.; $Q_1 = 0.1 m^3/s$; $H_1 = 10$ m; $N_2 = 500$ r.p.m.; $Q_2 = 0.05$ m^3/s

Design head of the second pump, H_2:

We know that, $$\frac{N_1\sqrt{Q_1}}{H_1^{3/4}} = \frac{N_2\sqrt{Q_2}}{H_2^{3/4}}$$

or, $$\frac{750 \times \sqrt{0.1}}{(10)^{3/4}} = \frac{500 \times \sqrt{0.05}}{(H_2)^{3/4}}$$

or, $$42.17 = \frac{111.8}{(H_2)^{3/4}}$$

$$\therefore \quad H_2 = \left(\frac{111.8}{42.17}\right)^{4/3} = \mathbf{3.67\ m}\ \textbf{(Ans.)}$$

(*ii*) Scale ratio, L_r:

We know that, $$\frac{Q_1}{Q_2} = (L_r)^{2.5}$$

or, $$\frac{0.1}{0.05} = (L_r)^{2.5}$$

or, $$L_r = \left(\frac{0.1}{0.05}\right)^{\frac{1}{2.5}} = \mathbf{1.32}\ \textbf{(Ans.)}$$

3.9. MINIMUM SPEED FOR STARTING A CENTRIFUGAL PUMP

When a centrifugal pump is started, it will start delivering liquid only if the pressure rise in the impeller is more than or equal to the manometric head (H_{mano}). In other words, there will be no flow of liquid until the speed of the pump is such that the required centrifugal head caused by the centrifugal force on rotating water when the impeller is rotating, but there is *no flow*

$$= \frac{u_2^2}{2g} - \frac{u_1^2}{2g} = \frac{u_2^2 - u_1^2}{2g}$$

Flow will commence only if $\dfrac{u_2^2 - u_1^2}{2g} \geq H_{mano}$

For *minimum speed,* we must have:

$$\frac{u_2^2 - u_1^2}{2g} = H_{mano} \qquad ...(3.17)$$

Also, $$\eta_{mano} = \frac{gH_{mano}}{V_{w2}u_2} \qquad \text{[Eqn (3.9)]}$$

$$\therefore \quad H_{mano} = \eta_{mano} \times \left(\frac{V_{w2}u_2}{g}\right)$$

Also, $$u_1 = \frac{\pi D_1 N}{60}; \quad u_2 = \frac{\pi D_2 N}{60}$$

Substituting the values in eqn. (3.17), we get:

$$\frac{1}{2g}\left[\left(\frac{\pi D_2 N}{60}\right)^2 - \left(\frac{\pi D_1 N}{60}\right)^2\right] = \eta_{mano} \times \frac{V_{w2}}{g} \times \left(\frac{\pi D_2 N}{60}\right)$$

Dividing both sides by $\dfrac{\pi N}{g \times 60}$, we have:

$$\frac{\pi N}{120}(D_2^2 - D_1^2) = \eta_{mano} \times (V_{w2} \times D_2)$$

$$\therefore \quad N\,(i.e.\ N_{min.}) = \frac{120 \times \eta_{mano} \times V_{w2} \times D_2}{\pi\,(D_2^2 - D_1^2)} \qquad ...(3.18)$$

Example 3.26. *A centrifugal pump working in a dock pump 1565 litres per second against a mean lift of 6.1m when the impeller rotates at 200 r.p.m.The impeller diameter is 1.22 m and the area at outer periphery is 6450 cm^2. If the vanes are set back at an angle of 26° at the outlet, determine:*

(*i*) *Hydraulic efficiency,*

(*ii*) *Power required to drive the pump, and*

(*iii*) *Minimum speed to start pumping if the ratio of external to internal diameter is 2.*

[PTU]

Solution. Discharge through the pump, Q = 1565 litres/ sec. or 1.565 m^3/s

Actual or manometric head = 6.1 m

Speed of the impeller, N = 200 r.p.m.

Diameter of the impeller at outlet, D_2 = 1.22 m

Area at the outer periphery = 6450 cm^2 = 0.645 m^2

Outlet vane angle, ϕ = 26°

(*i*) Hydraulic efficiency, η_h:

Peripheral or tangential velocity of impeller at outlet,

$$u_2 = \frac{\pi D_2 N}{60} = \frac{\pi \times 1.22 \times 200}{60} = 12.77 \text{ m/s}$$

$$V_{f2} = \frac{Q}{A} = \frac{1.565}{0.645} = 2.43 \text{ m/s}$$

Refer to Fig. 3.8. $V_{w2} = u_2 - \dfrac{V_{f2}}{\tan\phi} = 12.77 - \dfrac{2.43}{\tan 26°} = 7.79 \text{ m/s}$

$$\text{Euler head, } H_e = \frac{V_{w2}u_2}{g} = \frac{7.79 \times 12.77}{9.81} = 10.14 \text{ m}$$

If the effect of slip is *neglected*, ideal head equals Euler head.

$$\therefore \text{ Hydraulic (or manometric) efficiency, } \eta_h = \frac{\text{Actual or manometric head}}{\text{Ideal or Euler head}}$$

or, $\eta_h = \dfrac{6.1}{10.14} = 0.6016$ or **60.16 % (Ans.)**

(ii) Power required to drive the pump:

Power required to drive the pump (neglecting mechanical losses)

$$= \frac{wQ}{1000} \times \frac{V_{w2}u_2}{g} \text{ kW}$$

$$= \frac{9810 \times 1.565 \times 7.79 \times 12.77}{1000 \times 9.81} = \mathbf{155.68 \text{ kW}} \textbf{ (Ans.)}$$

(iii) Minimum speed to start pumping, N_{min}:

For *minimum speed*, we must have:

$$\frac{u_2^2 - u_1^2}{2g} = H_{mano}$$

Since, $D_2 = 2D_1$ (*Given*), $\therefore u_2 = 2u_1$

$$\therefore \quad \frac{u_2^2 - u_2^2/4}{2g} = 6.1, \text{ or, } \frac{3u_2^2}{8g}, \text{ or, } u_2 = \left(\frac{6.1 \times 8 \times 9.81}{3}\right)^{\frac{1}{2}} = 12.63 \text{ m/s}$$

Also, $u_2 = \dfrac{\pi D_2 N_{min}}{60}$, or, $12.63 = \dfrac{\pi \times 1.22 \times N_{min}}{60}$

$$\therefore \quad N_{min} = \frac{12.63 \times 60}{\pi \times 1.22} = \mathbf{197.7 \text{ r.p.m.}} \textbf{ (Ans.)}$$

Example 3.27. *A centrifugal pump impeller has diameters at inlet and outlet as 360 mm and 720 mm respectively. The flow velocity at outlet is 2.4 m/s and the vanes are set back at an angle of 45° at the outlet. If the manometric efficiency is 70 percent, calculate the minimum starting speed of the pump.*

Solution. Diameter of impeller at inlet, D_1 = 360 mm or 0.36 m

Diameter of impeller at outlet, D_2 = 720 mm or 0.72 m

The flow velocity at outlet, V_{f2} = 2.4 m/s

Outlet vane angle, ϕ = 45°

Manometric efficiency, η_{mano} = 70 %

Minimum starting speed of the pump, N_{min}:

Refer to Fig 3.8. From velocity triangle at *outlet*, we have:

$$\tan\phi = \frac{V_{f2}}{u_2 - V_{w2}}, \text{ or, } u_2 - V_{w2} = \frac{V_{f2}}{\tan\phi} = \frac{2.4}{\tan 45°} = 2.4 \text{ m/s}$$

$$\therefore \quad V_{w2} = u_2 - 2.4$$

But, $$u_2 = \frac{\pi D_2 N_{min}}{60} = \frac{\pi \times 0.72 N_{min}}{60} = 0.0377\, N_{min}$$

$$\therefore \quad V_{w2} = 0.0377 N_{min} - 2.4$$

Using eqn. (3.17) for minimum starting speed, we have:

$$N_{min} = \frac{120 \times \eta_{mano} \times V_{w2} \times D_2}{\pi .(D_2^2 - D_1^2)}$$

$$= \frac{120 \times 0.70 \times (0.0377 N_{min} - 2.4) \times 0.72}{\pi\,(0.72^2 - 0.36^2)}$$

or, $$N_{min} = 49.51\,(0.0377 N_{min} - 2.4)$$

$$= 1.866 N_{min} - 118.824$$

or, $$0.866 N_{min} = 118.824$$

$$\therefore \quad N_{min} = \frac{118.824}{0.866} = \mathbf{137.2\ r.p.m.}\ \textbf{(Ans.)}$$

3.10. EFFECT OF VARIATION OF DISCHARGE ON THE EFFICIENCY

When a centrifugal pump *runs and discharges at its 'designed speed', its efficiency is maximum.* If the discharge is either *increased or decreased its efficiency drops owing to loss of head due to shock at entry to impeller.*

Refer to Fig 3.22:

- *abc* is velocity triangle at *inlet* when the pump is operating under the normal conditions of flow rate (discharge). The relative velocity V_{r1} is along the contour of vane (θ is the inlet vane angle).
- With the decrease (or increase) in discharge, the flow velocity decreases (or increases) from *cb* to *cd.*
- With the pump running at the *same speed,* the peripheral velocity *ac remains unchanged* and the effective inlet velocity diagram becomes edc.
- The new relative velocity *ad* (V'_{r1}) *no longer remains parallel to the vane and consequently shock occurs at entry to impeller.* With fixed value of flow velocity *cd* (V'_{f1}) and flow taking place along the blade, the velocity vector should have the form *cde; de* is parallel to *ab. The sudden change in peripheral velocity ac brings about shock* and as a consequence of which there is a loss of head.

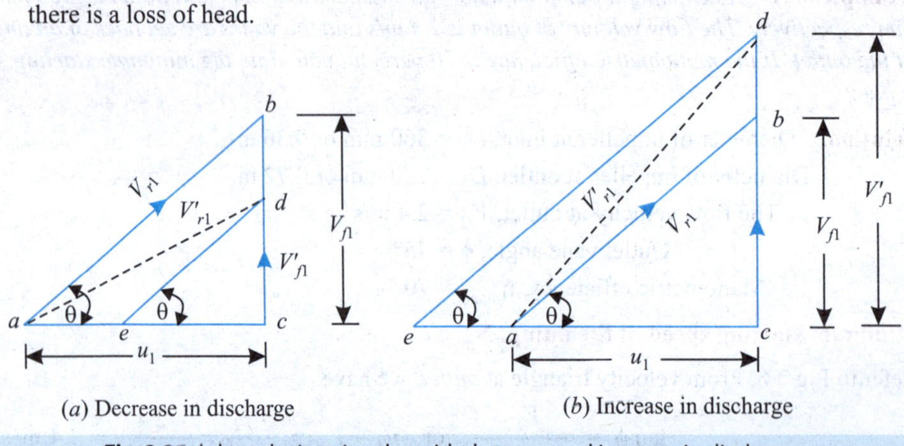

Fig. 3.22. Inlet velocity triangles with decrease and increase in discharge.

Loss of head (at entrance to impeller), $H_L = \dfrac{(\text{Change of velocity})^2}{2g} = \dfrac{(ae)^2}{2g}$

(*a*) *For decrease in discharge:* $$H_L = \frac{(ae)^2}{2g} = \frac{(ac - ec)^2}{2g} = \frac{(u_1 - V'_{f1}\cot\theta)^2}{2g} \quad ...(3.19)$$

(*b*) *For increase in discharge:* $$H_L = \frac{(ae)^2}{2g} = \frac{(ec - ac)^2}{2g} = \frac{(V'_{f1}\cot\theta - u_1)^2}{2g} \quad ...(3.20)$$

Example 3.28. *A centrifugal pump is delivering 0.216 m³/s of water against a head of 18 m; the speed of rotation of impeller being 600 r.p.m. The diameters at outer and inner periphery of the impeller are 600 mm and 300 mm respectively. The area of flow is constant at 0.084 m² from inlet to outlet of impeller. If the vanes of the impeller are bent at an angle of 35° to the tangent at exit, determine:*

(*i*) *Manometric efficiency,*
(*ii*) *Inlet vane angle, and*
(*iii*) *Loss of head at inlet to impeller when the discharge is reduced by 35 percent.*

Solution. Discharge through the pump, $Q = 0.216 \text{ m}^3/\text{s}$

Manometric head, $H_{mano} = 18$ m

Speed of rotation of impeller, $N = 600$ r.p.m.

Diameter of impeller at outlet, $D_1 = 600 \text{ mm} = 0.6$ m

Diameter of impeller at inlet, $D_2 = 300 \text{ mm} = 0.3$ m

The area of flow, $A_f = 0.084 \text{ m}^2$

Outlet vane angle, $\phi = 35°$

(*i*) **Manometric efficiency, η_{mano}:**

Peripheral or tangential velocities at inlet and outlet of the impeller are:

$$u_1 = \frac{\pi D_1 N}{60} = \frac{\pi \times 0.3 \times 600}{60} = 9.42 \text{ m/s}$$

$$u_2 = \frac{\pi D_2 N}{60} = \frac{\pi \times 0.6 \times 600}{60} = 18.85 \text{ m/s}$$

Flow velocity, $$V_{f1} = V_{f2} = V_f = \frac{Q}{A_f} = \frac{0.216}{0.084} = 2.57 \text{ m/s}$$

Refer to Fig 3.8. $$V_{w2} = u_2 - \frac{V_{f2}}{\tan\phi} = 18.85 - \frac{2.57}{\tan 35°} = 15.18 \text{ m/s}$$

Manometric efficiency, $$\eta_{mano} = \frac{gH_{mano}}{V_{w2}u_2} \quad ...[\text{Eqn. } (3.9)]$$

$$= \frac{9.81 \times 18}{15.18 \times 18.85} = 0.617 \text{ or } \mathbf{61.7\%} \textbf{ (Ans.)}$$

(*ii*) **Inlet vane angle, θ:**

Refer to Fig. 3.8. $$\tan\theta = \frac{V_{f1}}{u_1} = \frac{2.57}{9.42} = 0.2728$$

$$\therefore \quad \theta = \tan^{-1}(0.2728) = \mathbf{15.26°} \textbf{ (Ans.)}$$

(*iii*) Loss of head when the discharge is reduced by 35% :

When the discharge is reduced by 35 %, the velocity of flow,

$$V_f = V'_{f1} = 2.57 \times (1 - 0.35) = 1.67 \text{ m/s}$$

$$\text{Loss of head at inlet} = \frac{(u_1 - V'_{f1} \cot\theta)^2}{2g} \qquad \text{...[Eqn. (3.19)]}$$

$$= \frac{(9.42 - 1.67 \times \cot 15.26°)^2}{2 \times 9.81} = \mathbf{0.55 \text{ m of water}} \text{ (Ans.)}$$

3.11. EFFECT OF NUMBER OF VANES OF IMPELLER ON HEAD AND EFFICIENCY

The velocities indicated in velocity triangles (shown in Fig 3.5 and elsewhere) known as Euler's velocity triangles, can be obtained in practice only if the impeller has *very closely - spaced vanes.* Practically, it is impossible to have a very large number of vanes due to the two *reasons;* (*i*) The larger the number of vanes, the greater is the obstructed area owing to vanes thickness and consequently there is greater loss of head loss due to friction (*ii*) In order to fabricate the impeller easily, it is desirable that passages should be wider; it also minimises the possibility of pump being choked due to floating debris in the liquid.

Although, practically, the vanes are designed according to Euler's velocity triangles yet there is slight difference in the actually developed velocity triangles. The actual velocity of whirl (V_{w2}) at the outlet, *due to secondary or circulatory flow in the impeller,* is less than that in the Euler's velocity triangles. consequently, the actual head imparted (H_{actual}) by the impeller is less than the Euler head (H_e).

Thus, $$H_{actual} < H_e$$

The ratio $\left(\frac{H_{actual}}{H_e}\right)$ is known as *vane efficiency or effectiveness* ε (Greek 'epsilon')

i.e. $$\varepsilon = \frac{H_{actual}}{H_e}$$

It has been observed through experiments that as the *number of vanes is increased the* value of *ε increases and approaches unity.* The value of ε, in addition to number of vanes, depends on the shape of the vane and the outlet vane angle. In general, for radial flow pumps the value of ε varies from 0.6 to 0.8 as the number of vanes is increased from 4 to 12. However, for impeller with vanes more than 24 the value of ε may be taken as unity. Unless otherwise mentioned, the value of ε is taken as unity.

3.12. WORKING PROPORTIONS OF CENTRIFUGAL PUMPS

1. Speed ratio (K_u). *The speed ratio is the ratio of peripheral speed at exit (u_2) to the theoretical velocity of jet corresponding to manometric head (H_{mano})*. Thus,

$$K_u = \frac{u_2}{\sqrt{2gH_{mano}}}; \; K_u \text{ varies from 0.95 to 1.25}$$

2. Flow ratio, (K_f). *The flow ratio is the ratio of the velocity of flow at exit to the theoretical velocity of the jet corresponding to manometric head (H_{mano}).* Thus,

$$K_f = \frac{V_{f2}}{\sqrt{2gH_{mano}}}; \; K_f \text{ varies from 0.1 to 0.25}$$

Diameters of impellers and pipes:

(i) Outlet diameter of impeller (D_2):

Peripheral speed, $u_2 = \frac{\pi D_2 N}{60}$, where N is the speed of rotation of impeller.

Also, $$u_2 = K_u\,(\sqrt{2gH_{mano}}),$$ where K_u is the speed ratio.

$$\therefore \quad \frac{\pi D_2 N}{60} = K_u\sqrt{2gH_{mano}}, \text{ or, } D_2 = \frac{60\,K_u\sqrt{2gH_{mano}}}{\pi N}$$

or, $$D_2 = \frac{84.6K_u\sqrt{H_{mano}}}{N} \quad ...(3.21)$$

If D_2 and N are known, by using eqn. (3.21) we can find out the head which a pump can develop; it will serve as a cheak for existing pump.

(ii) Inlet diameter of impeller (D_1):

Depending upon specific speed (N_s) or total head (H_{mano}) the inlet diameter D_1 is kept in range; $D_1 = \frac{1}{3}D_2$ to $\frac{2}{3}D_2$. An average value of $D_1 = 0.5D_2$ is usually taken.

(iii) Least diameter of impeller:

The least or minimum diameter of an impeller can be determined on the basics of the fact that the pump will start delivering liquid only when *centrifugal head equals the total head H_{mano}*. Thus,

$$\frac{u_2^2 - u_1^2}{2g} = H_{mano}$$

or, $$\left(\frac{\pi D_2 N}{60}\right)^2 - \left(\frac{\pi D_1 N}{60}\right)^2 = 2gH_{mano}, \text{ or, } \left(\frac{\pi N}{60}\right)^2\left(D_2^2 - D_1^2\right) = 2gH_{mano}$$

Taking $D_1 = 0.5D_2$, we obtain:

$$\left(\frac{\pi N}{60}\right)^2[D_2^2 - (0{\cdot}5D_2)^2] = 2gH_{mano}, \text{ or, } \left(\frac{\pi N}{60}\right)^2 \times 0.75D_2^2 = 2 \times 9.81\,H_{mano}$$

$$\therefore \quad D_2 = \left(\frac{2 \times 9.81H_{mano}}{0.75}\right)^{1/2} \times \frac{60}{\pi N} = \frac{97.68\sqrt{H_{mano}}}{N} \quad ...(3.22)$$

(iv) Diameter of suction pipe (D_s) :

If D_s is the diameter of the suction pipe and V_s is the velocity of flow in suction pipe (usually V_s is 1.5 to 3 m/s), then the quantity of water to be pumped is given by:

$$Q = \frac{\pi}{4}D_s^2 \times V_s, \text{ or, } D_s^2 = \frac{4Q}{\pi V_s}$$

or, $$D_d = \sqrt{\frac{4Q}{\pi V_s}} \quad ...(3.23)$$

(v) Diameter of delivery pipe (D_d):

If D_d is the diameter of the delivery pipe and V_d is the velocity of flow in delivery pipe (usually V_d is 1.5 to 3.5 m/s), then the discharge,

$$Q = \frac{\pi}{4} \times D_d^2 \times V_d, \text{ or, } D_d^2 = \frac{4Q}{\pi V_d}$$

$$\therefore \qquad D_d = \sqrt{\frac{4Q}{\pi V_d}} \qquad \text{...(3.24)}$$

Note : The value of V_d is generally equal to or slightly higher than that V_s.

3.13. MULTI-STAGE CENTRIFUGAL PUMPS

A multi-stage centrifugal pump is one which has *two or more identical impellers* mounted on the same shaft or on different shafts. The important functions performed by a multi-stage pump are:

1. *To produce heads greater than that permissible with a single impeller, 'discharge remaining constant'. The task can be achieved by '**series arrangement**' where in the impellers are mounted on the same shaft and enclosed in the same casing.*
2. *To discharge a large quantity of liquid, 'head remaining same'.* This task is accomplished by '***parallel arrangement***' wherein impellers are mounted on *separate shafts*

3.13.1. Pumps in Series

For obtaining a high head, a number of impellers are mounted in *series* or on the same shaft. Fig. 3.23 shows such an arrangment for a two-stage pump. The discharge from impeller–1 passes through a guided passage and enters the impeller–2. At the outlet of impeller–2, the pressure of water will be more than the pressure of water at outlet of impeller–1. Thus if more number of impellers are mounted on the same shaft the pressure at outlet will be increased further. If in each stage, the manometric head imposed on the liquid is H_{mano}, then for *n identical* impellers the total head developed will be; $H_{total} = nH$, however, the *discharge* passing through each impeller is *same.*

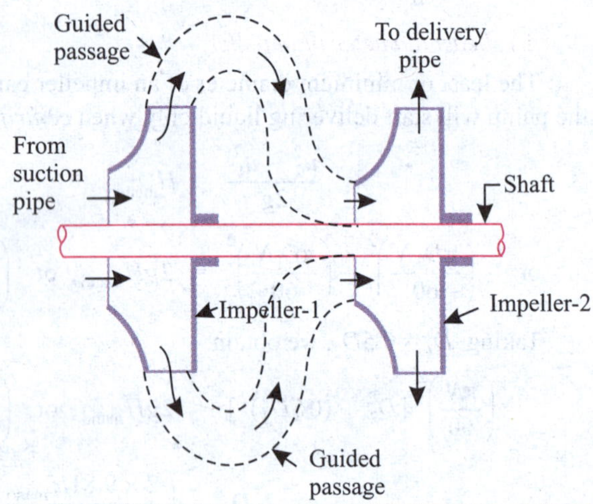

Fig. 3.23. Two-stage pump-impellers in series.

The series arrangement is employed for delivering a *relatively small quantity of liquid against very high heads.*

The *advantages* of multi–stage pumps–impellers in series over single-stage pumps are as follows:

1. Less loss due to friction.
2. Reduced stresses.
3. Small slip leakage.
4. The number of stages may be so chosen that the pump speed suits the driving motor speed.
5. By proper arrangement of impellers a thrust can be eliminated.
6. Owing to lower specific speed of individual impellers a higher suction lift is possible.

3.13.2. Pumps in Parallel

When a *large quantity of liquid* is required to be pumped against a *relatively small head* (which is impossible for a single pump to accomplish), two or more pumps are employed which are so arranged that each of these pumps working separately lifts thc liquid from a common sump and delivers it to a common collecting pipe through which it is carried to required height. This arrangement is known *as pumps in parallel* (since each pump delivers the liquid against the *same head*). If Q is the discharge capacity of one pump and there are *n identical* pumps (arranged in parallel) then total discharge will be, $Q_{total} = nQ$

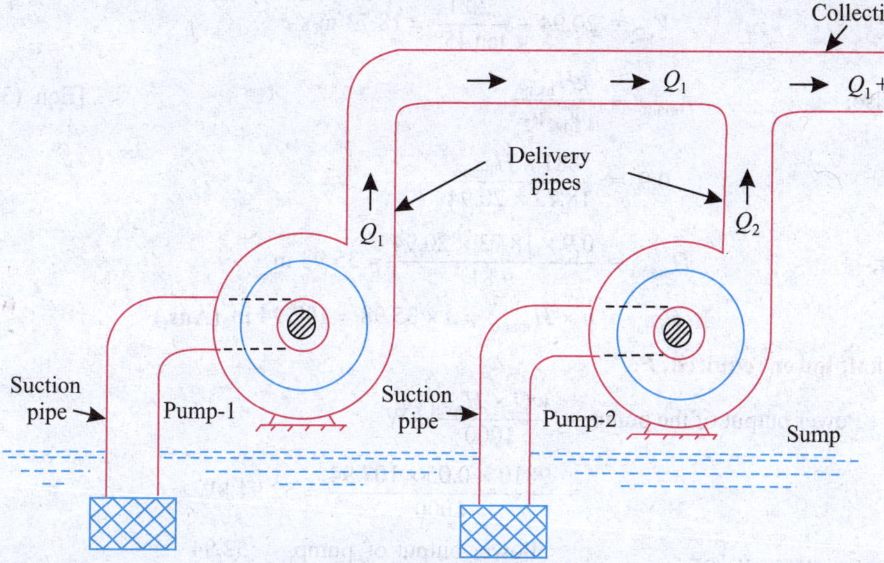

Fig. 3.24. Pumps in parallel.

Example 3.29. *A three stage centrifugal pump has impellers 400 mm in diameter and 20 mm wide at outlet. The vanes are curved back at the outlet at 45° and reduce the circumferential area by 10 percent. The manometric efficiency is 90 percent and the overall efficiency is 80 percent. The pump is running at 1000 r.p.m. and delivering 0.05 m³/s. Determine:*

(i) Head generated by the pump, and

(ii) Shaft power required to run the pump.

Solution. Number of stages, $n = 3$

Diameter of impeller at outlet, $D_2 = 400 \text{ mm} = 0.4 \text{ m}$

Width at outlet, $B_2 = 20 \text{ mm} = 0.02 \text{ m}$

Outlet vane angle, $\phi = 45°$

Reduction in area at outlet = 10%

Manometric efficiency, $\eta_{mano} = 90\%$

Overall efficiency, $\eta_0 = 80\%$; Speed, $N = 1000$ r.p.m.

Discharge through the pump, $Q = 0.05 \text{ m}^3\text{/s}$.

(i) Head generated by the pump, H_{total}:

$$\text{Area of flow at outlet} = 0.9\pi D_2 B_2 = 0.9 \times \pi \times 0.4 \times 0.02 = 0.02262 \text{ m}^2$$

$$\therefore \quad \text{Velocity of flow at outlet, } V_{f2} = \frac{Q}{\text{Area of flow}} = \frac{0.05}{0.02262} = 2.21 \text{ m/s}$$

Peripheral or tangential velocity of impeller at outlet,

$$u_2 = \frac{\pi D_2 N}{60} = \frac{\pi \times 0.4 \times 1000}{60} = 20.94 \text{ m/s}$$

Refer to Fig. 3.8. From velocity triangle at *outlet*, we have:

$$\tan\phi = \frac{V_{f2}}{u_2 - V_{w2}}, \text{ or, } u_2 - V_{w2} = \frac{V_{f2}}{\tan\phi}, \text{ or, } V_{w2} = u_2 - \frac{V_{f2}}{\tan\phi}$$

or,
$$V_{w2} = 20.94 - \frac{2.21}{\tan 45°} = 18.73 \text{ m/s}$$

Also,
$$\eta_{mano} = \frac{gH_{mano}}{V_{w2}u_2} \quad \text{...[Eqn (3.9)]}$$

$$0.9 = \frac{9.81 \times H_{mano}}{18.73 \times 20.94}$$

or,
$$H_{mano} = \frac{0.9 \times 18.73 \times 20.94}{9.81} = 35.98 \text{ m}$$

$$\therefore \quad H_{total} = n \times H_{mano} = 3 \times 35.98 = \mathbf{107.94\ m} \text{ (Ans.)}$$

(*ii*) Shaft lower required, *P*:

$$\text{Power output of the pump} = \frac{wQ \times H_{total}}{1000} \text{ kW}$$

$$= \frac{9810 \times 0.05 \times 107.94}{1000} = 52.94 \text{ kW}$$

$$\text{Now, Overall efficiency, } \eta_0 = \frac{\text{Power output of pump}}{\text{Power input to the pump}} = \frac{52.94}{P}$$

$$\therefore \quad P = \frac{52.94}{\eta_0} = \frac{52.94}{0.8} = \mathbf{66.17\ kW} \text{ (Ans.)}$$

Example 3.30. *It is required to pump water out of deep well under a total head of 90 m. A number of identical pumps of design speed 1000 r.p.m. and specific speed 30 with a rated capacity of 0.15 m³/s are available. How many pumps are required and how should they be connected whether in series or in parallel ?*

Solution. Total head, $H_{total} = 90$ m; Design speed, $N = 1000$ r.p.m.
Specific speed, $N_s = 30$; Discharge through each pump, $Q = 0.15 \text{ m}^3/\text{s}$

Number of pumps required, *n*:

Let, H_{mano} = Manometric head developed by each pump.

Now,
$$\text{specific speed, } N_s = \frac{N\sqrt{Q}}{(H_{mano})^{3/4}}$$

or,
$$30 = \frac{1000 \times \sqrt{0.15}}{(H_{mano})^{3/4}}, \text{ or, } H_{mano} = \left[\frac{1000 \times \sqrt{0.15}}{30}\right]^{4/3} = 30.28 \text{ m}$$

$$\therefore \quad \text{Number of pumps stages} = \frac{H_{total}}{H_{mano}} = \frac{90}{30.28} = 3$$

As the total head required to be developed is more that the head developed by each pump, the pumps should be connected in **series. (Ans.)**

3.14. SPECIFIC SPEED

The **specific speed** *of a centrifugal pump is defined as the speed of a geometrically similar pump which would deliver unit quantity (one cubic metre of liquid per second) against a unit head (one metre)*. It is denoted by N_s. The specific speed is a characteristic of pumps which can be used as a basis for *comparing the performance of different pumps.*

An expression for specific speed may be obtained as follows:

$$\text{Discharge, } Q = \text{Area} \times \text{velocity of flow}$$

$$= \pi DB \times V_f$$

$$\text{or,} \quad Q \propto D \times B \times V_f \qquad ...(i)$$

where, D and B are the diameter and width of the pump impeller respectively.

Now, $B \propto D$, therefore, from eqn. (*i*), we have:

$$Q \propto D^2 \times V_f \qquad ...(ii)$$

Also, the tangential velocity is given by:

$$u = \frac{\pi DN}{60}, \text{ or, } u \propto DN \qquad ...(iii)$$

Again, the tangential velocity (u) and velocity of flow (V_f) bear the following relationship with the manometric head (H_{mano}):

$$u \propto V_f \propto \sqrt{H_{mano}} \qquad ...(iv)$$

Substituting the value of u in eqn. (*iii*), we have:

$$\sqrt{H_{mano}} \propto DN, \text{ or, } D \propto \frac{\sqrt{H_{mano}}}{N}$$

Substituting the value of D in eqn. (*ii*), we obtain:

$$Q \propto \frac{H_{mano}}{N^2} \times V_f \propto \frac{H_{mano}}{N^2} \times \sqrt{H_{mano}} \propto \frac{(H_{mano})^{3/2}}{N^2} \quad \left[\because V_f \propto \sqrt{H_{mano}} \quad ...\text{eqn.}(iv)\right]$$

$$\therefore \quad Q = K\frac{(H_{mano})^{3/2}}{N^2} \qquad ...(v)$$

(where, K = constant of proportionality)

By definition: When $H = 1$ m, and $Q = 1$ m^3/s, then N = specific speed, N_s. That is,

$$1 = K\frac{(1)^{3/2}}{N_s^2}, \text{ or, } K = N_s^2$$

Eqn. (*v*) may then be written as:

$$Q = N_s^2 \frac{(H_{mano})^{3/2}}{N^2}, \text{ or, } N_s^2 = \frac{N^2 Q}{(H_{mano})^{3/2}}$$

$$\therefore \quad \text{Specific speed, } N_s = \frac{N\sqrt{Q}}{(H_{mano})^{3/4}} \qquad ...(3.25)$$

— The values of N, H and Q are those for normal operating condition (the design point) which would generally coincide with the optimum efficiency.

— In case of a multi-stage pump the value of H_{mano} to be used in eqn. (3.25) is obtained by *dividing the total head developed by the number of stages.* For a double suction pump *half* the actual discharge delivered by the pump is taken as Q.

The ranges of specific speeds for different types of pumps are tabulated below:

Type of pump	Slow speed radial flow	Medium speed radial flow	High speed radial flow	Mixed flow (or screw type)	Axial flow (or propeller type)
Specific speed	10 to 30	30 to 50	50 to 80	80 to 160	160 to 500

— Within *limits of net positive suction head* (NPSH) is the net head in metres of liquid that is required to make the liquid flow through the suction pipe from the sump to the impeller) available, the pump with the *highest specific speed* is generally the *best choice.* Such a pump can operate at the *highest rotational speed and is of the smallest size.*

Example 3.31. *The diameter and width of a centrifugal pump impeller are 300 mm and 60 mm respectively. The pump is delivering 144 litres of liquid per second with a manometric efficiency of 85 percent. The effective outlet vane angle is 30°. If the speed of rotation is 950 r.p.m. calculate specific speed of the pump.*

Solution. Diameter of impeller at outlet, $D_2 = 300 \text{ mm} = 0.3 \text{ m}$

Width of impeller at outlet, $B_2 = 60 \text{ mm} = 0.06 \text{ m}$

Discharge through the pump, $Q = 144 \text{ litres/s} = 0.144 \text{ m}^3/\text{s}$

Manometric efficiency, $\eta_{mano} = 85\ \%$

The effective outlet vane angle, $\phi = 30°$

Speed of rotation, $N = 950$ r.p.m.

Specific speed of the pump, N_s :

Peripheral or tangential velocity at outlet of impeller,

$$u_2 = \frac{\pi D_2 N}{60} = \frac{\pi \times 0.3 \times 950}{60} = 14.92 \text{ m/s}$$

$$\text{Velocity of flow, } V_{f2} = \frac{Q}{\pi D_2 B_2} = \frac{0.144}{\pi \times 0.3 \times 0.06} = 2.55 \text{ m/s}$$

Refer to Fig. 3.8. From velocity triangle at *outlet*, we have:

$$V_{w2} = u_2 - \frac{V_{f2}}{\tan\phi} = 14.92 - \frac{2.55}{\tan 30°} = 10.5 \text{ m/s}$$

$$\text{Now, manometric efficiency, } \eta_{mano} = \frac{gH_{mano}}{V_{w2}u_2} = \frac{9.81 \times H_{mano}}{10.5 \times 14.92}$$

$$\therefore \quad H_{mano} = \frac{\eta_{mano} \times 10.5 \times 14.92}{9.81} = \frac{0.85 \times 10.5 \times 14.92}{9.81} = 13.57 \text{ m}$$

$$\text{Specific speed, } N_s = \frac{N\sqrt{Q}}{(H_{mano})^{3/4}} \quad \text{...[Eqn. (3.25)]}$$

$$= \frac{950 \times \sqrt{0.144}}{(13.37)^{3/4}} \simeq \mathbf{51\ (Ans.)}$$

3.15. MODEL TESTING AND GEOMETRICALLY SIMILAR PUMPS

In order to know the performance of prototypes, the models of centrifugal pumps are tested. The performance of the prototype pump will be correctly predicted by its model test only if the following *conditions* are satisfied:

- Specific speed of model = Specific speed of prototype

i.e. $$(N_s)_m = (N_s)_p$$

or, $$\left[\frac{N\sqrt{Q}}{(H_{mano})^{3/4}}\right]_m = \left[\frac{N\sqrt{Q}}{(H_{mano})^{3/4}}\right]_p \quad ...(3.26)$$

- Peripheral tangential velocity, $u = \dfrac{\pi DN}{60}$, also $u \propto \sqrt{H_{mano}}$

$$\therefore \quad \sqrt{H_{mano}} \propto DN, \text{ or, } \frac{\sqrt{H_{mano}}}{DN} = \text{constant}$$

$\left[\dfrac{H_{mano}}{D^2N^2}\right.$ is called the *head or lift co-efficient* $\left.\right]$

or, $$\left(\frac{\sqrt{H_{mano}}}{DN}\right)_m = \left(\frac{\sqrt{H_{mano}}}{DN}\right)_p \quad ...(3.27)$$

- Also $$Q \propto D^2 \times V_f \quad \text{...(Eqn (iii) of Art. 3.7)}$$

But, $$V_f \propto u \propto DN$$

$$\therefore \quad Q \propto D^2 \times DN, \text{ or, } Q \propto D^3N, \text{ or, } \frac{Q}{D^3N} = \text{constant}$$

The factor $\left[\dfrac{Q}{D^3N}\right.$ is called **flow co-efficient** $\left.\right]$

or, $$\left(\frac{Q}{D^3N}\right)_m = \left(\frac{Q}{D^3N}\right)_p \quad ...(3.28)$$

- Power of the pump, $P = wQH_{mano}$ (neglecting losses)

$$\therefore \quad P \propto Q \times H_{mano}$$

or, $$P \propto D^3N \times H_{mano} \quad (\because Q \propto D^3N)$$

or, $P \propto D^3N \times D^2N^2$, or, $P \propto D^5N^3$, or, $\dfrac{P}{D^5N^3} = \text{constant}$ $\quad (\because \sqrt{H_{mano}} \propto DN)$

$\left[\dfrac{P}{D^5N^3}\right.$ is called the **power co-efficient** $\left.\right]$

or, $$\left(\frac{P}{D^5N^3}\right)_m = \left(\frac{P}{D^5N^3}\right)_p \quad ...(3.29)$$

In case of *geometrically* ***similar*** *pumps* the suffix '*m*' is replaced by '1' and suffix '*p*' is replaced by '2'.

Example 3.32. *In order to predict the performance of a large centrifugal pump, a scale model of one - sixth size was made with the following specifications: Power P = 25 kW; Head H_{mano} = 7 m; Speed N = 1000 r.p.m. If the prototype pump has to work against a head of 22 m, calculate its working speed, the power required to drive it, and the ratio of the flow rates handled by the two pumps.*

Solution. Scale ratio = one – sixth

Model:	**Prototype:**
Power, P_m = 25 kW	Power, P_p = ?
Head, $(H_{mano})_m$ = 7 m	Head, $(H_{mano})_p$ = 22 m
Speed, N_m = 1000 r.p.m.	Speed, N_p = ?

Speed of prototype, N_p:

Using eqn. (3.27), $\left(\dfrac{\sqrt{H_{mano}}}{DN}\right)_m = \left(\dfrac{\sqrt{H_{mano}}}{DN}\right)_p$, we have:

$$\frac{(\sqrt{H_{mano}})_m}{D_m N_m} = \frac{(\sqrt{H_{mano}})_p}{D_p N_p}, \text{ or, } N_p = \frac{(\sqrt{H_{mano}})_p}{(\sqrt{H_{mano}})_m} \times \frac{D_m}{D_p} \times N_m$$

or $$N_p = \frac{\sqrt{22}}{\sqrt{7}} \times \frac{1}{6} \times 1000 = \textbf{295.47 r.p.m. (Ans.)}$$

Power required to drive the prototype pump, P_p:

Using eqn. (3.29), $\left(\dfrac{P}{D^5 N^3}\right)_m = \left(\dfrac{P}{D^5 N^3}\right)_p$, we have:

$$\frac{P_m}{D_m^5 N_m^3} = \frac{P_p}{D_p^5 N_p^3}, \text{ or, } P_p = P_m \times \frac{D_p^5 N_p^3}{D_m^5 N_m^3} = P_m \times \left(\frac{D_p}{D_m}\right)^5 \times \left(\frac{N_p}{N_m}\right)^3$$

$$= 25 \times \left(\frac{6}{1}\right)^5 \times \left(\frac{295.47}{1000}\right)^3 = \textbf{5014.6 kW (Ans.)}$$

Ratio of the flow rates, $\dfrac{Q_p}{Q_m}$:

Using eqn. (3.28), $\left(\dfrac{Q}{D^3 N}\right)_m = \left(\dfrac{Q}{D^3 N}\right)_p$, we have :

$$\frac{Q_m}{D_m^3 N_m} = \frac{Q_p}{D_p^3 N_p}, \text{ or, } \frac{Q_p}{Q_m} = \frac{D_p^3 N_p}{D_m^3 N_m} = \left(\frac{D_p}{D_m}\right)^3 \times \frac{N_p}{N_m}$$

or, $$\frac{Q_p}{Q_m} = \left(\frac{6}{1}\right)^3 \times \frac{295.47}{1000} = \textbf{63.82 (Ans.)}$$

Example 3.33. *Two geometrically similar pumps are running at the same speed of 1000 r.p.m. One pump has an impeller diameter of 300 mm and lifts water at the rate of 0.02 m^3/s against a head of 15 m. Determine the head and impeller diameter of the other pump to deliver half the discharge.* **[M.U.]**

Solution. Pump -1: Speed, $N_1 = 1000$ r.p.m

Diameter, $D_1 = 300$ mm or 0.3 m

Head, $H_{mano_1} = 15$ m

Discharge, $Q_1 = 0.02\ m^3/s$

Pump -2: Speed, $N_2 = 1000$ r.p.m.

Diameter, $D_2 = ?$

Head, $H_{mano_2} = ?$

Discharge, $Q_2 = \frac{Q_1}{2} = \frac{0.02}{2} = 0.01 m^3/s$

Head, H_{mano_2}:

Using the eqn. (3.26), $\left[\frac{N\sqrt{Q}}{(H_{mano})^{3/4}}\right]_1 = \left[\frac{N\sqrt{Q}}{(H_{mano})^{3/4}}\right]_2$, we have:

$$\frac{N_1\sqrt{Q_1}}{(H_{mano_1})^{3/4}} = \frac{N_2\sqrt{Q_2}}{(H_{mano_2})^{3/4}}$$

or, $$\frac{1000 \times \sqrt{0.02}}{(15)^{3/4}} = \frac{1000 \times \sqrt{0.01}}{(H_{mano_2})^{3/4}}$$

or, $$(H_{mano_2})^{3/4} = \frac{1000 \times \sqrt{0.01} \times (15)^{3/4}}{1000 \times \sqrt{0.02}} = 5.389$$

or, $$H_{mano_2} = (5.389)^{4/3} = \mathbf{9.45\ m}$$ **(Ans.)**

Impeller diameter, D_2:

Using the eqn. (3.27), $\left[\frac{\sqrt{H_{mano}}}{DN}\right]_1 = \left[\frac{\sqrt{H_{mano}}}{DN}\right]_2$, we have:

$$\frac{\sqrt{H_{mano_1}}}{D_1N_1} = \frac{\sqrt{H_{mano_2}}}{D_2N_2}$$

or, $$\frac{\sqrt{15}}{0.3 \times 1000} = \frac{\sqrt{9.45}}{D_2 \times \sqrt{1000}}$$

$\therefore$ $$D_2 = \frac{0.3 \times 1000 \times \sqrt{9.45}}{1000 \times \sqrt{15}} = 0.238 \text{ m or } \mathbf{238\ mm}$$ **(Ans.)**

Example 3.34. *3 m^3 of water per second is lifted to a height of 30 m with an efficiency of 75 percent by single - stage centrifugal pump. The impeller diameter is 300 mm and it is rotating at 2000 r.p.m. Find the number of stages and diameter of each impeller of a similar multi-stage pump to lift 5 m^3 of water per second to a height of 200 m when rotating at 1500 r.p.m.*

[Allahabad University]

Solution. Single-stage pump:

Discharge, $Q_1 = 3\ m^3/s$

Manometric height, $H_{mano_1} = 30$ m

Diameter of impeller, $D_1 = 300$ mm or 0.3 m

Speed $N_1 = 2000$ r.p.m.

Multi-stage pump:

Discharge, $Q_2 = 5\ m^3/s$

Manometric height, $H_{mano_2} = ?$ (per stage)

Diameter of each impeller, $D_2 = ?$

Speed, $N_2 = 1500$ r.p.m.

Number of stage, n:

Since specific speed should be same, therefore, applying eqn. 3.26, we have:

$$\left[\frac{N\sqrt{Q}}{(H_{mano})^{3/4}}\right]_1 = \left[\frac{N\sqrt{Q}}{(H_{mano})^{3/4}}\right]_2$$

or, $$\frac{N_1\sqrt{Q_1}}{(H_{mano_1})^{3/4}} = \frac{N_2\sqrt{Q_2}}{(H_{mano_2})^{3/4}}, \text{ or, } \frac{2000 \times \sqrt{3}}{(30)^{3/4}} = \frac{1500 \times \sqrt{5}}{(H_{mano_2})^{3/4}}$$

or, $$(H_{mano_2})^{3/4} = \frac{1500 \times \sqrt{5} \times (30)^{3/4}}{2000 \times \sqrt{3}} = 12.411, \text{ or, } H_{mano_2} = (12.411)^{4/3} = 28.71 \text{ m}$$

$$\therefore \text{Number of stages} = \frac{\text{Total head}}{\text{Head per stage}} = \frac{200}{28.71} = \mathbf{6.966 \simeq 7} \textbf{ (Ans.)}$$

Diameter of each impeller, D_2:

Using eqn.3.27, $\left[\frac{\sqrt{H_{mano}}}{DN}\right]_1 = \left[\frac{\sqrt{H_{mano}}}{DN}\right]_2$, we have:

$$\frac{\sqrt{H_{mano_1}}}{D_1N_1} = \frac{\sqrt{H_{mano_2}}}{D_2N_2}, \text{ or, } \frac{\sqrt{30}}{0.3 \times 2000} = \frac{\sqrt{28.71}}{D_2 \times 1500}$$

or, $$D_2 = \frac{0.3 \times 2000 \times \sqrt{28.71}}{1500 \times \sqrt{30}} = 0.3913 \text{m or } \mathbf{391.3\,mm} \textbf{ (Ans.)}$$

Example 3.35. *A centrifugal pump is discharging 0·025 m³/s of water against a total head of 18 m. The diameter of the impeller is 0.4 m and it is rotating at 1400 r.p.m. Calculate the head, discharge and ratio of powers of a geometrically similar pump of diameter 0.25 m when it is running at 2800 r.p.m.*

Solution. Centrifugal pump:

Discharge, $Q_1 = 0{\cdot}025$ m³/s

Head, $H_{mano_1} = 18$ m

Diameter, $D_1 = 0.4$ m

Speed, $N_1 = 1400$ r.p.m

Geometrically similar pump:

Discharge, $Q_2 = ?$

Head, $H_{mano_2} = ?$

Diameter $D_2 = 0.25$ m

Speed, $N_2 = 2800$ r.p.m.

Head, H_{mano_2}:

Using eqn. (3.27) : $\left[\frac{\sqrt{H_{mano}}}{DN}\right]_1 = \left[\frac{\sqrt{H_{mano}}}{DN}\right]_2$, we have:

$$\frac{\sqrt{H_{mano_1}}}{D_1N_1} = \frac{\sqrt{H_{mano_2}}}{D_2N_2}, \text{ or, } \frac{\sqrt{18}}{0.4 \times 1400} = \frac{\sqrt{H_{mano_2}}}{0.25 \times 2800}$$

or, $$H_{mano_2} = \left(\frac{\sqrt{18} \times 0.25 \times 2800}{0.4 \times 1400}\right)^2 = \mathbf{28.125\,m} \textbf{ (Ans.)}$$

Discharge, Q_2:

Using eqn. (3.28) : $\left(\frac{Q}{D^3N}\right)_1 = \left(\frac{Q}{D^3N}\right)_2$, we have:

$$\frac{Q_1}{D_1^3N_1} = \frac{Q_2}{D_2^3N_2}, \text{ or, } \frac{0.025}{0.4^3 \times 1400} = \frac{Q_2}{0.25^3 \times 2800}$$

or, $$Q_2 = \frac{0.025 \times 0.25^3 \times 2800}{0.4^3 \times 1400} = \mathbf{0.0122\ m^3/s}\ \textbf{(Ans.)}$$

Ratio of power, $\frac{P_1}{P_2}$:

Using eqn. (3.29): $\left(\frac{P}{D^5 N^3}\right)_1 = \left(\frac{P}{D^5 N^3}\right)_2$, we have

$$\frac{P_1}{D_1^5 N_1^3} = \frac{P_2}{D_2^5 N_2^3}, \text{ or, } \frac{P_1}{P_2} = \frac{D_1^5 N_1^3}{D_2^5 N_2^3} = \frac{0.4^5 \times 1400^3}{0.25^3 \times 2800^3} = \mathbf{1.31}\ \textbf{(Ans.)}$$

3.16. CHARACTERISTICS OF CENTRIFUGAL PUMPS

Ordinarily a centrifugal pump is worked under its maximum efficiency conditions. However, when the pump is run at conditions different from the design conditions, it performs differently. Therefore, to predict the behaviour of the pump under varying conditions of speeds, heads, discharges or powers, tests are usually conducted. The results obtained from these tests are plotted in from of *characteristic curves;* these curves delineate useful information about the performance of a pump in its installation.

The following *four types of characteristic curves* are usually prepared for centrifugal pumps:

1. Main characteristic curves,
2. Operating characteristic curves,
3. Constant efficiency or Muschel curves, and
4. Constant head and constant discharge curves.

1. Main characteristic curves:

The main characteristic curves are obtained as follows:

- The pump is run at a *constant speed* and the *discharge is varied* over the desired range (by delivery valve)
- Measurements are taken for manometric head (H_{mano}) and shaft power (P) for each discharge(Q).
- Calculations are made for the pump overall efficiency, η_0
- The curves are plotted between Q and H_{mano}; Q and P; and Q and η_0 for that speed.
- The same procedure is repeated by running the pump at another speed.
- A family of curves is obtained as shown in Fig. 3.25.

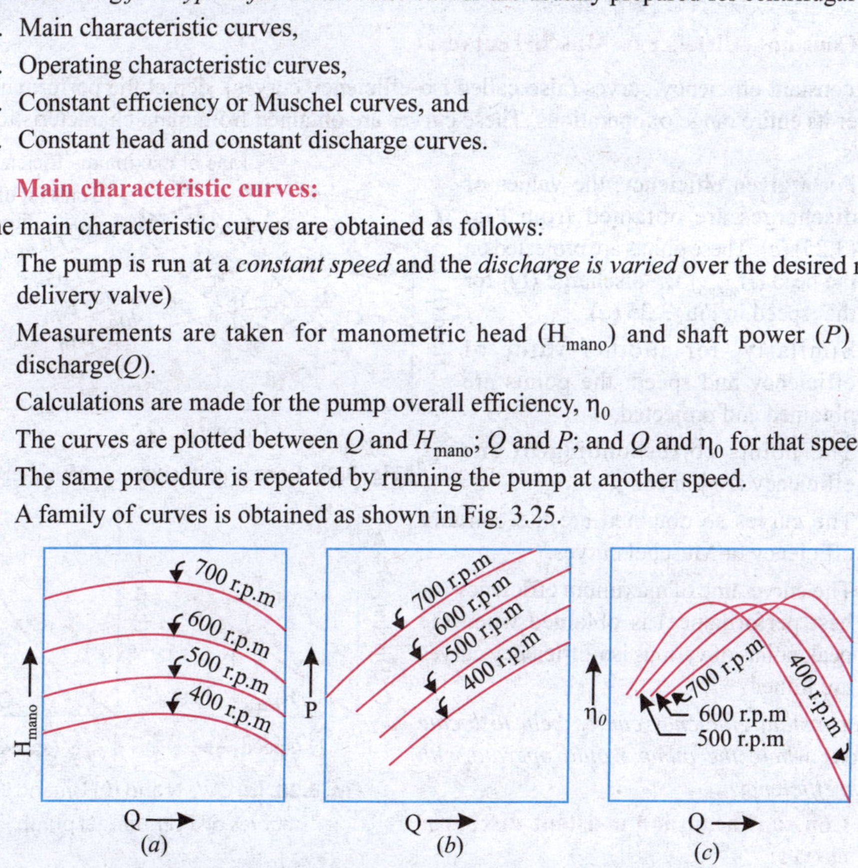

Fig. 3.25. Main characteristic curves.

2. Operating characteristic curves:

When a centrifugal pump operates at the *design speed* (same as speed of driving motor) the *maximum* efficiency occurs. Evidently for *optimum performance,* the pump needs to be operated at the design speed. To obtain *operating characteristic curves* the pump is run at the design speed and the discharge is varied, as in the case of main characteristic curves. The operating characteristic curves are shown in Fig. 3.26. The design discharge and head *are obtained from the corresponding* curves where the efficiency is maximum,

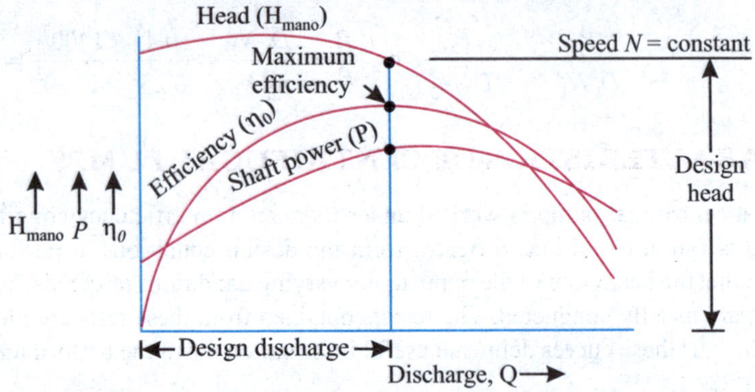

Fig. 3.26. Operating characteristic curves of a centrifugal pump.

3. Constant efficiency or Muschel curves:

The constant efficiency curves (also called iso-efficiency curves), depict the performance of a pump over its entire range of operations. These curves are obtained from main characteristic curves as follows:

- For a given efficiency, the values of discharges are obtained from Fig. (3.25) (*c*). These points are projected on the head (H_{mano}) *v/s* discharge (Q) for that speed in Fig. 3.25 (*a*).
- Similarly, for another value of efficiency and speed, the points are obtained and projected.
- The points corresponding to one efficiency are joined.
- The curves so obtained are the constant efficiency or Muschel curves.
- The curve/ line of maximum efficiency (or best performance) is obtained when the peak points of various iso-efficiency curves are joined.

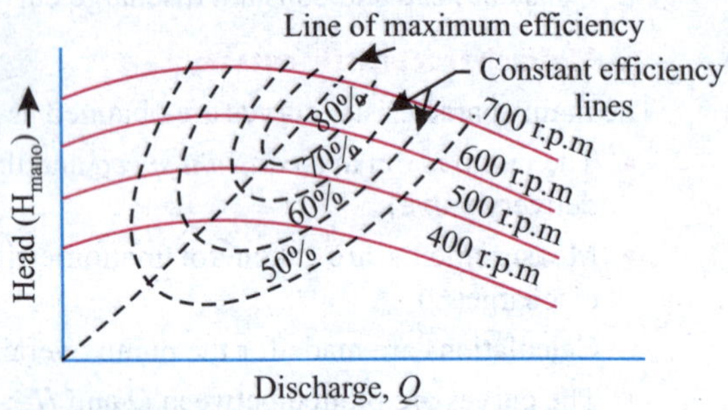

Fig. 3.27. Constant efficiency or Muschel curves

The constant efficiency curves help to locate the regions where the pump would operate with maximum efficiency.

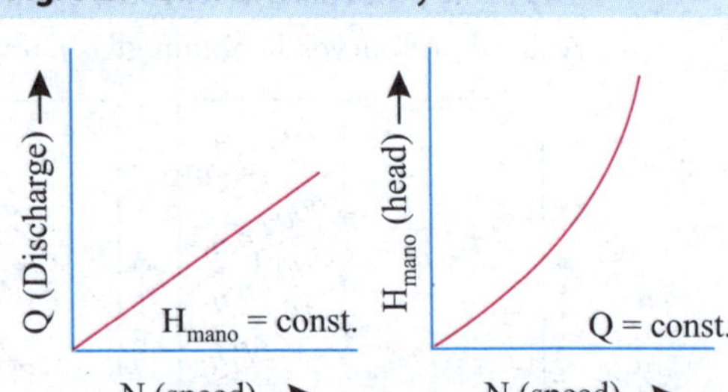

Fig. 3.28. (*a*) Qv/s N and **(*b*)** Hmano v/s N curves of a centrifugal pump

4. Constant head and constant discharge curves:

The performance of a variable speed pump for which the *speed constantly varies* can be determined by these curves. When the pump has a variable speed, the plots between Q and N, and

H_{mano} and N may be obtained. In the first case H_{mano} is kept constant and in the second case, Q is kept constant. The curves are shown in Fig. 3.28.

3.17. NET POSITIVE SUCTION HEAD (NPSH)

Fig. 3.29. shows a centrifugal pump drawing liquid from a sump open to a atmosphere.

Let, h_s = Vertical distance between the centre line of the pump and the free liquid surface of the sump,

V_s = Velocity of liquid in the suction pipe,

h_{fs} = Losses in the suction pipe upto the pump inlet (1),

p_1 = Absolute static pressure at pump inlet,

p_a = Absolute atmospheric pressure, and

p_v = Vapour pressure of the liquid for a given temperature.

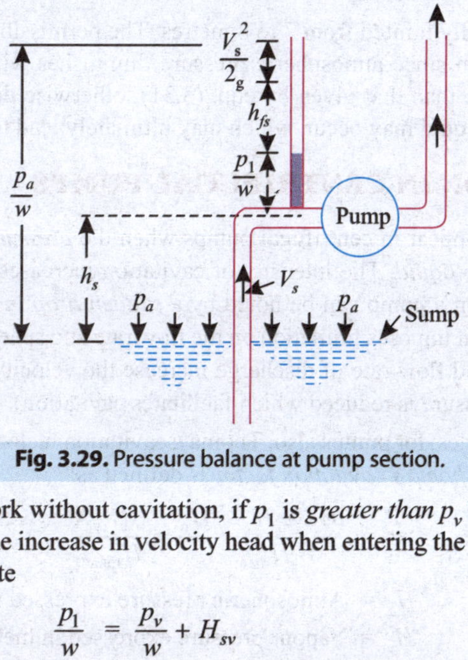

Fig. 3.29. Pressure balance at pump section.

Now the pump will work without cavitation, if p_1 is *greater than* p_v by an amount equal to that required by the liquid for the increase in velocity head when entering the impeller; if this amount be denoted by H_{sv}, we can write

$$\frac{p_1}{w} = \frac{p_v}{w} + H_{sv} \qquad ...(i)$$

Also,

$$\frac{p_1}{w} = \frac{p_a}{w} - \left(h_s + h_{fs} + \frac{V_s^2}{2g}\right) \qquad ...(ii)$$

From (*i*) and (*ii*), we have:

$$H_{sv} = \frac{p_a}{w} - \left(h_s + h_{fs} + \frac{V_s^2}{2g}\right) - \frac{p_v}{w}$$

or,

$$H_{sv} = H_a - H_s - H_v \qquad ...(3.30)$$

$$\left(H_a = \frac{p_a}{w}, \text{ and, } H_v = \frac{p_v}{w}\right)$$

where, H_s = Total suction head $= \left(h_s + h_{fs} + \frac{V_s^2}{2g}\right)$

This value of H_{sv} is frequently called the *net positive suction head* (NPSH). Thus the net positive suction head may *be defined "as the difference between the net inlet head and the head corresponding to the vapour pressure of the liquid"*. NPSH may also be defined as *"the net head (in metres of liquid) that is required to make the liquid flow through the suction pipe from the sump to the impeller."*

This *term has significance only when cavitating liquids are handled.*

NPSH is a parameter (diamensional) that *can be used to check cavitation in pump*. The term *NPSH* is a frequently used in *pump industry.* The minimum *NPSH* depends upon the pump design, its speed and the discharge.

From eqn. (*ii*), the **limiting value of suction lift** (h_s) is given by:

$$h_s = \left(\frac{p_a - p_v}{w}\right) - h_{fs} - \frac{V_s^2}{2g}, \text{ when } p_1 = p_v \quad ...(3.31)$$

Suction height is usually limited from 7 to 8 metres. The permissible suction lift would be less at elevated pump elevation since atmospheric pressure diminishes with altitude. The suction lift should in no case be more than that given by eqn. (3.31), otherwise due to reduction in pressure, rapid vaporization of the liquid may occur, which may ultimately lead to cavitation.

3.18. CAVITATION IN CENTRIFUGAL PUMPS

Cavitation begins to appear in centrifugal pumps when the *pressure at the suction falls below the vapour pressure of the liquid.* The intensity of cavitation increases with the decrease in value of NPSH. The cavitation in a pump can be noted by a *sudden drop in efficiency, head and power requirement.* The cavitation imposes limitation on the flow rate and speed of rotation of pump (since as the speed of rotation and flow rate of discharge increase the velocity of liquid at inlet increases due to which absolute pressure is reduced which facilitates cavitation).

As in the case of turbines, for pumps also, Thoma's cavitation factor is used to indicate the onset of cavitation. For pumps *Thoma's cavitation factor* is defined as:

$$\sigma = \frac{H_a - H_s - H_v}{H_{mano}} = \frac{H_{sv}}{H_{mano}} \quad ...(3.32)$$

where, H_a = Atmospheric pressure expressed in metres of water head,

H_v = Vapour pressure expressed in metres of water head,

H_s = Total suction head $\left(= h_s + h_{fs} + \frac{V_s^2}{2g}\right)$,

H_{sv} = Net positive suction head (NPSH), and

H_{mano} = Manometric head.

The cavitation will occur if the value of σ is less than the critical value σ_c at which the cavitation just begins. The cavitation parameter σ is a *function of specific speed, efficiency of the pump, and number of vanes.*

The *harmful effects of cavitation are:*

(*i*) Pitting and erosion of surface (due to continuous hammering action of collapsing bubbles)

(*ii*) Sudden drop in head, efficiency and the power delivered to the fluid.

(*iii*) Noise and vibration (produced by the collapse of bubbles)

The *factors which facilitate onset of cavitation are as follows:*

(*i*) Restricted suction, (*ii*) High runner speed,

(*iii*) Too high specific speed for optimum design parameters, and

(*iv*) Too high temperature of the flowing liquid.

Suction specific speed:

For geometrically similar machines (homologous),

$$Q \propto ND^3 \quad ...(i)$$

$$NPSH \propto \frac{V^2}{2g} \propto \frac{Q^2}{D^4} \quad ...(ii)$$

Eliminating the dimension D from the above expression, we have:

$$\frac{N\sqrt{Q}}{(NPSH)^{3/4}} = \text{constant} = (N_s)_{\text{suc.}} \quad ...(3.33)$$

The parameter $(N_s)_{suc}$. is called the **suction specific speed.** When the different machines have equal values of $(N_s)_{suc}$., it indicates that the machines are operating with similar degree of cavitation.

Now by eliminating $N\sqrt{Q}$ from the following expressions, we get:

$$(N_s)_{\text{suc.}} = \frac{N\sqrt{Q}}{(NPSH)^{3/4}} \quad \text{...Suction specific speed}$$

$$N_s = \frac{N\sqrt{Q}}{(H_{\text{mano}})^{3/4}} \quad \text{... pump specific speed (normal)}$$

$$\frac{NPSH}{H_{\text{mano}}} = \left[\frac{N_s}{(N_s)_{suc.}}\right]^{4/3}$$

or, $$\sigma = \left(\frac{N_s}{(N_s)_{suc.}}\right)^{4/3} \quad ...(3.34)$$

Example 3.36. *Tests on a pump model indicate a cavitation parameter $\sigma_c = 0.10$. A homologous unit is to be installed at a location where atmospheric pressure, $p_a = 0{\cdot}91$ bar and vapour pressure $p_v = 0{\cdot}035$ bar absolute and is to pump water against a head of 25 m. What is the maximum permissible suction head ?* **[P.E.C.]**

Solution. Cavitation paremeter, $\sigma_c = 0.10$

Atmospheric pressure, $p_a = 0.91$ bar, or, $H_a = \frac{0.91 \times 10^5}{9810} = 9.27$ m of water

Vapour pressure, $p_v = 0.035$ bar, or, $H_v = \frac{0.035 \times 10^5}{9810} = 0.356$ m of water

Manometric head, $H_{\text{mano}} = 25$ m

Maximum permissible suction head, h_s:

Using the relation: $\sigma = \frac{H_a - H_s - H_v}{H_{\text{mano}}}$, or, $\sigma_c = \frac{H_c - H_s - H_v}{H_{\text{mano}}}$, neglecting head lost due to friction

or, $$0.10 = \frac{9.27 - h_s - 0.356}{25}, \quad \text{or,} \quad 0.10 \times 25 = 8.914 - h_s$$

$\therefore$ $$h_s = 8.914 - 0.10 \times 25 = \mathbf{6.41\ m}\ \textbf{(Ans.)}$$

Example 3.37. *Find the height from the water surface at which a centrifugal pump may be installed in the following case to avoid cavitation:*

Atmospheric pressure = 1.01 bar; vapour pressure = 0.022 bar; inlet and other losses in suction pipe = 1.42 m; effective head of pump = 49 m; and cavitation parameter = 0.115.

Solution. *Given :*

$$p_a = 1.01 \text{ bar, or, } H_a = \frac{1.01 \times 10^5}{9810} = 10.29 \text{ m}$$

$$p_v = 0.022 \text{ bar, or, } H_v = \frac{0.022 \times 10^5}{9810} = 0.224 \text{ m}$$

Inlet and other losses in suction pipe, $h_{fs} = 1.42$ m,

Effective head of pump (manometric head), $H_{mano} = 49$ m.

Cavitation parameter, $\sigma = 0.115$.

Installation height above water surface, h_s:

Cavitation factor/ parameter is given by:

$$\sigma = \frac{H_a - H_s - H_v}{H_{mano}} = \frac{H_a - (h_s + h_{fs}) - H_e}{H_{mano}}$$

or,
$$0.115 = \frac{10.29 - (h_s + 1.42) - 0.224}{49} = \frac{8.646 - h_s}{49}$$

or,
$$h_s = 8.646 - 0.115 \times 49 = \mathbf{3.01 \text{ m}} \text{ (Ans.)}$$

Example 3.38. *A single-stage centrifugal pump runs at 600 r.p.m. and delivers 360 m³/min.. of water against a head of 144 m. The pump impeller is 2.4 m in diameter and it has a positive suction lift (including the velocity head and friction) of 3.6 m. Laboratory tests are to be conducted on a model with 0.54 m diameter impeller and on a reduced head of 114 m. Calculate the speed, discharge and suction lift for the laboratory tests. Assume atmospheric head = 10.18 m of water and vapour head = 0·32 m of water.*

Solution. Prototype pump:

Speed, $N_p = 600$ r.p.m.

Discharge, $Q_p = 360 \text{ m}^3/\text{min}$

Manometric head, $(H_{mano})_p = 144$ m

Diameter of impeller, $D_p = 2.4$ m

Positive suction lift = 3.6 m

Vapour head, $H_v = 0{\cdot}32$ m of water.

Model pump:

Speed, $N_m = ?$

Discharge, $Q_m = ?$

Manometric head, $(H_{mano})_m = 114$ m

Diameter of impeller, $D_m = 0.54$ m

Atmospheric head, $H_a = 10.18$ m of water.

(*i*) Speed of the model pump, N_m:

Using the relation:
$$\left(\frac{\sqrt{H_{mano}}}{DN}\right)_m = \left(\frac{\sqrt{H_{mano}}}{DN}\right)_p \quad \text{...[Eqn. (3.27)]}$$

or,
$$\frac{(\sqrt{H_{mano}})_m}{D_m N_m} = \frac{(\sqrt{H_{mano}})_p}{D_p N_p}, \text{ or, } N_m = \frac{(\sqrt{H_{mano}})_m}{(\sqrt{H_{mano}})_p} \times \frac{D_p}{D_m} \times N_p$$

or,
$$N_m = \sqrt{\frac{114}{144}} \times \frac{2.4}{0.54} \times 600 = \mathbf{2372.7 \text{ r.p.m.}} \text{ (Ans.)}$$

(ii) Discharge for the model pump, Q_m:

Using the relation: $\left(\frac{Q}{D^3N}\right)_m = \left(\frac{Q}{D^3N}\right)_p$ ...[Eqn. (3.28)]

$$\frac{Q_m}{D_m^3 N_m} = \frac{Q_p}{D_p^3 N_p}, \text{ or, } Q_m = Q_p \left(\frac{D_m}{D_p}\right)^2 \times \frac{N_m}{N_p}$$

or,
$$Q_m = 360 \times \left(\frac{0.54}{2.4}\right)^3 \times \frac{2372.7}{600} = 16.21 \text{ m}^3/\text{min}$$

(iii) Positive suction lift with which model should be tested, H_s:

Cavitation factor for the prototype,

$$\sigma_p = \frac{H_a - H_s - H_v}{(H_{mano})_p} = \frac{10.18 - 3.6 - 0.32}{144} = 0.0435$$

For cavitation similarity, $\sigma_m = \sigma_p$

$$\therefore \quad \sigma_m = \frac{10.18 - H_s - 0.32}{114}, \text{ or, } 0.0435 = \frac{9.86 - H_s}{114}$$

$\therefore H_s = 9.86 - 0.0435 \times 114 =$ **4.9 m** *(including velocity head and friction)* **(Ans.)**

3.19. PRIMING OF A CENTRIFUGAL PUMP

The operation of filling the suction pipe, casing of the pump and a portion of the delivery pipe completely from outside source with the liquid to be raised, before starting the pump, to remove any air, gas or vapour from these parts of the pump is called **priming** *of a centrifugal pump.* If a centrifugal pump is *not primed before starting, air pockets inside the impeller may give rise to vortices and cause discontinuity of flow. Further, dry running of the pump may result in rubbing and seizing of the wearing rings and cause serious damage.*

- **Small pumps** are usually primed by pouring liquid into the funnel provided for the purpose. While doing priming, the air-vent valve provided in the pump casing is opened; the air escapes through the valve. The priming is continued till all air from the suction pipe, impeller and casing has been removed.
- **Large pumps** are primed by evacuating the casing and the suction pipe by a *vacuum pump or by an ejector;* the liquid is thus drawn up the suction pipe from the sump and the pump is filled with liquid.
- The internal construction of some pumps is such that *special arrangements containing a supply of liquid are provided in the suction pipe due to which automatic priming of the pump occurs;* such pumps are known as **'self priming pumps'.**

3.20. SELECTION OF PUMPS

- The main criteria of the selection of the type of pump are values of discharge (Q), head (H) and speed (N). From these values the specific speed of the pump is calculated and subsequently the type of the pump can be decided.
- When the specific speed is low and it is possible to increase the pump speed, it is better to use multi-stage pump; the number of stages are decided on the basis of the head and the type of the pump to be used.
- The *type of impeller* is another aspect of pump selection:

(i) *Impeller shrounded type* ... for pumping fresh clean water.

(ii) *Impeller-unshrounded or propeller type* ... for pumping soild-liquid mixture or near plastic material.

(For pumping *molasses* etc. sometimes positive displacement screw pump or lobe pumps are employed.)

(iii) *Mixed flow impellers with diffuser vanes* ... used for *deep well or submersible pumps.*

- Axial flow pumps are employed for *very low heads* of about six metres and *for large discharges.*
- *Radial flow pumps* are used when the *head is high.*

3.21. OPERATIONAL DIFFICULTIES IN CENTRIFUGAL PUMPS

The type of operational *difficulties* commonly experienced in centrifugal pumps and their *remedies (*given in *parentheses*) are as given below:

I. Pump fails to start pumping:

1. Pump may not be properly primed – (Reprime the pump).
2. Total head against which the pump is working may be much higher than that for which the pump is designed – (Check the head with accurate gauges; reduce the head or change the pump).
3. Impeller may be clogged – (Clean the impeller).
4. The rotation of the impeller may be in the wrong direction. (Change the direction of rotation).
5. Too high suction lift (Reduce the suction lift).
6. Law speed – (Increase the speed).

II. Pump is not working upto capacity and pressure:

1. Leakage of air into the pump – (Plug the leakege).
2. Some of the parts are damaged due to excessive wear and tear – (Replace the worn out/ damaged parts).

III. Pump stops working:

1. Presence of air in suction line – (Remove the air by priming and plug the entry of air).
2. High suction lift (Reduce the suction lift).

IV. Pump has very low efficiency :

1. Speed may be too high – (Reduce the speed)
2. Head may be too low and the pump delivers the liquid in large quantity – (Reduce the discharge or change the pump).
3. Pump may be operating in wrong direction – (Correct the direction of rotation of impeller).
4. Shaft may be bent, the impeller may be touching the casing, stuffing boxes may be too tight, wearing rings may be worn – (Repair the affected parts).

HIGHLIGHTS

1. A pump is a contrivance which provides energy to a fluid in a fluid system; it assists to increase the pressure energy or kinetic energy, or both of the fluid by converting the mechanical energy.
2. (*a*) Work done per second per unit weight of liquid

$$= \frac{V_{w2}u_2}{g}, \text{ assuming flow at inlet to be } radial \qquad ...(i)$$

(*b*) If the flow is *not radial,* the expression for work done may be written as:

Work done per second per unit weight of liquid

$$= \frac{1}{g}(V_{w2}u_2 - V_{w1}u_1) \qquad ...(ii)$$

Eqn. (*ii*) is known as the *Euler momentum equation for centrifugal pumps.*

The term $\frac{1}{g}(V_{w2}u_2 - V_{w1}u_1)$ is referred to as Euler head (H_e)

(*c*) Work done per second per unit weight of liquid (or H_e)

$$= \frac{V_2^2 - V_1^2}{2g} + \frac{u_2^2 - u_1^2}{2g} + \frac{V_{r1}^2 - V_{r2}^2}{2g} \qquad ...(iii)$$

This equation is sometimes called the *fundamental equation of a centrifugal pump.*

3. *Suction head* (h_s). It is the vertical height of the centreline of pump shaft above the liquid surface in the sump from which the liquid is being raised.
Delivery head (h_d). It is the vertical height of the liquid surface in the tank/reservoir to which the liquid is delivered above the centreline of the pump shaft.
The sum of suction head and delivery head is known as *static head* (H_{stat}).
4. *Manometric head* (H_{mano}). The head against which a centrifugal pump has to work is known as manometric head. It is given as:

(*i*) $H_{mano} = \frac{V_{w2}u_2}{g}$ – loss of head in the pump (*i.e.* impeller and casing)

(*ii*) $H_{mano} = H_{stat} + \text{losses in pipe} + \frac{V_d^2}{2g}$

$$= (h_s + h_d) + (h_{fs} + h_{fd}) + \frac{V_d^2}{2g}$$

(*iii*) H_{mano} = Total head at outlet of the pump – total head at inlet of the pump

$$= \left(\frac{p_2}{w} + \frac{V_2^2}{2g} + z_2\right) - \left(\frac{p_1}{w} + \frac{V_1^2}{2g} + z_1\right)$$

5. The various efficiencies of the pump are

(*i*) Manometric efficiency, $\eta_{mano} = \frac{gH_{mano}}{V_{w2}u_2}$

(*ii*) Volumetric efficiency, $\eta_v = \frac{Q}{Q+q}$

where, Q = Actual liquid discharge at the pump outlet second, and

q = Leakage of liquid per second from the impeller (through the clearances between the impeller and casing)

(*iii*) Mechanical efficiency, $\eta_m = \dfrac{w(Q+q)\,(V_{w2}u_2/g)}{P}$

$$= \frac{P - P_{mach.loss}}{P} \quad \text{(where, } P = \text{shaft power)}$$

(*iv*) Overall efficiency, $\eta_0 = \dfrac{wQH_{mano}}{P}$

Also, $\eta_0 = \eta_{mano} \times \eta_v \times \eta_m$

6. The *minimum speed for starting a centrifugal pump* is given by:

$$N(i.e.\ N_{min}) = \frac{120 \times \eta_{mano} \times V_{w2} \times D_2}{\pi(D_2^2 - D_1^2)}$$

7. A multi-stage pump is one which has two or more identical impellers (mounted on the same shaft or on different shafts); to produce a *high head* the impellers are connected in *series* while to *discharge a large quantity of liquid,* the impellers are connected in *parallel.*

8. The *specific speed (N_s)* of a centrifugal pump is defined as the speed of a geometrically similar pump which would deliver unit quantity (one cubic metre of liquid per second) against a unit head (one metre). Thus.

$$N_s = \frac{N\sqrt{Q}}{(H_{mano})^{3/4}}$$

9. For complete similarity between the model and prototype/actual centrifugal pump the following conditions should be satisfied:

(*i*) $$\left[\frac{N\sqrt{Q}}{(H_{mano})^{3/4}}\right]_m = \left[\frac{N\sqrt{Q}}{(H_{mano})^{3/4}}\right]_p$$

(*ii*) $$\left(\frac{\sqrt{H_{mano}}}{DN}\right)_m = \left(\frac{\sqrt{H_{mano}}}{DN}\right)_p$$

(*iii*) $$\left(\frac{Q}{D^3N}\right)_m = \left(\frac{Q}{D^3N}\right)_p$$

(*iv*) $$\left(\frac{P}{D^5N^3}\right)_m = \left(\frac{P}{D^5N^3}\right)_p$$

10. The characteristics curves are used for predicting the behaviour and performance of a pump when it is working under different heads, speeds and rates of flow.

11. The net positive suction head (*NPSH*) may be defined as "The difference between the net inlet head and the head corresponding to the vapour pressure of the liquid"

12. *Cavitation* begins to appear in centrifugal pumps when the pressure at the suction falls below the vapour pressure of the liquid. It can be noted by *sudden drop in efficiency, head and power requirement.*

OBJECTIVE TYPE QUESTIONS

Choose the Correct Answer:

1. Which of the following types of impeller is used for centrifugal pumps dealing with muds?
   (*a*) One-side shrouded
   (*b*) Two -sides shrouded
   (*c*) Double section
   (*d*) Open.
2. Which of the following statements is *correct* with reference to an impeller with backward curved vanes?
   (*a*) It has a falling head - discharge characteristic.
   (*b*) It has rising head - discharge characteristic
   (*c*) It is easier to fabricate.
   (*d*) It cannot run at speeds other than the design speed.
3. The head developed by a centrifugal pump may be expressed as
   (*a*) $H = \dfrac{V_{w2}u_2 + V_{w1}u_2}{g}$
   (*b*) $H = \dfrac{V_{w2}u_2}{g}$
   (*c*) $H = \dfrac{V_{w1}u_1}{g}$
   (*d*) none of the above.
4. With reference to a centrifugal pump which of the following statements is *incorrect ?*
   (*a*) The discharge control valve is fitted in the delivery pipe.
   (*b*) The suction pipe is provided with a foot valve and a strainer.
   (*c*) The suction pipe has larger diameter as compared to the discharge pipe.
   (*d*) The discharge control valve is fitted in the suction pipe.
5. A centrifugal pump should be so installed above the water level in the sump such that
   (*a*) the negative pressures are not allowed to develop in the impeller
   (*b*) the negative pressures do not reach as low a value as the vapour pressure
   (*c*) its height is more than 10·28 m at ordinary temperature of liquid
   (*d*) none of the above.
6. The specific speed of a pump is defined as the speed of unit of such a size that it
   (*a*) produces unit power with unit head available
   (*b*) delivers unit discharge at unit head
   (*c*) requires unit power to develop unit head
   (*d*) delivers unit discharge at unit power.
7. Which of the following statements pertaining to a given centrifugal pump is *correct* ?
   (*a*) Discharge varies as the square of speed
   (*b*) Power varies as the square of speed
   (*c*) Discharge varies directly as speed
   (*d*) Head varies inversely as speed.
8. The delivery valve, while starting centrifugal pump, is kept
   (*a*) fully closed (*b*) fully open
   (*c*) half open (*d*) in any position.
9. Which of the following is *not* a dismensionless parameter ?
   (*a*) Friction factor
   (*b*) Specific speed
   (*c*) Thoma's cavitation parameter
   (*d*) Pressure co-efficient.
10. With reference to manometric head which of the following statements is *correct* ?
    (*a*) It is the head developed by the pump.
    (*b*) It is the height to which water is lifted by the pump measured above the pump centreline.
    (*c*) it is the difference in elevation between the water surface in the high level reservoir and the water level in the sump.
    (*d*) It is the difference in the piezometric heads between the points on the delivery and suction pipes as close to the pump as possible.
11. Higher specific speeds (160 to 500) of centrifugal pump indicate that the pump is of
    (*a*) radial flow type (*b*) axial flow type
    (*c*) mixed flow type (*d*) any of these types.
12. What will happen if requirements of net positive suction head (NPSH) for a given pump are not satisfied ?
    (*a*) The pump will get cavitated.
    (*b*) The pump will consume more power.
    (*c*) The pump will not develop head.
    (*d*) The pump will have a low efficiency.
13. The net positive suction head (*NPSH*) which represents the suction head at the impeller eye is given by
    (*a*) $\dfrac{p_a - p_v}{w} - h_s - h_{fs}$
    (*b*) $\dfrac{p_a - p_v}{w} - h_s + h_{fs}$

(c) $\frac{p_a - p_v}{w} + h_s + h_{fs}$

(d) $\frac{p_a - p_v}{w} + h_s - h_{fs}$.

14. To prevent cavitation, the suction lift must be considerably lower than the maximum limit expressed by

(a) $\frac{p_a - p_v}{w} + \frac{V_s^2}{2g} - h_{fs}$

(b) $\frac{p_a - p_v}{w} - \frac{V_s^2}{2g} + h_{fs}$

(c) $\frac{p_a - p_v}{w} - \frac{V_s^2}{2g} - h_{fs}$

(d) $\frac{p_a - p_v}{w} + \frac{V_s^2}{2g} + h_{fs}$.

15. In centrifugal pumps, cavitation is reduced by

(a) increasing the flow velocity
(b) reducing the discharge
(c) throttling the discharge
(d) reducing the suction head.

16. Which of the following statements pertaining to a centrifugal pump is *incorrect*?

(a) The suction lift of a pump can be upto 10·3 m or even more.
(b) The impellers of a multi-stage pump are arranged in parallel to discharge a large quantity of liquid.
(c) The volute casing of the pump maintains the velocity of flow constant, prevents eddies and converts velocity head to pressure head.
(d) The manometric head refers to the difference between the total energy of liquid at exit from and at inlet to the pump.

17. Regarding cavitation which of the following statements in *incorrect*?

(a) Cavitation affects the performance of a turbine to a lesser degree than that of a pump.
(b) Thoma's cavitation parameter has different expressions for turbines and pumps.
(c) With the increase in pump speed, there is increase in its minimum net positive suction head requirement.
(d) The leading edge of blades in pumps and the trailing edge of blades in water turbines are more susceptible to cavitation damage.

18. A centrifugal pump is taking much of power, the probable reason may be

(a) liquid being pumped is heavy
(b) speed of the pump is low
(c) there is leakage of air
(d) ineffective strainer and foot valve arrangement.

19. In rotodynamic pumps, the increase in energy level is due to

(a) centrifugal energy only
(b) pressure energy only
(c) kinetic energy only
(d) combination of *a*, *b* and *c*.

20. In a pump there is

(a) accelerating flow
(b) decelerated flow
(c) either of the above
(d) none of the above.

21. In a centrifugal pump the regulating valve is provided on

(a) the suction pipe (b) delivery pipe
(c) the casing (d) none of the above.

22. In a centrifugal pump the sum of suction head and delivery head is known as

(a) manometric head (b) total head
(c) static head (d) none of the above.

23. Regarding manometric head (H_{mano}) which of the following relations is *correct*?

(a) H_{mano} = head imparted by the impeller to the liquid – loss of head in the pump
(b) H_{mano} = static head + losses in pipes + $\frac{V_d^2}{2g}$
(c) H_{mano} = total head at outlet of the pump- total head at inlet of pump
(d) All of the above.

24. The ratio of power outlet of the pump to the power input to the pump is known as

(a) mechanical efficiency
(b) overall efficiency
(c) manometric efficiency
(d) none of the above.

25. The flow ratio in case of a centrifugal pump varies from

(a) 0.1 to 0.25 (b) 0.25 to 0.40
(c) 0.40 to 0.50 (d) 0.50 to 0.65.

ANSWERS

1. (*d*)	**2.** (*a*)	**3.** (*b*)	**4.** (*d*)	**5.** (*b*)	**6.** (*b*)
7. (*c*)	**8.** (*a*)	**9.** (*b*)	**10.** (*d*)	**11.** (*b*)	**12.** (*a*)
13. (*a*)	**14.** (*c*)	**15.** (*d*)	**16.** (*a*)	**17.** (*b*)	**18.** (*a*)
19. (*d*)	**20.** (*a*)	**21.** (*b*)	**22.** (*c*)	**23.** (*d*)	**24.** (*b*)
25. (*a*).					

THEORETICAL QUESTIONS

1. What is a pump ?
2. How are pumps classified ?
3. How are centrifugal pumps classified ?
4. State the advantages of a centrifugal pump over a displacement (reciprocating) pump.
5. List the main component parts of a centrifugal pump and explain them briefly.
6. Explain the working of a single - stage centrifugal pump with a neat sketch.
7. How does a volute casing differ from a vortex casing for the centrifugal pump ?
8. Explain briefly, with neat sketches, any two of the following types of casing
   (*i*) Volute casing; (*ii*) Vortex casing;
   (*iii*) Casing with guide blades/vanes.
9. Derive an expression for the work done by the impeller of a centrifugal pump on liquid per second per unit weight of liquid.
10. What is 'Euler head' ?
11. Define the following terms
    (*i*) Static head,
    (*ii*) Manometric head, and
    (*iii*) Total head.
12. Enumerate the losses which occur when a centrifugal pump operates.
13. Explain briefly the following efficiencies of a centrifugal pump:
    (*i*) Manometric efficiency,
    (*ii*) Volumetric efficiency,
    (*iii*) Mechanical efficiency, and
    (*iv*) Overall efficiency.
14. Discuss the influence of exit blade angle on the performance and efficiency of a centrifugal pump. Assume radial flow at entrance.
15. Derive an expression for the minimum speed for starting a centrifugal pump.
16. Explain briefly the effect of variation of discharge on the efficiency.
17. What is the effect of number of vanes of impeller on head and efficiency ?
18. State the difference between single stage and multi-stage pumps.
19. Discribe multi-stage pump with (*i*) impeller in series and (*ii*) impellers in parallel.
20. Define specific speed of a centrifugal pump. Derive an expression for the same.
21. How does the specific speed of a centrifugal pump differ from that of a turbine ?
22. Write down the ranges of specific speeds for the following types of pumps:
    (*i*) Slow speed radial flow,
    (*ii*) Medium speed radial flow,
    (*iii*) High speed radial flow, and
    (*iv*) Axial flow.
23. How is the model testing of the centrifugal pumps carried out ?
24. What do you mean by 'characteristics of centrifugal pumps ?
25. What is the significance of characteristic curves?
26. What do you mean by 'net positive suction head' (*NPSH*) ?
27. What is 'cavitation' ?
28. What are the effects of cavitation ? Give the necessary precautions against cavitation.
29. List the factors which facilitate onset of cavitation.
30. Define 'suction specific speed'.
31. What is priming ? Why is it necessary ?
32. How are small and large centrifugal pumps primed ?
33. How is the selection of pumps made ?
34. Give the operational difficulties commonly experienced in centrifugal pumps and their remedies.
35. Why are centrifugal pumps used sometimes in series and sometimes in parallel ? Draw the following characteristic curves for a centrifugal pump:
    Head, power and efficiency *versus* discharge with constant speed.

UNSOLVED EXAMPLES

1. The impeller of a centrifugal pump has an external diameter of 400 mm and internal diameter of 180 mm and it runs at 1440 r.p.m. Assuming a constant radial flow through the impeller at 2·5 m/s and that the vanes at the exit are set back at an angle of 25°, determine : (*i*) Inlet vane angle, (*ii*) The angle, absolute velocity of water at the exit makes with the tangent, and (*iii*) The work done per N of water.

   [**Ans.** (*i*) 10.42° ; (*ii*) 5.75° ; (*iii*) 76.18 Nm]

2. A centrifugal pump delivers water against a net head of 14·5 m and a design speed of 1000 r.p.m. The vanes are curved back to an angle of 30° with the periphery. The impeller diameter is 300 mm and outlet width 50 mm. Determine the discharge of the pump if manometric efficiency is 95 percent.

   [**Ans.** 0.1675 m^3/s]

3. The impeller of a centrifugal pump having external and internal diameters 500 mm and 250 mm respectively, width at outlet 50 mm and running at 1000 r.p.m. works against a head of 40 m. The velocity of flow through the impeller is constant and equal to 2.5 m/s. The vanes are set back at angle of 40°at outlet. Determine: (*i*) Inlet vane angle, (*ii*) Work done by the impeller on water per second, and (*iii*) Manometric efficiency.

   [**Ans.** (*i*) 10.81°; (*ii*) 119.2 kNm; (*iii*) 64.6%]

4. Determine the head imparted to a fluid as it passes through an impeller of 250 mm outlet diameter and 100 mm inlet diameter rotated at 1440 r.p.m. The outlet vane angle is set back at an angle of 20° to the tangent. Assume radial entrance and velocity of flow as 3 m/s.

   [**Ans.** 20.4 m]

5. A centrifugal pump impeller has an outer diamter of 300 mm and width at outer periphery 12·5 mm. The pump has radial inflow and delivers 0·08 m^3/s of water against a total of 40 m. if the speed of the pump is 1500 r.p.m. and its manometric efficiency is 80 percent find the blade angle at the exit. [**Ans.** $\phi = 69°$]

6. A centrifugal pump is running at 1000 r.p.m. and working against a head of 20 m. The rate of flow through the pump is 0.2 m^3/s. The outlet vane angle of impeller is 45° and velocity of flow at outlet is 2.5 m/s. If the manometric efficiency of the pump is 80 percent, calculate the diameter and width of impeller at outlet,

   [**Ans.** (*i*) 324 mm ; (*ii*) 78.6 mm]

7. A centrifugal pump (diffusion type) has a suction lift of 1.5 m and the delivery tank is 13.5 m above the pump. The velocity of water in the delivery pipe is 1.5 m/s. The radial velocity of flow through the wheel is 3 m/s and the tangent to the vane at exit from the wheel makes an angle of 120° with the direction of motion. Assuming that the water enters radially and neglecting friction and other losses, determine : (*i*) Velocity of wheel at exit , (*ii*) Velocity and pressure head at exit from the wheel, and(*iii*) Direction of fixed guide vanes.

   [**Ans.** (*i*) 13.07 m/s; (*ii*) 6.5 m of water; (*iii*) 14.89°]

8. A centrifugal pump impeller whose external and internal diameters are 400 mm and 200 mm respectively is running at 950 r.p.m. The rate of flow through the pump is 0.035 m^3/s. The suction and delivery heads are 5 m and 25 m respectively. The diameters of the suction and delivery pipe are 120 mm and 80 mm respectively. If the outlet vane angle is 45°, the flow velocity is constant and equal to 1.8 m/s and power required to drive the pump is 15 kW, determine:

   (*i*) Inlet vane angle, (*ii*) The overall efficiency, and (*iii*) The manometric efficiency.

   [**Ans.** (*i*) 10.26°; (*ii*) 61.76 % ; (*iii*) 73.65 %]

9. A centrifugal pump impeller whose external diameter and width at the outlet are 0.8 m and 0·1m respectively is running at 550 r.p.m. The angle of impeller vanes at outlet is 40°. The pump delivers 0.98 m^3 of water per second under an effective head of 35 m. If the pump is driven by a 500 kW motor, determine : (*i*) The manometric efficiency, (*ii*) The overall efficiency, and (*iii*) The mechanical efficiency, Assume water enters the vanes radially at inlet.

   [**Ans.** (*i*) 81 %; (*ii*) 67 %; (*iii*) 83 %]

10. A centrifugal pump impeller having external and internal diameters 400 mm and 200 mm respectively is running at 1200 r.p.m. The widths of impeller at outlet and inlet are 8 mm and 16 mm respectively. The rate of flow of water through the pump is 0.015 m^3/s. The outlet vane angle of the impeller is 30°. If the loss of head through the impeller is 1.15 m find the pressure rise in the impeller.

    Assume water enters the impeller radially at inlet. [**Ans.** 30.7 m]

11. A centrifugal pump is delivering 0.04 m^3 of water per second to a height of 20 m through a 150 mm diameter 100 m long pipeline. If the inlet losses in suction pipe are equal to 0.33 m and friction factor is 0.06 for the pipeline find the power required to drive the pump.

Assume overall efficiency of the pump as 70 percent. [**Ans.** 17.4 kW]

12. A centrifugal pump with 1.2 m diameter runs at 200 r.p.m. and discharges 1880 litres/sec, the average lift being 6 m. The angle which the vanes make at exit with the tangent to the impeller is 26° and the radial velocity of flow is 2.5 m/s. Determine the manometric efficiency and the least speed to start pumping against a head of 6 m; the inner diameter of the impeller being 0.6 m . [**Ans.** 199.36 r.p.m]

13. The diameter of a centrifugal pump impeller is 300 mm and its width is 600 mm. the pump delivers 120 litres/sec with a manometric efficinecy of 85 percent. The effective outlet vane angle is 30°. If the speed of the rotation is 1000 r.p.m. calculate the specific speed of the pump.

14. A multi - stage centrifugal pump has four identical impellers, keyed to the same shaft. The width and diameter of each impeller at outlet are 50 mm and 600 mm respectively. The vanes of each impeller are having outlet angle as 45°. The speed of the pump is 400 r.p.m. and the total manometric head developed is 40 m. If the discharge through the pump is 0.2 m^3/s, find the manometric efficiency. [**Ans.** 74.82 %]

15. Find the number of pumps required to take water from a deep well under a total head of 120 m. All the pumps are identical and are running at 800 r.p.m. The specific speed of each pump is given as 25 while the rated capacity of each pump is 0.16 m^3/s [**Ans.** 4]

16. A centrifugal pump is discharging 0.03 m^3/s of water against a total head of 20 m. The diameter of the impeller is 400 mm and it is rotating at 1500 r.p.m. Calculate the head, discharge and ratio of powers of a geometrically similar pump of diameter 250 mm when it is running at 3000 r.p.m. [**Ans.** 31.25 m; 0.01465 m^3/s ; 1.31]

17. A centrifugal pump (single stage) runs at 500 r.p.m. and delivers 300 m^3/min of water against a head of 120 m. The pump impeller is 2 m in diameter and it has a positive suction lift (including the velocity head and friction) of 3 m. Laboratory tests are to be conducted on a model with 450 mm diameter impeller and on a reduced head of 95 m. Assuming atmospheric head = 10.15 m of water and vapour head = 0.34 m of water calculate the speed, discharge and suction lift for the laboratory tests. [**Ans.** 1977 r.p.m.; 13.51 m^3/min ; 4.418 m]

Chapter 4

RECIPROCATING PUMPS

4.1. INRODUCTION

The reciprocating pump is a *positive displacement pump* as it sucks and raises the liquid by actually displacing it with a piston/plunger that executes a reciprocating motion in a closely fitting cylinder. The amount of liquid pumped is equal to the volume displaced by the piston.

The pumps designed with disk pistons create pressures upto 25 bar and the plunger pumps built up still higher pressures. Discharge from these pumps is almost wholly dependent on the pump speed.

The total efficiency of a reciprocating pump is about 10 to 20% higher than a comparable centrifugal pump.

Reciprocating pumps for industrial uses have almost become *obsolete owing to their high capital cost as well as maintenance cost as compared to that of centrifugal pumps*. However, small hand-operated pumps such as cycle pumps, football pumps, kerosene pumps, village well pumps and pumps used as important parts of hydraulic jack etc. still find wide applications. *The reciprocating pump is best suited for relatively small capacities and high heads*. This type of pump is very common in *oil drilling operations*.

The reciprocating pump is generally employed for:

(*i*) Light oil pumping,

(*ii*) Feeding small boilers condensate return, and

(*iii*) Pneumatic pressure systems.

4.2. CLASSIFICATION OF RECIPROCATING PUMPS

Reciprocating pumps are *classified* as follows:

1. According to the water being in contact with piston:

(*i*) *Single-acting pump* ...water is in contact with *one side* of the piston

(*ii*) *Double-acting pump* ...water is in contact with *both sides* of the piston.

2. According to number of cylinders:

(*i*) Single cylinder pump

(*ii*) Double cylinder pump (or two throw pump)

(*iii*) Triple cylinder pump (or three throw pump)

(*iv*) Duplex double-acting pump (or four throw pump)

(*v*) Quintuplex pump or (five throw pump).

In general the reciprocating pumps having more than one cylinder are known as *multi-cylinder pumps*.

4.3. MAIN COMPONENTS AND WORKING OF A RECIPROCATING PUMP

Refer to Fig. 4.1. The **main parts** *of a reciprocating pump are*:

1. Cylinder
2. Piston
3. Suction valve
4. Delivery valve
5. Suction pipe
6. Delivery pipe
7. Crank and connecting rod mechanism operated by a power source e.g. steam engine, internal combustion engine or an electric motor.

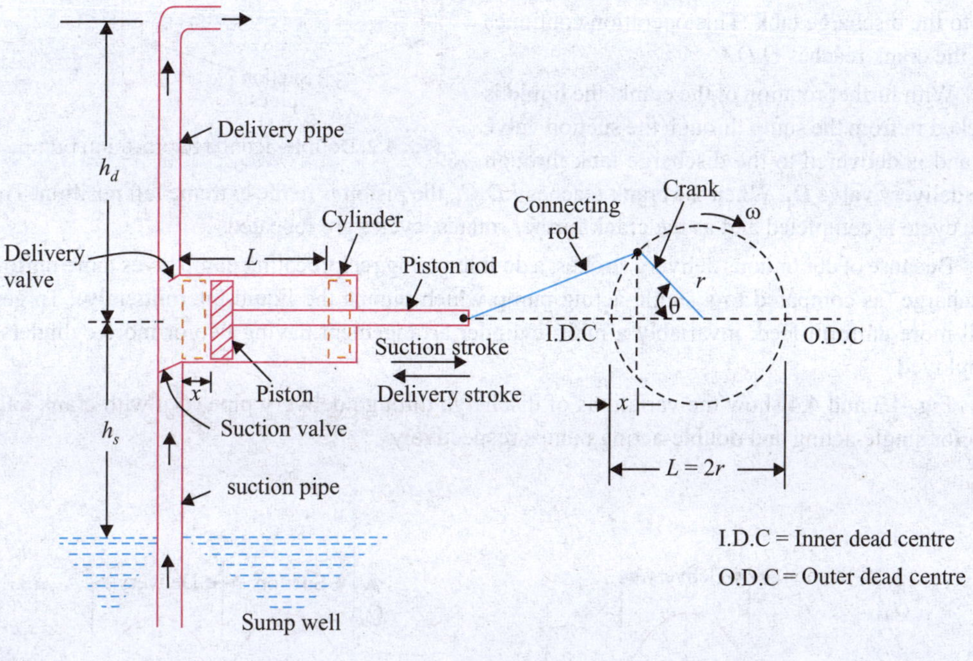

Fig. 4.1. Schematic view of single-acting reciprocating pump.

Working of a single-acting reciprocating pump:

As shown in Fig. 4.1, a single acting reciprocating pump has one suction pipe and one delivery pipe. It is usually placed above the liquid level in the sump. When the crank rotates the piston moves backward and forward inside the cylinder. The pump operates as follows:

— Let us suppose that initially the crank is at the inner dead centre (I.D.C.) and crank rotates in the clockwise direction. As the crank rotates, the piston moves towards right and a vacuum is created on the left side of the piston. This vacuum causes suction valve to open and consequently the liquid is forced from the sump into the left side of the piston. When the crank is at the outer dead centre (O.D.C) the suction stroke is completed and the left side of the cylinder is full of liquid.

— When the crank further turns from O.D.C to I.D.C., the piston moves inward to the left and high pressure is built up in the cylinder. The delivery valve opens and the liquid is forced into the delivery pipe. The liquid is carried to the discharge tank through the delivery pipe. At the end of delivery stroke the crank comes to the I.D.C and the piston is at the extreme left position.

Working of a double-acting reciprocating pump:

Refer to Fig. 4.2. In a double-acting reciprocating pump, suction and delivery *strokes occur simultaneously*. When the crank rotates from *I.D.C.* in the clockwise direction, a vacuum is created on the left side of piston and the liquid is sucked in from the sump through value S_1. At the same time, the liquid on the right side of the piston is pressed and a high pressure causes the delivery valve D_2 to open and the liquid is passed on to the discharge tank. This operation continues till the crank reaches *O.D.C.*

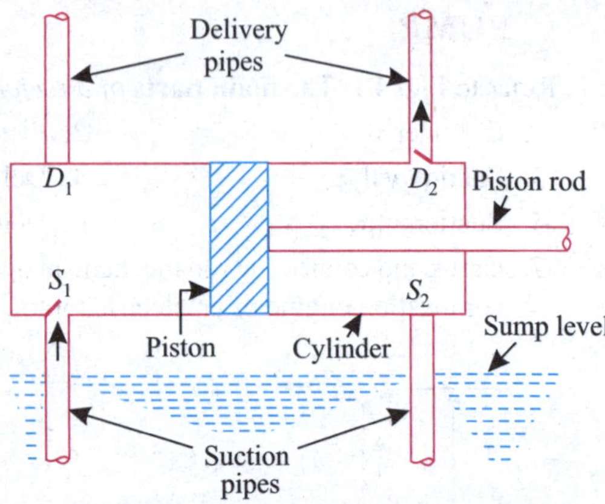

Fig. 4.2. Double-acting reciprocating pump.

With further rotation of the crank, the liquid is sucked in from the sump through the suction valve S_2 and is delivered to the discharge tank through the delivery valve D_1. When the crank reaches *I.D.C.*, the piston is in the extreme left position. Thus one cycle is completed and as the crank further rotates, cycles are repeated.

Because of continuous delivery strokes, a double-acting reciprocating pump gives more uniform discharge (as compared to a single-acting pump which pumps the liquid intermittently). To get a still more uniform feed, invariably a multi-cylinder arrangement having two or more cylinders is employed.

Fig. 4.3 and 4.4 show the variations of discharge through delivery pipe (Q_d) with crank angle (θ) for single-acting and double-acting pumps respectively.

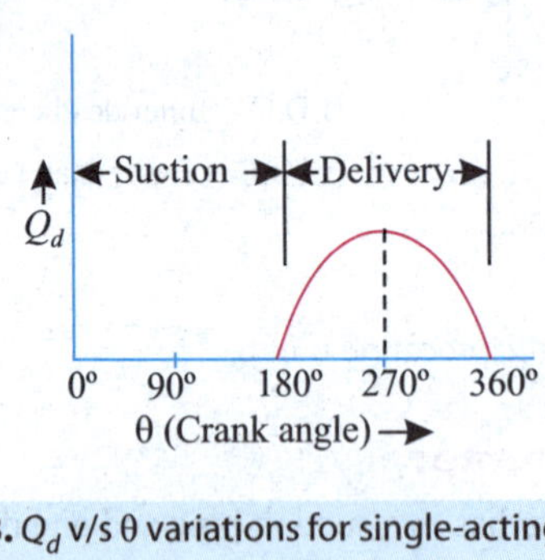

Fig. 4.3. Q_d v/s θ variations for single-acting pump.

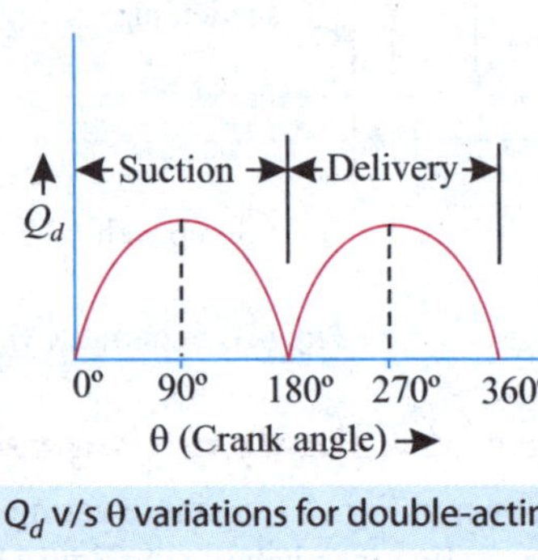

Fig. 4.4. Q_d v/s θ variations for double-acting pump.

4.4. DISCHARGE, WORK DONE AND POWER REQUIRED TO DRIVE RECIPROCATING PUMP

4.4.1. Single-acting reciprocating pump

Consider a single-acting reciprocating pump shown in Fig. 4.1

Let, D = Diameter of the cylinder, m

A = Cross-sectional area of the piston/cylinder = $\frac{\pi}{4} D^2$ m^2

r = Radius of crank, m

N = Speed of the crank, r.p.m.

L = Length of the stroke (= $2r$), m

h_s = Height of the centre of the cylinder above the liquid surface, m and

h_d = Height to which the liquid is raised above the centre of the cylinder, m.

Volume of liquid sucked in during suction stroke = $A \times L$

∴ *Discharge* of the pump per second, $Q = A \times L \times \frac{N}{60}$ ...(4.1)

Weight of water delivered per second, $W = wQ = \frac{wALN}{60}$ ...(4.2)

Work done per second = Weight of water lifted/sec. × total height through which liquid is lifted

$$= W(h_s + h_d) = \frac{wALN}{60}(h_s + h_d) \quad ...(4.3)$$

∴ *Power required* to drive the pump $= \frac{wALN}{60 \times 1000}(h_s + h_d)$ kW ...(4.4)

(where, w = weight density of liquid in N/m^3)

4.4.2. Double-acting Reciprocating Pump

Refer to Fig. 4.2.

Let, D = Diameter of the piston,

d = Diameter of the piston rod,

A_{pr} = cross-sectional area of the piston rod = $\frac{\pi}{4} d^2$

Area on one side of the piston, $A = \frac{\pi}{4} D^2$

Area on other side of the piston where piston rod is connected to the piston,

$$A' = A - A_{pr} = \frac{\pi}{4} D^2 - \frac{\pi}{4} d^2 = \frac{\pi}{4}(D^2 - d^2).$$

Volume of liquid delivered in one revolution of crank

$$= AL + A'L = (A + A')L = \left[\frac{\pi}{4} D^2 + \frac{\pi}{4}(D^2 - d^2)\right] L$$

∴ *Discharge* of the pump per second $= \left[\frac{\pi}{4} D^2 + \frac{\pi}{4}(D^2 - d^2)\right] L \times \frac{N}{60}$ ...(4.5)

If the diameter of the piston rod '*d*' is very small as compared to the diameter of the piston '*D*' then it can be neglected and hence *discharge* of the pump per second will become

$$Q = \left(\frac{\pi}{4}D^2 + \frac{\pi}{4}D^2\right) \times \frac{LN}{60} = 2 \times \frac{\pi}{4}D^2 \times \frac{LN}{60} = \frac{2\,A\,L\,N}{60} \quad \text{... (4.6)}$$

Evidently the output of a double acting pump is two-times that of a single acting pump.

Work done per second = Weight of water delivered × total height through which liquid is lifted

$$= \left(w \times \frac{2ALN}{60}\right) \times (h_s + h_d)$$

$$= \frac{2wALN}{60}(h_s + h_d) \quad \text{...(4.7)}$$

Power required to drive the pump, $P = \dfrac{2wALN}{60 \times 1000}(h_s + h_d)$ kW ...(4.8)

(where, w = weight density of liquid in N/m^3)

4.5. CO-EFFICIENT OF DISCHARGE AND SLIP OF RECIPROCATING PUMP

4.5.1. Co-efficient of Discharge

In a reciprocating pump, the actual discharge ($Q_{act.}$) is always slightly different from the theoretical discharge ($Q_{th.}$) due to following *reasons*:

(*i*) Leakage through the valves, glands and piston packing,

(*ii*) Imperfect operation of the valves (suction and discharge), and

(*iii*) Partial filling of cylinder by the liquid.

The ratio between actual discharge and theoretical discharge is known as the **co-efficient of discharge (C_d) of the pump**. That is,

$$C_d = \frac{\text{Actual discharge}}{\text{Theoretical discharge}} = \frac{Q_{act.}}{Q_{th.}} \quad \text{...(4.9)}$$

When the value of C_d is expressed in percentage, it is known as '**volumetric efficiency**' of the pump. Volumetric efficiency depends upon the dimensions of the pump and its value ranges from 85-98%.

4.5.2. Slip

The difference between the theoretical discharge and actual discharge is *called the* **slip** *of the pump*. That is

$$\text{Slip} = Q_{th.} - Q_{act.} \quad \text{...(4.10)}$$

But the slip is oftenly expressed in percentage which is given by,

$$\%\,\text{Slip} = \frac{Q_{th.} - Q_{act.}}{Q_{th.}} \times 100 = \left(1 - \frac{Q_{act.}}{Q_{th.}}\right) \times 100 = (1 - C_d) \times 100 \quad \text{...(4.11)}$$

The percentage of slip for the pumps maintained in *good condition is* of the order of 2% or even less.

Negative slip. In most of the reciprocating pumps $Q_{act.}$ is *less than* $Q_{th.}$; in such a case the value of C_d is *less than unity* and the *slip of the* pump is '*positive*'. However, in some cases $Q_{act.}$ may be *more* than $Q_{th.}$; in such a case C_d is more than *unity* and the *slip* will be '*negative*'. The slip *will be negative when there is a direct connection between the suction and delivery sides before the end of suction stroke. This happens if the momentum of liquid in the suction pipe is large enough to open* *the delivery valve before the beginning of delivery stroke. The negative slip is possible in case of pumps having long suction pipe and a short delivery pipe, especially when these are operating at high speeds.*

Example 4.1. *A single-acting reciprocating pump, running at 50 r.p.m. delivers 0.00736 m^3/s of water. The diameter of the piston is 200 mm and stroke length 300 mm. The suction and delivery heads are 3.5 m and 11.5 m respectively. Determine:*

(i) Theoretical discharge,

(ii) Co-efficient of discharge,

(iii) Percentage slip of the pump, and

(iv) Power required to run the pump.

Solution. Speed of the pump, $N = 50$ r.p.m.

Actual discharge, $Q_{act.} = 0.00736\ m^3/s$

Diameter of the piston, $D = 200\ mm = 0.2\ m$

$\therefore$ Area, $A = \frac{\pi}{4} \times 0.2^2 = 0.0314\ m^2$

Stroke length, $L = 300\ mm = 0.3\ m$

Suction head, $h_s = 3.5\ m$

Delivery head, $h_d = 11.5\ m$

(i) Theoretical discharge, $Q_{th.}$:

$$Q_{th.} = \frac{ALN}{60} = \frac{0.0314 \times 0.3 \times 50}{60} = \mathbf{0.00785\ m^3/s\ (Ans.)}$$

(ii) Co-efficient of discharge, C_d:

$$C_d = \frac{Q_{act.}}{Q_{th.}} = \frac{0.00736}{0.00785} = \mathbf{0.937\ (Ans.)}$$

(iii) Percentage slip of the pump:

$$\%\ \text{slip} = \frac{Q_{th.} - Q_{act.}}{Q_{th.}} \times 100 = \frac{0.00785 - 0.00736}{0.00785} \times 100 = \mathbf{6.24\%\ (Ans.)}$$

(iv) Power required to run the pump, P:

$$P = \frac{wALN}{60 \times 1000}(h_s + h_d)\,\text{kW} = \frac{9810 \times 0.0314 \times 0.3 \times 50}{60 \times 1000}(3.5 + 11.5) = \mathbf{1.155\ kW\ (Ans.)}$$

Example 4.2. *A single-acting reciprocating pump operating at 120 r.p.m. has a piston diameter of 200 mm and stroke of 300 mm. The suction and delivery heads are 4 m and 20 m, respectively. If the efficiency of both suction and delivery strokes is 75 percent, determine the power required by the pump.* **[UPTU]**

Solution. *Given:* $N = 120$ r.p.m.; $D = 200\ mm = 0.2\ m$; $L = 300\ mm = 0.3\ m$; $h_s = 4\ m$; $h_d = 20\ m$; $\eta_{\text{(suction and delivery strokes, each)}} = 75\%$

Power require by the pump, P :

Theoretical discharge of the pump,

$$Q = \frac{ALN}{60} = \frac{\frac{\pi}{4} \times (0.2)^2 \times 0.3 \times 120}{60} = 0.0188\ m^3/s$$

Power required to drive the pump,

$$P = \frac{wQ\,(h_s + h_d)}{\eta} = \frac{9810 \times 0.0188\,(4 + 20)}{0.75} = 5901.7 \text{ W or } \mathbf{5.9\ kW}\ \textbf{(Ans.)}$$

Example 4.3. *The discharge / crank-angle characteristic for a reciprocating pump has the likeness of the top half of a sine-curve as shown in Fig. 4.5.*

Draw the quantitative characteristic curve for a system consisting of three such pumps connected in parallel and discharging at a crank phase difference of 120° from each other. Identify the magnitude and position of two consecutive maxima on that curve. **[GATE]**

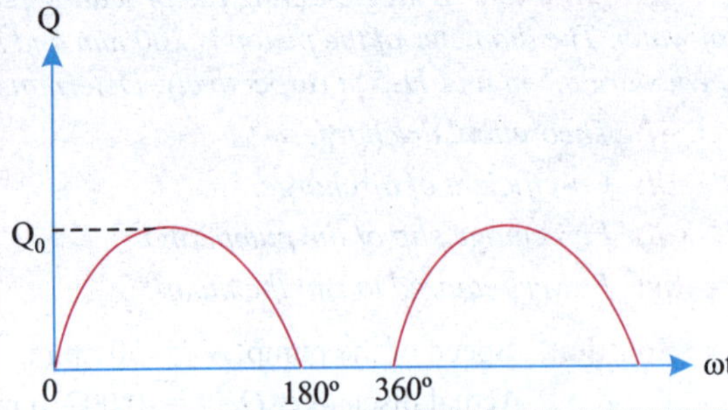

Fig. 4.5

Solution. The maximum discharge is equal to Q_0 (maximum discharge of one pump) and occurs at 60° intervals of 30°, 90°, 150°........ (See Fig. 4.6).

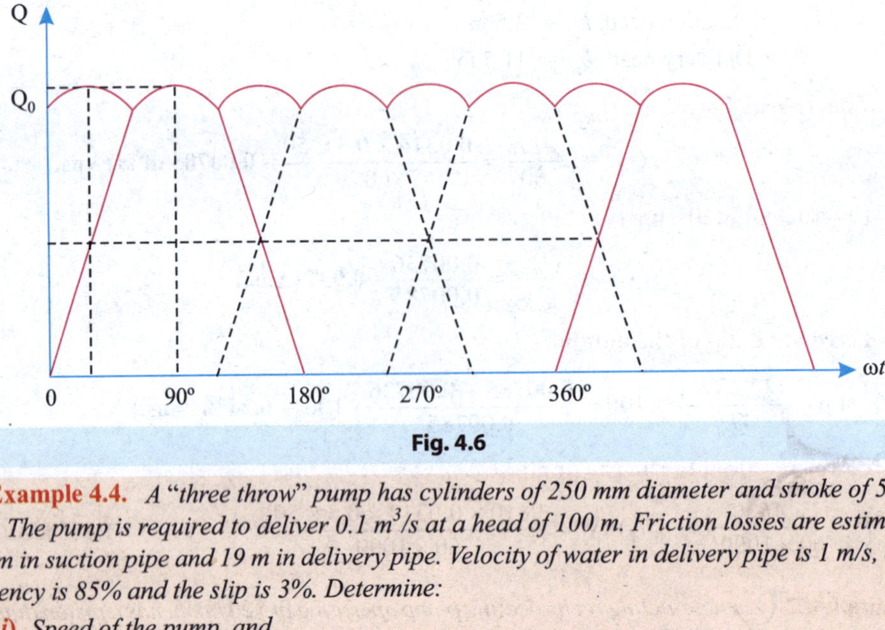

Fig. 4.6

Example 4.4. *A "three throw" pump has cylinders of 250 mm diameter and stroke of 500 mm each. The pump is required to deliver 0.1 m³/s at a head of 100 m. Friction losses are estimated to be 1 m in suction pipe and 19 m in delivery pipe. Velocity of water in delivery pipe is 1 m/s, overall efficiency is 85% and the slip is 3%. Determine:*

(i) Speed of the pump, and

(ii) Power required to run the pump. **[PTU]**

Solution. Diameter of each cylinder, D = 250 mm = 0.25 m

Stroke length of each cylinder, L = 500 mm = 0.5 m

Actual discharge, $Q_{act.}$ = 0.1 m^3/s

Static head, $(h_s + h_d)$ = 100 m

Friction loss in suction pipe, h_{fs} = 1 m

Friction loss in delivery pipe, h_{fd} = 19 m

Velocity of water in delivery pipe, V_d = 1 m/s

Overall efficiency of the pump, η_0 = 85%

Percentage slip = 3%

(i) Speed of the pump, N:

A three throw pump uses three equal cylinders with rams connected to cranks at 120° apart driven by a common shaft.

For a three throw pump, the theoretical discharge is given by:

$$Q_{th} = 3 \times \frac{ALN}{60} = 3 \times \frac{\pi}{4} \times 0.25^2 \times \frac{0.5 \times N}{60} = 0.001227 \text{ N}$$

$$\therefore \text{ Actual discharge, } Q_{act} = \left(1 - \frac{3}{100}\right) Q_{th.} = 0.97 \times 0.001227 \text{ N} = 0.00119 \text{ N}$$

But, $Q_{act} = 0.1 \text{ m}^3/\text{s}$ ...(*Given*)

$$\therefore \quad 0.1 = 0.00119\, N, \text{ or, } N = \frac{0.1}{0.00119} = \mathbf{84 \text{ r.p.m (Ans.)}}$$

(ii) Power required to run the pump, P:

Total head against which pump has to work,

$$H = (h_s + h_d) + (h_{fs} + h_{fd}) + \frac{V_d^2}{2g}$$

$$= 100 + (1 + 19) + \frac{1.0^2}{2 \times 9.81} = 120.05 \text{ m}$$

$$\therefore \quad \text{Water power} = \frac{wQ_{act.}H}{1000} \text{ kW} = \frac{9810 \times 0.1 \times 120.05}{1000} \text{ 117.77 kW}$$

Power required to drive the shaft,

$$P = \frac{\text{Water power}}{\text{Overall efficiency}} = \frac{117.77}{0.85} = \mathbf{138.55 \text{ kW (Ans.)}}$$

4.6. EFFECT OF ACCELERATION OF PISTON ON VELOCITY AND PRESSURE IN THE SUCTION AND DELIVERY PIPES

Refer to Fig. 4.1. If the crank rotates uniformly and the length of connecting rod is enough compared to the radius of crank, the piston makes simple harmonic motion. This causes acceleration during the first half of the stroke, and deceleration during the second half of the stroke.

Let, A = Area of the cylinder,
a = Area of the pipe (suction or delivery),
l = Length of pipe (suction or delivery),
r = Radius of the crank, and
ω = Angular speed of the crank in rad/s.

The crank is rotating with an angular velocity ω and let in time t seconds, the crank turns through angle θ (in radians) from I.D.C. (inner dead centre). The displacement of the piston in time t is x as shown in Fig. 4.1.

Now, angle turned by the crank in time t, $\theta = \omega t = \frac{2\pi N}{60} \times t$

(where, N = rotational speed of crank in r.p.m.)

The corresponding distance (x) travelled by the piston,

$$x = r - r\cos\theta = r(1 - \cos\theta) = r(1 - \cos\omega t) \quad ...(4.12)$$

$$\text{Velocity of the piston, } V = \frac{dx}{dt} = \frac{d}{dt}[r(1 - \cos\omega t)] = \frac{d}{dt}(r - r\cos\omega t)$$

or, $$V = \omega r \sin \omega t \qquad ...(4.13)$$

Acceleration of the piston, $$a_p = \frac{dV}{dt} = \frac{d}{dt}(\omega r \sin \omega t) = \omega^2 r \cos \omega t \qquad ...(4.14)$$

Now from continuity considerations, the *volume of liquid flowing from the pipe equals the volume of liquid flowing into the cylinder.*

∴ Velocity of liquid in the pipe (v) × area of pipe (a)

$$= \text{velocity of piston } (V) \times \text{area of cylinder } (A)$$

∴ $$v = \frac{AV}{a} = \frac{A}{a}\omega r \sin \omega t \qquad ...(4.15)$$

Acceleration of liquid in pipe $$= \frac{d}{dt}(v) = \frac{d}{dt}\left[\frac{A}{a}\omega r \sin \omega t\right]$$

$$= \frac{A}{a}\omega^2 r \cos \omega t \qquad ...(4.16)$$

Mass of water in pipe = Density × volume of liquid in pipe = ρal

Force required to accelerate the water in the pipe = Mass × acceleration

$$= \rho al \times \frac{A}{a}\omega^2 r \cos \omega t$$

∴ Intensity of pressure due to acceleration

$$= \frac{\text{Force required to accelerate the liquid}}{\text{Area of pipe}} = \frac{\rho al \times \frac{A}{a}\omega^2 r \cos \omega t}{a}$$

$$= \rho l \times \frac{A}{a}\omega^2 r \cos \theta$$

∴ Pressure head due to acceleration, $$h_a = \frac{\text{Intensity of pressure}}{\text{Weight density of liquid } (w)}$$

$$= \frac{\rho l \times \frac{A}{a}\omega^2 r \cos \theta}{\rho g} \qquad (\because w = \rho g)$$

$$= \frac{l}{g} \times \frac{A}{a}\omega^2 r \cos \theta \qquad ...(4.17)$$

The pressure head due to acceleration in the suction and delivery pipes is obtained by using subscripts 's' and 'd' respectively in the eqn. (4.17) as follows:

$$h_{as} = \frac{l_s}{g} \times \frac{A}{a_s}\omega^2 r \cos \theta \qquad ...(4.18)$$

$$h_{ad} = \frac{l_d}{g} \times \frac{A}{a_d}\omega^2 r \cos \theta \qquad ...(4.19)$$

Thus the pressure head due to acceleration given by eqn. (4.17), is a *function of angular displacement* θ.

Note : It may be noted that for any stroke, *the angular displacement θ is measured from the instant of commencement of that stroke.* In case of *suction stroke* the piston moves *outward* and θ is measured from I.D.C. (inner dead centre) and during *delivery stroke* the piston moves *inward* and θ is measured from *O.D.C* (outer dead centre).

The values of 'h_a' for different values of θ are:

(*i*) When $\theta = 0°$ (*i.e.* the *beginning* of the stroke), $h_a = \frac{l}{g}\frac{A}{a}\omega^2 r$ ...(4.20)

($\because \cos 0° = 1$)

(*ii*) When $\theta = 90°$ (*i.e.* the *middle* of the stroke), $h_a = 0$ ...(4.21)

($\because \cos 90° = 0$)

(*iii*) When $\theta = 180°$ (*i.e.* the end of the stroke), $h_a = -\frac{l}{g}\frac{A}{a}\omega^2 r$ ...(4.22)

($\because \cos 180° = -1$)

∴ Maximum pressure head due to acceleration, $(h_a)_{max} = \frac{l}{g} \times \frac{A}{a}\omega^2 r$ ...(4.23)

For $0° < \theta < 90°$, h_a has +ve values and for $90° < \theta < 180°$, h_a has –ve values, thereby indicating that for the first half of the stroke there is *acceleration head* development and in the later half of the stroke *retardation head* is developed.

In case the connecting rod is *not very long as compared to crank length* then it *cannot* be assumed that the piston has a simple harmonic motion and in that case the pressure head, h_a is given by:

$$h_a = \frac{l}{g}\frac{A}{a}\omega^2 r \cos\theta\left(\cos\theta + \frac{\cos 2\theta}{n}\right) \quad ...(4.24)$$

where, n = Ratio of the length of connecting rod to the crank length.

From eqn. (4.24), we have:

(*i*) When $\theta = 0°$ (*i.e.* the *beginning* of the stroke), $h_a = \frac{l}{g}\frac{A}{a}\omega^2 r\left(1 + \frac{1}{n}\right)$ ...(4.25)

(*ii*) When $\theta = 90°$ (*i.e.* the middle of stroke), $h_a = 0$ ...(4.26)

(*iii*) When $\theta = 180°$ (*i.e.* at the *end* of the stroke), $h_a = \frac{l}{g}\frac{A}{a}\omega^2 r\left(1 - \frac{1}{n}\right)$ ...(4.27)

Effect of variation of velocity on friction in pipes:

The liquid flowing through suction and delivery pipes causes loss of head due to friction, which is given by Darcy-Weisbach equation as:

$$h_f = \frac{4 f l v^2}{d \times 2g} \quad ...(i)$$

where,
f = Co-efficient of friction,
l = Length of the pipe,
d = Diameter of the pipe, and
v = Velocity of liquid in the pipe.

Also the velocity of liquid in the pipe, $v = \frac{A}{a}\omega r \sin \omega t = \frac{A}{a}\omega r \sin\theta$ ...[Eqn. (4.15)]

Substituting the value of 'v' in (*i*), we get:

$$h_f = \frac{4 fl}{d \times 2g}\left(\frac{A}{a}\omega r \sin\theta\right)^2 \quad ...(4.28)$$

The variation of h_f with θ is **parabolic.** The values of h_f for suction and delivery pipes are obtained from eqn. (4.28) by using subscripts '*s*' for suction pipe and '*d*' for delivery pipe as:

$$h_{fs} = \frac{4fl_s}{d_s \times 2g}\left(\frac{A}{a_s}\,\omega r \sin\theta\right)^2 \quad ...(4.29)$$

$$h_{fd} = \frac{4fl_d}{d_d \times 2g}\left(\frac{A}{a_d}\,\omega r \sin\theta\right)^2 \quad ...(4.30)$$

The loss of head due to friction (h_f) in pipes given by eqn. (4.28) varies with θ as:

(*i*) When θ = 0° (*i.e.* the *beginning* of the stroke), $h_f = 0$ ...(4.31)

(*ii*) When θ = 90° (*i.e.* the *middle* of the stroke), $h_f = \frac{4fl}{d \times 2g}\left(\frac{A}{a}\,\omega r\right)^2$ ...(4.32)

(*iii*) When θ = 180° (*i.e.* the *end* of the stroke), $h_f = 0$ ...(4.33)

∴ Maximum value of loss of head due to friction,

$$(h_f)_{max} = \frac{4fl}{d \times 2g}\left(\frac{A}{a}\,\omega r\right)^2 \quad ...(4.34)$$

4.7. INDICATOR DIAGRAMS

The **indicator diagram** *of a reciprocating pump is the diagram which shows the pressure head of the liquid in the cylinder corresponding to any position during the suction and delivery strokes. It is a graph between pressure head and stroke length of the piston for one complete revolution* (pressure head is taken as ordinate and stroke length as abscissa).

4.7.1. Ideal Indicator Diagram

The indicator diagram obtained by *neglecting* the loss of head due to friction in the suction and delivery pipes and the effect of acceleration of piston, is known as an *ideal indicator diagram*. Such diagram for a single-cylinder single-acting pump is shown is Fig. 4.7, the line *EF* represents the atmospheric pressure head $\left[H_{atm.} = \frac{p_a}{w}\right]$ equal to 10.3 m of water.

Let, h_s = Suction head, and

h_d = Delivery head.

- The pressure head in the cylinder (represented by line AB) during *suction stroke*, is *constant* and equal to suction head (h_s) which is below the atmospheric pressure head ($H_{atm.}$) by a height h_s. The *absolute pressure head* in cylinder during the suction stroke will be ($H_{atm.} - h_s$); it is shown by ordinate AS at the beginning of the stroke and by ordinate BT at the end of stroke and, is *uniform* throughout the stroke.
- During the *delivery stroke* the pressure head in the cylinder

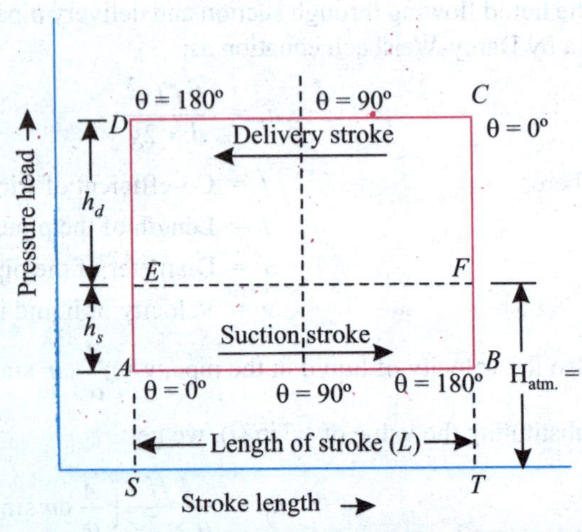

Fig. 4.7. Ideal indicator diagram.

(represented by line CD) is *constant* and equal to delivery head (h_d). The *uniform absolute pressure head* throughout the delivery stroke is ($h_d + H_{atm.}$) and is denoted by the ordinate TC or SD.

The work done by the pump per second is

$$= \frac{w \times ALN}{60} \times (h_s + h_d) \qquad \text{...[Eqn. (4.3)]}$$

$$= K \times L \times (h_s + h_d) \qquad \left(\text{where, } K = \frac{wAN}{60} = \text{const.}\right)$$

$$\propto L \times (h_s + h_d)$$

But from Fig. 4.7, the area of indicator diagram *ABCDA*

$$= AB \times BC = AB\,(BF + FC) = L\,(h_s + h_d) \qquad \text{...}(ii)$$

From (*i*) and (*ii*), we have:

Work done by the pump ∝ area of indicator diagram

Thus, work done by the pump per second $\frac{wAN}{60}$ × area of indicator diagram ...(4.35)

If the pump is double-acting, *neglecting* the area of the piston rod, work done per second is proportional to *twice* the area of the indicator diagram.

4.7.2. Effect of Acceleration in Suction and Delivery Pipes on Indicator Diagram

The effect of acceleration in suction and delivery pipes is discussed below:

(*a*) Effect of acceleration in the suction pipe:

As the piston (considering it as the beginning of the stroke) moves outward, it should create not only a negative pressure equal to the suction head (h_s) but it should also *accelerate the liquid*. If h_{as} is the acceleration head, then total negative pressure head at the *beginning* of the suction stroke is ($h_s + h_{as}$), the ordinate *EA'*, *absolute pressure head* at this point is denoted by ordinate *A'S*. So that separation *does not* take place, the absolute pressure at the beginning of stroke should not fall *below the vapour pressure*.

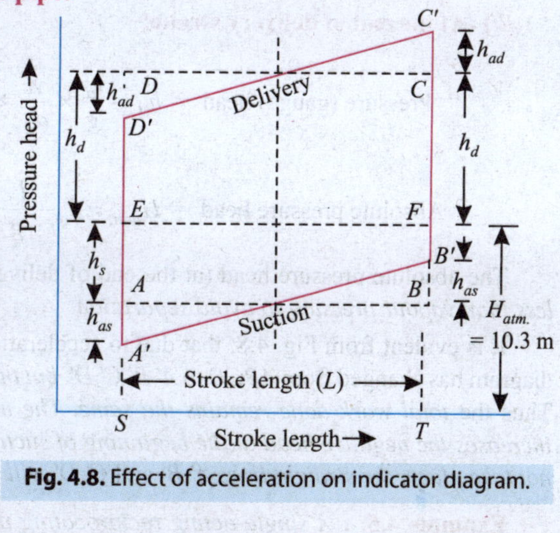

Fig. 4.8. Effect of acceleration on indicator diagram.

If l_s and a_s are length and cross-sectional area of the suction pipe respectively, then:

(*i*) At the beginning of the suction stroke:

$$\text{The accelerating head, } h_{as} = \frac{l_s}{g} \times \frac{A}{a_s}\,\omega^2 r \qquad \text{...[Eqn. 4.20]}$$

$$\text{Negative pressure (vacuum) head, } h_s + h_{as} = h_s + \frac{l_s}{g} \times \frac{A}{a_s}\,\omega^2 r \qquad \text{...(4.36)}$$

$$\text{Absolute pressure head} = H_{atm.} - \left(h_s + \frac{l_s}{g} \times \frac{A}{a_s}\,\omega^2 r\right)$$

(ii) At the middle of the suction stroke:

$$\text{The acceleration head, } h_{as} = 0 \qquad \text{...[Eqn. 4.20]}$$

$$\text{Negative pressure (vacuum) head} = h_s \qquad \text{...(4.37)}$$

$$\text{Absolute pressure head} = H_{atm.} - h_s$$

(iii) At the end of the suction stroke:

$$\text{The acceleration head, } h_{as} = -\frac{l_s}{g} \times \frac{A}{a_s}\,\omega^2 r \qquad \text{...[Eqn. 4.22]}$$

$$\text{Negative pressure (vacuum) head} = h_s + h_{as} = h_s - \frac{l_s}{g} \times \frac{A}{a_s}\,\omega^2 r \qquad \text{...(4.38)}$$

$$\text{Absolute pressure head} = H_{atm.} - \left(h_s - \frac{l_s}{g} \times \frac{A}{a_s}\,\omega^2 r\right)$$

(b) Effect of acceleration in the delivery pipe:

In the *beginning* of delivery stroke the liquid in the delivery pipe is *accelerated*, while at the *end* of delivery stroke the liquid is *retarded*.

If l_d and a_d are the length and cross-sectional area of the delivery pipe respectively, then:

(i) At the beginning of delivery stroke:

$$\text{Pressure head (gauge)} = h_d + h_{ad} = h_d + \frac{l_d}{g} \times \frac{A}{a_d}\,\omega^2 r \qquad \text{...(4.39)}$$

(ii) At the middle of delivery stroke:

$$\text{Pressure (gauge) head} = h_d \qquad (\because h_{ad} = 0) \qquad \text{...(4.40)}$$

(iii) At the end of delivery stroke:

$$\text{Pressure (gauge) head} = h_d - \frac{l_d}{g} \times \frac{A}{a_d} \times \omega^2 r \qquad \text{...(4.41)}$$

$$\text{Absolute pressure head} = H_{atm.} + h_d - \frac{l_d}{g} \times \frac{A}{a_d} \times \omega^2 r \qquad \text{...(4.42)}$$

The absolute pressure head (at the end of delivery stroke) given by eqn. (4.42) *should not be less than vapour pressure to avoid separation.*

It is evident from Fig. 4.8. that due to acceleration in suction and delivery pipes, the indicator diagram has changed from *ABCD* to *A' B' C' D' but the area of indicator diagram remains unaltered.* Thus the *total work done remains the same. The main effect of the acceleration head is that it increases the negative head at the beginning of suction stroke. If the simple harmonic motion does not take place, the straight lines A' B'* and *C' D' will become slightly curved.*

Example 4.5. *A single-acting reciprocating pump has a diameter (piston) of 150 mm and stroke length 350 mm. The centre of the pump is 3.5 m above the water surface in the sump and 22 m below the delivery water level. Both the suction and delivery pipes have the same diameter of 100 mm and are 5 m and 30 m long respectively. If the pump is working at 30 r.p.m., determine:*

(i) *The pressure heads on the piston at the beginning, middle and end of both suction and delivery strokes.*

(ii) *The power required to drive the pump.*

Take atmospheric pressure as 10.3 m of water.

Solution. Diameter of piston, D = 150 mm = 0.15 m

Stroke length L = 350 mm = 0.35 m

Suction head, h_s = 3.5 m

Delivery head, h_d = 22 m

Diameter of pipes, $d_s = d_d$ = 100 mm = 0.1 m

Length of suction pipe, l_s = 5 m

Length of delivery pipe, l_d = 30 m

Speed of pump, N = 30 r.p.m.

Atmospheric pressure, $H_{atm.}$ = 10.3 m

(*i*) The pressure heads on the piston:

(*a*) Suction stroke:

The pressure head *due to acceleration* in *suction pipe* is given by:

$$h_{as} = \frac{l_s}{g} \times \frac{A}{a_s} \times \omega^2 r \cos\theta \qquad \text{...Eqn. (4.18)}$$

At the *beginning* of the stroke ($\theta = 0°$),

$$h_{as} = \frac{l_s}{g} \times \frac{A}{a_s} \times \omega^2 r = \frac{5}{9.81} \times \left[\frac{(\pi/4) \times 0.15^2}{(\pi/4) \times 0.1^2}\right] \times \left(\frac{2\pi \times 30}{60}\right)^2 \times (0.175) = 1.98 \text{ m of water}$$

$$\left(\because r = \frac{L}{2} = \frac{0.35}{2} = 0.175 \text{ m}\right)$$

At the *middle* of the stroke ($\theta = 90°$), $h_{as} = 0$

At the *end* of the stroke ($\theta = 180°$), $h_{as} = -\frac{l_s}{g} \times \frac{A}{a_s} \times \omega^2 r = -1.98$ m of water (as calculated above)

∴ Pressure heads (on the piston) in metres of water *during suction stroke*:

At *beginning* = $(h_s + h_{as})$ = 3.5 + 1.98 = 5.48 m (vacuum)

= 10.3 – 5.48 = **4.82 m of water (absolute) (Ans.)**

At *middle* = $(h_s + 0)$ = 3.5 m (vacuum)

= 10.3 – 3.5 = **6.8 m of water (absolute) (Ans.)**

At *end* = 3.5 – 1.98 = 1.52 m (vacuum)

= 10.3 – 1.52 = **8.78 m of water (absolute) (Ans.)**

(*b*) Delivery stroke:

The pressure head *due to acceleration* in *delivery pipe* is given by:

$$h_{ad} = \frac{l_d}{g} \times \frac{A}{a_d} \times \omega^2 r \cos\theta$$

At the *beginning* of the stroke ($\theta = 0°$),

$$h_{ad} = \frac{l_d}{g} \times \frac{A}{a_d} \times \omega^2 r = \frac{30}{9.81} \times \left[\frac{(\pi/4) \times 0.15^2}{(\pi/4) \times 0.1^2}\right] \times \left(\frac{2\pi \times 30}{60}\right)^2 \times 0.175 = 11.88 \text{ m}$$

At the *middle* of the stroke ($\theta = 90°$), $h_{ad} = 0$

At the *end* of the stroke ($\theta = 180°$), $h_{ad} = -\frac{l_d}{g} \times \frac{A}{a_d} \times \omega^2 r = -11.088$ m (as calculated above)

∴ Pressure heads (on the piston) in the metres of water *during delivery stroke*:

At *beginning* $= h_d + h_{ad} = 22 + 11.88 = 33.88$ m (*gauge*)

$= 10.3 + 33.88 =$ **44.18 m of water (absolute) (Ans.)**

At *middle* $= h_d + 0 = 22$ m (*gauge*)

$= 10.3 + 22 =$ **32.3 m of water (absolute) (Ans.)**

At *end* $= h_d - h_{ad} = 22 - 11.8 = 10.2$ m (*gauge*)

$= 10.3 + 10.2 =$ **20.5 m of water (absolute) (Ans.)**

(*ii*) Power required to drive the pump, P:

Theoretical discharge of the pump,

$$Q = \frac{ALN}{60} = \frac{\pi}{4} \times (0.15)^2 \times 0.35 \times 30 \times \frac{1}{60} = 0.00309 \text{ m}^3/\text{s}$$

Work done by the pump

$$= wQ(h_s + h_d) = 9810 \times 0.00309\,(3.5 + 22) = 772.98 \text{ Nm/s or J/s}$$

∴ *Power required to drive the pump, P* = **772.98 W (Ans.)**

Example 4.6. *A single-acting reciprocating pump has a diameter (piston) of 100 mm and stroke length 200 mm. The length and diameter of the suction pipe are 6.5 m and 50 mm respectively. If the suction lift of the pump is 3.2 m and separation occurs when pressure in the pump falls below 2.5 m of water absolute and the manometer reads 763 mm of mercury, find the maximum speed at which pump can be run without separation in the suction pipe.*

Solution. Diameter of piston, $D = 100$ mm $= 0.1$ m

∴ Area, $A = \frac{\pi}{4} \times 0.1^2 = 0.00785$ m^2

Stroke length, $L = 200$ mm $= 0.2$ m

∴ Crank radius, $r = L/2 = 0.2 = 0.1$ m

Length of suction pipe, $l_s = 6.5$ m

Diameter of suction pipe, $d_s = 50$ mm $= 0.05$ m

∴ Area of suction pipe, $a_s = \frac{\pi}{4} \times 0.05^2 = 0.001963$ m^2

The suction lift of the pump, $h_s = 3.2$ m

Separation pressure head, $h_{sep} = 2.5$ m

Maximum speed at which pump can run without separation, *N*:

$$\text{Atmospheric head, } H_{atm.} = \frac{763}{1000} \times 13.6 = 10.377 \text{ m of water}$$

During suction stroke, the possibility of separation is only at the *beginning of the stroke*. At the beginning of suction stroke $\theta = 0°$ and $\cos\theta = 1$, that gives:

$$h_{as} = \frac{l_s}{g} \times \frac{A}{a_s} \times \omega^2 r = \frac{6.5}{9.81} \times \frac{0.00785}{0.001963} \times \omega^2 \times 0.1 = 0.265\,\omega^2$$

Pressure head in the cylinder at the beginning of suction stroke

$= (h_s + h_{as})$ vacuum

$= H_{atm.} - (h_s + h_{as})$ absolute.

This absolute pressure (at the beginning of suction stroke) should not fall *below* the vapour pressure head ($h_{sep.}$) *to avoid separation*, thus, in *limiting condition*,

$$H_{atm.} - (h_s + h_{as}) = h_{sep.}$$

or, $\quad 10.377 - (3.2 + 0.265\omega^2) = 2.5$, or, $\quad 10.377 - 3.2 - 0.265\omega^2 = 2.5$

or, $\quad \omega^2 = \dfrac{10\cdot377 - 3\cdot2 - 2\cdot5}{0\cdot265}$ $\therefore$ Angular velocity, $\omega = 4.2$ rad./s.

But, $\quad \omega = \dfrac{2\pi N}{60}$, or, $N = \dfrac{60\omega}{2\pi} = \dfrac{60 \times 4\cdot2}{2 \times \pi}$ **40.1 r.p.m. (Ans.)**

Example 4.7. *The cross-sectional area of a plunger of a reciprocating pump equals 1.5 times that of a delivery pipe. The delivery pipe is 60 m long and it rises upward at a slope of 1 in 6. If the plunger has an acceleration of 2.4 m/s² at the end of the stroke and separation pressure is 2.5 m of water, find whether separation will take place and, if so, at which section of the pipe.*

Assume simple harmonic motion, and take atmospheric pressure = 10.3 m of water.

Solution. Cross-sectional area of plunger, $A = 1.5\ a_d$ (area of delivery pipe)

Length of delivery pipe, $l_d = 60$ m

Slope of the delivery pipe = 1 in 6

Acceleration of the plunger at the end of the stroke = 2.4 m/s²

Separation pressure head, $h_{sep.} = 2.5$ m

Atmospheric pressure head, $H_{atm.} = 10.3$ m

Delivery head, h_d = Slope of the delivery pipe × length of the delivery pipe

$$= \frac{1}{6} \times 60 = 10 \text{ m}$$

The possibility of separation during delivery stroke is only at the *end of the stroke* where $\theta = 180°$ and $\cos\theta = -1$; that gives the pressure head due to acceleration in the delivery pipe,

$$h_{ad} = \frac{l_d}{g} \times \frac{A}{a_d} \times \omega^2 r \cos\theta$$

$$h_{ad} = -\frac{l_d}{g} \times \frac{A}{a_d} \times \omega^2 r = -\frac{60}{9\cdot81} \times 1.5 \times 2.4 = -22.02 \text{ m}$$

($\because$ Acceleration, $\omega^2 r = 2.4$ m/s²)

$\therefore$ Pressure head in the cylinder at the end of delivery stroke

$= (h_d + h_{ad})$ *gauge*

$= H_{atm} + (h_d + h_{ad})$ absolute

$= 10.3 + (10 - 22.02) = -1.72$ m of water

Since the absolute pressure head is *less than* the allowable separation pressure head of 2.5 m of water therefore, **separation will occur. (Ans.)**

Let, $\quad l$ = The length of pipe upto the section where the separation occurs.

Then, $\quad h_d = \dfrac{1}{6} \times l = 0.1667l$

$$h_{ad} = -\frac{l}{g} \times 1.5 \times 2.4 = -\frac{l}{9\cdot81} \times 1.5 \times 2.4 = -0.367l$$

From *limiting condition for separation*, we have:

$$H_{atm.} + (h_d + h_{ad}) = h_{sep.}$$

$$10.3 + (0.1667\,l - 0.367l) = 2.5$$

$$l = \frac{10.3 - 2.5}{(0.367 - 0.1667)} = \mathbf{38.94\ m\ (Ans.)}$$

Example 4.8. *The diameter and stroke length of a single-acting reciprocating pump are 75 mm and 150 mm respectively. It takes its supply of water from a sump 3 m below the pump through a pipe 5 m long and 40 mm in diameter. It delivers water to a tank 12 m above the pump through a pipe 30 mm in diameter and 15 m long. If separation occurs 75 kN/m² below the atmospheric pressure, find the maximum speed at which pump may be operated without separation. Assume that the piston has a simple harmonic motion.* **[Rajasthan University]**

Solution. Diameter of pump, $D = 75 \text{ mm} = 0.075 \text{ m}$

$$\therefore \quad \text{Area, } A = \frac{\pi}{4} \times 0.075^2 = 0.004418 \text{ m}^2$$

Stroke length, $L = 150 \text{ mm} = 0.15 \text{ m}$

$$\therefore \quad \text{Crank radius, } r = \frac{0.15}{2} = 0.075 \text{ m}$$

Suction head, $h_s = 3$ m

Delivery head, $h_d = 12$ m

Length of suction pipe, $l_s = 5$ m

Diameter of suction pipe, $d_s = 40 \text{ mm} = 0.04 \text{ m}$

$$\therefore \quad \text{Area of suction pipe, } a_s = \frac{\pi}{4} \times 0.04^2 = 0.001256 \text{ m}^2$$

Length of delivery pipe, $l_d = 15$ m

Diameter of delivery pipe, $d_d = 30 \text{ mm} = 0.03 \text{ m}$

$$\therefore \quad \text{Area of delivery pipe, } a_d = \frac{\pi}{4} \times 0.03^2 = 0.0007068 \text{ m}^2$$

$$\text{Separation head, } h_{sep.} = -\frac{75 \times 1000}{9810} = -7.645 \text{ m of water}$$

Maximum speed at which pump may be operated without separation, *N*:

Speed of pump without separation during 'Suction stroke', *N*:

The possibility of separation, during suction stroke, is only at the *beginning of the stroke*. At the beginning of the stroke $\theta = 0°$ and $\cos\theta = 1$, that gives:

$$h_{as} = \frac{l_s}{g} \times \frac{A}{a_s} \times \omega^2 r = \frac{5}{9.81} \times \frac{0.004418}{0.001256} \times \omega^2 \times 0.075 = 0.134\,\omega^2$$

$\therefore$ Pressure head in the cylinder at the beginning of suction stroke

$$= (h_s + h_{as}) \text{ vacuum}$$

$$= H_{atm.} - (h_s + h_{as}) \text{ absolute}$$

Limiting condition for *no separation* gives:

$$H_{atm.} - (h_s + h_{as}) = h_{sep.}$$
$$H_{atm.} - (h_s + h_{as}) = (H_{atm.} - 7.645) \text{ abs.}$$

or, $7.645 = h_s + h_{as}$, or, $7.645 = 3 + 0.134\omega^2$, or, $\omega^2 = \dfrac{7 \cdot 645 - 3}{0 \cdot 134}$, or, $\omega = 5.88$ rad/s

But, $\omega = \dfrac{2\pi N}{60}$, or, $N = \dfrac{60\,\omega}{2\pi} = \dfrac{60 \times 5.88}{2 \times \pi} = \mathbf{56.15\ r.p.m}$

Speed of pump without separation during delivery stroke, N:

The pressure head due to acceleration in the delivery pipe is given as:

$$h_{ad} = \frac{l_d}{g} \times \frac{A}{a_d} \times \omega^2\, r \cos\theta$$

During delivery stroke the possibility of separation is only at the end of the stroke. At the end of delivery stroke $\theta = 180°$ and $\cos\theta = -1$, that gives:

$$h_{ad} = -\frac{15}{9.81} \times \frac{0.004418}{0.0007068} \times \omega^2 \times 0.075 = -0.7168\,\omega^2$$

∴ Pressure head in the cylinder at the end of delivery stroke

$$= (h_d + h_{ad}) \text{ gauge}$$
$$= H_{atm.} + (h_d + h_{ad}) \text{ absolute}$$

Limiting condition for *no separation* gives:

$$H_{atm.} + (h_d + h_{ad}) = h_{sep.}$$
$$H_{atm.} + (h_d + h_{ad}) = (H_{atm.} - 7.645) \text{ abs.}$$

or, $7.645 = -(h_d + h_{ad})$, or, $7.645 = -(12 - 0.7168\omega^2)$, or, $0.7168\omega^2 = 19.645$

or, Angular velocity, $\omega = \left(\dfrac{19.645}{0.716}\right)^{1/2} = 5.235$ rad/s

But, $\omega = \dfrac{2\pi N}{60}$, or, $N = \dfrac{60\omega}{2\pi} = \dfrac{60 \times 5.235}{2\pi} \simeq 50$ r.p.m.

∴ Maximum permissible speed, N = **50 r.p.m.** (*minimum* of the two speeds obtained above)

(Ans.)

Example 4.9. *The bore and stroke of a reciprocating pump are 250 mm and 500 mm respectively. The pump delivers water through a 100 mm delivery pipe to a tank located at 14 m above it and 27 m horizontally from it. If separation occurs at a pressure of 22 kN/m² absolute, find the safe speed at which pump should run for the following arrangements of delivery pipe: (i) The delivery pipe is horizontal from the pump and then vertical upto the tank and (ii) The delivery pipe is vertical from the pump and then horizontal upto the tank.*

The atmospheric pressure at the pump side = 10.3 m of water and connecting rod-crank ratio = 5.

Solution. The bore of the pump, $D = 250 \text{ mm} = 0.25 \text{ m}$

∴ Area, $A = \dfrac{\pi}{4} \times 0.25^2 = 0.0491 \text{ m}^2$

Stroke length, $L = 500 \text{ mm} = 0.5 \text{ m}$

$\therefore$ Crank radius, $r = \frac{L}{2} = \frac{0.5}{2} = 0.25$ m

Diameter of delivery pipe, $d_d = 100$ mm $= 0.1$ m

$\therefore$ Area of delivery pipe, $a_d = \frac{\pi}{4} \times 0.1^2 = 0.007854 \text{ m}^2$

Length of delivery pipe, $l_d = 14 + 27 = 41$ m

Separation head, $h_{sep.} = \frac{22 \times 1000}{9810} = 2.242$ m

Safe speed at which pump should run, *N*:

(*i*) *The delivery pipe is horizontal from the pump and then vertical upto the tank*:

Acceleration head at the end of delivery stroke,

$$h_{ad} = -\frac{l_d}{g} \times \frac{A}{a_d} \times \omega^2 r \left(1 - \frac{1}{n}\right) \quad \text{...[Eqn.(4.27)]}$$

$$= -\frac{41}{9.81} \times \frac{0.0491}{0.007854} \times \omega^2 \times 0.25 \left(1 - \frac{1}{5}\right) = -5.22\,\omega^2$$

The separation possibility, if any, is at the bend after the horizontal portion of the delivery pipe. The entire delivery head is available at the bend.

Limiting condition for no separation gives:

$$H_{atm.} + (h_d + h_{ad}) = h_{sep.}$$

or, $10.3 + (14 - 5.22\,\omega^2) = 2.242$, or, $5.22\,\omega^2 = 22.058$

$$\therefore \quad \omega = \left(\frac{22.058}{5.22}\right)^{1/2} = 2.055 \text{ rad/s}$$

But, $\omega = \frac{2\pi N}{60}$, or, $N = \frac{60\omega}{2\pi} = \frac{60 \times 2.055}{2\pi} = $ **19.62 r.p.m. (Ans.)**

(*ii*) *The delivery pipe is vertical from the pump and then horizontal upto the tank*:

The possibility of separation is at the bend after the vertical portion of the delivery pipe. The delivery head becomes *zero* at the bend and the horizontal pipe after the bend has a considerable value of acceleration head. The *limiting condition for no separation gives*:

$$H_{atm.} + (h_d + h_{ad}) = h_{sep.}$$

$$10.3 + (0 - 5.22\omega^2) = 2.242, \text{ or, } 5.22\omega^2 = 8.058$$

$$\therefore \quad \omega = \left(\frac{8.058}{5.22}\right)^{1/2} = 1.242 \text{ rad/s}$$

But, $\omega = \frac{2\pi N}{60}$, or, $\mathbf{N} = \frac{60\omega}{2\pi} = \frac{60 \times 1.242}{2\pi} = $ **11.86 r.p.m. (Ans.)**

From above calculations, it is evident that the *pump can run at a higher speed without occurrence of separation with* **arrangement (*i*). (Ans.)**

4.7.3. Effect of Friction in Suction and Delivery Pipes on Indicator Diagram

The head lost due to friction in suction and delivery pipes is given by eqns. (4.29) and (4.30) as:

$$h_{fs} = \frac{4fl_s}{d_s \times 2g}\left(\frac{A}{a_s}\omega r \sin\theta\right)^2, \text{ and } h_{fd} = \frac{4fl_d}{d_d \times 2g}\left(\frac{A}{a_d}\omega r \sin\theta\right)^2$$

From the above equations, it is evident that the variation of h_{fs} or h_{fd} with θ is *parabolic*.

(*i*) At the *beginning of suction or delivery stroke*: θ = 0°, sin θ = 0 and therefore h_{fs} = 0, h_{fd} = 0 *i.e.* there is no loss of head due to friction.

(*ii*) At the *middle* of the *suction or delivery stroke*: θ = 90°, sin θ = 1 and, therefore,

$$h_{fs} = \frac{4fl_s}{d_s \times 2g}\left(\frac{A}{a_s}\omega r\right)^2, \text{ and } h_{fd} = \frac{4fl_d}{d_d \times 2g}\left(\frac{A}{a_d}\omega r\right)^2$$

(*iii*) At the *end of suction or delivery stroke*: θ = 180°, sin θ = 0 and therefore h_{fs} and h_{fd} = 0

These results, evidently, indicate that frictional losses are zero at the beginning and end of the strokes and maximum at the mid of the strokes. Fig. 4.9 shows the effect of friction on the indicator diagram.

The work done against friction in suction and delivery pipes is given by the areas of parabolas *AGB* and *CDI*.

$$\text{Area } AGB = AB \times \frac{2}{3} GH = AB \times \frac{2}{3} h_{fs}$$

$$= L \times \frac{2}{3} h_{fs}$$

$$\left[\text{where, } h_{fs} = \frac{4fl_s}{d_s \times 2g}\left(\frac{A}{a_s}\omega r\right)^2\right]$$

$$\text{Similarly, area } CDI = CD \times \frac{2}{3} \times IJ = L \times \frac{2}{3} h_{fd}$$

$$(\because CD = AB = L)$$

$$\left[\text{where, } h_{fd} = \frac{4fl_d}{d_d \times 2g}\left(\frac{A}{a_d}\omega r\right)^2\right]$$

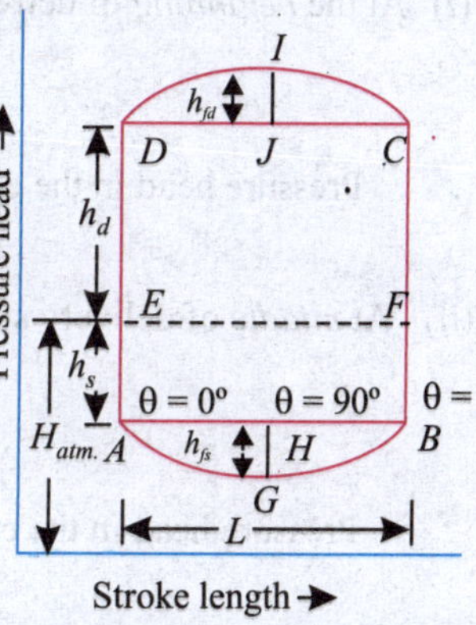

Fig. 4.9. Effect of friction on indicator diagram.

4.7.4. Effect of Acceleration and Friction in Suction and Delivery Pipes on Indicator Diagram.

The acceleration head (h_a) and friction head (h_f) at any instant of flow in the suction and delivery pipes of a reciprocating pump are given as:

$$h_a = \frac{l}{g}\frac{A}{a}\omega^2 r\cos\theta;\ h_f = \frac{4fl}{d \times 2g}\left(\frac{A}{a}\omega r \sin\theta\right)^2$$

(*a*) Suction stroke:

The pressure head on the piston during suction stroke for any angle θ of the crank = ($h_s + h_{as} + h_{fs}$)

(*i*) At the *beginning* of the suction stroke, θ = 0° and we have:

$$h_{as} = \frac{l_s}{g}\frac{A}{a_s}\omega^2 r \text{ and } h_{fs} = 0$$

∴ Pressure head in the cylinder = ($h_s + h_{as}$) *below* atmospheric head

$= H_{atm.} - (h_s + h_{as})$ absolute.

(*ii*) At *middle of* suction stroke, $\theta = 90°$ and we have:

$$h_{as} = 0, h_{fs} = \frac{4fl_s}{d_s \times 2g}\left(\frac{A}{a_s}\omega r\right)^2$$

$\therefore$ Pressure head in the cylinder $= (h_s + h_{fs})$ *below* atmospheric head

$= H_{atm.} - (h_s + h_{fs})$ absolute.

(*iii*) At the *end* of suction stroke, $\theta = 180°$ and we have:

$$h_{as} = -\frac{l_s}{g}\frac{A}{a_s}\omega^2 r, \text{ or } h_{fs} = 0$$

$\therefore$ Pressure head in the cylinder $= (h_s - h_{as})$ *below* atmospheric head

$= H_{atm.} - (h_s - h_{as})$ absolute.

(*b*) Delivery stroke:

The pressure head on the piston during delivery stroke for any angle θ of the crank $= (h_d + h_{ad} + h_{fd})$

(*i*) At the *beginning* of delivery stroke, $\theta = 0°$ and we have:

$$h_{ad} = \frac{l_d}{g}\frac{A}{a_d}\omega^2 r, \text{ and } h_{fd} = 0$$

$\therefore$ Pressure head in the cylinder $= (h_d + h_{ad})$ *above* atmospheric head

$= H_{atm.} + (h_d + h_{ad})$ absolute.

(*ii*) At *middle* of delivery stroke, $\theta = 90°$ and we have:

$$h_{ad} = 0 \text{ and } h_{fd} = \frac{4fl_d}{d_d \times 2g}\left(\frac{A}{a_d}\omega r\right)^2$$

$\therefore$ Pressure head in the cylinder $= (h_d + h_{fd})$ *above* atmospheric head

$= H_{atm.} + (h_d + h_{fd})$ absolute

(*iii*) At the *end* of delivery stroke, $\theta = 180°$ and we have:

$$h_{ad} = -\frac{l_d}{g}\frac{A}{a_d}\omega^2 r \text{ and } h_{fd} = 0$$

$\therefore$ Pressure head in the cylinder $= (h_d - h_{ad})$ *above* atmospheric head

$= H_{atm.} + (h_d - h_{ad})$ absolute.

Fig. 4.10 shows a complete indicator diagram including the effects of acceleration and friction.

Area of indicator diagram

$A'GB'C'ID'$ = Area $A'HB'C'JD'$ + area of parabola $A'GB'$ + area of parabola $C'ID'$

But, Area $A'HB'C'JD'$

$=$ area $ABCD = (h_s + h_d) \times L$

Area of parabola $A'GB'$

$= A'GB' = A'B' \times \frac{2}{3} \times HK$

$= \frac{2}{3} \times (A'B' \times HK)$

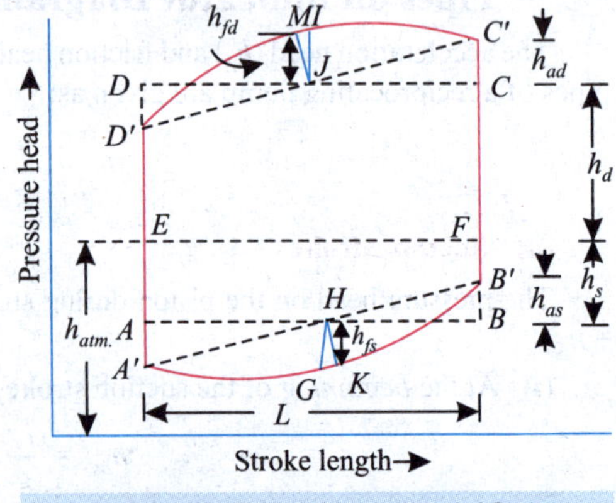

Fig. 4.10. Effect of acceleration and friction on indicator diagram.

$$= \frac{2}{3} \times (AB \times GH) = \frac{2}{3} \times L \times h_{fs}$$

Similarly, area of parabola $C'ID' = C'D' \times \frac{2}{3} JM = \frac{2}{3} (C'D' \times JM)$

$$= \frac{2}{3} (CD \times JI) = \frac{2}{3} \times L \times h_{fd}$$

$\therefore$ Area of indicator diagram $= (h_s + h_d) L + \frac{2}{3} \times L \times h_{fs} + \frac{2}{3} \times L \times h_{fd}$

$$= \left(h_s + h_d + \frac{2}{3} h_{fs} + \frac{2}{3} h_{fd} \right) L$$

As the area of the indicator diagram is proportional to work done by the pump, therefore,

Work done by pump per second $\propto \left(h_s + h_d + \frac{2}{3} h_{fs} + \frac{2}{3} h_{fd} \right) L$

$$= K \left(h_s + h_d + \frac{2}{3} h_{fs} + \frac{2}{3} h_{fd} \right) L$$

Where, K = a constant of proportionality

$$= \frac{wAN}{60} \quad \text{...for a single-acting pump}$$

$$= \frac{2wAN}{60} \quad \text{...for a double-acting pump.}$$

Hence, the work done per second by a *single-acting pump*

$$= \frac{wALN}{60} \left(h_s + h_d + \frac{2}{3} h_{fs} + \frac{2}{3} h_{fd} \right) \quad \text{...(4.43)}$$

and, for a double-acting pump

$$= \frac{2w\,ALN}{60} \left(h_s + h_d + \frac{2}{3} h_{fs} + \frac{2}{3} h_{fd} \right) \quad \text{...(4.44)}$$

Example 4.10. *A single-acting reciprocating pump has a stroke length of 150 mm, suction pipe is 7 m long and the ratio of suction pipe diameter to the piston diameter is 3/4. The water level in the sump is 2.5 m below the axis of the pump cylinder and the pipe connecting the sump and pump cylinder is 75 mm in diameter. If the crank is running at 75 r.p.m., determine the pressure head on the piston at the beginning, middle and end of the suction stroke. Take friction co-efficient, f = 0.01.* **[Delhi University]**

Solution. Stroke length, L = 150 mm = 0.15 m

$\therefore$ Crank radius, $r = L/2 = \frac{0 \cdot 15}{2} = 0.075$ m

Length of suction pipe, l_s = 7 m

Ratio of suction pipe dia. to the piston dia $\frac{d_s}{D} = 3/4$

The water level in the sump below the axis of the pump cylinder,

$$h_s = 2.5 \text{ m}$$

Diameter of suction pipe, d_s = 75 mm = 0.075 m

Speed of the crank, N = 75 r.p.m.

Friction co-efficient, f = 0.01

Pressure head on the piston at the beginning, middle and end of the suction stroke:

$$\frac{A}{a_s} = \frac{\text{Area of piston}}{\text{Area of suction pipe}} = \frac{\frac{\pi}{4}D^2}{\frac{\pi}{4}d_s^2} = \left(\frac{D}{d_s}\right)^2 = \left(\frac{4}{3}\right)^2 = \frac{16}{9}$$

$$\left(\because \frac{d_s}{D} = 3/4 \qquad \text{...Given}\right)$$

$$\text{Angular velocity, } \omega = \frac{2\pi N}{60} = \frac{2\pi \times 75}{60} = 7.85 \text{ rad/s}$$

The acceleration head (h_{as}) and friction head (h_{fs}) at any instant of flow through suction pipes are given as:

$$h_{as} = \frac{l_s}{g}\frac{A}{a_s}\omega^2 r\cos\theta = \frac{7}{9{\cdot}81} \times \frac{16}{9} \times 7.85^2 \times 0.075\cos\theta = 5.86\cos\theta$$

$$h_{fs} = \frac{4fl_s}{d_s \times 2g}\left(\frac{A}{a_s}\omega r\sin\theta\right)^2 = \frac{4 \times 0.01 \times 7}{0.075 \times 2 \times 9.81}\left(\frac{16}{9} \times 7.85 \times 0{\cdot}075 \times \sin\theta\right)^2$$

$$= 0.199\sin^2\theta$$

The pressure head on the piston during *suction stroke* for any angle θ of the crank = $h_s + h_{as} + h_{fs}$

(*i*) **At the beginning of suction stroke,** θ = 0° and we have:

$$h_{as} = 5.86 \text{ m, and, } h_{fs} = 0$$

∴ Pressure head on the piston $= H_{atm.} - (h_s + h_{as}) = H_{atm.} - (2.5 + 5.86)$

$= (H_{atm.} - 8.36 \text{ m})$ m of water *absolute*, or,

8.36 m of water vacuum. (Ans.)

(*ii*) At the *middle* of suction stroke, θ = 90° and we have:

$$h_{as} = 0 \text{ and } h_{fs} = 0.199 \text{ m}$$

∴ Pressure head on the piston $= H_{atm.} - (h_s + h_{fs}) = H_{atm.} - (2.5 + 0.199)$

$= (H_{atm.} - 2.699)$ m of water *absolute*, or,

2.699 m of water vacuum (Ans.)

(*iii*) At the *end* of suction stroke, θ = 180° and we have:

$$h_{as} = -5.86 \text{ m and } h_{fs} = 0$$

∴ Pressure head on the piston $= H_{atm.} - (h_s + h_{as}) = H_{atm.} - (2.5 - 5.86)$

$= H_{atm.} + 3.36$ m of *absolute*, or,

3.36 m of water gauge (Ans.)

Example 4.11. *The piston diameter and stroke length of a double-acting single cylinder reciprocating pump are 150 mm and 300 mm respectively. The centre of the pump is 4.5 m above the water level in the sump and 32 m below the delivery water level. Both the suction and delivery pipes have the same diameter of 75 mm and are 6 m and 36 m long respectively. If the pump is working at 30 r.p.m. determine:*

(*i*) *The pressure heads on the piston at the beginning, middle and end of both suction and delivery strokes,*

(*ii*) *The power required to drive pump if the mechanical efficiency is 80%,*

(*iii*) *The maximum head at any instant against which the pump has to work and its corresponding duty.*

Take atmospheric pressure head = 10.3 m of water, and Darcy's friction co-efficient for both the pipes as 0.01.

Solution. Diameter of the piston, $D = 150 \text{ mm} = 0.15 \text{ m}$

$\therefore$ Area, $A = \frac{\pi}{4} \times 0.15^2 = 0.01767 \text{ m}^2$

Stroke length, $L = 300 \text{ mm} = 0.3 \text{ m}$

$\therefore$ Crank radius, $r = L/2 = \frac{0.3}{2} = 0.15 \text{ m}$

Suction head, $h_s = 4.5 \text{ m}$

Delivery head, $h_d = 32 \text{ m}$

Diameters of suction and delivery pipes, $d_s = d_d = 75 \text{ mm} = 0.075 \text{ m}$

$\therefore$ Area, $a_s = a_d = \frac{\pi}{4} \times 0.075^2 = 0.00442 \text{ m}^2$

Length of suction pipe, $l_s = 6 \text{ m}$

Length of delivery pipe, $l_d = 36 \text{ m}$

Speed of the pump, $N = 30 \text{ r.p.m.}$

$\therefore$ Angular velocity, $\omega = \frac{2\pi N}{60} = \frac{2\pi \times 30}{60} = 3.14 \text{ rad/s}$

Mechanical efficiency, $\eta_{mech} = 80\%$

(*i*) Pressure heads on the piston:

Suction stroke:

$$\text{Acceleration head, } h_{as} = \frac{l_s}{g}\frac{A}{a_s}\omega^2 r \cos\theta$$

$$= \frac{6}{9 \cdot 81} \times \frac{0 \cdot 01767}{0 \cdot 00442} \times 3.14^2 \times 0.15 \cos\theta = 3.62 \cos\theta$$

$$\text{Friction head, } h_{fs} = \frac{4fl_s}{d_s \times 2g}\left(\frac{A}{a_s}\omega r \sin\theta\right)^2$$

$$= \frac{4 \times 0 \cdot 01 \times 6}{0 \cdot 075 \times 2 \times 9 \cdot 81}\left(\frac{0 \cdot 01767}{0 \cdot 00442} \times 3 \cdot 14 \times 0 \cdot 15 \times \sin\theta\right)^2 = 0.578 \sin^2\theta$$

The pressure head on the piston during *suction stroke* for any angle θ of the crank = $(h_s + h_{as} + h_{fs})$

(*a*) At the *beginning* of stroke, $\theta = 0°$ and we have:

$h_{as} = 3.62 \text{ m}$ and $h_{fs} = 0$

$\therefore$ Pressure head on the piston $= H_{atm.} - (h_s + h_{as})$

$= 10.3 - (4.5 + 3.67) =$ **2.13 m of water absolute** **(Ans.)**

(*b*) At the *middle* of stroke, $\theta = 90°$ and we have:

$$h_{as} = 0 \text{ and } h_{fs} = 0.578 \text{ m}$$

$\therefore$ Pressure head on the piston $= H_{atm.} - (h_s + h_{fs}) = 10.3 - (4.5 + 0.578)$

$= $ **5.22 m of water absolute (Ans.)**

(*c*) At the *end* of stroke, $\theta = 180°$ and we have:

$$h_{as} = -3.62 \text{ m and } h_{fs} = 0$$

$\therefore$ Pressure head on the piston $= H_{atm.} - (h_s + h_{as}) = 10.3 - (4.5 - 3.62)$

$= $ **9.42 m of water absolute (Ans.)**

Delivery stroke:

$$\text{Acceleration head, } h_{ad} = \frac{l_d}{g}\frac{A}{a_d}\omega^2 r\cos\theta$$

$$= \frac{36}{9.81} \times \frac{0.01767}{0.00442} \times 3.14^2 \times 0.15\cos\theta = 21.69\cos\theta$$

$$\text{Friction head, } \frac{h}{h_{fd}} = \frac{h}{h_{fd}} = \frac{4fl_d}{d_d \times 2g}\left(\frac{A}{a_d}\,\omega r\sin\theta\right)^2$$

$$= \frac{4 \times 0.01 \times 36}{0.075 \times 2 \times 9.81}\left(\frac{0.01767}{0.00442} \times 3.14 \times 0.15\sin\theta\right)^2 = 3.47\sin^2\theta$$

The pressure head on the piston during delivery stroke for an angle θ of the crank $= (h_d + h_{ad} + h_{fd})$

(*a*) At the *beginning* of the stroke, $\theta = 0°$ and we have:

$$h_{ad} = 21.69 \text{ m and } h_{fd} = 0$$

$\therefore$ Pressure head on the piston $= H_{atm.} + (h_d + h_{ad}) = 10.3 + (32 + 21.69)$

$= $ **63.99 m of water absolute (Ans.)**

(*b*) At the *middle* of stroke, $\theta = 90°$ and we have:

$$h_{ad} = 0, \text{ and } h_{fd} = 3.47 \text{ m}$$

$\therefore$ Pressure head on the piston $= H_{atm.} + (h_d + h_{fd}) = 10.3 + (32 + 3.47)$

$= $ **45.77 m of water absolute (Ans.)**

(*c*) At the *end* of stroke, $\theta = 180°$ and we have:

$$h_{ad} = -21.69 \text{ m, and, } h_{fd} = 0$$

$\therefore$ Pressure head on the piston $= H_{atm.} + (h_d + h_{ad}) = 10.3 + (32 - 21.69)$

$= $ **20.61 m of water absolute (Ans.)**

(*ii*) The power required to drive the pump:

Work done by the pump (double-acting)

$$= \frac{2wALN}{60}\left(h_s + h_d + \frac{2}{3}h_{fs} + \frac{2}{3}h_{fd}\right) \quad \text{...[Eqn. (4.44)]}$$

$$= \frac{2 \times 9810 \times 0.01767 \times 0.3 \times 30}{60}\left(4.5 + 32 + \frac{2}{3} \times 0.578 + \frac{2}{3} \times 3.47\right) = 2038.44 \text{ Nm/s}$$

$\therefore$ Power required to drive the pump $= \dfrac{\text{Work done by the pump}}{\eta_{mech.}}$

$= \dfrac{2038.44}{0.8} = 2548 \text{ W} = \mathbf{2.548 \text{ kW}}$ **(Ans.)**

(*iii*) The maximum head and power required to drive the pump:

The maximum head against which the pump has to work is larger of

(*a*) $(h_s + h_{fs}) + (h_d + h_{fd}) = (4.5 + 0.578) + (32 + 3.47)$

$= 40.548$ m (mid position of piston)

(*b*) $(h_s + h_{as}) + (h_d + h_{ad}) = (4.5 - 3.62) + (32 + 21.69)$

$= 54.57$ m (end position of piston)

Thus, *maximum head* = **54.57 m (Ans.)**

$\therefore$ The work done by the pump

$$= \frac{2wALN}{60} \times 54.57 = \frac{2 \times 9810 \times 0.01767 \times 0.3 \times 30}{60} \times 54.57 = 2837.8 \text{ N/ms.}$$

$\therefore$ *Power required to drive the pump*

$= \dfrac{2837.8}{0.8} = 3547 \text{ W} = \mathbf{3.547 \text{ kW}}$ **(Ans.)**

Example 4.12. *The bore and stroke of a double-acting single-cylinder reciprocating pump, running at 30 r.p.m., are 200 mm and 400 mm respectively. The pump draws water from a sump 1.2 m. below the pump through a suction pipe 100 mm in diameter and 3.0 m long. The water is delivered to a tank 28 m above the pump through a delivery pipe 100 mm in diameter and 38 m long. Assuming the motion of the piston to be simple harmonic determine the net force due to fluid pressure on the piston when it has moved through a distance of 100 mm from the I.D.C. (inner dead centre).*

Take friction co-efficient for both the suction and delivery pipes as 0.006. Neglect the size of the piston rod.

Solution. Bore of the pump, $D = 200 \text{ mm} = 0.2 \text{ m}$

Stroke length, $L = 400 \text{ mm} = 0.4 \text{ m}$

$\therefore$ Crank radius, $r = \dfrac{L}{2} = \dfrac{0.4}{2} = 0.2$ m

Area of piston, $A = \dfrac{\pi}{4} \times 0.2^2 = 0.0314 \text{ m}^2$

Area of suction and delivery pipes, $a_d = a_s = \dfrac{\pi}{4} \times 0.1^2 = 0.00785 \text{ m}^2$

Speed of the pump, $N = 30$ r.p.m.

$\therefore$ Angular velocity, $\omega = \dfrac{2\pi N}{60} = \dfrac{2\pi \times 30}{60} = 3.14$ rad/s

Suction head, $h_s = 1.2$ m

Delivery head, $h_d = 28$ m

Length of suction pipe, $l_s = 3.0$ m

Length of delivery pipe, $l_d = 38$ m

Friction co-efficient for both the pipes, $f = 0.006$.

Net force on the piston:

(a) Suction side:

Distance moved by the piston from *I.D.C.*, $x = 100$ mm $= 0.1$ m ..*(Given)*

But, $x = r(1 - \cos\theta)$

$\therefore$ $0.1 = 0.2(1 - \cos\theta)$, or, $\cos\theta = 0.5$ or $\theta = 60°$

The pressure head on the piston during suction stroke for any angle θ of the crank $= (h_a + h_{as} + h_{fs})$

Now, $h_s = 1.2$ m ...*(Given)*

Acceleration head,

$$h_{as} = \frac{l_s}{g}\frac{A}{a_s}\omega^2 r\cos\theta = \frac{3}{9.81}\times\frac{0.0314}{0.00785}\times(3.14)^2\times 0.2\times\cos 60° = 1.206 \text{ m}$$

$$\text{Friction head, } h_{fs} = \frac{4fl_s}{d_s\times 2g}\left(\frac{A}{a_s}\omega r\sin\theta\right)^2$$

$$= \frac{4\times 0.006\times 3.0}{0.1\times 2\times 9.81}\left(\frac{0.0314}{0.00785}\times 3.14\times 0.2\times\sin 60°\right)^2 = 0.1737 \text{ m}$$

$\therefore$ Pressure head on the piston $= (h_s + h_{as} + h_{fs}) = 1.2 + 1.206 + 0.1737$

$= 2.579$ m of water below atmospheric head

$= 10.3 - 2.579 = 7.721$ m of water absolute

$\therefore$ Force on the piston from *suction side*

= Specific weight (w) × pressure head (h) × area of piston (A)

$= 9810 \times 7.721 \times 0.0314 = 2378.3$ N

(b) Delivery side:

The angular displacement from the O.D.C. (outer dead centre) for delivery stroke (corresponding to angular displacement of 60° from the I.D.C.), $\theta = 180° - 60° = 120°$

The pressure head on the piston during delivery stroke for any angle θ of the crank $= (h_d + h_{ad} + h_{fd})$

Acceleration head,

$$h_{ad} = \frac{l_d}{g}\frac{A}{a_d}\omega^2 r\cos\theta = \frac{38}{9.81}\times\frac{0.0314}{0.00785}\times 3.14^2\times 0.2\times\cos 120° = -15.27 \text{ m}$$

$$\text{Friction head, } h_{fd} = \frac{4fl_d}{d_d\times 2g}\left(\frac{A}{a_d}\omega r\sin\theta\right)^2$$

$$= \frac{4\times 0.006\times 38}{0.1\times 2\times 9.81}\left(\frac{0.0314}{0.00785}\times 3.14\times 0.2\times\sin 120°\right)^2 = 2.2 \text{ m}$$

$\therefore$ Pressure head on the piston $= (h_d + h_{ad} + h_{fd}) = 28 - 15.27 + 2.2$

$= 14.93$ m of water above atmospheric head

$= (10.3 + 14.93) = 25.23$ m of water absolute

$\therefore$ Force on the piston from delivery side

= Specific weight (w) × pressure head (h) × area of piston (A)

$= 9810 \times 25.23 \times 0.0314 = 7771.7$ N

$\therefore$ *Net force on the piston* $= 7771.7 - 2378.3 =$ **5393.4 N (Ans.)**

4.8. AIR VESSELS

An **air vessel** *is a closed chamber containing compressed air in the upper part and liquid being pumped in the lower part. One air vessel is fixed on the suction pipe just near the suction valve and one is fixed on the delivery pipe near the delivery valve. The air vessels are used for the following purposes:*

1. *To get continuous supply of liquid at a uniform rate (whatever fluctuations take place, they occur between the air vessels and the pump).*
2. *To save the power required to drive the pump* (By the use of air vessels the acceleration and friction heads are considerably reduced, thereby the work is also reduced).
3. *To run the pump at much higher speed without any danger of separation* (By fitting the air vessels as close to the pump as possible, the length of the pipe in which acceleration takes place is reduced due to which acceleration head is reduced, and pump can run at a high speed without separation).

Fig. 4.11 shows a reciprocating pump fitted with air vessels. When the liquid level in the air vessel rises, the air above is compressed, this compressed air forces the liquid as soon as the pressure in the pipe falls. The variation in air pressure may be reduced by increasing the capacity of the air vessels.

In a delivery pipe the liquid beyond the air vessel is assumed to flow with a uniform velocity (V_d). When the piston forces the liquid into the delivery pipe with a velocity *greater* than mean velocity V_d, *the additional liquid moves into the air vessel*. When the velocity is *less* than mean velocity, V_d, the liquid *flows out of the air vessel and makes up the deficiency*. The volume of liquid present in the portion of the delivery pipe between the cylinder and air vessel is *accelerated*. The same reasoning is applicable to the air vessel fitted on the suction pipe.

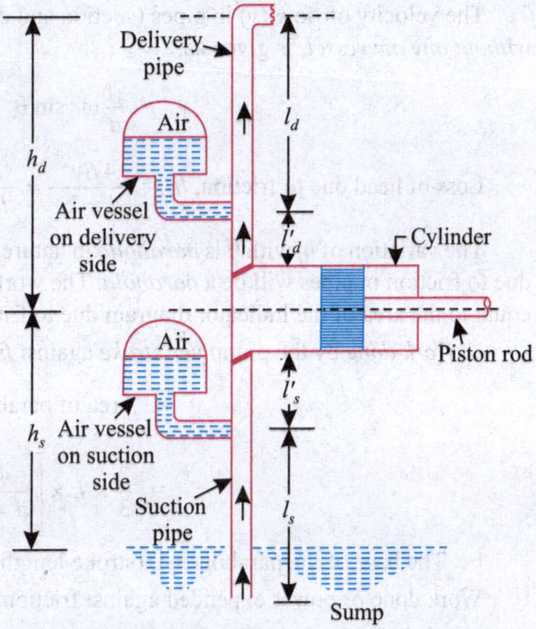

Fig. 4.11. Reciprocating pump with air vessels.

Fig. 4.12 shows the indicator diagrams without and with adequate air vessels.

Let,
A = Area of cross-section of the cylinder,
a = Area of cross-section of suction or delivery pipe,
l_d = Length of delivery pipe beyond the air vessel,
l_d' = Length of delivery pipe between cylinder and air vessel,
l_s = Length of suction pipe below air vessel,
l'_s = Length of suction pipe between cylinder and air vessel,
h_{ad} = Pressure head due to acceleration in delivery pipe,
h_{as} = Pressure head due to acceleration in suction pipe,
h_{fd} = Loss of head due to friction in delivery pipe beyond the air vessel,

h'_{fd} = Loss of head due to friction in delivery pipe between cylinder and air vessel,

h_{fs} = Loss of head due to friction in suction pipe below the air vessel, and

h'_{fs} = Loss of head due to friction in suction pipe between cylinder and air vessel.

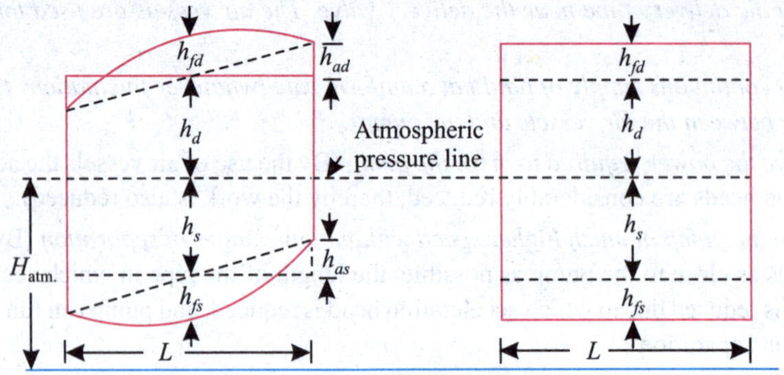

Fig. 4.12. Indicator diagrams without and with air vessels.

Case I. Work done or power expended against friction '*without air vessels*':

The velocity of flow (v) in pipes (suction and delivery), for a *single-acting* reciprocating pump *without any air vessel,* is given as:

$$v = \frac{A}{a}\omega r \sin\theta \qquad \text{...[Eqn. (4.15)]}$$

$$\text{Loss of head due to friction, } h_f = \frac{4flv^2}{d \times 2g} = \frac{4fl}{d \times 2g}\left(\frac{A}{a}\omega r \sin\theta\right)^2 \qquad \text{...(4.45)}$$

The variation of h_f with θ is *parabolic* in nature and hence indicator diagram for the loss of head due to friction in pipes will be a *parabola.* The work done by the pump against friction per stroke is equal to the area of the indicator diagram due to friction.

∴ *Work done* by the pump *per stroke* against *friction*

$$= \text{Area of parabola} = \frac{2}{3} \times \text{base} \times \text{height}$$

$$= \frac{2}{3} \times L \times \left[\frac{4fl}{d \times 2g} \times \left(\frac{A}{a}\omega r\right)^2\right] \qquad \text{...(4.46)}$$

[∵ The base of the parabola = L (stroke length), and the height = h_f at $\theta = 90°$]

Work done or power expended against friction,

$$P_1 = \frac{wAN}{60} \times \frac{2}{3} \times L \times \frac{4fl}{d \times 2g}\left(\frac{A}{a}\omega r\right)^2$$

or,

$$P_1 = \frac{wANL}{60} \times \frac{4fl}{3d \times g}\left(\frac{A}{a}\omega r\right)^2 \qquad \text{...(4.47)}$$

Case II. Work done or Power Expended Against Friction '*with air vessels*':

When air vessels having adequate capacity are fitted on the suction and delivery pipes, the velocity of flow in pipes (suction and delivery) may be *assumed constant* and equal to the *average/ mean flow velocity.*

$$\text{Mean velocity of flow, } \bar{v} = \frac{\text{Discharge}}{\text{Area of pipe}} = \frac{A \times L \times (N/60)}{a} = \frac{A \times 2r \times (\omega/2\pi)}{a} = \frac{A}{a} \times \frac{\omega r}{\pi}$$

$$\text{Loss of head due to friction, } h_f = \frac{4fl\bar{v}^2}{d \times 2g} = \frac{4fl}{d \times 2g}\left(\frac{A}{a} \times \frac{\omega r}{\pi}\right)^2 \quad ...(4.48)$$

Eqn. (4.48) shows that with the fitting of air vessels, the loss of head due to friction is independent of θ and hence indicator diagram is a *rectangle.*

∴ Work done by the pump per stroke against friction

$$= \text{Area of rectangle} = \text{base} \times \text{height}$$

$$= L \times \frac{4fl}{d \times 2g}\left(\frac{A}{a} \times \frac{\omega r}{\pi}\right)^2 \quad ...(4.49)$$

Work done or power expended against friction,

$$P_2 = \frac{wAN}{60} \times L \times \frac{4fl}{d \times 2g}\left(\frac{A}{a} \times \frac{\omega r}{\pi}\right)^2 \quad ...(4.50)$$

$$\text{Now, ratio, } \frac{P_2}{P_1} = \frac{\text{Work done or power expended against friction } \textit{with air vessels}}{\text{Work done or power expended against friction } \textit{without air vessels}}$$

$$= \frac{\frac{wANL}{60} \times \frac{4fl}{d \times 2g}\left(\frac{A}{a} \times \frac{\omega r}{\pi}\right)^2}{\frac{wANL}{60} \times \frac{4fl}{3d \times g}\left(\frac{A}{a}\omega r\right)^2} = \frac{3}{2\pi^2}$$

Percentage of work saved in pipe friction by fitting air vessels

$$= \frac{P_1 - P_2}{P_1} = 1 - \frac{P_2}{P_1} = \left(1 - \frac{3}{2\pi^2}\right) = \mathbf{0.848 \text{ or } 84.8\%}$$

Work saved in a double-acting reciprocating pump:

The work done or power expended against friction in case of double-acting reciprocating pump *without air vessel* is the same as given in case of single-acting reciprocating pump, *i.e*

$$P_1 = \frac{wANL}{60} \times \frac{4fL}{3d \times g}\left(\frac{A}{a}\omega r\right)^2$$

The mean velocity of flow, v for double-acting pump is given by:

$$v = \frac{\text{Discharge}}{\text{Area of pipe}} = \frac{2AL \times (N/60)}{a} = \frac{2A \times 2r \times (\omega/2\pi)}{a} = \frac{2A}{a} \times \frac{\omega r}{\pi}$$

$$\text{Loss of head due to friction,} \quad h_f = \frac{4flv^2}{d \times 2g} = \frac{4fl}{d \times 2g}\left(\frac{2A}{a} \times \omega r\right)^2 \quad ...(4.51)$$

Work done by the pump per stroke against friction

$$= L \times \frac{4fl}{d \times 2g}\left(\frac{2A}{a} \times \frac{\omega r}{\pi}\right)^2 \quad ...(4.52)$$

Work done or power expended against friction,

$$P_2 = \frac{wAN}{60} \times L \times \frac{4fl}{d \times 2g}\left(\frac{2A}{a} \times \frac{wr}{\pi}\right)^2 \quad ...(4.53)$$

Now ratio, $$\frac{P_2}{P_1} = \frac{\frac{wANL}{60} \times \frac{4fl}{d \times 2g}\left(\frac{2A}{a} \times \frac{wr}{\pi}\right)^2}{\frac{wANL}{60} \times \frac{4fl}{3d \times g}\left(\frac{A}{a}\omega r\right)^2} = \frac{6}{\pi^2}$$

Percentage of work saved in pipe friction by fitting air vessels

$$= \frac{P_1 - P_2}{P_1} = 1 - \frac{P_2}{P_1} = \left(1 - \frac{6}{\pi^2}\right) = 0.392 \text{ or } \mathbf{39.2\%}$$

Example 4.13. *A single-acting reciprocating pump is to raise a liquid of density 1200 kg/m³ through a vertical height of 11.5 m, from 2.5 m below pump axis to 9 m above it. The plunger moves with simple harmonic motion, has diameter 125 mm and stroke 225 mm. The suction and delivery pipes are of 75 mm diameter and 3.5 m and 13.5 m long respectively. There is a long vessel placed on the delivery pipe near the pump axis but there is no air vessel on the suction pipe. If separation takes place 0.88 bar below atmospheric pressure find:*

(i) Maximum speed with which the pump can run without separation taking place, and

(ii) Power required to drive the pump, if f = 0.02

Neglect slip for the pump. **[AMIE-Fluid Power Engg.]**

Solution. Density of liquid, $\rho = 1200 \text{ kg/m}^3$

Total vertical height $= 11.5$ m

Suction head, $h_s = 2.5$ m

Delivery head, $h_d = 9$ m

Diameter of plunger, $D = 125 \text{ mm} = 0.125$ m

$\therefore$ Area of plunger, $A = \frac{\pi}{4} \times 0.125^2 = 0.0123 \text{ m}^2$

Stroke length, $L = 225 \text{ mm} = 0.225$ m, $\therefore$ Crank radius, $r = \frac{0.225}{2} = 0.1125$ m

Diameter of each pipe (suction and delivery),

$$d_s = d_d = 75 \text{ mm} = 0.075 \text{ m}$$

$$\text{Area, } a_a = a_d = \frac{\pi}{4} \times 0.075^2 = 0.00442 \text{ m}^2$$

Length of suction pipe, $l_s = 3.5$ m

Length of delivery pipe, $l_d = 13.5$ m.

Friction co-efficient, $f = 0.02$.

Since air vessel is placed on delivery side only therefore, the *velocity in the delivery pipe will be uniform* and there will be *no accelerating head* (on delivery side).

Separation pressure $= 0.88$ bar below atmospheric pressure

$\therefore$ Separation pressure head $= \frac{0.88 \times 10^5}{1200 \times 9.81} = 7.47$ m $\left[\because h = \frac{p}{w}, \text{ and, } w = \rho g\right]$

(i) Maximum speed with which pump can run without separation, N:

The separation can take place only *at the beginning of suction stroke*. Since no air vessel has been fitted on suction side, therefore, there will be accelerating head (on suction side).

Pressure head at the beginning of suction stroke $= (h_s + h_{as})$ below atmosphere.

Limiting condition for *no separation* gives,

$$h_s + h_{as} = 7.47, \quad \text{or,} \quad 2.5 + h_{as} = 7.47, \quad \text{or,} \quad h_{as} = 4.97 \text{ m}$$

But h_{as} (at the beginning of stroke) $= \frac{l_s}{g} \times \frac{A}{a_s} \omega^2 r$

$$\therefore \quad 4.97 = \frac{3.5}{9.81} \times \frac{0.0123}{0.00442} \times \omega^2 \times 0.1125$$

$$\text{or,} \quad \omega = \left(\frac{4.97 \times 9.81 \times 0.00442}{3.5 \times 0.0123 \times 0.1125} \right)^{1/2} = 6.67 \text{ rad/s}$$

$$\text{But,} \quad \omega = \frac{2\pi N}{60}, \text{ or, } N = \frac{60\omega}{2\pi} = \frac{60 \times 6.67}{2\pi} \quad \textbf{63.69 r.p.m. (Ans.)}$$

(*ii*) Power required to drive the pump, *P*:

In case of a single-acting pump the discharge (Q) is given by :

$$Q = \frac{ALN}{60} = \frac{0.0123 \times 0.225 \times 63.69}{60} = 0.00294 \text{ m}^3\text{/s}$$

Velocity of liquid in the delivery pipe, $v = \frac{Q}{a_d} = \frac{0.00294}{0.00442} = 0.665$ m/s

Loss of head due to friction in delivery pipe,

$$h_{fd} = \frac{4 f l_d v^2}{d_d \times 2g} = \frac{4 \times 0.02 \times 13.5 \times 0.665^2}{0.075 \times 2 \times 9.81} = 0.324 \text{ m}$$

The maximum value of h_{fs}, during suction stroke, is given by:

$$h_{fs} = \frac{4 f l_s}{d_s \times 2g} \times \left(\frac{A}{a_s} \omega r \right)^2 = \frac{4 \times 0.02 \times 3.5}{0.075 \times 2 \times 9.81} \times \left(\frac{0.0123}{0.00442} \times 6.67 \times 0.1125 \right)^2 = 0.829 \text{ m}$$

Now, power required to drive the pump,

$$P = wQ \left(h_s + h_d + \frac{2}{3} h_{fs} + h_{fd} \right)$$

$$= (1200 \times 9.81) \times 0.00294 \left(2.5 + 9 + \frac{2}{3} \times 0.829 + 0.324 \right)$$

$$= \textbf{428.3 W (Ans.)}$$

Example 4.14. *The plunger diameter and stroke length of a single-acting reciprocating pump are 300 mm and 500 mm respectively. The speed of the pump is 50 r.p.m. The diameter and length of delivery pipe are 150 mm and 55 m respectively. If the pump is equipped with an air vessel on the delivery side at the centre line of the pump, find the power saved in overcoming friction in the delivery pipe.*

Take friction co-efficient, f = 0.01

Solution. Diameter of plunger, D = 300 mm = 0.3 m

$$\therefore \quad \text{Area, } A = \frac{\pi}{4} \times 0.3^2 = 0.07068 \text{ m}^2$$

Stroke length, L = 500 mm = 0.5 m,

$\therefore$ Crank radius, $r = \dfrac{0.5}{2} = 0.25$ m

Speed of the pump, N = 50 r.p.m.

$\therefore$ Angular velocity, $\omega = \dfrac{2\pi N}{60} = \dfrac{2\pi \times 50}{60} = 5.23$ rad/s

Diameter of delivery pipe, d_d = 150 mm = 0.15 m

$\therefore$ Area of delivery pipe, $a_d = \dfrac{\pi}{4} \times 0.15^2 = 0.01767\ \text{m}^2$

Length of delivery pipe, l_d = 55 m

Friction co-efficient, f = 0.01

Power saved in overcoming friction in the delivery pipe:

Maximum velocity of water in delivery pipe,

$$v_d = \frac{A}{a_d}\,\omega r = \frac{0.07068}{0.01767} \times 5.23 \times 0.25 = 5.23 \text{ m/s}$$

Maximum loss of head due to friction,

$$h_{fd} = \frac{4 f l_d v_d^2}{2g \times d_d} = \frac{4 \times 0.01 \times 55 \times 5.23^2}{2 \times 9.81 \times 0.15} = 20.45 \text{ m}$$

$\therefore$ Power required to overcome friction

$$= \frac{wALN}{60} \times \left(\frac{2}{3} h_{fd}\right) = \frac{9810 \times 0.07068 \times 0.5 \times 50}{60} \times \left(\frac{2}{3} \times 20.45\right) = 3938.7 \text{ W} = \mathbf{3.938\ kW}$$

With air vessel fitted, the velocity in the delivery pipe becomes constant and is given by,

$$v_d = \frac{A}{a_d} \times \frac{\omega r}{\pi} = \frac{0.07068}{0.01767} \times \frac{5.23 \times 0.25}{\pi} = 1.66 \text{ m/s}$$

Loss of head due to friction,

$$h_{fd} = \frac{4 f l_d v_d^2}{d_d \times 2g} = \frac{4 \times 0.01 \times 55 \times 1.66^2}{0.15 \times 2 \times 9.81} = 2.06 \text{ m}$$

$\therefore$ Power required to overcome friction

$$= \frac{wALN}{60} \times h_{fd} = \frac{9810 \times 0.07068 \times 0.5 \times 50}{60} \times 2.06 = 595 \text{ W or } 0.595 \text{ kW}$$

Hence, the *power saved* by fitting an air vessel

= 3.938 – 0.595 = **3.343 kW (Ans.)**

Example 4.15. *The diameter and stroke of a single-acting reciprocating pump are 300 mm and 500 mm respectively. The pump takes its supply of water from sump 3.2 m below the pump axis through a pipe 9 m long and 200 mm diameter. If separation occurs at 2.4 m of water absolute, determine:*

(i) The speed at which separation may take place at the beginning of suction stroke, and

(ii) The speed of the pump if an air vessel is fitted on the suction side 2.4 m above the sump water level.

Take atmospheric pressure head = 10.3 m of water, and friction co-efficient, f = 0.01.

Solution. Diameter of piston, $D = 300 \text{ mm} = 0.3 \text{ m}$

$$\therefore \quad \text{Area, } A = \frac{\pi}{4} \times 0.3^2 = 0.07068 \text{ m}^2$$

Stroke length, $L = 500 \text{ mm} = 0.5 \text{ m}$

$$\therefore \quad \text{Crank radius, } r = \frac{0.5}{2} = 0.25 \text{ m}$$

Suction head, $h_s = 3.2 \text{ m}$

Diameter of suction pipe, $d_s = 200 \text{ mm} = 0.2 \text{ m}$

$$\therefore \quad \text{Area of suction pipe, } a_s = \frac{\pi}{4} \times 0.2^2 = 0.0314 \text{ m}^2$$

Length of suction pipe, $l_s = 9 \text{ m}$

Separation head, $h_{sep} = 2.4$ m of water absolute

Atmospheric pressure head, $H_{atm.} = 10.3$ m of water

Friction co-efficient, $f = 0.01$

(*i*) The speed at which separation may take place (no air vessel fitted), *N*:

The pressure head due to acceleration in the suction pipe,

$$h_{as} = \frac{l_s}{g} \times \frac{A}{a_s} \omega^2 r \cos\theta$$

At the beginning of the suction stroke, $\theta = 0°$ and we have:

$$h_{as} = \frac{l_s}{g} \times \frac{A}{a_s} \omega^2 r = \frac{9}{9.81} \times \frac{0.07068}{0.0314} \times \omega^2 \times 0.25 = 0.516\,\omega^2$$

$\therefore$ Pressure head in the cylinder at the beginning of suction stroke

$$= (h_s + h_{as}) = (3.2 + 0.516\,\omega^2) \text{ below atmospheric head}$$

Limiting condition for *no separation* gives:

$$H_{atm.} - (h_s + h_{as}) = h_{sep.} \text{ or, } 10.3 - (3.2 + 0.516\,\omega^2) = 2.4$$

or,
$$\omega = \left(\frac{10.3 - 2.4 - 3.2}{0.516}\right)^{1/2} = 3.02 \text{ rad./s}$$

But,
$$\omega = \frac{2\pi N}{60}, \text{ or, } N = \frac{60\omega}{2\pi} = \frac{60 \times 3.02}{2\pi} = \mathbf{28.8 \text{ r.p.m.}} \textbf{ (Ans.)}$$

(*ii*) The speed of pump when an air vessel is fitted on suction side, *N*:

Since the air vessel is installed 2.4 m above the sump water level, therefore: (*i*) there will be a loss of head due to friction in the suction pipe for the length of $9 \times \frac{2.4}{3.2} = 6.75$ m; (*ii*) the acceleration pressure head will be restricted in the remaining $(9 - 6.75) = 2.25$ m length of suction pipe.

The pressure head due to acceleration (h_{as}) in the suction pipe at the beginning of suction stroke ($\theta = 0°$) is given by:

$$h_{as} = \frac{l_s}{g} \times \frac{A}{a_s} \times \omega^2 r = \frac{2.25}{9.81} \times \frac{0.07068}{0.0314} \times \omega^2 \times 0.25 = 0.129\,\omega^2$$

The velocity of water in the suction pipe *fitted with air vessel,*

$$v_s = \frac{A}{a_s} \times \frac{\omega r}{\pi} = \frac{0.07068}{0.0314} \times \frac{\omega \times 0.25}{\pi} = 0.179\ \omega$$

Loss of head due to friction, $$h_{fs} = \frac{4fl_s v_s^2}{d_s \times 2g} = \frac{4 \times 0.01 \times 6.75 \times (0.179\omega)^2}{0.2 \times 2 \times 9.81} = 0.0022\ \omega^2$$

Limiting condition for *no separation* gives:

$$H_{atm.} - (h_s + h_{as} + h_{fs}) = h_{sep.}$$

$$10.3 - (3.2 + 0.129\omega^2 + 0.0022\omega^2) = 2.4$$

or, $$10.3 - 3.2 - 0.1312\omega^2 = 2.4, \text{ or, } \omega = \left(\frac{10.3 - 3.2 - 2.4}{0.1312}\right)^{1/2} = 5.98 \text{ rad./s}$$

But, $$\omega = \frac{2\pi N}{60}, \text{ or, } N = \frac{60\omega}{2\pi} = \frac{60 \times 5.98}{2\pi} = \mathbf{57.1\ r.p.m.\ (Ans.)}$$

Evidently by *fitting an air vessel, the pump can be run at higher speeds without any chance of separation.*

Example 4.16. *A double-acting reciprocating pump is running at 30 r.p.m. Its bore and stroke are 250 mm and 400 mm respectively. The pump lifts water from a sump 3.8 m below and delivers it to tank at a height 65 m above the cylinder axis. The length of suction and delivery pipes are 6 m and 150 m respectively. The diameter of the delivery pipe is 100 mm. If an air vessel of adequate capacity has been fitted on the discharge side, determine:*

(*i*) *The minimum diameter of suction pipe to prevent cavitation assuming 2.5 m as the minimum head to prevent separation of flow which causes cavitation.*

(*ii*) *The maximum gross head against which pump has to work and the corresponding power of motor. Assume mechanical efficiency = 78% and slip = 1.5%.*

Take atmospheric pressure head (at the pump site) = 10.0 m, and friction co-efficient, f = 0.012.

Solution. The speed of the pump, $N = 30$ r.p.m.

$\therefore$ Angular velocity, $\omega = \frac{2\pi N}{60} = \frac{2\pi \times 30}{60} = 3.14$ rad/s

Bore of the pump, $D = 250$ mm $= 0.25$ m

$\therefore$ Area, $A = \frac{\pi}{4} \times 0.25^2 = 0.0491\ m^2$

Stroke length, $L = 400$ mm $= 0.4$ m

$\therefore$ Crank radius, $r = \frac{0.4}{2} = 0.2$ m

Suction head, $h_s = 3.8$ m

Delivery head, $h_d = 65$ m

Length of suction pipe, $l_s = 6$ m

Length of delivery pipe, $l_d = 150$ m

Diameter of delivery pipe, $d_d = 100$ mm $= 0.1$ m

$\therefore$ Area of delivery pipe, $a_d = \frac{\pi}{4} \times 0.1^2 = 0.00785\ m^2$

Separation head, $h_{sep.}$ = 2.5 m of water absolute
Mechanical efficiency, $\eta_{mech.}$ = 78%; slip = 1.5%
Atmospheric pressure head, $H_{atm.}$ = 10.0 m; friction co-efficient, f = 0.012.

(*i*) The minimum diameter of suction pipe, to prevent cavitation, d_s:

It is at the beginning of the suction stroke ($\theta = 0°$) where cavitation is likely to occur and at that instant the pressure head due to acceleration is given by,

$$h_{as} = \frac{l_s}{g}\frac{A}{a_s}\omega^2 r\cos\theta = \frac{l_s}{g}\frac{A}{a_s}\omega^2 r \quad [\because \text{when } \theta = 0°, \cos\theta = 1]$$

$$= \frac{6}{9.81} \times \frac{0.0491}{\frac{\pi}{4}\times d_s^2} \times 3.14^2 \times 0.2 = \frac{0.0754}{d_s^2} \quad ...(i)$$

$$\text{Friction head, } h_{fs} = \frac{4fl_s}{d_s \times 2g}\left(\frac{A}{a_s}\omega r \sin\theta\right)^2 = 0 \quad (\because \sin\theta = \sin 0° = 0)$$

Limiting condition for *no separation* gives,

$$H_{atm.} - (h_s + h_{as}) = h_{sep.}$$

$$\text{or,} \quad 10.0 - \left(3.8 + \frac{0.0754}{d_s^2}\right) = 2.5, \text{ or, } \frac{0.0754}{d_s^2} = 10.0 - 2.5 - 3.8$$

$$\text{or,} \quad d_s = \left(\frac{0.0754}{10.0 - 2.5 - 3.8}\right)^{1/2} = 0.1427 \text{ m or } \mathbf{142.7\ mm\ (Ans.)}$$

(*ii*) Gross head and power of motor:

$$\text{The discharge, } Q = \frac{2ALN}{60} = \frac{2 \times 0.0491 \times 0.4 \times 30}{60} = 0.01964 \text{ m}^3/\text{s}$$

The total head (H) against which the water has to be lifted is equal to sum of various heads at *I.D.C.* (inner dead centre) position of the piston,

i.e., $$H = (h_s + h_{as} + h_{fs}) + (h_d + h_{ad} + h_{fd})$$

(*a*) Suction side: h_s = 3.8 m (*Given*)

At the beginning of the stroke, $\theta = 0°$ and we have:

$$h_{as} = \frac{l_s}{g} \times \frac{A}{a_s}\omega^2 r = \frac{0.0754}{d_s^2} \quad [\text{as eqn. } (i)]$$

$$= \frac{0.0754}{0.1427^2} = 3.7 \text{ m}$$

$$\text{Friction head, } h_{fs} = \frac{4fl_s}{d_s \times 2g}\left(\frac{A}{a_s}\omega r \sin\theta\right)^2 = 0$$

$$(\because \sin\theta = \sin 0° = 0)$$

(*b*) Delivery side: h_d = 65 m (*Given*)

As an air vessel of adequate capacity has been installed on the delivery side, therefore acceleration is zero and velocity is no longer fluctuating (except in short length between the cylinder and air vessel). Also, since this shaft length has not been specified, the acceleration head can be assumed to be negligible.

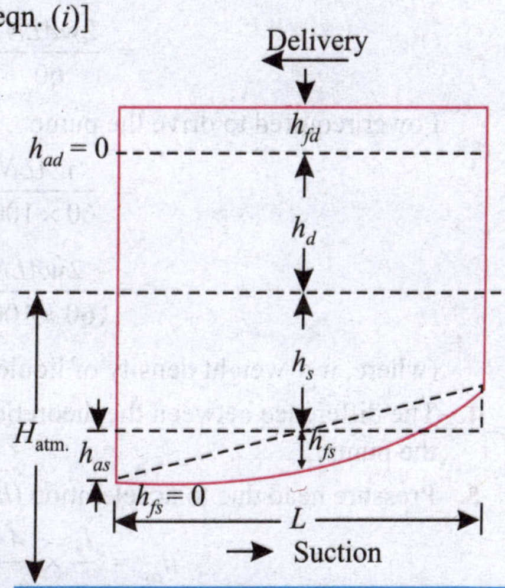

Fig. 4.13

$\therefore \qquad h_{ad} = 0$

Velocity in the delivery pipe,

$$v_d = \frac{Q}{a_d} = \frac{0.01964}{0.00785} = 2.5 \text{ m/s}$$

$$\therefore \qquad \text{Friction head, } h_{fd} = \frac{4 f l_d v_d^2}{d_d \times 2g} = \frac{4 \times 0.0012 \times 150 \times 2.5^2}{0.1 \times 2 \times 9.81} = 22.93 \text{ m}$$

$\therefore$ Gross head against which pump has to work,

$$H = (3.8 + 3.7 + 0) + (65 + 0 + 22.93) = 95.43 \text{ m}$$

Gross volume of water with a slip of 1.5%

$$= 0.01964 \times 1.015 = 0.01993 \text{ m}^3/\text{s}$$

$$\therefore \qquad \text{Power required} = wQH = 9810 \times 0.01993 \times 95.43 = 18657.8 \text{ W} \simeq 18.66 \text{ kW}$$

$$\text{Power of motor, } P = \frac{18.66}{\eta_{mech.}} = \frac{18.66}{0.78} = \mathbf{23.92 \text{ kW}} \textbf{ (Ans.)}$$

HIGHLIGHTS

1. The reciprocating pump is a positive displacement pump and consists of a cylinder, a piston a suction valve, a delivery valve, a suction pipe, a delivery pipe and crank and connecting rod mechanism operated by a power source e.g. steam engine, I.C. engine or an electric motor.
2. Discharge through a pump per second is given as

$$Q = \frac{ALN}{60} \qquad \text{...for a single-acting pump}$$

$$Q = \frac{2ALN}{60} \qquad \text{...for a double-acting pump}$$

3. Work done by reciprocating pump per second is given as

$$= \frac{wALN}{60}(h_s + h_d) \qquad \text{...for a single-acting pump}$$

$$= \frac{2wALN}{60}(h_s + h_d) \qquad \text{...for a double-acting pump}$$

Power required to drive the pump

$$= \frac{wALN}{60 \times 1000}(h_s + h_d) \; kW \qquad \text{...for a single-acting pump}$$

$$= \frac{2wALN}{60 \times 1000}(h_s + h_d) \; kW \qquad \text{...for a double-acting pump.}$$

(where, w = weight density of liquid in N/m^3.)

4. The difference between the theoretical discharge and actual discharge is called the '*slip*' of the pump.
5. Pressure head due to acceleration (h_a) in the suction and delivery pipes is given as:

$$h_{as} = \frac{l_s}{g} \times \frac{A}{a_s} \omega^2 r \cos\theta \qquad \text{...For suction pipe;}$$

$$h_{ad} = \frac{l_d}{g} \times \frac{A}{a_d} \omega^2 r \cos \theta \qquad \text{...For delivery pipe.}$$

6. The *indicator diagram* of a reciprocating pump is the diagram which shows the pressure head of the liquid in the pump cylinder corresponding to any position during the suction and delivery strokes. It is a graph between pressure head and stroke length of the piston for one complete revolution.
7. Work done by the pump is proportional to the area of the indicator diagram.
8. Work done by the pump per second due to acceleration and friction in suction and delivery pipes

$$= \frac{wALN}{60}\left(h_s + h_d + \frac{2}{3}h_{fs} + \frac{2}{3}h_{fd}\right) \qquad \text{...For a single-acting pump}$$

$$= \frac{2wALN}{60}\left(h_s + h_d + \frac{2}{3}h_{fs} + \frac{2}{3}h_{fd}\right) \qquad \text{...For a double-acting pump.}$$

9. An *air vessel* is a closed chamber containing compressed air in the upper part and liquid being pumped in the lower part. The air vessels are used: (*i*) To get continuous supply of liquid at a uniform rate, (*ii*) To save the power required to drive the pump and (*iii*) To run the pump at a much higher speed without any danger of separation.

OBJECTIVE TYPE QUESTIONS

Choose the correct Answer

1. With respect to a reciprocating pump which of the following statements is *incorrect* ?
   (*a*) The limiting value of separation pressure head for water is 6·8 m (absolute).
   (*b*) During suction, the separation may take place at the beginning of suction stroke.
   (*c*) During delivery, the separation may take place at the end of delivery stroke.
   (*d*) Indicator diagram shows variation of pressure head in the cylinder for one revolution of crank.
2. Reciprocating pumps are most suited where
   (*a*) constant heads are required on mains despite fluctuation in discharge
   (*b*) operating speeds are much high
   (*c*) constant supplies are required regardless of pressure fluctuations
   (*d*) none of the above.
3. Which of the following statements is *incorrect* for a reciprocating pump ?
   (*a*) The reciprocating pump is essentially a low speed machine.
   (*b*) The percentage of power saved by fitting air vessels is more in a double-acting than in a single-acting pump.
   (*c*) The reciprocating pumps can handle only low viscosity liquids free from impurities.
   (*d*) none of the above.
4. In a reciprocating pump the air vessels are used for which of the following purposes ?
   (*a*) To get continuous supply of liquid at a uniform rate.
   (*b*) To save the power required to drive the pump.
   (*c*) To run the pump at much higher speed without any danger of separation.
   (*d*) All of the above.
5. The pressure head due to acceleration in the suction pipe (h_{as}) is given as:
   (*a*) $\frac{l_s}{g} \times \frac{A}{a_s} \times \omega r \cos \theta$
   (*b*) $\frac{l_s}{2g} \times \frac{A^2}{a_s} \times \omega r \cos \theta$
   (*c*) $\frac{l_s}{g} \times \frac{A}{a_s} \times \omega^2 r \cos \theta$
   (*d*) none of the above.
6. Discharge through a double-acting reciprocating pump is given as
   (*a*) $\frac{ALN}{60}$ (*b*) $\frac{ALN}{120}$
   (*c*) $\frac{2ALN}{60}$ (*d*) $\frac{3ALN}{120}$.

ANSWERS

1. (*a*) 2. (*c*) 3. (*b*) 4. (*d*) 5. (*c*) 6. (*c*).

THEORETICAL QUESTIONS

1. Describe the principle and working of a reciprocating pump.
2. How are reciprocating pumps classified ?
3. Define slip, percentage slip and negative slip of a reciprocating pump.
4. What is negative slip in reciprocating pump ? Explain with neat sketches the function of air vessels in a reciprocating pump ?

   [AMIE, Fluid Power Engg.]
5. Obtain an expression for the pressure head due to acceleration in the suction and delivery pipes.
6. Define indicator diagram. Prove that work done by the pump is proportional to the area of indicator diagram.
7. Draw an indicator diagram, considering the effect of acceleration and friction in suction and delivery pipes. Find an expression for the work done per second in case of a single-acting reciprocating pump.
8. What is an air vessel ?
9. What are the uses of air vessels ?
10. Show from the first principles that work saved in a single-acting reciprocating pump, by fitting an air vessel, is 84.8 percent.

UNSOLVED EXAMPLES

1. A single-acting reciprocating pump, running at 60 r.p.m., delivers 0.53 m^3 of water per minute. The diameter of the piston is 200 mm and stroke length 300 mm. The suction and delivery heads are 4 m and 12 m respectively. Determine: (*i*) Theoretical discharge, (*ii*) Co-efficient of discharge, (*iii*) Percentage slip of the pump, and (*iv*) Power required to run the pump.

   [Ans. (*i*) 0.00942 m^3/s; (*ii*) 0.937; (*iii*) 6.26%; (*iv*) 1.47 kW]
2. A single-acting reciprocating pump having a bore of 150 mm and a stroke of 300 mm is raising water to height of 20 m above the sump level. The pump has an actual discharge of 0.0052 m^3/s. The efficiency of the pump is 70%. If the speed of pump is 60 r.p.m. determine: (*i*) Theoretical discharge, (*ii*) Theoretical power, (*iii*) Actual power, and (*iv*) Percentage slip.

   [Ans. (*i*) 0.0053 m^3/s; (*ii*) 1.04 kW, (*iii*) 1.48 kW; (*iv*) 1.88%]
3. A single-acting reciprocating pump has a piston diameter of 150 mm and stroke length 350 mm. The centre of the pump is 3 m above the water surface in the sump and 20 m below the delivery water level. Both the suction and delivery pipes have the same diameter of 100 mm and are 5 m and 30 m long respectively. If the pump is working at 35 r.p.m., determine:

   (*i*) Pressure heads due to acceleration at the beginning of suction and delivery strokes,

   (*ii*) Pressure heads in the cylinder at the beginning of suction and delivery strokes, and

   (*iii*) Pressure heads in the cylinder at the end of suction and delivery strokes.

   [Ans. (*i*) 2.695 m, 16.17 m; (*ii*) 4.605 m of water (abs.), 46.47 m of water (abs.), (*iii*) 9.99 m of water (abs.), 14.13 m (abs.)]
4. The diameter and stroke of a single-acting reciprocating pump are 125 mm and 300 mm. The pump is fed by a suction pipe 75 mm in diameter and 7 m long; the suction lift being 4 m. The separation occurs if the absolute pressure head in the cylinder during suction stroke falls below 2·5 m of water. What is the maximum speed at which pump can be run without separation in the suction pipe ?

   Take atmosphere pressure head = 10.3 m of water. **[Ans.** 34.1 r.p.m.]
5. The diameter and stroke length of single-acting reciprocating pump are 100 mm and 200 mm respectively. It takes its supply of water from a sump 4 m below the pump through a pipe 6 m long and 40 mm in diameter. It delivers water to a tank 14 m above the pump through a pipe 30 mm in diameter and 18 m long. If the separation occurs at 78.48 kN/m^2 below the atmospheric pressure, find the maximum speed at which pump may be operated without separation. Assume plunger has a simple harmonic motion. **[Ans.** 30.9 r.p.m.]

6. The cross-sectional area of plunger equals 1.65 times that of a delivery pipe. The delivery pipe is 55 m long and it rises upward at a slope of 1 in 5. If the plunger has an acceleration of 2.5 m/s^2 at the end of the stroke and separation pressure is 2.5 m of water find whether separation will take place and, if so, at which section of the pipe. Assume simple harmonic motion and take atmospheric pressure = 10·3 m of water

[**Ans.** Separation will take place; 35.45 m]

7. The bore and stroke of a reciprocating pump are 250 mm and 500 mm respectively. The pump delivers water through 100 mm delivery pipe to a tank located at 12 m above it and 25 m horizontally from it. If the separation occurs at a pressure of 22.5 kN/m^2 absolute, find the safe speed at which the pump should run for the following arrangements of delivery pipe:

   (*i*) The delivery pipe is horizontal from the pump and then vertical upto the tank and

   (*ii*) The delivery pipe is vertical from the pump and then horizontal upto the tank.

   Take: Atmospheric pressure (at the pump site) = 10.3 m of water, connecting rod-crank ratio = 5.

   [**Ans.** (*i*) 19.68 r.p.m; (*ii*) 12.45 r.p.m.]

8. The piston diameter and stroke length of a single acting reciprocating pump are 150 mm and 300 mm respectively. The centre of the pump is 5·0 m above the water level in the sump and 33 m below the delivery water level. Both the suction and delivery pipes have the same diameter of 75 mm and are 65 m and 39 m long respectively. If the pump is working at 30 r.p.m. determine:

   (*i*) The pressure head on the piston at the beginning, middle and end of both suction and delivery strokes, and

   (*ii*) The power required to drive the pump.

   Take atmospheric pressure head = 10.3 m of water and friction co-efficient, f = 0.01 for both the pipes.

   [**Ans.** (*i*) 1.38 m (abs.); 4.672 m (abs.), 9.22 m (abs.); 66.84 m (abs.); 47.067 m (abs.); 19.76 m (abs.); (*ii*) 1.064 kW]

9. The bore and stroke of a double-acting single-cylinder reciprocating pump, running at 35 r.p.m., are 200 mm and 400 mm respectively. The pump draws water from a sump 1·0 m below the pump through a suction pipe 100 mm in diameter and 2·5 m long. The water is delivered to a tank 30 m above the pump through a delivery pipe 100 mm in diameter and 40 m long. Assuming the motion of the piston to be simple harmonic determine the net force due to fluid pressure on the piston when it has moved through a distance of 100 mm from inner dead centre (*I.D.C.*).

   Take friction co-efficient for both the suction and delivery pipes as 0.0075. Neglect the size of the piston rod. [**Ans.** 2.522 kN]

10. The plunger diameter and stroke length of a single-acting reciprocating pump are 300 mm and 500 mm respectively. The speed of the pump is 60 r.p.m. The diameter and length of delivery pipe are 150 mm and 60 m respectively. If the pump is equipped with an air vessel on the delivery side at the centre line of the pump find the power saved in overcoming friction in delivery pipe.

    Take friction co-efficient, f = 0.01.

    [**Ans.** 6.3 kW]

11. The diameter and stroke of a single-acting reciprocating pump are 300 mm and 500 mm respectively. The pump takes its supply of water from sump 3.5 m below the pump axis through a pipe 10 m long and 200 mm diameter. If separation occurs at 2.5 m of water absolute, determine:

    (*i*) The speed at which separation may take place at the beginning of suction stroke, and

    (*ii*) The speed of the pump if an air vessel is fitted on the suction side 2.5 m above the sump water level.

    Take atmospheric pressure head = 10.3 m of water and friction co-efficient, f = 0.01.

    [**Ans.** (*i*) 26.16 r.p.m; (*ii*) 48.56 r.p.m.]

Chapter 5

MISCELLANEOUS HYDRAULIC MACHINES

5.1. INTRODUCTION

There are many hydraulic devices/machines which are based on the principles of *fluid statics* and *fluid kinematics* and are used for either *storing the hydraulic energy and then transmitting when needed or magnifying the hydraulic energy several times and transmitting the same*. In all such machines power is transmitted with the help of a fluid which may be a liquid (water or oil). In this chapter following hydraulic devices/machines will be discussed.

1. Hydraulic accumulator,
2. Hydraulic intensifier,
3. Hydraulic press,
4. Hydraulic crane,
5. Hydraulic lift,
6. Hydraulic ram,
7. Hydraulic coupling,
8. Hydraulic torque converter,
9. Air lift pump, and
10. Jet pump.

5.2. HYDRAULIC ACCUMULATOR

Hydraulic accumulator *is a device used to store the energy of liquid under pressure and make this energy available* (*as a quick secondary source of power*) *to hydraulic machines, such as presses, lifts and cranes*. In case of hydraulic crane or lift, the liquid under pressure needs to be supplied during upward motion of the load only. This energy is supplied from hydraulic accumulator. But when the lift is moving downward, no large external energy is required and during that period the energy from the pump is stored in the accumulator.

The function of hydraulic accumulator is analogous to that of the flywheel of a reciprocating engine and an electric storage battery. It damps out pressure surges and shocks in the hydraulic system, and thus it functions as a *pressure regulator*.

5.2.1. Simple Hydraulic accumulator

Construction and working. Fig. 5.1 shows a *simple hydraulic accumulator*. It consists of a fixed vertical cylinder, containing a sliding ram/plunger. A load/weight is placed on the top, to create pressure in the cylinder chamber. One side of the cylinder is connected to the pump and the other side to the machine.

In the beginning, the ram is at the lowermost position. During idle periods of driven machine (say crane or lift) high pressure liquid supplied by the pump is admitted in the hollow space of the cylinder, it raises the ram, on which the heavy load is placed. Flow of more liquid continues till the ram is at its uppermost position; at this position, the cylinder is full of water and the maximum amount of pressure energy is accumulated. This accumulated energy is later discharged to the driven machine, during its working stroke (*i.e.* when it requires maximum amount of energy).

Capacity of accumulator. The *maximum amount of energy that the accumulator can store* is known as the *capacity of the accumulator*.

Let, A = Area of the sliding ram,

L = Stroke or lift of the ram,

p = Intensity of pressure of liquid supplied by the pump, and

W = Total weight of the ram (including the weight of the load on the ram).

Then, $W = P \times A$

The work done in lifting the ram = $W \times$ lift of ram

$= W \times L$

$= p \times A \times L \qquad (\because W = p \times A)$

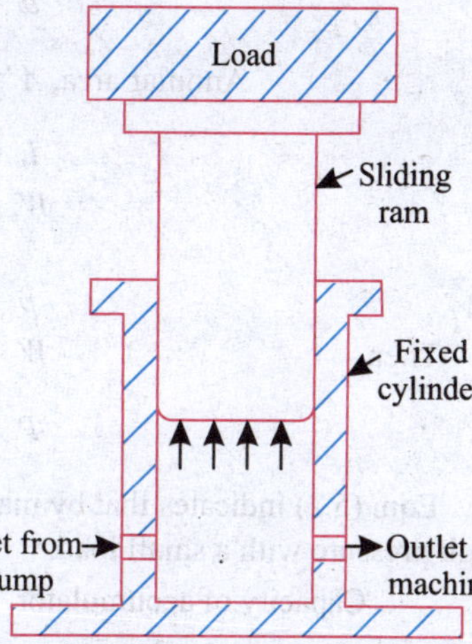

Fig. 5.1 Simple hydraulic accumulator.

But the work done in lifting the ram = Energy stored in the accumulator = Capacity of the accumulator

$\therefore$ Capacity of the accumulator = $p \times A \times L$

$= p \times$ volume of accumulator ...(5.1)

$(\because A \times L = \text{volume of accumulator})$

5.2.2. Differential Hydraulic Accumulator

Fig. 5.2 shows a differential hydraulic accumulator. *The advantage of this accumulator is that liquid can be stored at a high pressure by a comparatively small load on the ram.* It consists of a fixed vertical ram/plunger inside which is provided a central liquid passage of small diameter. This fixed cylindrical ram/plunger is surrounded by closely fitting brass bush, which is surrounded by an inverted *sliding* cylinder having a circular projected collar on which weights are placed. Passages for liquid to enter and leave the unit are provided in the fixed ram, and connected to the inlet and outlet pipes, as shown in the Fig. 5.2.

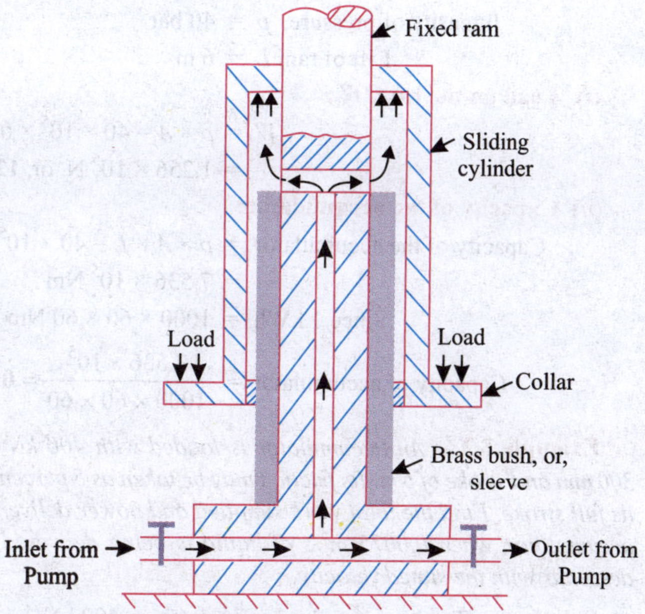

Fig. 5.2 Differential hydraulic accumulator.

The liquid supplied from the pump enters the cylinder through central vertical hole provided in the fixed ram and causes the loaded cylinder to move upwards, thus storing the hydraulic energy. When the liquid is drawn by the machine from the accumulator, the liquid leaves the cylinder through the same central hole. The liquid entering the cylinder exerts pressure on the *annular area of the cylinder which is equal to cross-sectional area (horizontal) of the brass bush or sleeve.*

Let, D = External diameter of the bush,

d = Diameter of fixed ram,

$\therefore$ Annular area, $A = \frac{\pi}{4}(D^2 - d^2)$,

L = Vertical lift of the sliding cylinder,

W = Total weight of sliding cylinder (including the weight placed on the cylinder), and

p = Intensity of pressure of liquid by pump.

Then, $W = p \times A$

$$\therefore \quad P = \frac{W}{A} \qquad ...(5.2)$$

Eqn. (5.2) indicates that by making the area of the bush small, it is possible to store liquid at a high pressure with a small load.

$$\text{Capacity of accumulator} = W \times L$$
$$= p \times A \times L = p \times \text{volume} \qquad ...(5.3)$$

Example 5.1. *An accumulator has a ram of 200 mm diameter and a lift of 6 m. If the liquid is supplied at a pressure of 40 bar, find: (i) load on the ram, and (ii) capacity of the accumulator.* **[Anna University]**

Solution. Diameter of ram, $D = 200 \text{ mm} = 0{\cdot}2 \text{ m}$

$\therefore$ Area, $A = \frac{\pi}{4} \times 0.2^2 = 0.0314 \text{ m}^2$

Intensity of pressure, $p = 40$ bar

Lift of ram $L = 6$ m

(*i*) Load on the ram, *W*:

$$W = p \times A = 40 \times 10^5 \times 0.0314$$
$$= 1.256 \times 10^5 \text{ N, or, } \mathbf{125.6 \text{ kN (Ans.)}}$$

(*ii*) Capacity of the accumulator:

$$\text{Capacity of the accumulator} = p \times A \times L = 40 \times 10^5 \times 0.0314 \times 6 \text{ Nm}$$
$$= 7.536 \times 10^5 \text{ Nm}$$

$$\text{Since 1 kWh} = 1000 \times 60 \times 60 \text{ Nm}$$

$$\therefore \quad \text{Capacity of accumulator} = \frac{7.536 \times 10^5}{1000 \times 60 \times 60} = \mathbf{0.209 \text{ kWh (Ans.)}}$$

Example 5.2. *An accumulator is loaded with 400 kN weight. The ram has a diameter of 300 mm and stroke of 6 m. Its friction may be taken as 5 percent. It takes two minutes to fall through its full stroke. Find the total work supplied and power delivered to the hydraulic appliance by the accumulator, when 0.0075 m³/s of liquid is being delivered by a pump, while the accumulator descends with the stated velocity.* **[AMIE, Fluid Power Engg.]**

Solution. Total load on the accumulator = 400 kN

Diameter of ram, $D = 300 \text{ mm} = 0.3 \text{ m}$

$\therefore$ Area of ram, $A = \frac{\pi}{4} \times 0.3^2 = 0.07068 \text{ m}^2$

Stroke of ram, $L = 6$ m

Friction = 5 percent.

Total work supplied and power delivered:

Net load on accumulator when it descends = $400 \times 0.95 = 380$ kN

Time taken by ram to fall through full stroke, $t = 2$ min, or, 120 s

$$\therefore \quad \text{Distance moved by ram per sec.} = \frac{1}{t} = \frac{6}{120} = 0.05 \text{ m/s}$$

Liquid supplied by pump = $0.0075 \text{ m}^3/\text{s}$ (*Given*)

Work done by accumulator per second

= Net load on ram × distance moved by ram per sec.

= $380 \times 0.05 = 19$ kN m/s

$$\text{Intensity of pressure of water, } p = \frac{\text{Net load}}{\text{Area}} = \frac{380 \times 10^3}{0.07068} = 5376.3 \times 10^3 \text{ N/m}^2$$

$$\text{Pressure head, } H = \frac{p}{w} = \frac{5376.3 \times 10^3}{9810} = 548 \text{ m}$$

Work supplied by pump per second = Weight of water supplied per sec. × pressure head

= $wQH = 9810 \times 0.0075 \times 548$

= 40319 Nm/s, or, 40.319 kN m/s

$\therefore$ *Total work supplied to hydraulic machine*

= Work supplied by accumulator + work supplied by the pump

= 19 + 40.319 = **59.319 kN m/s** **(Ans.)**

Power delivered to the hydraulic machine = **59.319 kW** **(Ans.)**

Example 5.3. *It is required to transmit 25 kW power from an accumulator through a pipeline 100 mm diameter and 1500 m long. The ram is loaded with a weight of 1250 kN and the friction loss in the pipeline equals 2.5 per cent of the total power being transmitted. Determine the diameter of the ram.*

Take friction co-efficient = 0.01.

Solution. Power to be transmitted, $P = 25$ kW

Diameter of the pipe, $d = 100$ mm = 0.1 m

Length of the pipe, $l = 1500$ m

Load on the ram, $W = 1250$ kN

Friction loss in the pipeline = 2.5 percent of the total power being transmitted

$$= \frac{2.5}{100} \times 25 = 0.625 \text{ kW}$$

Diameter of the ram, D:

The loss of head due to friction in pipeline is given by,

$$h_f = \frac{4flV^2}{d \times 2g} = \frac{4 \times 0.01 \times 1500 \times V^2}{0.1 \times 2 \times 9.81} = 30.58\, V^2$$

(where, V = velocity of flow through the pipeline)

$$\text{Power lost due to friction} = \frac{wQh_f}{1000} \text{ kW (where, } w = 9810 \text{ N/m}^3 \text{ for water)}$$

or,
$$0.625 = \frac{9810 \times (\pi/4) \times 0.1^2 \times V \times 30.58V^2}{1000} = 2.356V^3$$

$$\therefore \quad = \left(\frac{0.625}{2.356}\right)^{1/3} = 0.64 \text{ m/s}$$

$$\text{Discharge in the pipeline} = \frac{\pi}{4} \times 0.1^2 \times 0.64 = 0.005026 \text{ m}^3/\text{s}$$

The same discharge flows through the accumulator.

$$\text{Power developed by accumulator} = \frac{wQH}{1000} \text{kW}$$

(where, H = pressure head of water in the accumulator.)

$$\therefore \quad 25 = \frac{9810 \times 0.005026 \times H}{1000},$$

or,
$$H = \frac{25 \times 1000}{9810 \times 0.005026} = 507 \text{ m of water}$$

Pressure intensity in the accumulator, $p = wH$, or, $p = 9810 \times 507 = 4973670 \text{ N/m}^2$

But,
$$\text{Pressure intensity} = \frac{\text{Load on the ram}}{\text{Area of the ram}} = \frac{1250 \times 10^3}{\frac{\pi}{4} \times D^2}$$

$$\therefore \quad 4973670 = \frac{1250 \times 10^3}{\frac{\pi}{4} \times D^2}, \text{ or, } D^2 = \frac{1250 \times 10^3}{\frac{\pi}{4} \times 4973670} = 0.3199$$

$\therefore$ $D = 0.565$ m or **565 mm (Ans.)**

Example 5.4. *The diameters of two portions of the ram of a differential accumulator are 150 mm and 140 mm respectively, the stroke being 1·25 m. If the accumulator is supplied with water at pressure 1200 m of water, find load on the ram and the capacity.*

Solution. *Given* : $D = 150$ mm $= 0.15$ m; $d = 140$ mm $= 0.14$ m;

Stroke length = 1.25 m; Pressure head, $h = 1200$ m.

Load on the ram and the capacity:

$$\text{Pressure on the ram, } p = wh = 9810 \times 1200 = 117.72 \times 10^5 \text{ N/m}^2$$

$$\textit{Load on the ram} = \text{Pressure intensity} \times \text{annular area}$$

$$= 117.72 \times 10^5 \times \frac{\pi}{4}(0.15^2 - 0.14^2) = 0.268 \times 10^5 \text{ N or } \mathbf{26.8\ kN\ (Ans.)}$$

$$\textit{Capacity of ram} = \text{Load on the ram} \times \text{stroke length}$$

$$= 26.8 \times 1.25 = 33.5 \text{ kN m}$$

$$\therefore \quad \text{Capacity in kWh} = \frac{33.5 \times 10^3}{1000 \times 60 \times 60} = \mathbf{0.0093\ kWh\ (Ans.)}$$

Example 5.5. *The diameters of the two parts of the ram of a differential accumulator are 150 mm and 120 mm, and stroke length is 1.25 m. If the pressure of water is 7850 kN/m² when the load is at rest at the upper end of stroke or when the load is moving with uniform velocity, what will be the weight of the loaded cylinder? How* ***much*** *energy can be stored in the accumulator?*

Find also the diameter of the ram of an ordinary accumulator to move the same load with the help of the same water pressure.

Solution. $D = 150$ mm = 0.15 m; $d = 120$ mm = 0.12 m;

Stroke length = 1.25 m; Pressure of water = 7850 kN/m^2.

Weight of loaded cylinder:

$$\text{Weight of loaded cylinder} = \text{Pressure} \times \text{annular area}$$

$$= 7850 \times \frac{\pi}{4}(0.15^2 - 0.12^2) = \mathbf{49.94\ kN\ (Ans.)}$$

Energy stored in accumulator:

$$\text{Energy stored in the accumulator} = \text{Load} \times \text{displacement}$$

$$= 49.94 \times 1.25 = \mathbf{62.42\ kNm\ (Ans.)}$$

Diameter of the ram of an ordinary accumulator, D′:

$$\text{Load on the ram} = \text{Pressure intensity} \times \text{area}$$

$$49.94 = 7850 \times \frac{\pi}{4}(D')^2$$

$$\therefore \quad D' = \left(\frac{49.94 \times 4}{7850 \times \pi}\right)^{1/2} = 0.09 \text{ m or } \mathbf{90\ mm\ (Ans.)}$$

Example 5.6. *A hydraulic accumulator has sliding ram of 400 mm diameter which slides through 7.5 m in 3 minutes during its working stroke, while weight on the ram including its self weight is equivalent to 300 kN. The pump supplies water at 0.009 m^3/s rate and packing friction amounts to 4 percent of total load. Determine:*

(*i*) *Pressure intensity of water,*

(*ii*) *Power delivered to machine supplied by accumulator,*

(*iii*) *Power required to drive the pump having efficiency 72 percent.*

Solution. Diameter of sliding ram, D = 400 mm = 0·4 m

$$\therefore \quad \text{Area, } A = \frac{\pi}{4} \times 0.4^2 = 0.1256 \text{ m}^2$$

Weight on ram including self weight, W = 300 kN

Packing friction = 4% of total load

$$= \frac{4}{100} \times 300 = 12 \text{ kN}$$

∴ Discharge of water supplied by pump, q = 0.009 m^3/s

Efficiency of pump, η = 72%

(*i*) Pressure intensity of water, *p*:

$$\text{Net load on the sliding ram} = 300 - 12 = 288 \text{ kN}$$

$$\therefore \quad \text{Pressure intensity of water} = \frac{288}{0.1256} \simeq \mathbf{2293\ kN/m^2\ (Ans.)}$$

(*ii*) Power delivered to machine supplied by accumulator, *P*:

Let Q = Total discharge of water supplied to the machine,

q = Discharge of water supplied by the pump, (= 0.009 m^3/s), and

L = Stroke length of the accumulator ram, (= 7.5 m),

Then, $AL = (Q - q)\, t$, or, $0.1256 \times 7.5 = (Q - 0.009) \times (3 \times 60)$

or, $$Q = \frac{0.1256 \times 7.5}{3 \times 60} + 0.009 = 0.0142 \text{ m}^3/\text{s}$$

$\therefore$ Power delivered to the machine $= p \times Q$

$= 2293 \times 0.0142$ kW $=$ **32.56 kW (Ans.)**

(*iii*) Power required to drive the pump:

Power required to drive the pump $= \dfrac{p \times q}{\eta} = \dfrac{2293 \times 0.009}{0.72}$ kW $=$ **28.66 kW (Ans.)**

Example 5.7. *An accumulator maintains a pressure of 6000 kN/m² in a 50 mm diameter hydraulic main. A hydraulic crane situated at a distance of 250 m from the accumulator is supplied with pressure water from this main. The ram of the hydraulic crane is of 220 mm diameter. Velocity ratio of the crane hook to ram is 4 : 1. A pressure of 280 kN/m² may be assumed on the ram to account for mechanical friction of ram, pulleys etc. Calculate the load lifted when it is raised with a speed of 0.6 m/s.*

Assume a co-efficient of friction for the hydraulic main as 0.01. **[UPTU]**

Solution. Pressure in the accumulator $= 6000$ kN/m²

Diameter of hydraulic main, $d = 50$ mm $= 0.05$ m

Length of hydraulic main, $l = 250$ m

Diameter of the ram, $D = 220$ mm $= 0.22$ m

Velocity ratio of the crane hook to ram $= 4 : 1$

Pressure lost due to mechanical friction $= 280$ kN/m²

Velocity of crane hook $= 0{\cdot}6$ m/s

Co-efficient of friction for the hydraulic main, $f = 0.01$

Load lifted, *W*:

$$\text{Velocity of ram} = \frac{0.6}{4} = 0.15 \text{ m/s}$$

Let V be the velocity of water in the hydraulic main. Since the quantity of water per second flowing through the main is equal to the quantity per second in the ram cylinder, therefore,

$$\left(\frac{\pi}{4} \times 0{\cdot}05^2\right) \times V = \frac{\pi}{4} \times 0.22^2 \times 0.15, \text{ or, } V = 2.9 \text{ m/s}$$

$$\text{Head of water in accumulator} = \frac{6000}{9.81} = 611.6 \text{ m of water} \qquad (\because w = 9.81 \text{ kN/m}^3)$$

$$\text{Head lost in friction in main, } h_f = \frac{4 f l V^2}{d \times 2g} = \frac{4 \times 0.01 \times 250 \times 2.9^2}{0.05 \times 2 \times 9.81} = 85.7 \text{ m of water}$$

Head lost due, to mechanical friction of ram, pulleys etc.

$$= \frac{280}{9.81} = 28.5 \text{ m of water}$$

Net head available on the ram $= 611.6 - (85.7 + 28.5) = 497.4$ m of water

$\therefore$ Net intensity of pressure on ram

$= 497.4 \times 9.81$ kN/m² $= 4879.5$ kN/m² ($\because w = 9.81$ kN/m³)

$$\text{Load on the ram} = 4879.5 \times \left(\frac{\pi}{4} \times 0.22^2\right) = 185.48 \text{ kN}$$

$$\therefore \quad \textit{Load lifted} \text{ by the crane hook} = \frac{185.48}{4} = \mathbf{46.37\ kN} \ \textbf{(Ans.)}$$

Example 5.8. *A weight loaded accumulator operates certain machinery through a pipe 100 mm in diameter and 600 m long. The accumulator has a ram 300 mm diameter and 3.5 m stroke, loaded with 320 kN; it is supplied with water by a three throw pump running at 45 r.p.m., the plungers of pump having a diameter of 45 mm and a stroke of 360 mm. The slip of the pump has been estimated as 5 per cent. If the power absorbed by the machinery is 37 kW, calculate the longest period during which it may be operated continuously.*

Take co-efficient of friction for the pipe as 0·0075. **[PTU]**

Solution.

Diameter of the pipe, d = 100 mm = 0.1 m

Length of the pipe, l = 600 m

Diameter of this ram, D = 300 mm = 0.3 m

Stroke of the ram, L = 360 mm = 0.36 m

Load on the ram, W = 320 kN

Diameter of each plunger of the pump = 45 mm = 0.045 m

Stroke of each plunger = 360 mm = 0.36 m

Speed of pump = 45 r.p.m.

Slip of the pump = 5%

Power absorbed by the machinery, P = 37 kW

Co-efficient of friction for the pipe, f = 0.0075.

Longest period during which machinery can be operated continuously, *t*:

$$\text{Pressure head in the accumulator} = \frac{W}{\left(\frac{\pi}{4} \times D^2\right) \times w} = \frac{320}{\frac{\pi}{4} \times 0.3^2 \times 9.81} = 461.5 \text{ m}$$

$$(\because w = 9.81 \text{ kN/m}^3)$$

$$\text{Loss of head due to friction in the pipe, } h_f = \frac{4flV^2}{d \times 2g}$$

$$= \frac{4 \times 0.0075 \times 600 \times V^2}{0.1 \times 2 \times 9.81} = 9.17\,V^2$$

$$\therefore \quad \text{Effective head at the machine, } H = (461.5 - 9.17\,V^2)$$

$$\text{Power supplied to machinery} = wQH$$

$$37 = 9.81 \times \left(\frac{\pi}{4} \times 0.1^2 \times V\right) \times (461.5 - 9.17V^2)$$

$$= 0.077V(461.5 - 9.17V^2) = 35.53\,V - 0.706V^3$$

or, $$0.706V^3 - 35.53V + 37 = 0$$

or, $$V^3 - 50.32V + 52.4 = 0$$

Solving by trial, we get V = 1.064 m/s

$\therefore$ Discharge through the pipe, $Q = \frac{\pi}{4} \times 0.1^2 \times 1.064 = 0.008357 \text{ m}^3/\text{s}$

This discharge, Q is the same as the discharge leaving the accumulator.

Also, the discharge entering the accumulator = Discharge of the three-throw pump.

$$\text{Discharge of the pump} = 3 \times \left[\frac{\pi}{4} \times 0.045^2 \times 0.36 \times \frac{45}{60}\right] = 0.001288 \text{ m}^3/\text{s}$$

$\therefore$ Water supplied from the accumulator = 0.008357 – 0.001288 = 0.00707 m³/s

$$\text{Volume of accumulator} = \frac{\pi}{4} \times 0.3^2 \times 3.5 = 0.247 \text{ m}^3$$

$\therefore$ The longest period during which machinery may be operated continuously (t) = time in which accumulator will be emptied

$$\therefore \quad t = \frac{0.247}{0.00707} = \mathbf{34.94\ s}\ \textbf{(Ans.)}$$

5.3. HYDRAULIC INTENSIFIER

Hydraulic intensifier *is a device which increases the intensity of pressure of a given liquid with the help of low pressure liquid of large quantity*. It finds its application at places where a liquid of very high pressure is to be developed from available low pressure. It is located between the pump and the machine (*e.g.* press, crane, lift) that needs high pressure liquid for its operation. Its action is similar to that of a step-up electrical transformer.

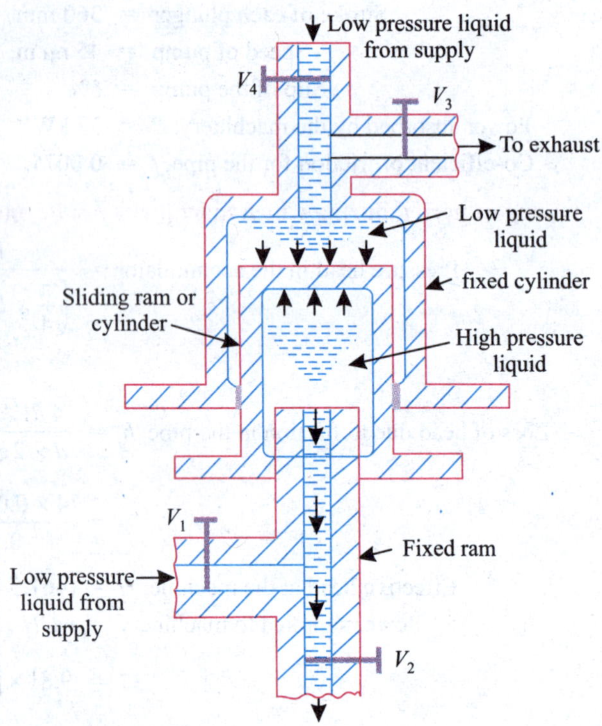

Fig. 5.3. Hydraulic intensifier.

Construction and working. A hydraulic intensifier consists of a fixed ram surrounded by the sliding cylinder, which is itself encased in a bigger and fixed cylinder (Fig. 5·3). The sliding cylinder contains water at high pressure (which is supplied to the machine through fixed ram) whereas the fixed cylinder contains water from the main supply at a low pressure.

— Initially when the sliding cylinder lies at the bottom of the stroke, the fixed cylinder is full of low pressure liquid. The valves V_2 and V_4 are then closed, the valve V_1 is opened thus admitting the low pressure liquid into the sliding cylinder; the valve V_3 is also opened which permits the low pressure liquid from the fixed cylinder to be discharged to the exhaust and the sliding cylinder to move upward. When the sliding cylinder reaches its topmost position, the inside of the sliding cylinder is full of low pressure liquid.

— The valves V_1 and V_3 are then closed and the valves V_2 and V_4 are opened. The low pressure liquid (from supply) then enters the fixed cylinder, and forces the sliding cylinder to move downward; pressure of liquid beneath is raised and the high pressure liquid is supplied to the driven machine.

The above cycle of operation is repeated.

The intensifier described above is *single-acting* (which gives supply during downward stroke only); however, *double-acting* intensifiers are also made, which give continuous supply of high pressure liquid.

— By means of an intensifier, it is possible to raise the intensity of pressure as high as 160 MN/m^2.

— Depending upon fluid used, intensifiers may also be of the following types:

(*i*) *Hydro-pneumatic intensifier*—Here air is supplied to the fixed cylinder instead of low pressure liquid.

(*ii*) *Steam intensifier*—Here steam under pressure is supplied to the fixed cylinder instead of low pressure liquid.

Let, p_1 = Pressure intensity of low pressure liquid (from supply) in the fixed cylinder,

A_1 = Cross-sectional area of sliding cylinder,

p_2 = Intensity of high pressure liquid in the fixed ram, and

A_2 = Cross-sectional area of the fixed ram.

The force exerted by low pressure liquid on the sliding cylinder in the downward direction $= p_1 \times A_1$

The force exerted by high pressure liquid on the sliding cylinder in the upward direction $= p_2 \times A_2$

For the equilibrium of the sliding cylinder at any position, we have

$$p_1 A_1 = p_2 A_2$$

∴ The intensity of high pressure liquid, $p_2 = \dfrac{p_1 A_1}{A_2}$ (neglecting friction effects) ...(5.4)

Example 5.9. *A hydraulic intensifier gets the low pressure liquid at a pressure of 40 bar and delivers it to a machine at a pressure of 160 bar. If the intensifier has a capacity of 0.021 m^3 and stroke 1.2 m, calculate the diameters of the fixed ram and the sliding cylinder to be used for this intensifier.*

Solution. Intensity of pressure of low pressure liquid, $p_1 = 40 \text{ bar} = 40 \times 10^5 \text{ N/m}^2$

Intensity of pressure of high pressure liquid, $p_2 = 160 \text{ bar} = 160 \times 10^5 \text{ N/m}^2$

Capacity of intensifier = 0.021 m^3

Stroke length = 1.2 m

Diameters of fixed ram (D_2) and the sliding cylinder (D_1):

Capacity of intensifier = Area of fixed ram × stroke length

$$0.021 = A_2 \times 1.2, \quad \text{or,} \quad A_2 = \frac{0.021}{1.2} = 0.0175 \text{ m}^2$$

$$\therefore \quad \frac{\pi}{4} D_2^2 = 0.0175, \text{ or, } D_2 = \left(\frac{4 \times 0.0175}{\pi}\right)^{1/2} = 0.149 \text{ m or } \mathbf{149\ mm}\ \textbf{(Ans.)}$$

Considering equilibrium of the sliding cylinder (neglecting friction effects), we have:

$$p_1 A_1 = p_2 A_2, \text{ or, } 40 \times 10^5 \times A_1 = 160 \times 10^5 \times 0.0175, \text{ or, } A_1 = 0.07 \text{ m}^2$$

$$\therefore \quad \frac{\pi}{4} D_1^2 = 0.07, \text{ or, } d_1 = \left(\frac{4 \times 0.07}{\pi}\right)^{1/2} = 0.298 \text{ m or } \mathbf{298\ mm}\ \textbf{(Ans.)}$$

Example 5.10. *An intensifier has a ram diameter of 150 mm and a sliding cylinder diameter of 750 mm. Calculate the pressure of water on the low pressure side of the intensifier if the pressure of water on high pressure side is 21000 kN/m². The loss due to friction at each of the packings of the intensifier is 5% of the total force on each of the packings.*

Solution. Sliding cylinder diameter, D_1 = 750 mm = 0.75 m

Ram diameter, D_2 = 150 mm = 0.15 m

Pressure of water on high pressure side, p_2 = 21000 kN/m²

Loss due to friction, k = 5% of the total force on each of the packings.

Pressure of water on low pressure side, p_1:

Considering equilibrium of the sliding cylinder, we have:

$$p_1 A_1\left(1-\frac{k}{100}\right) = \frac{p_2 A_2}{\left(1-\frac{k}{100}\right)}$$

or, $$p_1 \times \frac{\pi}{4} \times 0\cdot75^2\left(1-\frac{5}{100}\right) = 21000 \times \frac{\pi}{4} \times 0.15^2 \times \frac{1}{(1-5/100)}$$

or, $$p_1 \times \frac{\pi}{4} \times (0.75)^2 \times 0.95 = 21000 \times \frac{\pi}{4} \times 0.15^2 \times \frac{1}{0.95}$$

or, $$0.4197 p_1 = 390.6 \therefore P_1 = \frac{390.63}{0.4197} = \mathbf{930.7\ kN/m^2}\ \textbf{(Ans.)}$$

Example 5.11. *An intensifier receives water from an overhead tank through a pipeline 60 mm in diameter and 110 m long and conveys high pressure water to a hydraulic press which has 250 mm ram diameter and exerts a force of 350 kN. The diameters of ram and sliding piston of intensifier are 0·1 m and 1 m respectively. If the level of water in the overhead tank is 15 m above the inlet to the low pressure side of the intensifier, calculate the speed with which the ram of the hydraulic press moves to exert the force.*

Take friction co-efficient, f = 0.0075 for the pipeline between the supply reservoir and the intensifier.

Solution. Diameter of sliding piston of intensifier, D_1 = 1 m

Diameter of ram of intensifier, D_2 = 0.1 m

Diameter of pipeline, d_p = 60 mm = 0.06 m

Length of pipeline, l_p = 110 m

Diameter of ram of hydraulic press, D = 250 mm = 0.25 m

Friction co-efficient, f = 0.0075

Velocity of the ram of hydraulic press, V_r:

Let, V_p = Velocity of water in the pipeline.

Loss of head due to friction,

$$h_f = \frac{4 f l_p V_p^2}{d_p \times 2g} = \frac{4 \times 0.0075 \times 110 \times V_p^2}{0.06 \times 2 \times 9.81} = 2.8 V_p^2$$

Pressure head on the low pressure side of the intensifier,

$$h_1 = (15 - 2.8\, V_p^2) \text{ m of water}$$

$\therefore$ Corresponding intensity of pressure, $p_1 = w h_1 = 9810\,(15 - 2.8\, V_p^2)$ N/m²

Pressure intensity on the high pressure side of the intensifier,

$$p_2 = p_1 \times \frac{A_1}{A_2} = 9810\,(15 - 2.8V_p^2) \times \left(\frac{\frac{\pi}{4} \times 1^2}{\frac{\pi}{4} \times 0.1^2} \right) = 9810 \times (1500 - 280V_p^2)$$

Neglecting loss of pressure between the intensifier and hydraulic press, force on the ram of hydraulic press

$$= 9810\,(1500 - 280\,V_p^2) \times \frac{\pi}{4} \times 0.25^2$$; this force *equals the load on press.*

$$\therefore\ 9810\,(1500 - 280\,V_p^2) \times \frac{\pi}{4} \times 0.25^2 = 350 \times 1000$$

or, $1500 - 280\,V_p^2 = 726.8$, or, $V_p = \left(\frac{1500 - 726.8}{280} \right)^{1/2} = 1.66$ m

Discharge through the pipeline $= \frac{\pi}{4} \times 0.06^2 \times 1.66 = 0.00469\ \text{m}^3/\text{s}$

The same discharge flows to the low pressure side of the intensifier.

Discharge on the high pressure side of the intensifier

$$= 0.00469 \times \left(\frac{0.1}{1} \right)^2 = 4.69 \times 10^{-5}\ \text{m}^3/\text{s}$$

$$\because \left[\frac{Q_1}{Q_2} = \frac{A_1 L}{A_2 L} = \frac{A_1}{A_2},\ \text{or,}\ Q_2 = Q_1 \times \left(\frac{A_2}{A_1} \right) = Q_1 \left(\frac{D_2}{D_1} \right)^2 \right]$$

where,
Q_1 = Rate of discharge of low pressure liquid,
Q_2 = Rate of discharge of high pressure liquid to machine, and
L = Stroke length.

From continuity considerations, we have:

$$\frac{\pi}{4} \times 0.25^2 \times V_r = 4.69 \times 10^{-5},\ \text{or,}\ V_r = \frac{4.69 \times 10^{-5}}{\frac{\pi}{4} \times 0.25^2} = 0.000955\ \text{m/s}$$

$$= \mathbf{0.0573\ m/min}$$ **(Ans.)**

5.4. HYDRAULIC PRESS

The **hydraulic press** *is a device used for lifting heavy loads by the application of much smaller force. It is based on Pascal's law, which states that intensity of pressure is transmitted equally in all directions through a mass of fluid at rest.*

Working principle. The working principle of a hydraulic press may be explained with the help of Fig. 5.4. Consider a ram and plunger, operating in two cylinders of different diameters, which are inter-connected at the bottom, through a chamber, which is filled with some liquid.

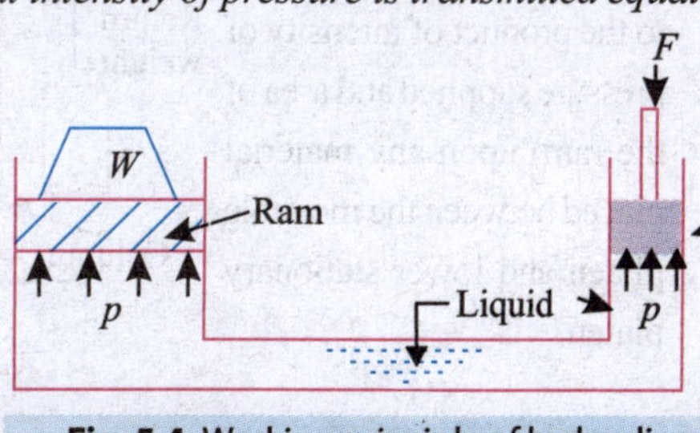

Fig. 5.4. Working principle of hydraulic press.

Let,

W = Weight to be lifted,

F = Force applied on the plunger,

A = Area of ram, and

a = Area of plunger.

Pressure intensity produced by the force F, $p = \dfrac{F}{\text{Area of plunger}} = \dfrac{F}{a}$

As per Pascal's law, the above intensity p will be equally transmitted in all directions.

∴ The pressure intensity on ram $= p = \dfrac{F}{a} = \dfrac{W}{A}$

or,
$$W = F \times \frac{A}{a} \quad \text{...(5.5)}$$

Eqn. (5.5) indicates that by applying a small force F on the plunger, a large force W may be developed by the ram.

Mechanical advantage of press $= \dfrac{A}{a}$

If the force in the plunger is applied by a lever which has a mechanical advantage $\dfrac{L}{l}$ (Fig. 5.5), then total mechanical advantage of machine

$$= \frac{L}{l} \times \frac{A}{a}$$

The ratio $\dfrac{L}{l}$ is known as *leverage of press*.

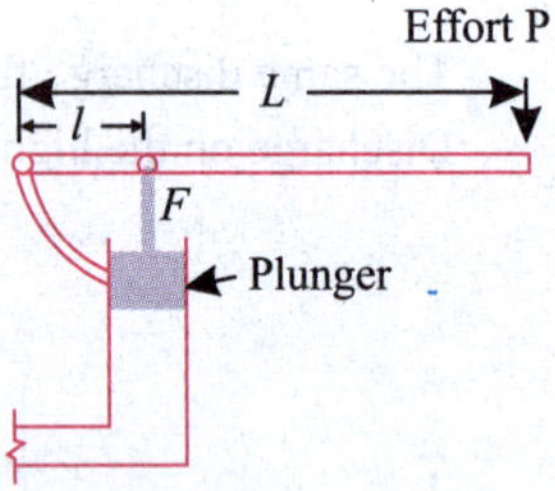

Fig. 5.5

Fig. 5.6 shows an '*Elementary Inverted Hydraulic Press*'.

— It consists of a cylinder (fixed) in which the ram slides. The lower end of the ram carries a movable platen which moves up and down with the ram. The upper and lower stationary platens are joined by columns.

— When liquid under high pressure is supplied to the cylinder, the ram moves downward and applies tremendous pressure (equal to the product of intensity of pressure supplied and area of the ram) upon any material placed between the movable platen and lower stationary platen.

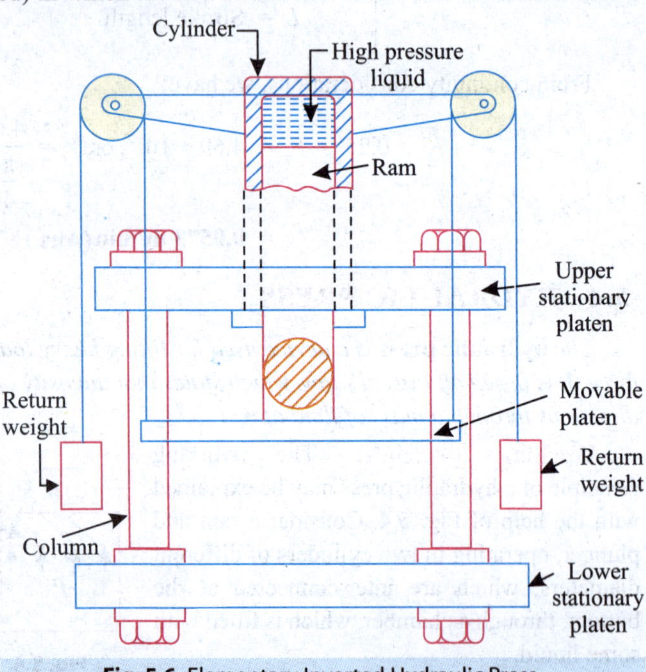

Fig. 5.6. Elementary Inverted Hydraulic Press.

For bringing the ram back in position, the liquid from the cylinder is taken out, subsequently the ram (along with movable platen) moves up by the action of return weights.

It may be noted that in some large presses it is possible to produce total thrust ranging from about 50 MN to 100 MN.

Hydraulic presses may be employed for the following jobs:

(*i*) Metal press work (to press sheet metal to any required shape).

(*ii*) Drawing and pushing rods.

(*iii*) Bending and straightening any metal piece.

(*iv*) Packing press.

(*v*) Cotton press.

(*vi*) Autoclave vulcanising press.

(*vii*) To prepare moulds and casting of bakelite (Bakelite press).

(*viii*) Forgoing press.

(*ix*) Plate press etc.

Example 5.12. *The diameters of ram and plunger of a hydraulic press are 100 mm and 12.5 mm respectively. Find the force required to be applied on the plunger to raise a load of 24 kN on the ram. If the plunger has a stroke of 200 mm, how many strokes will be required to lift the load by 500 mm. Also calculate the volume of additional liquid required. Further if the time taken to lift the load is 12 minutes, what will be power required to drive the plunger? Neglect frictional effects.*

Solution. Diameter of ram, D = 100 mm = 0.1 m.

$$\therefore \quad \text{Area of ram, } A = \frac{\pi}{4} \times 0.1^2 = 0.00785 \text{ m}^2$$

Diameter of plunger, d = 12.5 mm = 0.0125 m

$$\therefore \quad \text{Area of plunger, } a = \frac{\pi}{4} \times 0.0125^2 = 0.0001227 \text{ m}^2$$

Load to be raised, W = 24 kN

Stroke of plunger = 200 mm = 0.2 m

Distance through which load is to be lifted = 500 mm = 0.5 m

Time taken to lift the load = 12 minutes

Force required to raise a load of 24 kN, *F*:

Since intensity of pressure is same throughout a static mass of fluid,

$$\frac{F}{a} = \frac{W}{A}, \text{ or, } \frac{F}{0.0001227} = \frac{24}{0.00785}, \text{ or, } F = \frac{24 \times 0.0001227}{0.00785} = \mathbf{0.375\ kN\ (Ans.)}$$

Number of strokes, *n*:

Number of strokes required to lift the load by 0.5 m,

$$n = \frac{\text{Total volume of liquid to be displaced}}{\text{Volume of liquid displaced in one stroke of plunger}}$$

$$\text{or,} \quad n = \frac{\frac{\pi}{4} \times (0.1)^2 \times 0.5}{\frac{\pi}{4} \times (0.0125)^2 \times 0.2} = \mathbf{160\ (Ans.)}$$

$$\textit{Volume of additional liquid} = \frac{\pi}{4} \times (0.1)^2 \times 0.5 = \mathbf{0.00392\ m^3}\ \textbf{(Ans.)}$$

Power required to drive motor, *P*:

$$\text{Work done by the press} = 24 \times 0.5 = 12 \text{ kNm (in 12 minutes)}$$

$$\text{Work done per sec.} = \frac{12}{12 \times 60} = 0.01666 \text{ kNm/s}$$

$$\therefore \quad \text{Power required, } P = 0.01666 \text{ kW or } \mathbf{16.66\ W}\ \textbf{(Ans.)}$$

Example 5.13. *A hydraulic press has a ram of 180 mm diameter and plunger of 36 mm diameter, with stroke length of 300 mm. Weight exerted by press ram amounts to 7 kN and distance moved is 0·9 m in 15 minutes. Determine:*

(*i*) *The force applied on plunger,*

(*ii*) *The number of strokes performed by the plunger,*

(*iii*) *Work done by the press ram, and*

(*iv*) *Power required to drive the plunger.*

Solution. Diameter of ram, $D = 180 \text{ mm} = 0.18 \text{ m}$

$$\therefore \quad \text{Area of ram, } A = \frac{\pi}{4} \times 0.18^2 = 0.0254 \text{ m}^2$$

Diameter of plunger, $d = 36 \text{ mm} = 0.036 \text{ m}$

$$\therefore \quad \text{Area of plunger, } a = \frac{\pi}{4} \times 0.036^2 = 0.001018 \text{ m}^2$$

Weight exerted by press ram, $W = 7 \text{ kN}$

Stroke length of plunger, $x = 300 \text{ mm} = 0.3 \text{ m}$

Distance moved by the ram, $y = 0.9 \text{ m}$

(*i*) The force applied on plunger, *F*:

We know,
$$\frac{F}{a} = \frac{W}{A}, \text{ or, } F = \frac{a}{A} \times W = \frac{0.001018}{0.0254} \times 7 = \mathbf{0.28\ kN}\ \textbf{(Ans.)}$$

(*ii*) The number of strokes performed by the plunger, *n* :

$$\text{Number of strokes, } n = \frac{A}{a} \times \frac{y}{x} = \frac{0.0254}{0.001018} \times \frac{0.9}{0.3} = 74.85 \simeq \mathbf{75}\ \textbf{(Ans.)}$$

(*iii*) Work done by the press ram:

Work done by the press ram $= 7 \times 0.9 = \mathbf{6.3\ kNm}$ **(Ans.)**

(*iv*) Power required to drive the plunger, *P*:

Power required to drive the plunger = Work done by the ram per sec.

$$= \frac{6.3}{15 \times 60} \text{ kW} = 0.007 \text{ kW or } \mathbf{7\ W}\ \textbf{(Ans.)}$$

Example 5.14. *The ram and plunger of a hydraulic press are 250 mm and 30 mm respectively, and the leverage of handle is 12 : 1. With a plunger stroke of 250 mm, the press is able to lift 180 kN through 1. 25 m in 2 minutes.*

Determine:

(i) Force applied at the end of lever,

(ii) Number of strokes to be performed by the plunger in one second, and

(iii) Power required to drive the plunger.

Take the packing friction of the plunger as well as the ram as 5% of the load.

Solution. Diameter of ram, $D = 250 \text{ mm} = 0.25 \text{ m}$

$\therefore$ Area of ram, $A = \frac{\pi}{4} \times 0.25^2 = 0.04908 \text{ m}^2$

Diameter of plunger, $d = 30 \text{ mm} = 0.03 \text{ m}$

$\therefore$ Area of plunger, $a = \frac{\pi}{4} \times 0.03^2 = 0.0007068 \text{ m}^2$

Leverage of handle $= 12 : 1$

Plunger stroke $= 250 \text{ mm} = 0.25 \text{ m}$

Load lifted, $W = 180$ kN

Packing friction, $k = 5\%$ of the load.

(i) Force applied at the end of lever, F':

The effective force transmitted to create pressure on the liquid is reduced by the amount lost in overcoming friction. Also, pressure intensity in a static mass of fluid is same throughout.

$$\therefore \quad p \times a = F\left(1 - \frac{k}{100}\right), \text{ and, } W = p \times A\left(1 - \frac{k}{100}\right)$$

$$\text{or, Intensity of pressure, } p = \frac{F\,(1 - k/100)}{a} = \frac{W}{A\,(1 - k/100)}$$

$$\text{or, } \quad F = \frac{Wa}{A\left(1 - \frac{k}{100}\right)^2} = \frac{180 \times 0.0007068}{0.04908 \times \left(1 - \frac{5}{100}\right)^2} = 2.87 \text{ kN}$$

Effort to be applied at the end of lever,

$$F' = \frac{F}{\text{Leverage of handle}} = \frac{2.87}{12} = \mathbf{0.239 \text{ kN}} \text{ (Ans.)}$$

(ii) Number of strokes per second:

$$\text{Number of strokes} = \frac{A}{a} \times \frac{\text{Distance moved by ram } (y)}{\text{Stroke length of plunger } (x)} = \frac{0.04908}{0.0007068} \times \frac{1.25}{0.25} = 348$$

$$\therefore \text{ Strokes per second} = \frac{348}{2 \times 60} \simeq \mathbf{3} \text{ (Ans.)}$$

(iii) Power required to drive the plunger, P:

Work done at plunger $= 2.87 \times 0.25 \times 349 = 248.69$ kNm (in 2 minutes)

$$\text{Work done per second} = \frac{249.69}{2 \times 60} = 2.08 \text{ kNm/s}$$

$\therefore$ *Power required,* $P = \mathbf{2.08 \text{ kW}}$ **(Ans.)**

5.5. HYDRAULIC CRANE

Hydraulic crane *is a device which is used for lifting heavy loads (upto 25 MN)*. It is widely used in docks for loading and unloading ships, warehouses, foundary workshops and heavy industries.

Construction. A hydraulic crane (Fig. 5.7) consists of a **crane** and a hydraulic **jigger**. The crane has a central mast from which a *tie* and a *jib* are fixed. The mast is fixed on a pedestal which can revolve. The load can be lifted and moved around the crane area depending upon the length of the jib. The load is suspended by a wire rope which passes over a pulley at the end of the jib and over the tie-rod to the hydraulic jigger. The hydraulic jigger consists of a cylinder and a ram at the end of which pulleys are attached. The pulleys enable to increase the velocity ratio between the wire rope and the ram (A four sheave pulley block system will have a velocity ratio 4 : 1, thus the load suspended at one end of the wire rope will move with 4 times the speed of the ram).

Working. When the load is to be *lifted* by the crane, liquid under pressure is admitted to the cylinder of the jigger; the liquid forces the sliding ram to move vertically up. Due to the movement of the ram in the vertically upward direction, the movable pulley block (attached to the ram) also moves upward. The *distance* between the two pulley blocks *increases*, the wire or rope is pulled and the load is lifted up. *Lowering* of the load is achieved by removing the liquid from the cylinder by Jigger through the outlet valve. As the liquid leaves the cylinder, the distance between the two sets of pulleys *decreases* which results in releasing more length of the wire rope and the load gets lowered.

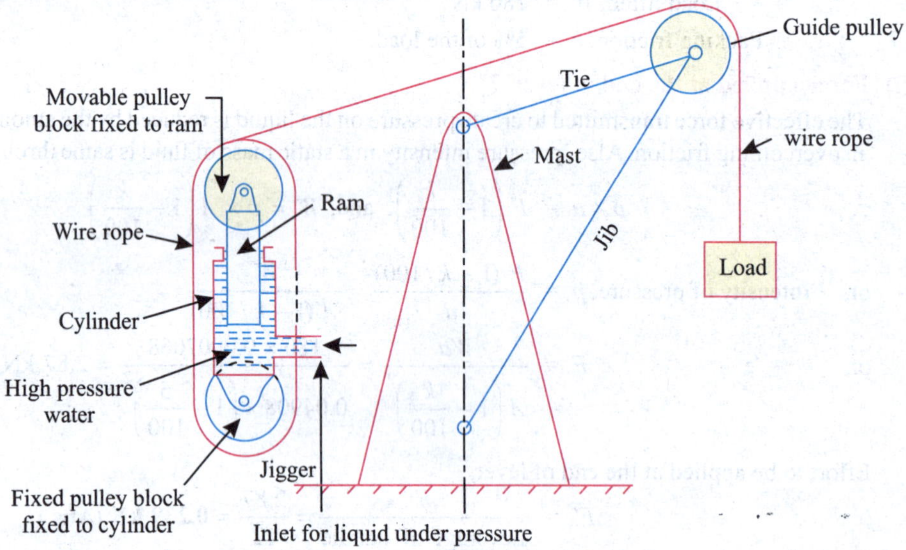

Fig. 5.7. Hydraulic crane.

The lifting speed of a modern hydraulic crane may be about 75 m per minute. However, hydraulic cranes have been replaced by the electric cranes these days.

Example 5.15. *A hydraulic crane ram has a diameter of 150 mm and the ratio between the movement of the load and the ram is 6 to 1. Water is supplied through a 40 mm diameter pipe having a length of 500 m, the pressure at the inlet end of the pipe being 7550 kN/m^2. The co-efficient of friction for the pipe is 0·01. A pressure of 440 kN/m^2 on the ram is required to overcome the mechanical losses. Determine:*

(*i*) *The maximum speed with which a load of 11 kN can be lifted, and*

(*ii*) *The load and speed of lifting which correspond to the maximum power obtained from the crane.*

Solution. Diameter of ram, D = 150 mm = 0.15 m.

The ratio between the movement of the load and the ram = 6 : 1

Diameter of pipe, d = 40 mm = 0.04 m

Length of pipe, l = 500 m

$$\text{Pressure at inlet end of pipe} = 7550 \text{ kN/m}^2$$

$$\text{The co-efficient of friction for the pipe, } f = 0.01$$

$$\text{Pressure required on the ram to overcome mechanical losses} = 440 \text{ kN/m}^2$$

$$\text{Load to be lifted, } W = 11 \text{ kN}$$

(*i*) The maximum speed with which a load of 11 kN can be lifted:

$$\text{Pressure head at inlet end of pipe} = \frac{7550}{9.81} = 769.6 \text{ m} \qquad (\because w = 9.81 \text{ kN/m}^3)$$

$$\text{Pressure head required to overcome mechanical losses} = \frac{440}{9.81} = 44.8 \text{ m}$$

$$\text{Frictional loss in pipe} = \frac{4flV^2}{d \times 2g} = \frac{4 \times 0.01 \times 500 \times V^2}{0.04 \times 2 \times 9.81} = 25.48V^2$$

(where, V = velocity of water in pipe)

$$\therefore \quad \text{Pressure head on the ram} = (769.6 - 44.8 - 25.48V^2) \text{ m}$$

$$\text{Pressure head required on the ram} = \frac{11 \times 6}{\frac{\pi}{4} \times 0.15^2 \times 9.81} = 380.7 \text{ m}$$

$$\therefore \quad 769.6 - 44.8 - 25.48V^2 = 380.7, \text{ or, } V^2 = \frac{769.6 - 44.8 - 380.7}{25.48} = 13.5$$

$$\therefore \quad V = 3.67 \text{ m/s}$$

$$\text{Velocity of pipe} \times \text{area} = \text{Velocity of ram} \times \text{ram area}$$

$$3.67 \times \frac{\pi}{4} \times 0.04^2 = \text{Velocity of ram} \times \frac{\pi}{4} \times 0.15^2$$

$$\text{or,} \quad \text{Velocity of ram} = 3.67 \times \left(\frac{0.04}{0.15}\right)^2 = 0.26 \text{ m/s}$$

$\therefore$ The maximum speed of the load = $0.26 \times 6 =$ **1.56 m/s (Ans.)**

(*ii*) The load and speed of lifting:

For maximum transmission of power, the pressure losses in the pipeline should be $\frac{1}{3}$rd of the available head. Therefore,

$$25.48\,V^2 = \frac{1}{3} \times 769.6, \text{ or, } V = \left(\frac{769.6}{3 \times 25.48}\right)^{1/2} = 3.17 \text{ m/s}$$

$$\text{Speed of lifting the head} = 3.17 \times \left(\frac{0.04}{0.15}\right)^2 \times 6 = \mathbf{1.35 \text{ m/s}} \textbf{ (Ans.)}$$

$$\text{Available head at the ram} = \left(769.6 - 44.8 - \frac{769.6}{3}\right) = 468.3 \text{ m}$$

$$= 9.81 \times 468.3 \text{ kN/m}^2 = 4594 \text{ kN/m}^2$$

$$\therefore \quad \text{The load that can be lifted} = \frac{4594}{6} \times \left(\frac{\pi}{4} \times 0.15^2\right) = \mathbf{13.53 \text{ kN}} \textbf{ (Ans.)}$$

Example 5.16. *In an installation of six hydraulic cranes, the working cycle of each of which takes 90 seconds in hoisting and lowering, each crane is fed with water at a pressure of 4900 kN/m^2 and is required to lift a load of 50 kN at a speed of 18 m/min, through a total height of 12 m, the jigger system giving a velocity ratio of 6. Estimate the diameter and stroke of the rams assuming an efficiency of 60 percent.*

It is assumed that all six cranes are making the working stroke at the same time. Calculate the minimum capacity of the pump feeding the installation and that of the accumulator.

[Anna University]

Solution. Number of hydraulic cranes = 6

Time taken by working cycle of each of the six cranes, t = 90 s

Pressure of water fed to each crane, p = 4900 kN/m^2

Load to be lifted, W = 50 kN

Speed with which load is lifted, V = 18 m/min

Total height through which load is lifted, H = 12 m

Velocity ratio = 6

Efficiency of each ram, η = 60%

Diameter (D) and stroke of each ram (L):

Load on each ram × velocity of ram × η = Load lifted × velocity of lifting the load

$$\therefore \quad \text{Load on each ram} = \frac{W \times \text{velocity ratio}}{\eta} = \frac{50 \times 6}{0.6} = 500 \text{ kN}$$

$$\text{Area of arm} = \frac{\text{Load on ram}}{\text{Pressure } (p)} = \frac{500}{4900} = 0.102 \text{ m}^2$$

$$\therefore \quad \frac{\pi}{4} \times D^2 = 0.102, \text{ or, } D = \left(\frac{0.102 \times 4}{\pi}\right)^{1/2} = 0.36 \text{ m} = \mathbf{360 \text{ mm}} \text{ (Ans.)}$$

$$\text{Stroke of ram, } L = \frac{\text{Distance } (H)}{\text{Velocity ratio}} = \frac{12}{6} = \mathbf{2 \text{ m}} \text{ (Ans.)}$$

Minimum capacity of the pump:

$$\text{Volume swept by ram} = \text{Area of ram} \times \text{stroke}$$
$$= 0.102 \times 2 = 0.204 \text{ m}^3$$

Since there are 6 cranes, volume of water supplied by the pump = 0.204 × 6 = 1.224 m^3

$$\textit{Minimum capacity of the pump} = \frac{\text{Volume of water}}{\text{Time taken}} = \frac{1.224}{90} = \mathbf{0.0136 \text{ m}^3/\text{s}} \text{ (Ans.)}$$

Capacity of accumulator:

$$\text{Operating time of cranes} = \frac{H}{V} = \frac{12}{18} \times 60 = 40 \text{ s}$$

$$\therefore \quad \text{Idle time of crane} = 90 - 40 = 50 \text{ s}$$

But the accumulator is continuously fed by the pumps,

∴ Accumulator volume

$$= 0.0136 \times 50 = \mathbf{0.68 \text{ m}^3} \text{ (Ans.)}$$

$$\text{Accumulator capacity} = p \times \text{volume}$$
$$= 4900 \times 0.68 = \mathbf{3332 \text{ kNm}} \text{ (Ans.)}$$

5.6. HYDRAULIC LIFT

Hydraulic lift *is a device used for carrying persons and loads from one floor to another, in a multi-storeyed building*. The hydraulic lifts are of the following two types:

1. Direct acting hydraulic lift, and
2. Suspended hydraulic lift.

1. Direct acting hydraulic lift. Refer to Fig. 5.8.

It consists of a ram sliding in a cylinder. A platform or a cage is fitted to the top end of ram on which goods may be placed or the persons may stand. As the liquid under pressure is admitted to the cylinder, the ram moves up and the cage is lifted. The lift of the cage is equal to the stroke of the ram. The cage moves in the downward direction when the liquid from the fixed cylinder is removed.

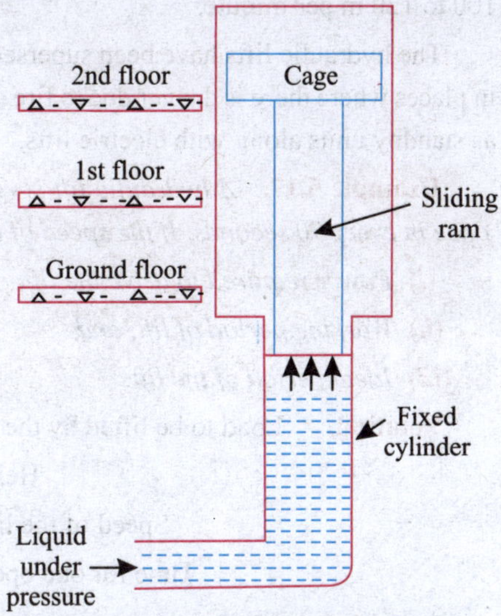

Fig. 5. 8 Direct acting hydraulic lift.

2. Suspended hydraulic lift. Refer to Fig. 5.9.

The suspended hydraulic lift is a modified form of the direct acting hydraulic lift. It is fitted with a *jigger* which is exactly same as in the case of a hydraulic crane (For the construction and operation of jigger, refer to Art. 5.5). The cage is suspended by ropes. It runs between guides of hard wood or round steel. In order to balance the weight of the cage, sliding balance weights are provided.

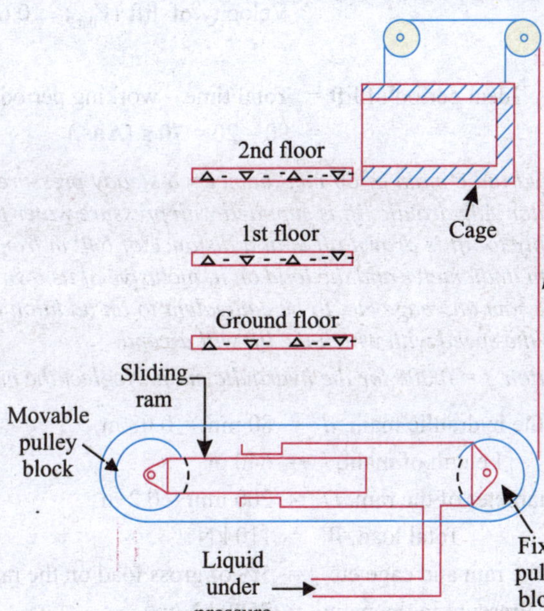

Fig. 5.9. Suspended hydraulic lift.

Modern lifts are of suspended type and have a *high velocity ratio;* the lifting speeds range from 100 to 120 m per minute.

The hydraulic lifts have been superseded by electric lifts. However hydraulic lifts are preferred in places where there is danger due to fire or explosions. The hydraulic lifts are also usually provided as standby units along with electric lifts.

Example 5.17. *A hydraulic lift is required to lift a load of 60 kN through a height of 12 m, once in every 90 seconds. If the speed of the lift is 0.6 m/s, determine:*

(i) Power required to drive the lift,

(ii) Working period of lift, and

(iii) Ideal period of the lift.

Solution. Load to be lifted by the lift, $W = 60$ kN

Height, $H = 12$ m

Speed of the lift, $V_{lift} = 0.6$ m/s

Time for one operation $= 90$ s.

(i) Power required to drive the lift, *P*:

Work done in lifting the load in 90 s $= W \times H = 60 \times 12 = 720$ kNm

$\therefore$ Work done/sec. $= \dfrac{720}{90} = 8$ kNm/s

$\therefore$ Power required to drive the lift, $P =$ **8 kW (Ans.)**

(ii) Working period of lift:

$$\textit{Working period of lift} = \frac{\text{Height of lift } (H)}{\text{Velocity of lift } (V_{\text{lift}})} = \frac{12}{0.6} = \mathbf{20\ s.}$$

(iii) Ideal period of lift:

Ideal period of lift = Total time – working period of lift

$= 90 - 20 =$ **70 s (Ans.)**

Example 5.18. *In a hydraulic main of 80 mm diameter a steady pressure of 7500 kN/m^2 is maintained by an accumulator. A hydraulic lift is supplied with pressure water from the main, and the point at which the supply to lift is drawn off is at a distance of 640 m from the accumulator. The ram of the lift is 200 mm in diameter and the load on it, inclusive of its own weight, is 110 kN. Assuming the friction of the ram and cage etc. to be equivalent to an addition of 5% of the gross load on the ram, determine the speed with which the lift will ascend.*

Take co-efficient of friction, f = 0.008 for the hydraulic main. Neglect the minor losses.

Solution. Diameter of the hydraulic main, $d = 80$ mm $= 0.08$ m

Length of main, $l = 640$ m

Diameter of the ram, $D = 200$ mm $= 0.2$ m

Total load, $W = 110$ kN

Friction of the ram and cage etc. = 5% of gross load on the ram

Pressure in the main = 7500 kN/m^2

Co-efficient of friction for the main, $f = 0.008$.

Speed with which the lift will ascend, V_{lift}:

Let, V = Velocity of water in hydraulic main.

$$\text{Loss of head due to friction, } h_f = \frac{4flV^2}{d \times 2g} = \frac{4 \times 0.008 \times 640 \times V^2}{0.08 \times 2 \times 9.81} = 13.05V^2$$

$$\text{Corresponding pressure intensity, } p = wh_f = 9.81 \times 13.05V^2 = 128V^2 \text{ kN/m}^2$$

$$(\because w = 9{\cdot}81 \text{ kN/m}^3)$$

$$\therefore \quad \text{Pressure intensity at the lift} = (7500 - 128V^2) \text{ kN/m}^2$$

$$\text{Force on the ram of hydraulic lift} = (7500 - 128V^2) \times \frac{\pi}{4} \times 0.2^2 \text{ kN}$$

This force equals the load to be lifted during the upward motion of the lift which is

$$= \left(110 + \frac{5}{100} \times 110\right) = 115.5 \text{ kN}$$

$$\therefore \quad (7500 - 128V^2) \times \frac{\pi}{4} \times 0.2^2 = 115.5, \text{ or, } 7500 - 128V^2 = \frac{115.5 \times 4}{\pi \times 0.02^2} = 3676.48$$

$$\text{or} \quad V = \left(\frac{7500 - 3676.48}{128}\right)^{1/2} = 5.46 \text{ m/s}$$

$$\text{Discharge through the pipeline} = \frac{\pi}{4} \times 0.08^2 \times 5.46 = 0.02744 \text{ m}^3/\text{s}$$

Since the same discharge flows to the ram of hydraulic lift, therefore, the velocity of lift,

$$V_{lift} = \frac{0.02744}{\frac{\pi}{4} \times D^2} = \frac{0.02744}{\frac{\pi}{4} \times 0.2^2} = \mathbf{0.873 \text{ m/s}} \textbf{ (Ans.)}$$

Example 5.19. *A load of 120 kN is required to be lifted by a hydraulic lift through a height of 16 metres once in every 2 minutes. The lift travels up at the rate of 1.25 m /s. During working stroke of the lift, the water is supplied to it from the accumulator and the pump at a pressure intensity of 3200 kN/m². If the efficiency of the pump is 82% and that of the lift is 77% determine:*

(i) Power required to the drive the pump, and

(ii) Minimum capacity of the accumulator.

Neglect friction losses in the pipe.

Solution. Load to be lifted, W = 120 kN

Height, H = 16 m

Speed of weight, V = 1.25 m/s

Pressure intensity of water, p = 3200 kN/m²

Efficiency of the pump, η_{pump} = 82%

Efficiency of the lift, η_{lift} = 77%.

(i) Power required to drive the pump, P_{pump}:

Work done by water in raising the lift = Load lifted × distance moved per sec.

$$= W \times V = 120 \times 1.25 = 150 \text{ kNm/s}$$

Power used at the lift = 150 kW

$$\text{Power supplied to the lift} = \frac{150}{0.77} = 194.8 \text{ kW}$$

This power has been supplied to the lift by water supplied from the pump and accumulator.

$$\text{Working period of the lift} = \frac{H}{V} = \frac{16}{1.25} = 12.8 \text{ s}$$

$$\text{Ideal period of lift} = \text{Total time} - \text{working period of lift}$$
$$= (2 \times 60) - 12.8 = 107.2 \text{ s}$$

During idle period the energy will be stored in the accumulator and during the working period the energy will be supplied to the lift from the accumulator.

Let P_{pump} be the *output of pump in kW*, then,

Energy stored in accumulator during the idle period

$$= P'_{pump} \times 107.2 \text{ kJ} \quad \text{...}(i)$$

The above energy is supplied to the lift during the working period of 12·8 s.

$$\text{Energy supplied by the accumulator per sec.} = \frac{P'_{pump} \times 107.2}{12.8} = 8.37\ P'_{pump} \text{ kJ/s}$$

Power supplied by the accumulator = $8.37\ P'_{pump}$ kW

∴ Total power supplied by the pump and accumulator

$$= P'_{pump} + 8.37\ P'_{pump} = 9.37\ P'_{pump}$$

But power supplied to the lift = 194.8 kW

$$\therefore \quad 9.37\ P'_{pump} = 194.8, \text{ or, } P'_{pump} = \frac{194.8}{9.37} = 20.79 \text{ kW}$$

$$\therefore \text{ Power required to drive the pump, } P_{pump} = \frac{P'_{pump} \text{ (output of pump)}}{\eta'_{pump} \text{ (efficiency of pump)}}$$

$$= \frac{20.79}{0.82} = \mathbf{25.35 \text{ kW}} \textbf{ (Ans.)}$$

***(ii)* Minimum capacity of accumulator:**

From expression (*i*), the energy stored in accumulator during the idle period

$$= P'_{pump} \times 107.2 = 20.79 \times 107.2 = 2228.69 \text{ kNm}$$

Also, the energy stored in the accumulator = p × volume of cylinder of the accumulator

∴ p × volume of cylinder of the accumulator = 2228.69

$$\therefore \text{ Volume (capacity) of the accumulator} = \frac{2228.69}{3200} = \mathbf{0.6965 \text{ m}^3} \textbf{ (Ans.)}$$

5.7. HYDRAULIC RAM

Hydraulic ram *is device with which small quantities of water can be pumped to higher levels from the available large quantity of water of low head.* It works on the *principle of water hammer.*

Construction. Fig. 5.10 shows the hydraulic ram. It consists of a valve box wherein low head water flows. The box contains a waste valve V_1 which opens inwards, and a delivery valve V_2, which opens outwards. Both the valves V_1 and V_2 are *non-return valves* that allow the flow only in one direction. Valve V_2 communicates with an air vessel which is connected to the delivery tank through a delivery pipe as shown in the Fig. 5.10.

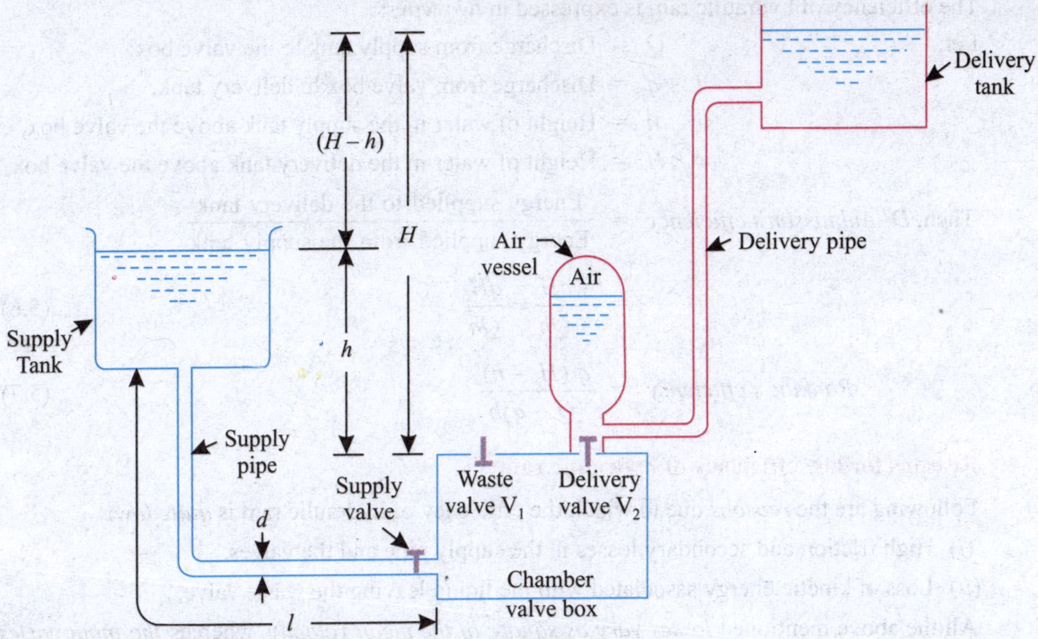

Fig. 5.10. Hydraulic ram.

Working. It works on the principle of '*water hammer*'. *When a flowing liquid is suddenly brought to rest, the change in momentum of liquid mass causes a sudden rise in pressure. This rise in pressure is utilised to raise a portion of the liquid to higher levels.*

— Initially the water flows down the supply pipe, the valve V_1 is open and water escapes through it to waste.

— As the velocity of flow in the supply pipe increases, dynamic pressure on the underside of valve V_1 becomes amply high so as to lift the valve V_1 and ultimately close it.

— Due to quick closure of valve V_1 water in the supply pipe is suddenly brought to rest and consequently pressure in the valve box increases.

— The increased high pressure lifts the valve V_2 and a part of water enters the air vessel. Subsequently air pressure inside the air vessel increases and that forces water to delivery tank through the delivery pipe.

As soon as momentum of water gets destroyed in the valve box, valve V_2 closes and the valve V_1 opens (the pressure in the valve box falls below the atmospheric pressure momentarily); the flow of water from supply tank recommences and the cycle is repeated. *The air vessel provides storage and helps to regulate the flow at the delivery end.*

Efficiency. The *efficiency* of a hydraulic ram depends upon the following:

(*i*) Losses in the pipe, (*ii*) losses in the valve box, and (*iii*) ratios $\frac{l}{h}$ and $\frac{h}{H}$.

From experiments it has been observed that, for *optimum performance*, values of $\frac{l}{H}$ and $\frac{h}{H}$ should be about 2.5 and 5 respectively. It is possible to obtain efficiencies as high as 75% *under these conditions*.

The performance of the hydraulic ram is influenced by the number of beats (openings of waste valve) which can be varied by varying the length of the travel of valve and weight of valve.

The efficiency of hydraulic ram is expressed in *two ways*:

Let,
Q = Discharge from supply tank to the valve box,
q = Discharge from valve box to delivery tank,
h = Height of water in the supply tank above the valve box,
H = Height of water in the delivery tank above the valve box,

Then, *D' Aubuisson's efficiency* $= \dfrac{\text{Energy supplied to the delivery tank}}{\text{Energy supplied from the supply tank}}$

$$= \frac{wqH}{wQh} = \frac{qH}{Qh} \quad \text{...(5.6)}$$

$$\textit{Rankine's efficiency} = \frac{q\,(H-h)}{(Q-q)h} \quad \text{...(5·7)}$$

Reasons for low efficiency of hydraulic ram:

Following are the *reasons* due to which the efficiency of hydraulic ram is *quite low*:

(*i*) High friction and secondary losses in the supply pipe and the valves.

(*ii*) Loss of kinetic energy associated with the liquid leaving the waste valve.

All the above mentioned *losses vary as square of the mean velocity*, whereas *the input varies directly as the mean velocity*.

The efficiency of the ram can be improved by reducing the mean velocity. It can be accomplished by *reducing the lift of the waste valve* (this limits the maximum velocity in the supply pipe and hence the mean velocity of flow):

Working cycle of a hydraulic ram:

Let,
V_{wv} = Velocity of liquid passing through waste valve just before its closure, (*wv* stands for waste valve)
d_{wv} = Diameter of waste valve,
b_{wv} = Lift of the waste valve,
W_{wv} = Weight of the waste valve,
h_{wv} = Dynamic pressure head acting on the waste valve,
d = Diameter of supply pipe,
V_{max} = Maximum velocity in the supply pipe just before the closure of the waste valve
l_s = Length of supply pipe,
t_1 = Time during which the velocity in the supply pipe builds up from zero to V_{max}, and
t_2 = Time during which the waste valve remains closed, *i.e.* the time during which delivery valve remains open in one beat (*i.e.* one complete cycle).

The dynamic pressure head, $h_{wv} = \dfrac{p_{wv}}{w} = \dfrac{V_{wv}^2}{2g}$ ...(5.8)

Also, $W_{wv} = \left(\dfrac{\pi}{4} \times d_{wv}^2\right) p_{wv}$, or, $h_{wv} = \dfrac{p_{wv}}{w} = \dfrac{4W_{wv}}{w\pi d_{wv}^2}$ ...(5.9)

Invoking continuity equation, we have:

$$(\pi \times d_{wv} \times b_{wv})V_{wv} = \frac{\pi}{4} \times d^2 \times V_{max} \qquad ...(5.10)$$

The time t_1 during which the velocity in the supply pipe builds up from zero to V_{max} is also equal to the time during which the waste valve remains open in one *beat* for gradual closure of valve in rigid pipe, the rise in pressure is given by,

$$h = \frac{l_s V_{max}}{g\, t_1}, \text{ or, } t_1 = \frac{l_s V_{max}}{hg} \qquad ...(5.11)$$

From water hammer equation, we have:

$$(H - h) = \frac{l_s \times V_{max}}{g \times t_2}, \text{ or, } t_2 = \frac{l_s V_{max}}{(H-h)\, g} \qquad ...(5.12)$$

$\therefore$ Total time for one cycle is given by,

$$t = t_1 + t_2 = \frac{l_s V_{max}}{h\, g} + \frac{l_s V_{max}}{(H-h)\, g} = \frac{l_s V_{max}}{g}\left(\frac{1}{h} + \frac{1}{H-h}\right) \qquad ...(5.13)$$

The number of beats per minute N, is equal to $(60/t)$ where, t is the time for one beat in seconds.

Let, q = Rate of discharge of water actually lifted by the ram, and

Q_{wv} = Rate of discharge of water flowing past the waste valve, then,

$$q = \left(\frac{\pi}{4} d^2\right)\left(\frac{V_{max}}{2}\right)\frac{t_2}{t} \qquad ...(5.14)$$

$$Q_{wv} = \left(\frac{\pi}{4} d^2\right)\left(\frac{V_{max}}{2}\right)\frac{t_1}{t} \qquad ...(5.15)$$

Fig. 5.11 shows the working cycle of hydraulic ram graphically. In Fig. 5.11 (*a*) is shown the velocity in supply pipe plotted against time, whereas Fig. 5.11 (*b*) shows the pressure head in the valve box plotted against time.

Eqns. 5.11 and 5.12 have been derived on the assumption that the loss of head due to friction is negligible. However, if loss of head due to friction in supply pipe (h_{fs}) and that in delivery pipe (h_{fd}) are taken into account, then eqns. 5.11 and 5.12 are modified as follows:

$$(h - h_{fs}) = \frac{l_s}{g} \times \frac{V_{max}}{t_1} \qquad ...(5.16)$$

$$[(H - f) + h_{fd}] = \frac{l_s}{g} \times \frac{V_{max}}{t_2} \qquad ....(5.17)$$

Eqn. 5.13, then, becomes:

$$t = t_1 + t_2$$

$$= \frac{l_s V_{max}}{g}\left[\frac{1}{(h - h_{fs})} + \frac{1}{\{(H-h) + h_{fd}\}}\right] \qquad ...(5.18)$$

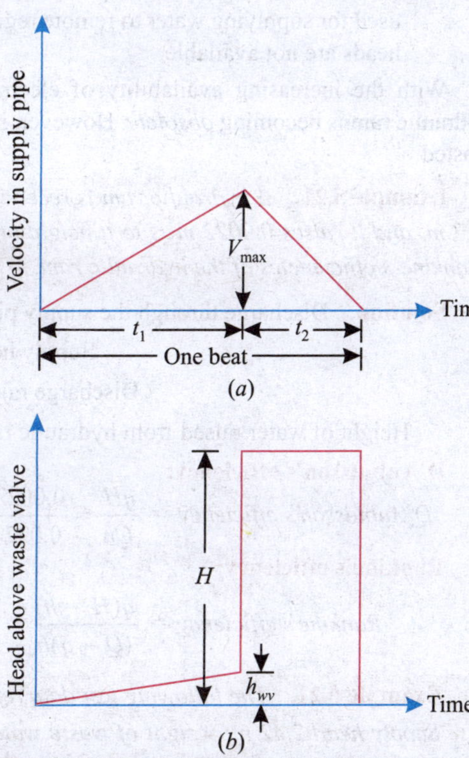

Fig. 5.11. Graphical representation of a working cycle of hydraulic ram.

Performance of hydraulic ram:

In Fig. 5.12 are shown the characteristic curves of hydraulic ram working under conditions of *constant waste-valve lift, constant supply head and varying delivery head.*

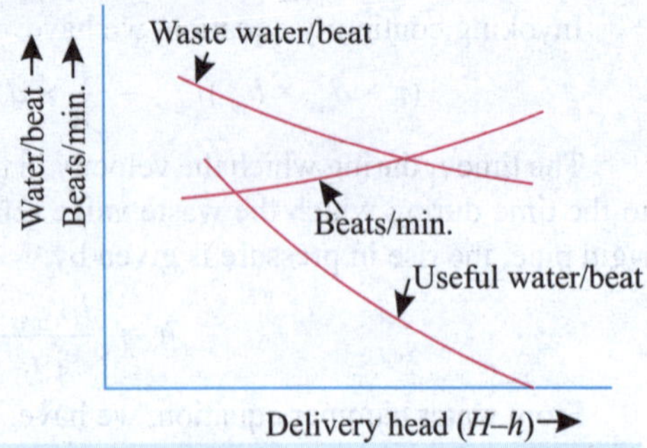

Fig. 5.12. Characteristic curves of hydraulic ram.

— With the *increase of delivery head the number of beats increases*. This is owing to the fact that as the delivery head is increased, more rapid retardation is impressed on the liquid column in the supply pipe, as a consequences of which the time taken per beat decreases and hence the number of beats per minute increases.

— Keeping the input unaltered, as *the head increases the quantity of useful water per beat decreases. With increase in head the waste water per beat decreases slightly.*

Characteristic features of hydraulic ram:

The characteristic features of hydraulic ram are as follows:

1. It works automatically and requires very little maintenance.
2. It does not need any external energy to pump water, but it works at the cost of large quantity of water.
3. Negligible running cost.
4. Due to absence of moving parts, frequent oiling is not required.
5. It is particularly suitable for pumping water from a rivulet for irrigation purposes. It can be used for supplying water to remote regions where other means of pumping water to higher heads are not available.

With the increasing availability of electric power to drive pumps even at remote places, hydraulic ram is becoming *obsolete*. However, it is used in those place where plenty of water can be wasted.

Example 5.21. *A hydraulic ram is receiving water at the rate of 0.022 m³/s from a height of 3.3 m, and it raises 0.0022 m³/s to a height of 5 m from the ram. Determine D'Aubuisson's and Rankine's efficiencies of the hydraulic ram.*

Solution. Discharge through the supply pipe, $Q = 0.022 \text{ m}^3\text{/s}$

Supply head, $h = 3.3$ m

Discharge raised, $q = 0.0022 \text{ m}^3\text{/s}$

Height of water raised from hydraulic ram, $H = 5$ m

D'Aubuisson's efficiency:

$$\text{D'Aubuisson's efficiency} = \frac{qH}{Qh} = \frac{0.0022 \times 21}{0.022 \times 3.3} = 0.6364 \text{ or } \mathbf{63.64\%} \textbf{ (Ans.)}$$

Rankine's efficiency:

$$\text{Rankine's efficiency} = \frac{q(H - h)}{(Q - q)h} = \frac{0.0022\,(21 - 3.3)}{(0.022 - 0.0022) \times 3.3} = \mathbf{0.596} \text{ or } \mathbf{59.6\%} \textbf{ (Ans.)}$$

Example 5.21. *The following test data refer to the hydraulic ram:*

Supply head 2.42 m; weight of waste water per minute 180 N; weight of water pumped per minute 5.4 N; and net head from the ram 44 m. Calculate D'Aubuisson's and Rankine's efficiencies.

Solution. Supply head, $h = 2.42$ m

Weight of waste water per minute $= 180$ N

Weight of water pumped per minute, $W' = 5.4$ N

$\therefore$ Total weight of water flowing per minute into the valve box,

$$W = 180 + 5.4 = 185.4 \text{ N}$$

Net head pumped from ram, $H = 44$ m

D'Aubuisson's efficiency:

$$\text{D'Aubuisson's efficiency} = \frac{qH}{Qh} = \frac{wqH}{wQh} = \frac{W'H}{Wh} = \frac{5.4 \times 44}{185.4 \times 2.42} = 0.529 \text{ or } \mathbf{52.9\%} \text{ (Ans.)}$$

[where, W' (= wq) = weight of water pumped per minute, w being weight density of water.]

$$\textit{Rankine's efficiency} = \frac{q(H-h)}{(Q-q)h} = \frac{wq\,(H-h)}{w\,(Q-q)h} = \frac{W'(H-h)}{(W-W')h}$$

$$= \frac{5.4\,(44-2.42)}{(185.4-5.4) \times 2.42} = 0.515 \text{ or } \mathbf{51.5\%} \text{ (Ans.)}$$

Example 5.22. *A hydraulic ram is being supplied water at the rate of 0.05 m³/s from a height of 5 m, and it raises 0.005 m³/s to a height of 35 m from the ram. The length and diameter of the pipe are 120 m and 70 mm respectively. If the co-efficient of friction is 0.009, calculate D'Aubuisson's and Rankine's efficiencies.*

Solution. Discharge through the supply pipe, $Q = 0.05 \text{ m}^3/\text{s}$

Supply head, $h = 5$ m

Discharge raised, $q = 0.005 \text{ m}^3/\text{s}$

Height of water raised from hydraulic ram, $H = 35$ m

Length of the pipe, $l = 120$ m

Diameter of the pipe, $d = 75 \text{ mm} = 0.075$ m

Co-efficient of friction, $f = 0.009$.

Efficiency of the ram:

Head lost due to friction in the delivery pipe,

$$h_f = \frac{4flV^2}{d \times 2g} = \frac{4 \times 0.009 \times 120 \times V^2}{0.075 \times 2 \times 9.81}$$

But, V = Velocity of water in delivery pipe $= \dfrac{q}{\frac{\pi}{4}d^2} = \dfrac{0.005}{\frac{\pi}{4} \times 0.075^2} = 1.13$ m/s

$$\therefore \quad h_f = \frac{4 \times 0.009 \times 120 \times 1.13^2}{0.075 \times 2 \times 9.81} = 3.75 \text{ m}$$

$\therefore$ Effective head developed by the ram, $H_e = H + h_f = 35 + 3.75 = 38.75$ m

$$\textit{D'Aubuisson's efficiency} = \frac{q \times H_e}{Q \times h} = \frac{0.005 \times 38.75}{0.05 \times 5} = 0.775 \text{ or } \mathbf{77.5\%} \text{ (Ans.)}$$

$$\textit{Rankine's efficiency} = \frac{q\,(H_e - h)}{(Q-q)\,h} = \frac{0.005\,(38.75-5)}{(0.05-0.005) \times 5} = 0.75 \text{ or } \mathbf{75\%} \text{ (Ans.)}$$

Example 5.23. *Water is supplied to hydraulic ram from a height of 1.5 m by a pipe 75 mm in diameter and 12 m long. The waste valve, which is 125 mm in diameter and of weight 15 N, lifts through 6.5 mm. Determine:*

(i) The number of beats per minute, and

(ii) The quantity of water delivered per minute to a tank 10 m above the waste valve.

Solution. Supply head, $h = 1.5$ m

Diameter of the supply pipe, $d_s = 75$ mm $= 0.075$ m

Length of the supply pipe, $l_s = 12$ m

Diameter of waste valve, $d_{wv} = 125$ mm $= 0.125$ m

Weight of the waste valve, $W_{wv} = 15$ N

Lift of the waste valve, $b_{wv} = 6.5$ mm $= 0.0065$ m

Delivery head above the waste valve, $H = 10$ m

(i) Number of beats per minute:

The dynamic pressure head, $h_{wv} = \dfrac{4W_{wv}}{w\pi d_{wv}^2}$ ...[Eqn. (5.9)]

$$= \frac{4 \times 15}{9810 \times \pi \times 0.125^2} = 0.1246 \text{ m of water}$$

Maximum velocity past waste valve just before closure,

$$V_{wv} = \sqrt{2gh_{wv}} = \sqrt{2 \times 9.81 \times 0.1246} = 1.56 \text{ m/s}$$

If V_{max} is the maximum velocity in the supply pipe, then from continuity consideration, we have:

$$(\pi \times d_{wv} \times b_{wv})\, V_{wv} = \left(\frac{\pi}{4} d_s^2\right) V_{max} \quad \text{...[Eqn. (5.10)]}$$

or, $$(\pi \times 0.125 \times 0.0065) \times 1.56 = \left(\frac{\pi}{4} \times 1.075^2\right) \times V_{max}$$

or, $$V_{max} = \frac{(\pi \times 0.125 \times 0.0065) \times 1.56}{\left(\dfrac{\pi}{4} \times 0.075^2\right)} = 0.9 \text{ m/s}$$

Let, t_1 = The time during which the velocity in the supply pipe builds up from zero to V_{max}, or, the time during which the waste valve remains *open* in one beat, and

t_2 = The time during which the waste valve remains *closed* or the time during which the delivery valve remains open in one beat and $(H - h)$ is the level of water in the delivery tank above that in the supply tank.

Then, $$t_1 = \frac{l_s \times V_{max}}{h \times g} = \frac{12 \times 0.9}{1.5 \times 9.81} = 0.734 \text{ s} \quad \text{...[Eqn (5.11)]}$$

and, $$t_2 = \frac{l_s \times V_{max}}{(H - h)\, g} = \frac{12 \times 0.9}{(10 - 1.5) \times 9.81} = 0.129 \text{ s}$$

∴ Total time for one cycle (one beat),

$$t = t_1 + t_2 = 0{\cdot}734 + 0{\cdot}129 = 0{\cdot}863 \text{ s}$$

Number of beats per minute $= \dfrac{60}{0.863} \simeq$ **69 (Ans.)**

(*ii*) **The quantity of water delivered per second to delivery tank:**

Discharge delivered to tank per second,

$$q = \left(\frac{\pi}{4} d_s^2\right) \times \frac{V_{max}}{2} \times \frac{t_2}{t} = \left(\frac{\pi}{4} \times 0.075^2\right) \times \frac{0.9}{2} \times \frac{0.129}{0.863}$$

$$= 0.000297 \text{ m}^3/\text{s}$$

$$= 0.000297 \times 60 \text{ m}^3/\text{min}$$

$$= \mathbf{0.01782 \text{ m}^3/\text{min.}} \text{ (Ans.)}$$

5.8. HYDRAULIC COUPLING

Hydraulic (or fluid) coupling *is a device which is employed for transmission of power from one shaft to another through a liquid medium.* It has no mechanical connection or face to face contact. *The magnitudes of input and output torques are equal.*

Construction. Refer to Fig. 5.13. It consists of the following two *rotating elements:*

(*i*) *Pump impeller*. It is attached to a driving shaft of the prime mover which may be an I. C. engine, a steam engine or an electric motor.

(*ii*) *Turbine runner*. It is attached to a driven shaft.

Both the above units are enclosed in a single housing filled with a liquid, usually oil, because of its lubricating power, availability and stability. This oil serves to transmit torque from the pump impeller to the turbine runner. There is no direct contact between the driving and driven parts.

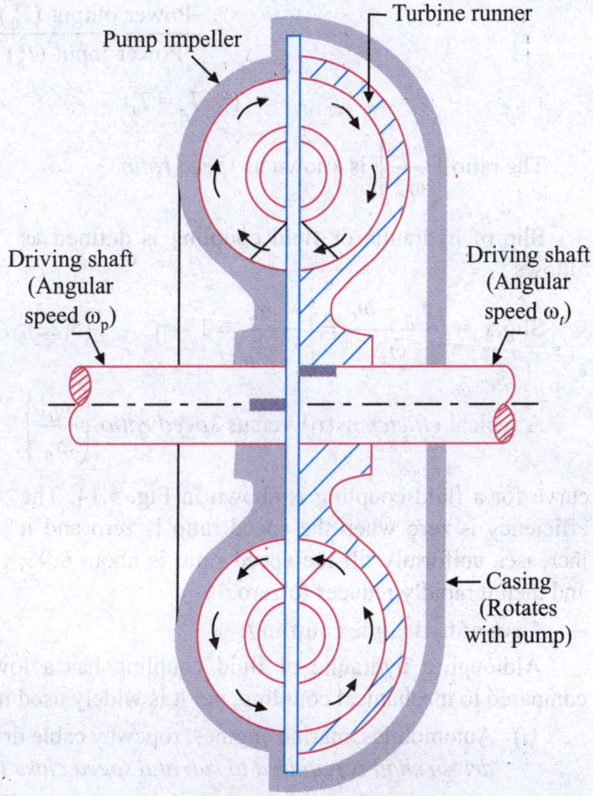

Fig. 5.13. Hydraulic coupling.

Working. As soon as the prime mover starts rotating, the pump impeller also starts rotating and throws the oil outward by centrifugal action. The oil then enters the turbine runner and exerts a force on the runner blades. The magnitude of the torque increases with an increase in the speed of the driving shaft and eventually when this torque *overcomes the inertia effects*, the turbine runner and the driven shaft begin to rotate. The oil from the runner then flows back into the pump impeller, thus a complete hydraulic (oil) circuit is established.

Expression for efficiency:

In a hydraulic coupling the power is transmitted *hydraulically* from the driving shaft to driven shaft and is free from engine vibrations. The efficiency of power transmission may be as high as 98 per cent.

Let, T_p = Torque on the pump or driving shaft '1' (*i.e.* input torque),

ω_p = Angular speed of the pump or the driving shaft '1',

T_t = Torque on the turbine or driven shaft '2' (*i.e.* output torque), and

ω_t = Angular speed of the turbine shaft or driven shaft '2'.

$$\therefore \quad \text{Power input, } P_i = T_p\,\omega_p, \qquad ...(5.19)$$

$$\text{Power output, } P_o = T_t\,\omega_t \qquad ...(5.20)$$

Since the tangential momentum due to the velocity of whirl suffers no change as the oil passes from impeller blades to the runner blades, therefore, the torque on the driven shaft (T_t) equals that on the driving shaft (T_p), *i.e.* $T_p = T_t$.

The efficiency (η) of the hydraulic coupling is defined as the *ratio of power output to power input, i.e.*

$$= \frac{\text{Power output } (P_0)}{\text{Power input } (P_i)} = \frac{T_t\,\omega_t}{T_p\,\omega_p} = \frac{\omega_t}{\omega_p} \qquad ...(5.21)$$

$$(\because T_p = T_t)$$

The ratio $\left(\frac{\omega_t}{\omega_p}\right)$ is known as *speed ratio.*

Slip of hydraulic or fluid coupling is defined as follows :

$$\text{Slip, } s = \frac{\omega_p - \omega_t}{\omega_p} = 1 - \frac{\omega_t}{\omega_p} = 1 - \eta \qquad ...(5.22)$$

A typical *efficiency* (η) versus *speed ratio* $\left(\frac{\omega_t}{\omega_p}\right)$ curve for a fluid coupling is shown in Fig. 5.14. The efficiency is zero when the speed ratio is zero and it increases uniformly till the speed ratio is about 0.95, and then it rapidly reduces to zero.

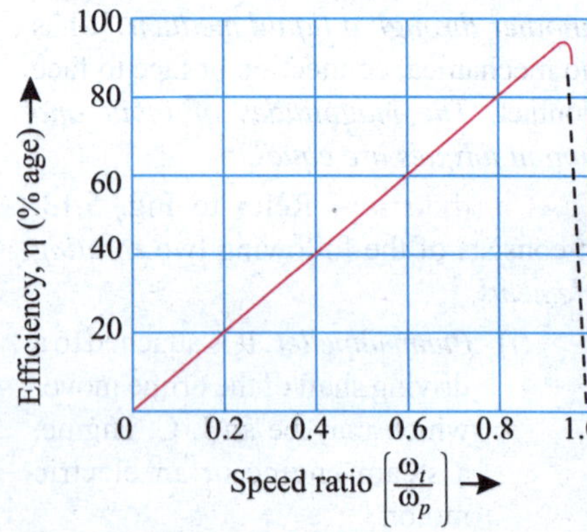

Fig. 5.14. Efficiency-speed ratio curve for hydraulic coupling.

Uses of hydraulic coupling:

Although a hydraulic or fluid coupling has a low value of transmission efficiency when compared to mechanical coupling, yet it is widely used in the following fields:

(*i*) Automobiles, marine engines, ropeway cable drive units and such other applications *where driven shaft is required to run at a speed close to that of the driving shaft.*

(*ii*) These couplings are particularly useful where *large initial loads are involved and smooth shock-free operations are required.*

5.9. HYDRAULIC TORQUE CONVERTER

Hydraulic torque converter *is a device used for transmitting increased or decreased power from one shaft to another.* A variable torque is impressed on the driven member without the use of a gear train or clutch. The torque at the driven shaft may be increased about five times the torque available at the driving shaft with an efficiency of about 90 percent.

Construction and working:

Refer to Fig. 5.15. The hydraulic torque converter consists of the following:

(*i*) Pump impeller coupled to the driving shaft,

(*ii*) Turbine runner coupled to the driven shaft, and

(*iii*) Stationary/fixed guide vanes (also known as reaction members) provided between the impeller and the turbine runner.

The liquid flowing from the pump impeller to turbine runner exerts a torque on the stationary guide vanes which change the direction of liquid, thereby making possible the transformation of torque and speed. Thus by suitably designing the stationary guide vanes the torque transmitted to the driving unit can be either increased or decreased. The torque relationship is given as :

$$T_t = T_p + T_v \qquad ...(5.23)$$

where, T_t = Torque transmitted to the *turbine shaft*,

T_p = Torque of *pump impeller*, and

T_v = Variation of torque caused by *fixed guide vanes*.

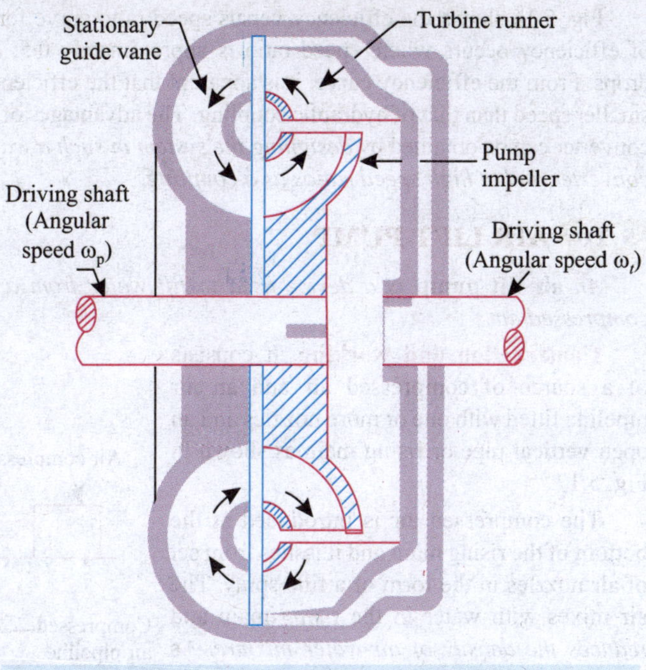

Fig. 5.15. Hydraulic torque converter.

$$\text{Power input, } P_i = T_p\,\omega_p$$

$$\text{Power output, } P_o = T_t\,\omega_t = (T_p + T_v)\,\omega t$$

$$\therefore \text{ Efficiency of torque converter, } \eta = \frac{\text{Power output } (P_o)}{\text{Power input } (P_i)} = \frac{(T_p + T_v)\,\omega_t}{T_p\omega_p}$$

$$= \frac{\omega_t}{\omega_p}\left(1 + \frac{T_v}{T_p}\right) \qquad ...(5.24)$$

From equation (5.22) we find that, when there are no guide vanes, torque converter reduces to flange coupling with $T_v = 0$ and then $\eta = \frac{\omega_t}{\omega_p} = 1 - s$.

The torque T_v depends upon the design of the stationary guide vanes; in eqn. (5.23) it can be +ve or –ve. If the guide vanes are designed to receive a torque from the fluid which is in *opposite direction* to that exerted on the turbine (or driven) shaft, an *increased output* results. On the other hand if the shape of the guide vanes is such that they receive the torque in the *same* sense as that of driven shaft, a *torque reduction* results. Thus, since the torque in case of a hydraulic torque converter can be increased or decreased (in case of the hydraulic coupling $T_p = T_t$), therefore, it is comparable to an electric transformer. Usually the *torque converters are used for increasing the torque*.

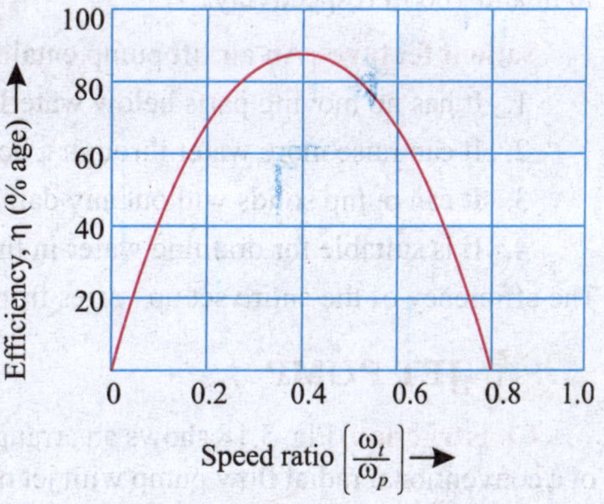

Fig. 5. 16. Efficiency v/s speed ratio curve for the hydraulic torque converter.

When a large reduction in speed ($\omega_t << \omega_p$), and a large torque is required, the hydraulic torque converters are designed which utilize two or more sets of turbine runners, and fixed guide vanes located between the turbine runners.

Fig. 5.16 shows the efficiency versus speed ratio curve for a torque converter, maximum value of efficiency occurs where speed ratio is approximately 0·5; at higher speed ratios the efficiency drops. From the efficiency curve, it is apparent that the efficiency of the torque converter is better at smaller speed than that of hydraulic coupling. The advantages of the hydraulic coupling and the torque converter can be obtained by *designing the system in such a way that at low speed ratios it acts as a converter and at high speed ratios as a coupling.*

5.10. AIR LIFT PUMP

An **air lift pump** *is a device used to lift water from a deep well or sump by utilizing the compressed air.*

Construction and working. It consists of a source of compressed air and an air pipeline fitted with one or more nozzles and an open vertical pipe or rising main as shown in Fig. 5.17.

The compressed air is introduced at the bottom of the rising main and it issues from set of air nozzles in the form of a fine spray. The air mixes with water in the rising main and *reduces the density of air-water mixture.* As soon as the pressure of the column of air-water mixture in the rising main of height H becomes less than the pressure due to the height of water column h in the bore well, the water begins to flow at the outlet of the rising main. The flow rate depends upon the density of the 'mixture' in the rising main/delivery pipe. It has been observed that best results are obtained when the value of the ratio $\left(\frac{h}{H-h}\right)$ is in between 4 and 1 for the values of h between 30 m and 100 m respectively.

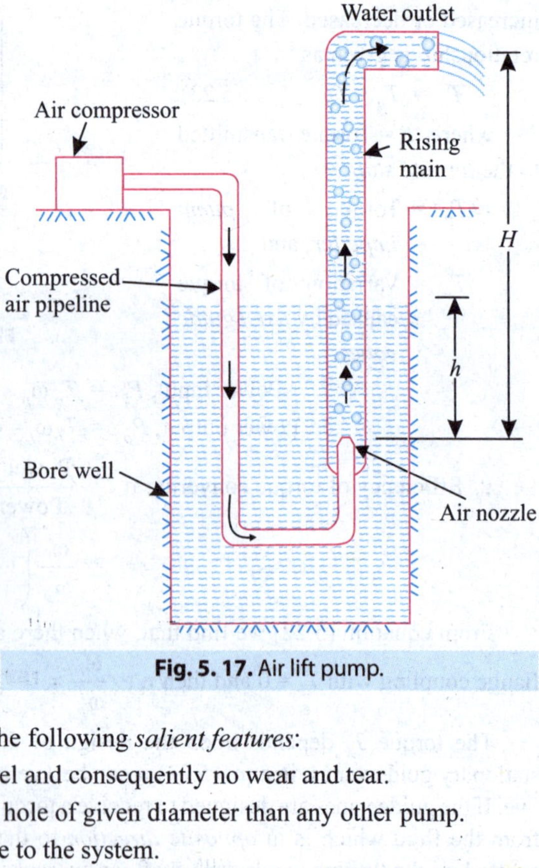

Fig. 5. 17. Air lift pump.

Salient features. An air lift pump entails the following *salient features*:

1. It has no moving parts below waterlevel and consequently no wear and tear.
2. It can raise more water through a bore hole of given diameter than any other pump.
3. It can pump solids without any damage to the system.
4. It is suitable for draining water in the mines where compressor units are already installed.

The efficiency of the entire set up varies from 20 to 40%.

5.11. JET PUMP

Construction. Fig. 5.18 shows an arrangement of a jet pump in a bore well. A jet pump consists of a conventional radial flow pump with jet nozzle at the suction end. It helps to increase the suction lift beyond the normal limit of about 8 metres of water head. With the use of jet assembly it is possible to increase the suction lift upto 60 m.

Working. The working operation of the jet pump is as follows:

—The suction side is completely filled with water and the pump is started.

— A stream of high pressure water from the delivery pipe of the pump is allowed to flow through the suction jet nozzle. The pressure energy of water is converted into kinetic energy due to which a local drop in the pressure takes place. Due to this pressure drop suction is created and water is sucked from the bore well. This action ensures a considerably large supply of low pressure water.

— When the streams with different velocities mix (in the mixing zone), some pressure rise takes place in the mixing zone.

— After the mixing zone, there is a diverging section where *further rise* of pressure occurs *due to decrease in velocity*.

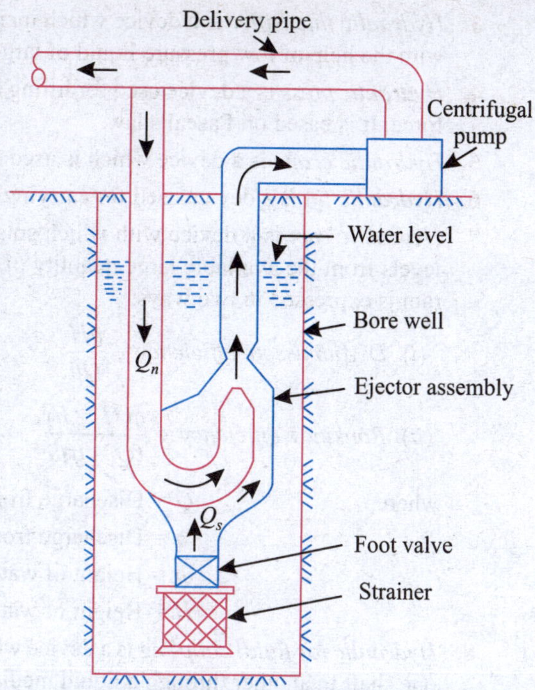

Fig. 5.18. Jet pump.

Efficiency of jet pump:

The efficiency of a jet pump is defined as:

$$\eta = \frac{Q_s (H_s + H_d)}{Q_n (H'_d - H_d)} \qquad ...(5.25)$$

where,
Q_s = Discharge through the suction pipe,
Q_n = Discharge through the nozzle,
H_s = Suction head,
H_d = Delivery head, and
H'_d = Pressure head on delivery side.

The efficiency of a jet pump is very low and is of the order of 25 percent. The major losses take place in the mixing zone.

Uses. The fields of application of a jet pump are:

1. A jet pump can be used to take out muddy water from excavation trenches.
2. To lift water from wells of smaller bores.
3. Employed in mining and for pumping oil.

HIGHLIGHTS

1. The *hydraulic accumulator* is a device used to store the energy of fluid under pressure and make this energy available to hydraulic machines such as presses, lifts and cranes. Its action is similar to that of an electrical storage battery.

   Capacity of hydraulic accumulator = $p \times A \times L$

   [where, p = Liquid pressure supplied by pump, A = area of the sliding ram, and L = Stroke or lift of the ram.]

2. A *differential accumulator* is a special type of accumulator that is used for storing energy at high pressure by comparatively small load on the ram.

3. *Hydraulic intensifier* is a device which increases the intensity of pressure of a given liquid with the help of low pressure liquid of large quantity.
4. *Hydraulic press* is a device used for lifting heavy loads by the application of much smalller force. It is based on Pascal's law.
5. *Hydraulic crane* is a device which is used for lifting heavy loads (upto 25 MN).
6. *Hydraulic lift* is a device used for carrying persons and loads from one floor to another.
7. *Hydraulic ram* is a device with which small quantities of water can be pumped to higher levels from the available large quantity of water of low head. The efficiency of hydraulic ram is expressed in two ways:

   (*i*) $D'Aubuisson' efficiency = \frac{qH}{Qh}$

   (*ii*) $Rankine's\ efficiency = \frac{q(H-h)}{(Q-q)h}$

   where, Q = Discharge from supply tank to the valve box,
   q = Discharge from the valve box to delivery tank,
   h = Height of water in the supply tank above the valve box, and
   H = Height of water in the delivery tank above the valve box.
8. *Hydraulic (or fluid) coupling* is a device which is employed for transmission of power from one shaft to another through a liquid medium.

   Efficiency of hydraulic coupling, $\eta = \frac{\omega_t}{\omega_p}$

   (where ω_t and ω_p are the angular speeds of the turbine shaft and pump shaft respectively) The magnitudes of input and output torque are equal.
9. *Hydraulic torque converter* is device used for transmitting increased or decreased torque from one shaft to another.

   Efficiency of torque converter, $\eta = \frac{\omega_t}{\omega_p}\left(1 + \frac{T_v}{T_p}\right)$

   (where T_v = variation of torque caused by fixed guide vanes; T_p = torque of pump impeller).
10. *Air lift pump* is a device used to lift water from a deep well or sump by utilizing the compressed air.

OBJECTIVE TYPE QUESTIONS

Choose the Correct Answer:

1. ..... is a device which increases the intensity of pressure of a given liquid with the help of low pressure liquid of large quantity.
   (*a*) Hydraulic press
   (*b*) Hydraulic crane
   (*c*) Hydraulic accumulator
   (*d*) Hydraulic intensifier.
2. Which of the following devices is used to store energy of liquid under pressure and make this energy available to hydraulic machines?
   (*a*) Hydraulic coupling
   (*b*) Hydraulic accumulator
   (*c*) Hydraulic ram
   (*d*) Hydraulic press.
3. .... is a device which is employed for transmission of power from one shaft to another through a liquid medium.
   (*a*) Hydraulic intensifier
   (*b*) Hydraulic torque converter
   (*c*) Hydraulic coupling
   (*d*) Hydraulic press.

4. Which of the following devices is used for transmitting increased or decreased torque from one shaft to another?
(a) Hydraulic ram
(b) Hydraulic coupling
(c) Hydraulic intensifier
(d) Hydraulic torque converter.

5. Efficiency of hydraulic torque converter is given by

(a) $\frac{\omega_t}{\omega_p}\left(2+\frac{T_v}{T_p}\right)$ (b) $\frac{\omega_t}{\omega_p}\left(1+\frac{T_v}{T_p}\right)$

(c) $\frac{\omega_p}{\omega_t}\left(1+\frac{T_v}{T_p}\right)$ (d) $\frac{\omega_t}{\omega_p}\left(1+\frac{T_p}{T_v}\right)$.

where, ω_t = Angular speed of turbine shaft;
ω_p = Angular speed of pump shaft;
T_v = Variation of torque caused by fixed guide vanes;
T_p = Torque of pump impeller.

6. ..... is a device with which small quantities of water can be pumped to higher levels from the available large quantity of water of low head.
(a) Hydraulic accumulator
(b) Hydraulic intensifier
(c) Hydraulic ram
(d) Air lift pump.

7. The function of which of the following hydraulic devices is analogous to that of the flywheel of a reciprocating engine and an electric storage battery?
(a) Hydraulic ram
(b) Hydraulic accumulator
(c) Hydraulic intensifier
(d) Hydraulic coupling.

8. Which of the following statements with regard to the purposes served by hydraulic units is *incorrect*?
(a) Hydraulic torque converter is used for transmitting increased or decreased torque to the driven shaft.
(b) Hydraulic ram is a pump that is used to lift small quantities of water to greater heights from large quantity of water available at small heights.
(c) Hydraulic intensifier stores the energy of fluid in the form of pressure energy and is used for lifting heavy weights.
(d) Hydraulic accumulator stores the energy of a fluid in the form of pressure energy; that helps to reduce the capacity of power house where there is intermittent demand for power.

9. The working of which of the following hydraulic units is based on Pascal's law?
(a) Air lift pump
(b) Jet pump
(c) Hydraulic coupling
(d) Hydraulic press.

10. With regard to hydraulic ram, which of the following statements is *incorrect*?
(a) It works automatically
(b) It requires very little maintenance
(c) Its running cost is high
(d) It has no moving parts.

ANSWERS

1. (d)	2. (b)	3. (c)	4. (d)	5. (b)	6. (c)
7. (b)	8. (c)	9. (d)	10. (c).		

THEORETICAL QUESTIONS

1. Describe with sketches the working of any two of the following hydraulic devices:
(i) Hydraulic crane
(ii) Hydraulic lift
(iii) Hydraulic press
(iv) Hydraulic coupling.
2. Describe with the aid of neat sketch the working of a hydraulic intensifier.
3. Obtain an expression for the capacity of a hydraulic accumulator.
4. What is the difference between a hydraulic accumulator and a hydraulic intensifier?
5. Describe with the aid of neat sketch the construction and working of a hydraulic ram.
6. What is the difference between a hydraulic coupling and a hydraulic torque converter?
7. Explain with neat sketch, the working of an air lift pump. Mention its advantages.
8. What is a jet pump?

9. What is the difference between a hydraulic ram and a centrifugal pump?

10. Write short notes on the following:

(*i*) Hydraulic lift,

(*ii*) Hydraulic press, and

(*iii*) Hydraulic coupling.

UNSOLVED EXAMPLES

1. An accumulator has a ram of 200 mm diameter and a lift of 6 m. If the liquid is supplied at 60 bar, find the necessary load on the ram and capacity of the accumulator.

[**Ans.** 188.49 kN; 0.314 kWh]

2. It is required to transmit 36·76 kW power from an accumulator through a pipeline 100 mm diameter and 1500 m long. The ram is loaded with a weight of 1226.25 kN and the friction loss in the pipeline equals 2 per cent of the total power being transmitted. Determine the diameter of the ram if friction co-efficient, $f = 0.01$.

[**Ans.** 493 mm]

3. The diameters of the two parts of the ram of a differential accumulator are 150 mm and 130 mm respectively, and stroke is 1.2 m. If the pressure of water is 98.1 bar when the load is either at rest at the upper end of the stroke or the load is moving with uniform velocity, what will be the weight of the loaded cylinder? How much energy can be stored in the accumulator?

Determine the diameter of the ram of an ordinary accumulator to move the same load with the help of the same water pressure?

[**Ans.** 43.16 kN; 51.8 kNm; 74.8 mm]

4. The pressure intensity of liquid supplied to an intensifier is 50 bar while the pressure intensity of liquid leaving the intensifier is 150 bar. If the intensifier has a capacity of 0.025 m^3 and stroke 1.25 m, calculate the diameters of the fixed ram and sliding cylinder to be used for this intensifier. [**Ans.** 159.6 mm; 319 mm]

5. A hydraulic press has a ram of 165 mm diameter and plunger of 33 mm diameter, with stroke length of 250 mm. Weight exerted by press ram amounts to 5.5 kN and distance moved is 1.2 m in 20 minutes. Determine:

(*i*) The force applied on plunger,

(*ii*) Number of strokes performed by the plunger,

(*iii*) Work done by the press ram, and

(*iv*) Power required to drive the plunger.

[**Ans.** (*i*) 0.22 kN; (*ii*) 120 (app.); (*iii*) 6.6 kNm; (*iv*) 5.5 W]

6. An intensifier receives water from an overhead tank through a pipeline 50 mm in diameter and 100 long and conveys high pressure water to a 600 kN hydraulic press having a ram 0·2 m diameter. The intensifier has low pressure piston diameter 1 m and high pressure ram diameter 0.1 m. If the static head on the low pressure side of the intensifier is 21 m, calculate the rate of movement of the ram when exerting its maximum force.

Take co-efficient of friction, $f = 0.0075$ for the pipeline (50 mm dia). [**Ans.** 0.026 m/min.]

7. In a hydraulic crane, the diameter of crane ram = 300 mm; length of supply pipe from accumulator = 150 m; diameter of the supply pipe = 50 mm; pressure at accumulator = 54 bar; mechanical friction of ram, pulleys etc. are equivalent to a pressure of 4.9 bar on the ram. Determine a relationship between the W kN lifted and the speed of lifting V and hence find V for $W = 50$ kN.

Take co-efficient of friction for the pipe = 0.01.

[**Ans.** $V = 0.1389\,(81.83 - 1.18\,W)^{1/2}$; 0.664 m/s]

8. A hydraulic crane is lifting a load of 11.772 kN through a height of 12 m with a speed of 0.3 m/s once in every two minutes. The crane is working under a pressure of 4905 kN/m^2 of water and has an efficiency of 65 percent. The crane is fed from an accumulator to which water is supplied by a pump. Determine: (*i*) Capacity of cylinder of the jigger, (*ii*) Capacity of accumulator, and (*iii*) Minimum power required to drive the pump.

[**Ans.** (*i*) 0.0443 m^3/s; (*ii*) 0.0295 m^3/s; (*iii*) 1.81 kW]

9. The following data refers to a hydraulic ram: Supply head 2.2 m; weight of waste water per minute 170 N; weight of water pumped per minute 5N and net head pumped from the ram 40 m.

Calculate D'Aubuisson's and Rankine's efficiencies. [**Ans.** 51.8%; 50.5%]

10. A hydraulic ram has a supply head of 2 m, the pipe being 60 mm in diameter and 5 m long. The waste valve, which is 150 mm in diameter and of weight 14.7 N, lifts through 6 mm. Determine:

(*i*) The number of beats per minute, and

(*ii*) The quantity of water delivered per minute to a tank 10 m above the ram.

[**Ans.** (*i*) 175, (*ii*) 0.0183 m^3/min.]

Chapter

6

WATER POWER DEVELOPMENT

6.1. Hydrology—definition—the hydrologic cycle—measurement of run-off—hydrograph—low duration curve, mass curve.

6.2. Hydro-power plant—introduction—application of hydroelectric power plants—advantages and disadvantages of hydroelectric power plants—average life of hydroplant components—hydroplant controls—safety measures in hydroelectric power plants—preventive maintenance of hydroplant—calculation of hydropower—cost of hydro-power plant—hydro-power development in India—combined hydro and steam power plants—comparison of hydro-power station with thermal-power stations

Highlights

Objective Type Questions

Theoretical Questions

Unsolved Examples.

6.1. HYDROLOGY

6.1.1. Definition

Hydrology *may be defined as the science which deals with the depletion and replenishment of water resources.* It deals with the surface water as well as the ground water. It is also concerned with the *transportation of water from one place to another, and from one form to another*. It helps us in determining the occurrence and avilability of water.

6.1.2. The hydrologic Cycle

The earth's water sources such as rivers, lakes, oceans and underground sources etc. mostly get their supplies from rains while the rain water itself is the evaporation from these sources. Water is lost to the atmosphere as vapour from earth, which is then precipitated back in the form of rain, snow, hail, dew, sleet or frost, etc. This evaporation and precipitation continues for ever, and thereby, a balance is maintained between the two. This process is known as *Hydrologic cycle*. It can be represented graphically as shown in Fig. 6.1 Hydrologic equation is expressed as follows:

$$P = R + E \qquad \text{...(6.1)}$$

where, P = Precipitaion,

R = Run-off, and

E = Evaporation.

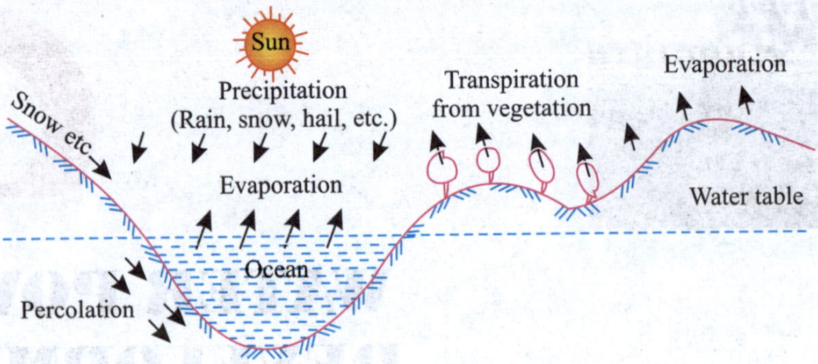

Fig. 6.1. Hydrologic cycle.

Precipitation. It includes all the water that falls from atmosphere to earth surface. Precipitation is of two types: (*i*) *Liquid precipitaion* (rain fall). (*ii*) *Solid precipitaion* (snow, hail).

Run-off and surface run-off. Run-off and surface run-off are two different terms. *Run-off* includes all the water flowing in the stream channel at any given section, while the *surface run-off* includes only the water that reaches the stream channel without first percolating down to the water table.

Run-off can, therefore, also be named as '*Discharge*' or '*Stream flow*'. *Rainfall duration, its intensity and a real distribution influence the rate and volume of run-off.*

Evaporation. Transfer of water from liquid to vapour state is called *evaporation.*

Transpiration. The process by which water is released to the atmosphere by the plants is called *transpiration.*

6.1.3. Measurement of Run-off

Run-off can be measured daily, monthly, seasonal or yearly. It can be measured by the following methods:

1. From rainfall records
2. Empirical formulae
3. Run-off curves and tables
4. Discharge observation method.

1. From rainfall records:

In this method consistent rainfall record for a sufficiently long period is taken and then average depth of rainfall over the catchment is determined. Then considering all the factors which affect run-off process, a co-efficient is arrived at for that catchment. Now a simple equation can be used to find out the run-off over the catchment.

$$\text{Run-off} = \text{Rainfall} \times \text{co-efficient.} \qquad ...(6.2)$$

2. Empirical formulae:

In this method, an attempt is made to derive a direct relationship between the rainfall and subsequent run-off. For this purpose some constants are established which give fairly accurate result for a specified region. Some important formulae are given below:

***(a) Khosla's formula*:**

$$R = P - 4.811\ T$$

where,
R = Annual run-off in mm,
P = Annual rainful in mm, and
T = Mean temperature in 0°C.

(b) Inglis formulae for hilly and plain areas of Maharashtra:

For *Ghat region*

$$R = 0.88\,P - 304.8$$

For *Plain region*

$$R = \frac{(P - 177.8) \times P}{2540}$$

(c) Lacey's formula:

$$R = \frac{P}{1 + \dfrac{3084\,F}{PS}}$$

where, R = Monsoon run-off in mm,
P = Monsoon rainfall in mm,
S = Catchment area factor, and
F = Monsoon duration factor.

Values of S for various types of catchment are given below:

Type of catchment	*Value of S*
Flat, cultivated and black cotton soils	0.25
Flat, partly cultivated, various soils	0.6
Average catchment	1.00
Hills and places with little cultivation	1.70
Very hilly and steep, with hardly any cultivation	3.45

Values of F for various duration of monsoon are given below:

Class of monsoon	*Value of F*
Very short	0.50
Standard length	1.00
Very long	1.50

3. Run-off curves and tables:

Each region has its own catchment area and rainfall characteristics. Thus formulae given above and co-efficients derived there in cannot be applied universally. However, for the same region the characteristics mostly remain unchanged. Based on this fact, the run-off co-efficients are derived once for all. Then a graph is plotted in which one axis represents rainfall and the other run-off. The curves obtained are called *run-off curves.* Alternatively a table can be prepared to give the run-off for a certain value of rainfall for a particular region.

4. Discharge observation method:

The run-off over a catchment can be computed by actual measurement of discharge at an outlet of a drainage basin. The complication in this method is that the discharge of the stream at the outlet comprises surface run-off as well as sub-surface flow. To find out the sub-surface run-off it is essential to *separate* the sub surface flow from the total flow. The separation can be done on an approximate basis but with correct analysis.

Factors affecting the run-off:

The following factors affect run-off :

1. Rainfall pattern
2. Character of catchment area

3. Topography
4. Shape and size of the catchment area
5. Vegetation
6. Geology of the area
7. Weather conditions.

6.1.4. Hydrograph

Hydrograph *is defined as a graph showing discharge (run-off) of flowing water with respect to time for a specified time.* Discharge graphs are known as *flood or run-off graphs.* Each hydrograph has a reference to a particular river site. The time period for discharge hydrography may be hour, day, week or month.

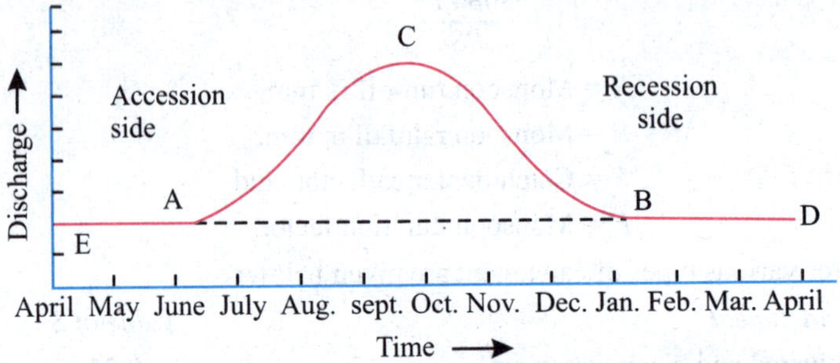

Fig. 6.2. Typical hydrograph.

Hydrograph of stream of river will depend on the characteristics of the catchment and precipitation over the catchment. Hydrograph will access the flood flow of rivers hence it is essential that anticipated hydrograph could be drawn for river for a given storm.

Hydrograph indicates the power available from the stream at different times of day, week or year.

Typical hydrographs are shown in Figs. 6.2 and 6.3.

The unit hydrograph:

The peak flow alone does not give sufficient information about the run-off since it (peak flow) represents a momentary value. Therefore it is necessary to understand the full hydrograph of flow. *The basic concept of unit hydrograph is that the hydrographs of run-off from two identical storms would be the same.* In practice identical storms occur very rarely. The rainfall generally varies in duration, amount and areal distribution. This makes it necessary to construct a typical hydrograph for a basin which could be used as a unit of measurement of run-off.

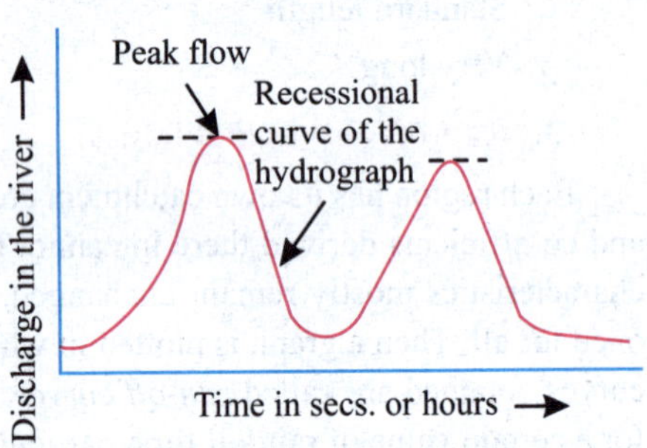

Fig. 6.3. Typical hydrograph.

A **unit hydrograph** *may be defined as a hydrograph which represents unit run-off resulted from an intense rainfall of unit duration and specific areal distribution.*

The following *steps* are used for the *construction of unit hydrograph:*

1. Choose an isolated intense rainfall of unit duration from past records.
2. Plot the discharge hydrograph for outlet from the rainfall records.
3. Deduct the base flow from stream discharge hydrograph to get hydrograph of surface run-off.

4. Find out the volume of surface run-off and convert this volume into cm of run-off over the catchment area.
5. Measure the ordinates of surface run-off hydrograph.
6. Divide these ordinates of obtained run-off in cm to get ordinates of unit hydrograph. Thus for any catchment unit hydrograph can be prepared once. Then whenever peak flow is to be found out, *multiply the maximum ordinate of unit hydrograph by the run-off value expressed in cm.* Similarly to obtain run-off hydrograph by the storm of same unit duration multiply the ordinates of the unit hydrograph by the run-off value expressed in cm. If the storm is of longer duration calculate the run-off in each unit duration of the storm. Then superimpose the run-off hydrographs in the same order giving a lag of unit period between each of them. Finally draw a summation hydrograph by adding all the overlapping ordinates. Generally the computations are done in a tabular form before the hydrograph is plotted.

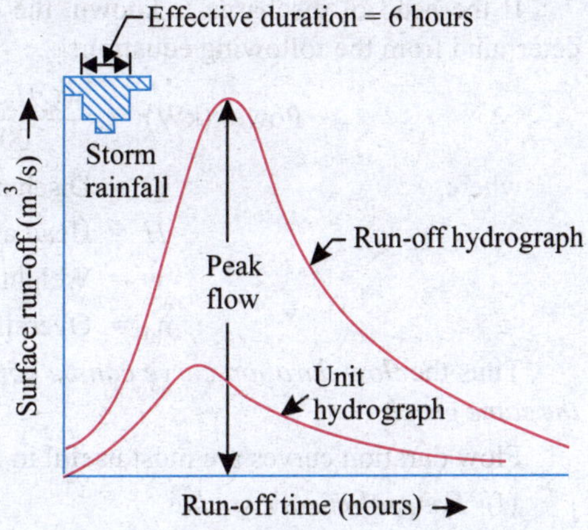

Fig. 6.4.

Fig. 6.4 shows how a run-off hydrograph is constructed from a unit hydrograph.

Limitations to the use of unit hydrograph:

1. Its use is limited to areas about 5000 sq. kilometres since similar rainfall distribution over a large area from storm to storm is rarely possible.
2. The odd-shaped basins (particularly long and narrow) have very uneven rainfall distribution. Therefore, unit hydrograph method is not adopted to such basins.
3. In mountain areas, the areal distribution is very uneven, even then unit hydrograph method is used because the distribution pattern remains same from storm to storm.

6.1.5. Flow Duration Curve

Refer to Fig. 6.5. Flow duration curve is another useful form to represent the run-off data for the given time. This curve is plotted between flow available during a period versus the fraction of time. If the magnitude on the ordinate is the potential power contained in the stream flow, the curve is known as *"power duration curve."* This curve is a very useful tool in the analysis for the development of water power.

The flow duration curve is drawn with the help of a hydrograph from the available run-off data and here it is necessary to find out the lengths of time duration for which certain flows are available. This information either from run-off data or from hydrograph is tabulated. Now the flow duration curve taking 100 percent time on *X*-axis and run-off on *Y*-axis can be drawn.

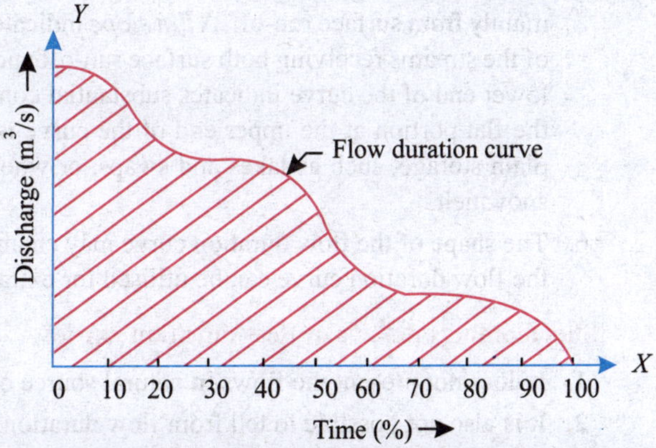

Fig. 6.5. Flow duration curve.

The area under the flow duration curve (Fig. 6.5) gives the *total quantity of run-off during that period* as the flow duration curve is representation of graph with its flows arranged in order of descending magnitude.

If the head of discharge is known, the possible power developed from water in kW can be determind from the following equation:

$$\text{Power (kW)} = \frac{wQH \times \eta_0}{1000}$$

where, Q = Discharge, m^3/s,

H = Head available, m,

w = Weight density of water, N/m^3, and

η_0 = Overall efficiency.

Thus the *flow duration curve can be converted to a power duration with some other scale on the same graph.*

Flow duration curves are most useful in the following cases:

(*i*) *For preliminary studies.*

(*ii*) *For comparison between streams.*

Uses of flow duration curve:

1. A flow duration curve allows the evaluation of low level flows.
2. It is highly useful in the planning and design of water resources projects. In particular, for hydropower studies, the flow duration curve serves to determine the potential for firm power generation. In the case of run-of the river plant, with no strogage facilities, the firm power is usually computed on the basis of flow available 90 to 97 precent of the time. The firm power is also known as the *primary power. Secondary power* is the power generated at the plant utilising water other than that used for the generation of firm power.
3. If a sediment rating curve is available for the given stream, the flow duration curve can be converted into cumulative sediment transport curve by multiplying each flow rate by its rate sediment transport. The area under this curve represents the *total amount of sediment transported.*
4. The flow duration curve also finds use in the design of drainage systems and in flood control studies.
5. A flow duration curve plotted on a log-log paper provides a qualitative description of the run-off variability in the stream. If the curve is having steep slope throughout, it indicates a stream with highly variable discharge. This is typical of the conditions where the flow is mainly from surface run-off. A *flat slope* indicates small variability which is a characteristic of the streams receiving both surface run-off and ground water run-off. A *flat-portion* at the lower end of the curve indicates substantial contribution from ground water run-off, while the flat portion at the upper end of the curve is characteristic of streams with large flood plain storage, such as lakes and swaps, or where the high flows are mainly derived from snowmelt.
6. The shape of the flow duration curve may change with the length of record. This aspect of the flow duration curve can be utilised for extrapolation of the short records.

Shortcomings/defects of flow duration curve:

1. It does not present the flows in natural source of occurrence.
2. It is also not possible to tell from flow duration curve whether the lower flows occurred in consecutive periods or were scattered throughout the considered period.

6.1.6. Mass Curve

A **mass curve** *is the graph of the cumulative values of water quantity (run-off) against time.* A mass curve is an integral curve of the hydrograph which expresses the area under the hydrograph from one time to another.

It is convenient device *to determine storage requirement* that is needed to produce a certain dependable flow from fluctuating discharge of a river by a reservoir.

Mass curve can also be used to solve *reserve problem* of determining the maximum demand rate that can be maintained by a given storage volume. However, it is a trial and error procedure.

The mass curve *will always have a positive shape* but of a greater of less degree depending upon the variations in the quantity of inflow water available. The negative inclination of mass curve would show that the amount of water flowing in the reservoir was less than the loss due to evaporation and seepage.

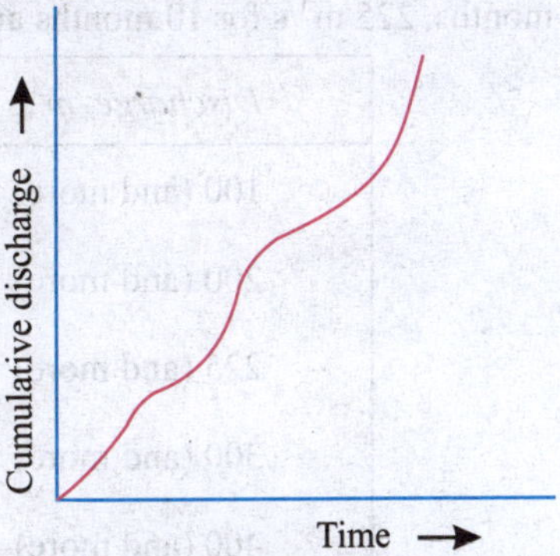

Fig. 6.6. Mass curve.

Example 6.1. *At a particular site the mean monthly discharge is as follows:*

Month	*Discharge m^3/s*	*Month*	*Discharge, m^3/s*
January	*100*	*July*	*1000*
February	*225*	*August*	*1200*
March	*300*	*September*	*900*
April	*600*	*October*	*600*
May	*750*	*November*	*400*
June	*800*	*December*	*200*

Draw the following :

(i) Hydrograph; *(ii) Flow duration curve.*

Solutions. (*i*) The **hydrogarph** is plotted between discharge (m^3/s) and time (months) as shown in Fig. 6.7.

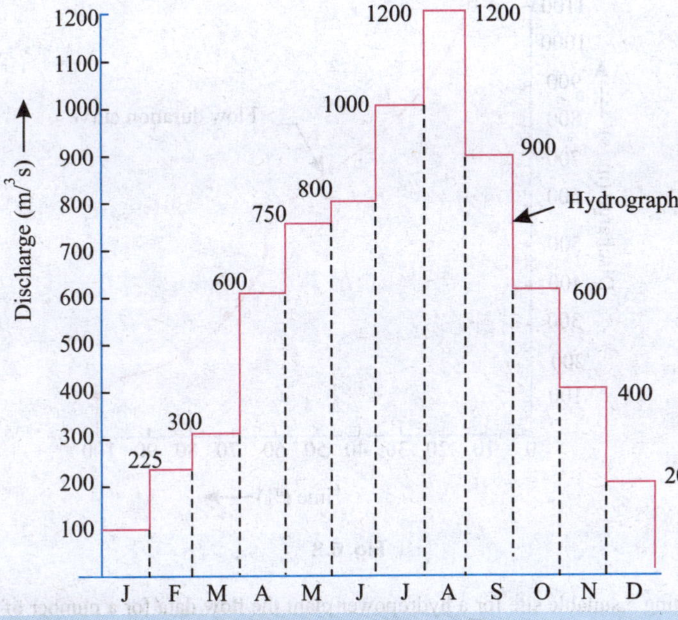

Fig. 6.7. Hydrograph.

***(ii)* Flow duration curve:**

In order to draw flow duration curve it is essential to find the length of time during which centain flows are available, e.g. 100 m^3/s is available for all 12 months, flow of 200 m^3/s for 11 months, 225 m^3/s for 10 months and so on. This information is indicated in the table below:

Discharge, m^3/s	*Length of time, months*	*%age time*
100 (and more)	12	100
200 (and more)	11	91.7
225 (and more)	10	83.3
300 (and more)	9	75.0
400 (and more)	8	66.7
600 (and more)	7	58.3
750 (and more)	5	41.7
800 (and more)	4	33.3
900 (and more)	3	25.0
1000 (and more)	2	16.7
1200 (and more)	1	8.3

The flow duration curve is then plotted as shown in Fig. 6.8

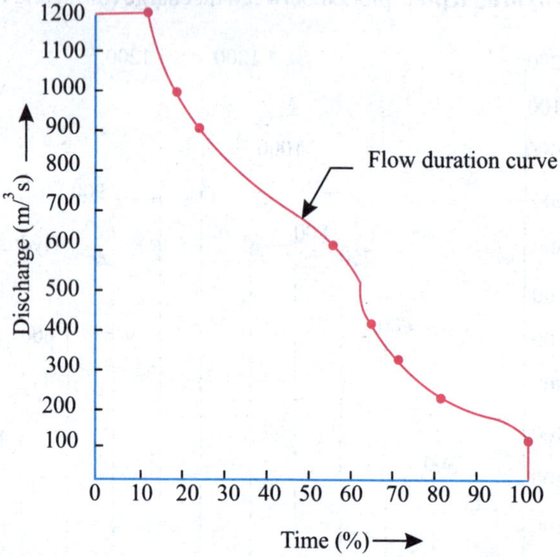

Fig. 6.8

Note: When selecting a suitable site for a hydropower plant the flow data for a number of years is collected and hydrographs and flow duration curves and the various periods are determined.

Example 6.2. *The run-off data of a river at a particular site is tabulated below:*

Month	*Mean discharge per month (millions of m^3)*	*Month*	*Mean discharge per month (millions of m^3)*
January	*40*	*July*	*75*
February	*25*	*August*	*100*
March	*20*	*September*	*110*
April	*10*	*October*	*60*
May	*0*	*November*	*50*
June	*50*	*December*	*40*

(*i*) *Draw a hydrograph and find the mean flow,*

(*ii*) *Also draw the flow duration curve, and*

(*iii*) *Find the power in MW available at mean flow if head available is 80 m and overall efficiency of generation is 85 %. Take each month of 30 days.*

Solution. (*i*) Hydrograph :

The hydrograph for the given data is drawn as shown in Fig. 6.9.

The mean discharge for the given data

$$= \frac{40 + 25 + 20 + 10 + 0 + 50 + 75 + 100 + 110 + 60 + 50 + 40}{12}$$

$$= \frac{580}{12} = 48.33 \text{ millions of m}^3\text{/month.}$$

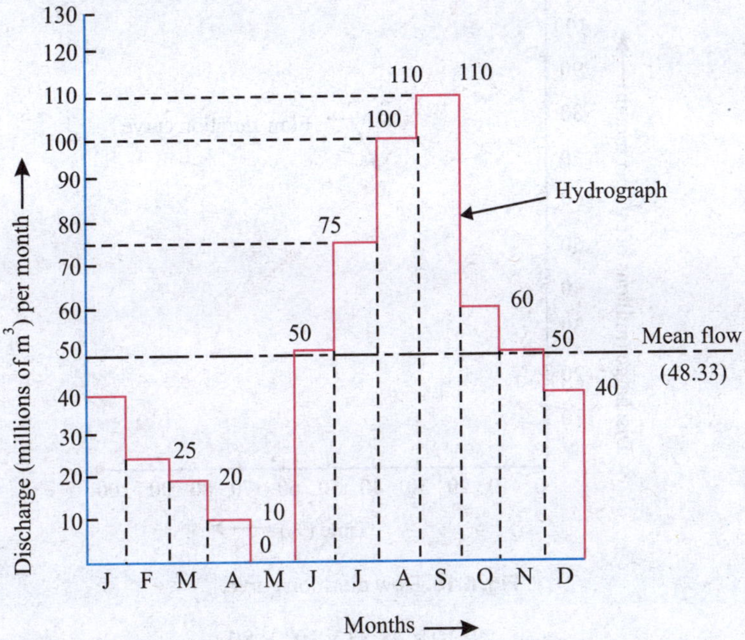

Fig. 6.9

(*ii*) Flow duration curve:

To obtain the flow duration curve, it is necessary to find the lengths of time during which certain flows are available. This information is tabulated using the hydrograph in the table as under:

Discharge per month (millions of m^3)	*Total number of months during which flow is available*	*Percentage time*
0	12	100
10	11	91.7
20	10	83.3
25	9	75
40	8	66·7
50	6	50
60	4	33·3
75	3	25·0
100	2	16·7
110	1	8·3

The flow duration curve can be drawn using the data tabulated as shown in Fig. 6.10

(*iii*) Average MW energy available :

$$\text{Average MW energy available} = \frac{wQH\eta_0}{1000}$$

$$\left[\text{where, } Q \text{ (discharge in m}^3\text{/s)} = \frac{48.33 \times 10^6}{30 \times 24 \times 3600}\right]$$

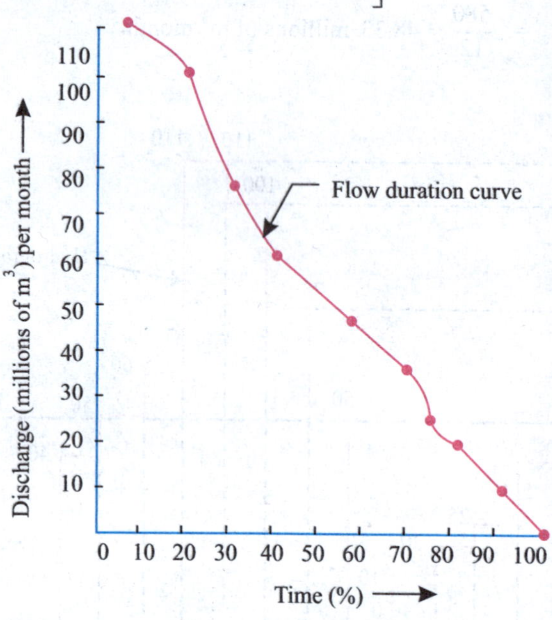

Fig. 6.10. Flow duration curve.

$$= \frac{9810 \times 48.33 \times 10^6 \times 80}{(30 \times 24 \times 3600)} \times 0.85 \times \frac{1}{1000} \text{ kW}$$

$= 12438$ kW, or, **12.438 MW** **(Ans.)**

6.2. HYDRO-POWER PLANT

6.2.1. Introduction

In hydroelectric plants, energy of water is utilised to move the turbines which in turn run the electric generators. The energy of water utilised for power generation may be kinetic or potential. The kinetic energy of water is its energy in motion and is a function of mass and velocity, while the potential energy is a function of the difference in level/head of water between two points. In either case continuous availability of water is a basic necessity; to ensure this, water collected in natural lakes and reservoirs at high altitudes may be utilised or water may be artificially stored by constructing dams across flowing stream. The ideal site is one in which a good system of natural lakes with substantial catchment area, exists at a high altitude. Rainfall is the primary source of water and depends upon such factors as temperature, humidity, cloudiness, wind etc. The usefulness of rainfall for power purposes further depends upon several complex factors which include its intensity, time distribution, topography of land etc. However it has been observed that only a small part of the rainfall can actually be utilised for power generation. A significant part is accounted for by direct evaporation, while another similar quantity seeps into the soil and forms the underground storage. Some water is also absorbed by vegetation. Thus only a part of water falling as rain actually flows over the ground surface as direct run off and forms the streams which can be utilised for hydroschemes.

First hydroelectric station was probably started in America in 1882 and thereafter development took place very rapidly. In India, the first major hydroelectric development of 4·5 MW capacity named as Sivasamudram Scheme in Mysore was commissioned in 1902. In 1914, a hydropower plant named Khopoli project of 50 MW capacity was commissioned in Maharashtra. The hydropower capacity, upto 1947, was nearly 500 MW.

Hydro (water) power is a conventional renewable source of energy which is clean, free from pollution and generally has a good environmental effect. However the following facts are *major obstacles* in the utilisation of hydropower resources :

(*i*) Large investments

(*ii*) Long gestation period

(*iii*) Increased cost of power transmission.

Next to thermal power, hydropower is important in regard to power generation. The hydroelectric power plants provide 30 percent of the total power of the world. The total hydropotential of the world is about 5000 GW. In some countries (like Norway) almost total power generation is hydrobased.

6.2.2. Application of Hydro-electric power plants

Earlier hydro-electric plants have been used as exclusive source of power, but the trend is towards use of hydro-power in an inter-connected systen with thermal stations. As a self-contained and independent power source, a hydroplant is most effective with adequate storage capacity otherwise the maximum load capacity of the station has to be based on the year. This increases the per unit cost of installation. By interconnecting hydropower with thermal (stream) power, a great deal of saving in cost can be effected due to:

(*i*) reduction in necessary reserve capacity,

(*ii*) diversity in construction programmes,

(*iii*) higher utilisation factors on hydroplants, and

(*iv*) higher capacity factors on efficient steam plants.

In an inter-connected system the base load is supplied by hydropower when the maximum flow demand is less than the stream flow while steam supplies the peak. When stream flow is lower than the maximum demand the hydroplant supplies the peak load and steam plant the base load.

6.2.3 Advantages and Disadvantages of Hydro-electric Power Plants

Advantages :

1. No fuel charges.
2. A hydroelectric plant is highly reliable.
3. Maintenance and operation charges are very low.
4. Running cost of the plant is low.
5. The plant has no standby losses.
6. The plant efficiency does not change with age.
7. It takes a few minutes to run and synchronise the plant.
8. Less supervising staff is required.
9. No fuel transportation problem.
10. No ash problem and atmosphere is not polluted since no smoke is produced in the plant.
11. In addition to power generation, these plants are also used for flood control and irrigation purposes.
12. Such a plant has comparatively a long life (100 to 125 years as against 20-45 years of a thermal plant.)
13. The number of operations required is considerably small compared with thermal power plants.
14. The machines used in hydro-electric plants are more robust and generally run at low speeds at 300 to 400 r.p.m. where as the machines used in thermal plants run at a speed 3000 to 4000 r.p.m. Therefore, there are no specialised mechanical problems or special alloys required for construction.
15. The cost of land is not a major problem since the hydroelectric stations are situated away from the developed areas.

Disadvantages :

1. The initial cost of the plant is very high.
2. It takes considerably long time for the erection of such plants.
3. Such plants are usually located in hilly areas far away from the load centre and as such they require long transmission lines and losses in them will be more.
4. Power generation by the hydro-electric plant is only dependent on the quantity of water available which in turn depends on the natural phenomenon of rain. So, if the rainfall is in time and proper and the required amount of it can be collected, the plants will function satisfactorily otherwise not.

6.2.4. Average Life of Hydroplant Components

The average life (approximate) of various components of hydro-electric power plant is given as follows:

Components	*Average life (years)*
1. Reservoirs	70–80
2. Dams	
(*i*) Earthen, concrete or masonry	150
(*ii*) Loose rock	60
3. Water ways	50–100
(*i*) Canals, tunnels	
(*ii*) Penstocks	
(*a*) Steel	40–50
(*b*) Concrete	25–50
4. Power house and equipment	
(*i*) Building	35–50
(*ii*) Generators	25
(*iii*) Transformers	30
(*iv*) Turbines (hydraulic)	5
(*v*) Pumps	20–25

6.2.5. Hydroplant controls

The various controls which are provided in a hydro-electric power plant are:

1. Hydraulic controls
2. Machine controls—starting and stopping
3. Machine controls—loading and frequency
4. Voltage control of generator and system
5. Machine protection.

6.2.6. Safety measures in hydroelectric power plants

Following safety measures need to be taken for the safe operation of a hydroelectric power plant:

1. Surge tanks
2. Screens
3. Sand traps
4. Jet dispersers
5. Pressure regulator.

Surge tanks. A surge tank is used to prevent sudden increase of pressure in the supply line or the penstock. It is placed as near as possible to the turbine. The tank may be open at the top or closed. In case, it is open at the top, it must not be lower than the level of the water in the reservoir.

Screens. These are provided to prevent logs, fishes, ice blocks and other obstructive elements from entering the pipelines and turbines.

Sand traps. Sand traps are provided to prevent the sand flowing with water in pipes since sand blast action of solid matter in the water causes rapid wear of nozzles, spears, blades etc. of the turbine.

Jet dispersers. The discharged water at the bottom of the high dams possesses large amount of energy which is likely to cause scouring of the channel below the dam and consequent damage to the dam foundation unless some means are provided to dissipate it. The possible remedies for this are either to discharge water into a cushion pool or to provide a jet disperser at the end of outlet pipe so that the end of the outlet pipe is such that the jet is broken up into a conical shower of drops and their energy is absorbed by air.

Pressure regulator. It is usually operated by a governor of the turbine. It is provided on the pipeline near the turbine inlet so that when the turbine gates are suddenly closed, pressure surges so produced are kept within the safe limits of the pipeline. The water discharge from the regulator is passed on to tailrace through a separate pipeline.

6.2.7. Preventive Maintenance to Hydroplant

The purpose of preventive maintenance is to minimise breakdown and excessive depreciation resulting from neglect. In a hydroplant (using reaction turbines) monthly, quarterly halfyearly and yearly inspection and maintenance are carried out on the following parts :

Inspection / Maintenance	*Parts*
Monthly	*Turbine cover parts* (e.g. leakage unit, drainage, holes, servomotor connections, turbine shaft and cover, oil pump etc.) *Operating ring of turbines.*
	Guide vane mechanism.
Quarterly	*Servomotor*
	Ejector cabinet
	Feedback system.
Half-yearly	*Governor machanism*
	Gauges
	Grease pumps for guide vanes and guide bearings.
	Grease pipes connected to grease pumps.
Yearly	*Turbine auxiliaries* (e.g. oil pressure tank, turbine guide bearing, turbine instruments)
	Scroll casing runner with guide vanes
	Emergency slide valve
	Pit liner
	Draft tube
	Runner blades checked for cavitational effects, cracks and wearing out.

6.2.8. Calculation of available hydro-power

The theoretical power (P_{th}) available from falling water can be calculated using the following formula:

$$P_{th} = \frac{wQH}{1000} \text{ kW} \quad \text{...(6.3)}$$

where, w = Weight density of water in N/m^3,

Q = Flow through turbine (or quantity or water available for hydropower generation) in m^3/s, and

H = Head available in metres.

The actual useful or effective output depends upon the efficiency of the various parts of the installation.

If, η_1 = Efficiency of pipelines, intake etc., and

η_2 = Efficiency of hydraulic turbine.

Then, overall efficiency $\eta_0 = \eta_1 \times \eta_2$. ...(6.4)

Since the turbine and the genertor are directly coupled on common shaft the hydro-electrical power available will be given by the equation:

or, $$P_{actual} = P_{th.} \times \eta_0 \quad ...(6.5)$$

or, $$P_{actual} = \frac{wQH}{1000} \times \eta_0 \text{ kW} \quad ...(6.5a)$$

6.2.9. Cost of Hydro-power Plant

The initial cost of any hydroplant is very high but the power produced by it is the cheapest. The following costs are included in development of a hydroplant :

1. Cost of land and riparian rights.
2. Cost of railways and highways required for the construction work.
3. Cost of construction.
4. Cost of engineering supervision of the project.
5. Cost of building etc.
6. Cost of equipment.
7. Cost of equipment used for power transmission.

6.2.10. Hydro-power Development in India

Hydropower is a renewable source of energy which entails many intrinsic advantages. In India, the scope of water power development is tremendous. The first hydropower station in India dates back to year 1897 when a *small* power station of 200 kW capacity was constructed at Darjeeling. Since them many big and small hydropower stations have been installed in the country.

Important hydroplants in India

State/Name of Power plant	*Installed capacity* **(MW)**
Andhra Pradesh	
Machkand (Stage I and II)	114
Upper silern	120
Lower silern	600
Srisailam	770
Nagarjun sagar pumped storage	100
Assam	
Umiam	54
Gujarat	
Ukai	300
Himachal Pradesh	
Baira suil	200
Jammu and Kashmir	
Salal	270
Karnataka	
Tungabhadra	72
Sharavati	890
Kailindi	395

Kerala	
Parambikulam-Aliyar	185
Sabarigiri	300
Idikki (Stage I)	390
Maharashtra	
Koyna (Stages I, II and III)	860
Manipur	
Loktak	70
Orissa	
Hirakud (Stage I and II)	270
Balimela	480
Punjab	
Bhakra Nangal	1084
Beas-Sutlej Link	780
Rajasthan	
Chambal	287
Uttar Pradesh	
Rihand	300
Yamuna (Stage I and II)	424
Tamil Nadu	
Kundah (Stages, I , II and III)	425
Kodiar	100

Although the present utilization of hydropower in over country is relatively small with the persent tempo of development and need for power resources it would not be long before the available potential is fully harnessed. Hydrofield provides immense scope for sophisticated study requiring application of modern mathematical and operational research techniques with the help of computers.

6.2.11. Combined Hydro and Steam Power Plants

A electrical power system should fulfil the following *objectives:*

1. To ensure an adequate and reliable electric power supply at all loads and at all times.
2. The source of energy should be such as to give the minimum overall cost of the system as a whole.

The above objectives (unless a country/region is rich either in abundant supply of cheap fuel or ample water power resources which can be developed at suitable site) can be best realised by a judicious combination of both hydro and thermal power. Hydropower represents a renewable source of energy which enjoys many intrinsic advantages as compared to thermal power. Although the cost of construction of hydropower plant is nearly same as that of a coal based steam power plant in terms of investment for MW, but hydropower plant uses water for power generation which is available in abundance in nature.

It is known that hydroplant can meet the demands of load variations more rapidly and easily. Thus, when the rate of flow of water is low, the steam plant can work at constant load producing a better efficiency and the hydroplant will work most effectively as peak load plant and its output can be varied to meet the load fluctuations.

The steam and hydroplants reverse their functions (steam plant providing the peal load and the hydroplant providing base load) when high rate of water flow is available. But even under this condition, *the steam plant output will remain constant and the hydroplant output will be varied to meet the load fluctuations.*

6.2.12. Comparison of Hydro-power Station with Thermal Power Station

The comparison between hydropower station and thermal power station is given below:

S. No.	Aspects	Hydropower station	Thermal power station
1.	*Raw material consumption,*	Nil. Water power is inexhaustible and is continuously replenished by the direct agency of sun.	Huge quantity of coal consumed, thereby exhausting "fuel reserves"
2.	*Cost of energy*	Cheaper	Costlier
3.	*Cost of energy generation.*	Immune to inflation	Very much influenced by the increase in the cost of fuel.
4.	*Life of plant*	Long useful life.	Not so long comparatively. The component parts deteriorate and become obsolete at a faster rate.
5.	*Pollution*	Free from problems of pollution.	Cause pollution and subsequently create health hazards.
6.	*Design, construction and reliability.*	Simple in design, robust in construction and reliable in operation.	Comparatively more complicated in design, less robust in construction and less reliable in operation .
7.	*Running below a certain minimum load factor.*	Can be run.	Cannot be run.
8.	*Reserve capacity and variation in power demands.*	Particularly suited to provide reserve capacity as well as meeting the exact needs of daily variation in power demands.	Comparatively not suited for the mentioned requirements.
9.	*Employment potential.*	More. Affords a relatively high employment potential and better utilization of the available local talent and resources.	Less
10.	*Man power required*	Small	Large
11.	*Labour problem*	Less	More
12.	*Foreign exchange requirement for equipment.*	Less	More
13.	*Construction time required.*	Almost same as thermal power station.	Almost same as hydropower station.
14.	*Overall capital expenditure requirements.*	Low	High

Example 6.3. *At a proposed site of hydroelectric power plant the available discharge and head are 330 m³/s and 28 m respectively. The turbine efficiency is 86%. The generator is directly coupled to the turbine. The frequency of generator is 50 Hz and number of poles used are 24. Find the least number of machines required if,*

(*i*) *A Francis turbine with a specific speed of 260 is used;*

(*ii*) *A Kaplan turbine with a specific speed of 700 is used.*

Solution. Available discharge, $Q = 330\ m^3/s$

Head, $H = 28$ m

Turbine efficiency, $\eta = 86\%$

Frequency of generation, $f = 50$ Hz

Number of poles used, $p = 24$.

As the generator is directly coupled to the turbine, the speed of turbine used must be equal to the synchronous speed of the generator.

$$N = \frac{120 f}{p} = \frac{120 \times 50}{24} = 250 \text{ r.p.m}$$

$$P = \eta \times wQH = 0.86 \times 9.81 \times 330 \times 28 = 77954 \text{ kW}$$

$$(\because w = 9.81 \text{ kN/m}^3)$$

(*i*) The power capacity of each Francis turbine (P_1) can be calculated by using the following formula:

$$N_s = \frac{N\sqrt{P_1}}{H^{5/4}}, \text{ or, } 260 = \frac{250\sqrt{P_1}}{(28)^{5/4}}$$

$$\therefore \quad P_1 = \left[\frac{260 \times (28)^{5/4}}{250}\right]^2 = 4487 \text{ kW}$$

$\therefore$ *Number of Francis turbines required* $= \frac{P}{P_1} = \frac{77954}{4487} = \mathbf{17.37 \simeq 18.}$ **(Ans.)**

(*ii*) The power capacity of each Kaplan turbine can be calculated by using the following formula:

$$N_s = \frac{N\sqrt{P_2}}{H^{5/4}}, \text{ or, } 700 = \frac{250\sqrt{P_2}}{(28)^{5/4}}$$

$$\therefore \quad P_2 = \left[\frac{700 \times (28)^{5/4}}{250}\right]^2 = 32524.5 \text{ kW}$$

$\therefore$ *Number of Kaplan turbines required*

$$= \frac{P}{P_2} = \frac{77954}{32524{\cdot}5} = \mathbf{2.4 \simeq 3.} \text{ (Ans.)}$$

Example 6.4. *The following data relate to a proposed hydro-electric station:*

Available head = 28 m; catchment area = 420 sq. km; rainfall = 140 cm/year; percentage of total rainfall utilised = 68%; penstock efficiency = 94%; turbine efficiency = 80%; generator efficiency = 84% and load factor = 44%.

(*i*) *Calculate the power developed, and*

(*ii*) *Suggest suitable machines and specify the same.*

Solution. Head available, $H = 28$ m

Catchment area, $A = 420$ sq. km ($= 420 \times 10^6$ m^2)

Rainfall $= 140$ cm /year ($= 1.4$ m/year)

Rainfall utilised, $h = 68\%$ of the total rainfall

$= (0.68 \times 1.4)$ m per year

Penstock efficiency $\eta_p = 94\%$

Turbine efficiency, $\eta_t = 80\%$

Generator efficiency, $\eta_g = 84\%$

Load factor $= 44\%$.

(*i*) Power developed, *P*:

Quantity of water available per year $= A \times h$

$$= (420 \times 10^6) \times (0.68 \times 1.4) = 399.84 \times 10^6 \text{ m}^3$$

Hence the quantity of water available per second,

$$Q = \frac{399.84 \times 10^6}{(365 \times 24) \times 3600} = 12{\cdot}6 \text{ m}^3/\text{s}.$$

$$\therefore \quad P = \eta_0 \times wQH \text{ (where } \eta_0 = \text{overall efficiency} = \eta_p \times \eta_t \times \eta_g)$$

$$P = \eta_p \times \eta_t \times \eta_g \times wQH$$

$$= 0.94 \times 0.8 \times 0.84 \times 9.81 \times 12.6 \times 28 = 2186.2 \text{ kW} \quad (\because w = 9.81 \text{ kN/m}^3)$$

Hence, average output of generating units = **2186.2 kW (Ans.)**

Example 6.5. *The following data is available for a hydropower plant :*

Available head = 140 m; catchment area = 200sq. km; annual average rainfall = 145 cm; turbine efficiency = 85 % ; generator efficiency = 90 % ; Percolation and evaporation losses = 16 %.

Determine the following :

(i) Power developed, and

(ii) Suggest type of turbine to be used if runner speed is to be kept below 240 r.p.m.

Solution. Head available, $H = 140$ m

Catchment area, $A = 200$ sq. km ($= 200 \times 10^6$ m^2)

Annual average rainfall, $h = 145$ cm ($= 1.45$ m)

Turbine efficiency, $\eta_t = 85\%$

Generator efficiency, $\eta_g = 90\%$

Percolation and evaporation losses, $z = 16\% = 0.16$

(*i*) Power developed, *P* :

Quantity of water avilable for power generation per year

$$= A \times h \times (1 - z)$$

$$= 200 \times 10^6 \times 1.45 \times (1 - 0.16) = 2.436 \times 10^8 \text{ m}^3 \text{/year}$$

Hence, quantity of water available for power generation per second,

$$Q = \frac{2.436 \times 10^8}{(365 \times 24) \times 3600} = 7.72 \text{ m}^3/\text{s}$$

$$P = \eta_0 \times wQH$$

$$= \eta_t \times \eta_g \times wQH$$
$$= 0.85 \times 0.9 \times 9.81 \times 7.72 \times 140 = 8111 \text{ kW } (w = 9.81 \text{ kN/m}^3)$$

or, **8.111 MW (Ans.)**

***(ii)* Type of turbine to be used:**

$$\text{Specific speed, } N_s = \frac{N\sqrt{P}}{H^{5/4}} = \frac{240\sqrt{8111}}{(140)^{5/4}} = \mathbf{44.88 \text{ r.p.m.}} \text{ (Ans.)}$$

Single pelton turbine with 4 jets can be used. Further since head available is large and discharge is low, pelton turbine will work satisfactorily.

Example 6.6. *From the investigation of hydrosite the following data is avilable:*

Available head	*45 m*
Total catchment area	*60 sq. km*
Rainfall per annum	*140 cm*
Percentage of rainfall utilized	*68%*
Turbine efficiency	*82%*
Generator efficiency,	*90%*
Penstock efficiency	*74%*

Calculate the suitable capacity of a turbo- generator.

Solution. Head available, $H = 45$ m

Catchment area, $A = 60$ sq. km ($= 60 \times 10^6 \text{m}^2$)

Availabe rainfall, $h = (0.68 \times 1.4)$ m

Turbine efficiency, $\eta_t = 82$ %

Generator efficiency, $\eta_g = 90$ %

Penstock efficiency, $\eta_p = 74$ %

Quantity of water available per annum

$$= A \times h = 60 \times 10^6 \times 0.68 \times 1.4 = 57.12 \times 10^6 \text{ m}^3/\text{annum}$$

Hence, quantity of water available per second

$$Q = \frac{57.12 \times 10^6}{(365 \times 24) \times 3600} = 1.81 \text{ m}^3/\text{s}$$

$$\text{Now overall efficiency, } \eta_0 = \eta_p \times \eta_t \times \eta_g = 0.74 \times 0.82 \times 0.9 = 0.546$$

$$\therefore \quad \text{Power developed, } P = \eta_0 \times wQH$$
$$= 0.546 \times 9.81 \times 1.81 \times 45 = 436 \text{ kW}$$

If a load factor of 55 percent is assumed, then,

$$\text{Maximum kW} = \frac{436}{0{\cdot}55} \simeq 793 \text{ kW}$$

So a *generator of* **800 kW maximum rating** *can be selected.*

$$\therefore \text{ Power of the turbine } \frac{793}{0.82} = \mathbf{967 \text{ kW}} \text{ (Ans.)}$$

For a head of 45 m, which is low, *a vertical shaft Francis or Kaplan trubine may be employed.*

Example 6.7. *A hydro-electric power plant produces 27 MW under a head of 15 metres. If the overall efficiency of the plant is 72%, determine :*

(i) Type of turbine; *(ii) Synchronous speed of the generator.*

Solution. Power developed, $P = 27$ MW ($= 27 \times 10^3$ kW)

Head, $H = 15$ m

Overall efficiency, $\eta_0 = 72\%$.

(i) Type of turbine :

$$P = \eta_0 \times wQH$$

$$27 \times 10^3 = 0.72 \times 9.81 \times Q \times 15$$

$$\therefore \quad Q = \frac{27 \times 10^3}{0.72 \times 9.81 \times 15} = 188.8 \text{ m}^3/\text{s}$$

As the head is low and discharge is high so a propeller type of turbine should be used **(Ans.)**

(ii) Synchronous speed of the generator, N_{syn} :

$$\text{Specific speed, } N_s = \frac{1150}{H^{1/4}} \text{ (approx.)}$$

$$= \frac{1150}{(15)^{1/4}} = 584.3 \text{ r.p.m.}$$

$$\text{Speed of rotation, } N = \frac{N_s \times H^{5/4}}{\sqrt{P}} \qquad \left(\because N_s = \frac{N\sqrt{P}}{H^{5/4}}\right)$$

$$= \frac{584.3 \times (15)^{5/4}}{\sqrt{27 \times 10^3}} \simeq 105 \text{ r.p.m}$$

$$\text{For generator, } N = \frac{120 f}{p}$$

$$105 = \frac{120 \times 50}{p}$$

[where, f = frequency (= 50 Hz)]

$$\therefore \quad \text{Number of poles, } p = \frac{120 \times 50}{105} = 57.14 = 60 \text{ (say)}$$

(as the number of poles is necessarily an *even* number)

$$\text{Again,} \quad N_{syn} = \frac{120 f}{p} = \frac{120 \times 50}{60} = \mathbf{100 \text{ r.p.m.}} \quad \textbf{(Ans.)}$$

Example 6.8. *Calculate the power developed in MW from a hydro-electric power plant with the following data :*

Available head	*50 m*
Catchment area	*250 sq. km*
Average annual rainfall	*120 cm*
Rainfall lost due to evaporation	*20%*
Turbine efficiency	*82%*
Generator efficiency	*84%*
Head lost in penstock	*4%.*

Solution. Head available, $H = 50$ m

Catchment area, $A = 250$ sq. km ($= 250 \times 10^6$ m^2)

Average annual rainfall, = 120 cm (= 1.2 m)

Evaporation, = 20%

∴ Average annual rainfall available, $h = (1 - 0.2) \times 1.2 = 0.96$ m

Turbine efficiency, η_t = 82%

Generator efficiency, η_g = 84%

Penstock efficiency, $\eta_p = 100 - 4 = 96\%$

Quantity of water available *per annum*

$$= A \times h = 250 \times 10^6 \times 0.96 = 2.4 \times 10^8 \text{ m}^3$$

Hence, quantity of water available *per second,*

$$Q = \frac{2.4 \times 10^8}{(365 \times 24) \times 3600} = 7.61 \text{ m}^3/\text{s}$$

Overall efficiency, $\eta_o = \eta_p \times \eta_t \times \eta_g$

$$= 0.96 \times 0.82 \times 0.84 = 0.66$$

Power developed, $P = \eta_0 \times wQH$

$$= 0.66 \times 9.81 \times 7.61 \times 50 \text{ kW}$$

$$= 2463.6 \text{ kW or } 2.463 \text{ MW}$$

Hence, power developed = **2.463 MW. (Ans.)**

Example 6.9. *In an hydro-electric power plant the reservoir is 225 m above the turbine house. The annual replenishment of reservoir is 3.5 × 10^{12} N. Calculate the energy available at the generating station bus bars if the loss of head in the hydraulic system is 25 m and the over all efficiency of the system is 85%.*

If maximum demand of 45 MW is to be supplied, determine the diameter of two steel penstocks.

Solutuion. Actual head available, $H = 225 - 25 = 200$ m

Overall efficiency, η_0 = 85%

Annual replenishment, $W = 3.5 \times 10^{12}$ N.

(*i*) Energy output :

E = Energy available at the turbine house

$$= WH = 3.5 \times 10^{12} \times 200 = 7 \times 10^{14} \text{ Nm or J}$$

$$= \frac{7 \times 10^{14}}{36 \times 10^5} = 1.944 \times 10^8 \text{ kWh} \quad [\because 1 \text{ kWh} = 36 \times 10^5 \text{ J}]$$

Energy output $= \eta_0 \times E$

$$= 0.85 \times 1.944 \times 10^8 = \mathbf{1.652 \times 10^8 \text{ kWh. (Ans.)}}$$

(*ii*) Diameter of steel penstock, D :

Kinetic energy of water = Loss of potential energy

$$\therefore \quad \frac{1}{2}mC^2 = mgH$$

$$\therefore \quad C = \sqrt{2gH} = \sqrt{2 \times 9.81 \times 200} = 62.64 \text{ m/s}$$

(where, C = Velocity of water in each penstock, m = mass of water in kg)

Now, $\frac{1}{2}mC^2$ = Energy to be supplied

$$\frac{1}{2} m \times (62.64)^2 = 45 \times 10^6 \text{ W}$$

$$\therefore \qquad m = \frac{45 \times 10^6 \times 2}{(62.64)^2} = 22937 \text{ kg}$$

Let, A = Area of two penstocks, m^2,

A_1 = Area of each penstock = $\frac{A}{2}$ and

D = Diameter of each penstock.

Then, $m = A \times C \times \rho$

(where, ρ = Mass density of water)

$$22973 = A \times 62.64 \times 1000 \qquad \left(\rho = \frac{w}{g} = \frac{9810}{9.81} = 1000 \text{ kg/m}^3\right)$$

$$A = \frac{22973}{62.64 \times 1000} = 0.366 \text{ m}^2$$

and, $$A_1 = \frac{A}{2} = \frac{0.366}{2} = 0.183 \text{ m}^2$$

Now, $$0.183 = \frac{\pi}{4} D^2$$

$$\therefore \qquad D = \left(\frac{0.183 \times 4}{\pi}\right)^{1/2} = \mathbf{0.483 \text{ m.}} \text{ (Ans.)}$$

Example 6.10. *It is observed that a run-of-river plant operates as peak load plant with a weekly load factor of 25%, all this capacity being firm capacity. Determine the minimum flow in river so that power plant may act as a base load plant. The following data is supplied : Rated installed capacity of generating plant = 10 MW; operating head = 16 m; Plant efficiency = 86%. If the stream flow is 15 m^3/s, find the daily load factor of the plant.*

Solution. Weekly load factor = 25%

Rated installed capacity of generating plant = 10 MW (= 10000 kW)

Operating head H = 16 m

Plant efficiency, η_0 = 86%

Minimum flow in river in m^3/sec, Q :

$$\because \qquad \text{Load factor} = \frac{\text{Average load}}{\text{Maximum demand}}$$

$$\therefore \qquad \text{Average load} = \text{Load factor} \times \text{maximum demand}$$

$$= 0.25 \times 10000 = 2500 \text{ kW}$$

$$E = \text{Total energy generated in one } week$$

$$= 2500 \times 24 \times 7 = 42 \times 10^4 \text{ kWh}$$

Now, $$\text{Power developed, } P = \eta_0 wQH \text{ kW}$$

$$= 0.86 \times 9.81 \times Q \times 16 \text{ kW} = 134.98\, Q \text{ kW}$$

$$\therefore \qquad E_1 = \text{Total energy generated in one } week$$

$$= 134.98\,Q \times 24 \times 7 = 22676.6\,Q \text{ kWh}$$

Now,

$$E = E_1$$

$$42 \times 10^4 = 22676.6\,Q$$

$$\therefore \quad Q = \frac{42 \times 10^4}{22676.6} = 18.52 \text{ m}^3/\text{s}$$

Hence, minimum flow rate = **18.52 m³/s. (Ans.)**

Power developed when stream flow is 15 m³/s,

$$P_1 = 134.98 \times 15$$

$$= 2024.7 \text{ kW}$$

Energy generated per day,

$$E_2 = P_1 \times \text{time} = 2024.7 \times 24$$

$$= 48592.8 \text{ kWh}$$

$$\therefore \quad \textit{Daily load factor} = \frac{\text{Average load}}{\text{Maximum load}}$$

$$= \frac{48592.8}{10000 \times 24} = \mathbf{0.2025 \text{ or } 20.25\%} \textbf{ (Ans.)}$$

Example 6.11. *Calculate the firm capacity of a run-of-river hydro-power plant to be used as 8 hours peak plant assuming daily flow in a river to be constant at 15 m³/s. Also calculate pondage factor and pondage if the head of the plant is 11 m and overall efficienty is 85%.*

Solution. Discharge, $Q = 15 \text{ m}^3/\text{s}$

Plant head, $H = 11 \text{ m}$

Overall efficiency, $\eta_0 = 85\%$

Specific weight of water, $w = 9.81 \text{ kN/m}^3$

P = Firm capacity without pondage

$$= \eta_0 \times wQH = 0.85 \times 9.81 \times 15 \times 11$$

$$= 1375.8 \text{ kW}$$

$$PF = \text{Pondage factor} = \frac{t_1}{t_2}$$

where, t_1 = Total hours in one day = 24, and

t_2 = Number of hours for which plant runs = 8

[***Pondage factor*** *is the ratio of total inflow hours in a given period to the total number of hours for which plant runs during the same period.*]

$$PF = \frac{24}{8} = \mathbf{3.} \textbf{ (Ans.)}$$

$$Q_1 = 15 \times 3 = 45 \text{ m}^3/\text{s}$$

P_1 = Firm power with pondage

$$= 1375.8 \times 3 = 4127.4 \text{ kW}$$

Pondage (*magnitude*) = (24 – 8) = 16 hours flow

$$= 16 \times 60 \times 60 \times 15 = \mathbf{8.64 \times 10^5 \text{ m}^3.} \textbf{ (Ans.)}$$

Example 6.12. *The following data relate to a pump storage power plant :*

Gross head	*280 m*
Dia. of head-race tunnel	*4.0 m*
Length of head-race tunnel	*620 m*
Flow velocity	*6.5 m/s*
Friction factor	*0.018*
Pumping efficiency	*85%*
Generation efficiency	*90%*

If the power plant discharges directly in the lower reservoir determine the plant efficiency.

Solution. Head, H = 280 m

Dia. of head-race tunnel, D = 4.0 m

Length of head-race tunnel, L = 620 m

Flow velocity C = 6.5 m/s

Friction factor f = 0.018

Pumping efficiency η_p = 85%

Generation efficiency η_g = 90%

Plant efficiency, η_{plant} :

Loss of head due to friction (h_f) is given by the equation :

$$h_f = \frac{fLC^2}{2gD} = \frac{0.018 \times 620 \times 6.5^2}{2 \times 9.81 \times 4.0} = 6.0 \text{ m}$$

Now, $h_f = xH$

or, $6 = x \times 280 \quad \therefore x = \frac{6}{280} = 0.0214$

$$\therefore \quad \eta_{plant} = \frac{1-x}{1+x} \times \eta_p \times \eta_g = \frac{(1-0.0214)}{(1+0.0214)} \times 0.85 \times 0.9$$

$= \textbf{0.7329 or 73.29\%}$ **(Ans.)**

Example 6.13. *The nature of load required for 24 hours and thermal efficiencies of the plant at the respective loads are given in the table below :*

Time period	*Load (MW)*	*Thermal efficiency (%age)*
10 A.M. to 6 P.M.	*120*	*32%*
6 P.M. to 8 P.M.	*60*	*24%*
8 P.M. to 12 A.M.	*30*	*15%*
12 A.M. to 6 A.M.	*15*	*10%*
6 A.M. to 10 A.M.	*75*	*25%*

(*i*) *Find the total input to the thermal plant if the load is supplied by the single thermal plant only.*

(*ii*) *If the above load is taken by combined thermal and pump storage plant, then find the percentage saving in the input to the plant. Thermal efficiency at full load = 32%.*

(*iii*) *The overall efficiencies in both cases.*

In pump storage plant, the pump and turbine are separate. The efficiency of pump is 82% and water turbine is 92%.

Solution. The load curve, drawn as per data given, is shown in Fig. 6.11.

Total output per day $= 75 \times 4 + 120 \times 8 + 60 \times 2 + 30 \times 4 + 15 \times 6$

$= 300 + 960 + 120 + 120 + 90 = 1590$ MWh

(i) Total input to the thermal plant :

The input to the thermal plant

$$= \frac{75 \times 4}{0.25} + \frac{120 \times 8}{0.32} + \frac{60 \times 2}{0.24} + \frac{30 \times 4}{0.15} + \frac{15 \times 6}{0.1}$$

$= 1200 + 3000 + 500 + 800 + 900 =$ **6400 MWh. (Ans.)**

(ii) Percentage saving in the input to plant :

The overall efficiency of the pump storage plant

$= 0.82 \times 0.92 = 0.7544$, or, 75.44%.

Assume that the capacity of the thermal plant is x MW when it is working in combination with pump-storage plant.

The energy used from the thermal plant to pump the water of pump storage plant during off-peak period *must be equal to* the energy supplied by the pump-storage plant during peak period.

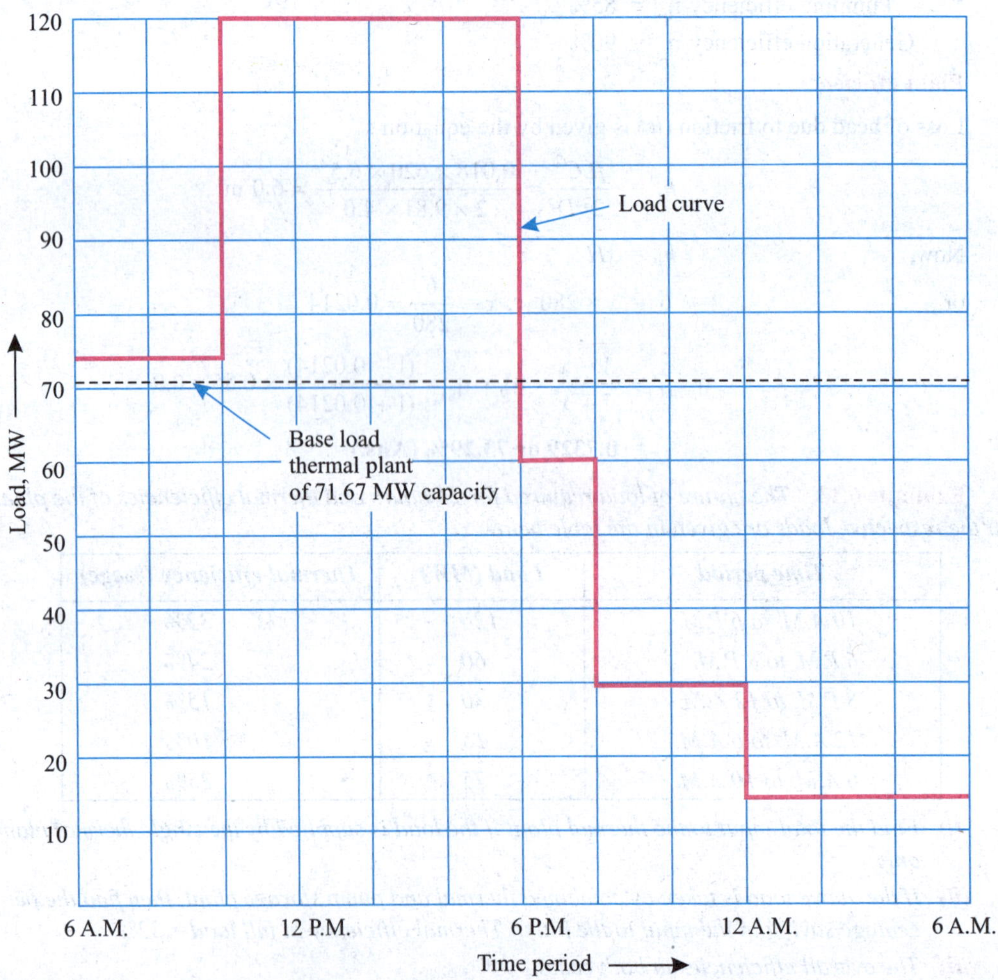

Fig. 6.11

From the Fig. 6.11, we have:

$[(x-60)\times 2+(x-30)\times 4+(x-15)\times 6]\times 0.7544=(75-x)\times 4+(120-x)\times 8$

or, $[(2x-120)+(4x-120)+(6x-90)]\times 0.7544=(300-4x)+(960-8x)$

or, $(12x-330)\times 0.7544 = 1260-12x$

or, $9.053x-248.95 = 1260-12x$

$$\therefore \quad x = \frac{1260+248.95}{(9.053+12)} = 71.67 \text{ MW}$$

The energy supplied in the second case

$$= \frac{71.67\times 24}{0.32} = 5375 \text{ MWh}$$

The percentage saving in input if the load is taken by combined thermal and pump storage plant

$$= \frac{6400-5375}{6400} = 0.16 \text{ or } \mathbf{16\%.}$$ **(Ans.)**

(*iii*) The overall efficiency in the first case :

$$= \frac{1590}{6400} = 0.2484 = \mathbf{24.84\%.}$$ **(Ans.)**

The overall efficiency in the *second case*

$$= \frac{1590}{5375} = 0.2958 \text{ or } \mathbf{29.58\%.}$$ **(Ans.)**

Example 6.14. *At a particular site of a river, the mean monthly discharge for 12 months is tabulated below :*

Month	Discharge (millions of m^3 per month)	Month	Discharge (millions of m^3 per month)
April	*250*	*Oct.*	*1000*
May	*100*	*Nov.*	*750*
June	*750*	*Dec.*	*750*
July	*1250*	*Jan.*	*500*
Aug.	*1500*	*Feb.*	*400*
Sep.	*1200*	*Mar.*	*300*

(*i*) *Draw hydrograph for the given discharges and find the average monthly flow.*

(*ii*) *Also draw the flow duration curve.*

(*iii*) *The power available at mean flow of water if available head is 90 metres at the site and overall efficiency of the generation is 82 percent.*

Take 30 days in a month.

Solution.

(*i*) Hydrograph:

The hydrograph, drawn as per data given, is shown in Fig. 6.12.

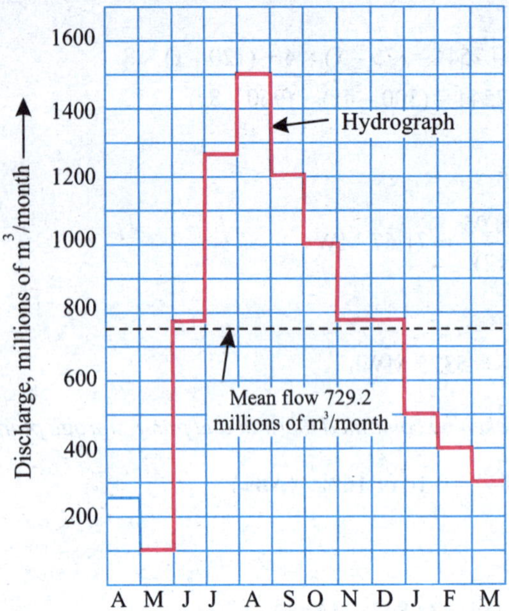

Fig. 6.12. Hydrograph.

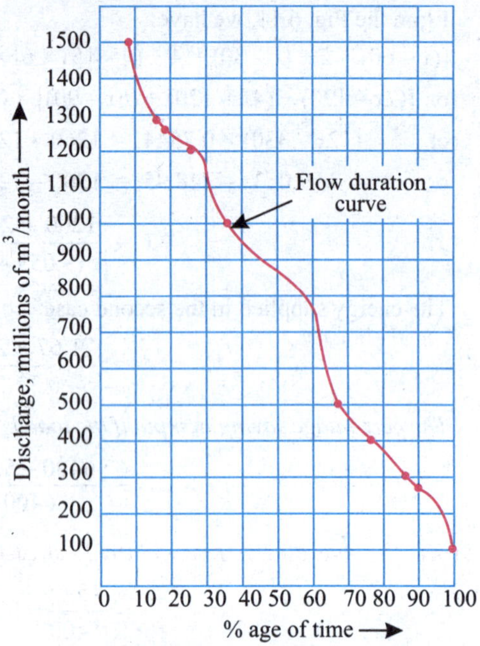

Fig. 6.13. Flow duration curve.

The average monthly flow (Refer to Fig. 6.12)

$$= \frac{250 + 100 + 750 + 1250 + 1500 + 1200 + 1000 + 750 + 750 + 500 + 400 + 300}{12}$$

$= \textbf{729.2 millions of m}^3\textbf{/month.}$ **(Ans.)**

(*ii*) Flow duration curve :

In order to obtain the *flow duration curve* it is necessary to find the lengths of time during which certain flows are available. This information is tabulated, using the hydrograph, in the following table:

Discharge per month millions of m^3	*Total number of months during which flow is available*	*Percentage time during which flow is available*
100	12	100
250	11	91.8
300	10	83.40
400	9	76.00
500	8	66.60
750	7	58.40
1000	4	33.30
1200	3	25.00
1250	2	16.65
1500	1	8.325

By using the above tabulated data, the *flow duration curve* can be shown in Fig. 6.13.

(iii) Power available at mean flow of water :

The mean/average flow available per second

$$= \frac{729.2 \times 10^6}{30 \times 24 \times 3600} = 281.3 \text{ m}^3/\text{s}$$

Average kW available at the site

$$= \frac{wQH}{1000} \times \eta_g \text{ MW} \quad (\because w = \rho g = 1000 \times 9.81 = 9.81 \text{ kN/m}^3)$$

$$= \frac{9.81 \times 281.3 \times 90}{1000} \times 0.82 = \mathbf{203.6\ MW}. \text{ (Ans.)}$$

Example 6.15. *The data for a weekly flow at a particular site is given below for 12 weeks :*

Week	*Weekly flow, m³/s*	*Week*	*Weekly flow, m³/s*
1	*3000*	*7*	*600*
2	*2000*	*8*	*2250*
3	*2700*	*9*	*4000*
4	*1000*	*10*	*2000*
5	*750*	*11*	*1500*
6	*500*	*12*	*1000*

With the help of mass curve, find the size of the reservoir and the possible rate of available flow after the reservoir has been built.

Solution. In order to draw *mass curve,* we need to find the cumulative volume of water that can be stored week after week. This is done as tabulated in the table below:

Week (*a*)	*Weekly flow in m³/s* (*b*)	*Weekly flow in day-sec-metres* (*c*) = (*b*) × 7	*Cumulative volume in day-sec-metres* (*d*)
1	3000	21000	21000
2	2000	14000	35000
3	2700	18900	53900
4	1000	7000	60900
5	750	5250	66150
6	500	3500	69650
7	600	4200	73850
8	2250	15750	89600
9	4000	28000	117600
10	2000	14000	131600
11	1500	10500	142100
12	1000	7000	149100

If the *mean flow* is available in the week at the given rate, the the total flow in the week = $7 \times \text{day} \times \text{m}^3/\text{s} = 7 \times$ day-sec-metres.

By using the above tabulated data, the mass curve can be drawn as shown in Fig. 6.14.

- Draw the tangent at the highest point on the mass curve from '*p*' and *measure the highest distance between the tangent drawn and mass curve which gives the capacity of the reservoir.*

In this **case, capacity of the reservoir = 18 × 10^3 day-sec-metres (Ans.)**

- The *slope of the line 'pq' gives the flow rate available for the given capacity reservoir.*

$$\therefore \quad \textbf{Flow rate available} = \frac{qr}{pr} = \frac{54 \times 10^3 \text{ (day-sec-metres)}}{5.5 \times 7 \text{ (days)}} = \textbf{1402.6 m}^3 \textbf{/ s. (Ans.)}$$

Example 6.16. *The following run-off data is collected for twelve months at a particular site :*

Month	*Flow per month, millions of m^3*	*Month*	*Flow per month, millions of m^3*
1	50	7	95
2	25	8	20
3	10	9	15
4	40	10	100
5	5	11	85
6	5	12	40

Determine the following :

(*i*) *The required capacity for the uniform flow of 25 millions m^3 per month throughout the year.*

(*ii*) *Spill-way capacity.*

(*iii*) *Average flow capacity if whole water is used and required capacity of the reservoir for this condition.*

Solution. In order to draw the ***mass curve,*** we need to find the cumulative volume of water that can be stored month after month. This is done as shown in the following table :

Month	*Flow per month (millions of m^3)*	*Cumulative volume, (millions of m^3)*
1	50	50
2	25	75
3	10	85
4	40	125
5	5	130
6	5	135
7	95	230
8	20	250
9	15	265
10	100	365
11	85	450
12	40	490

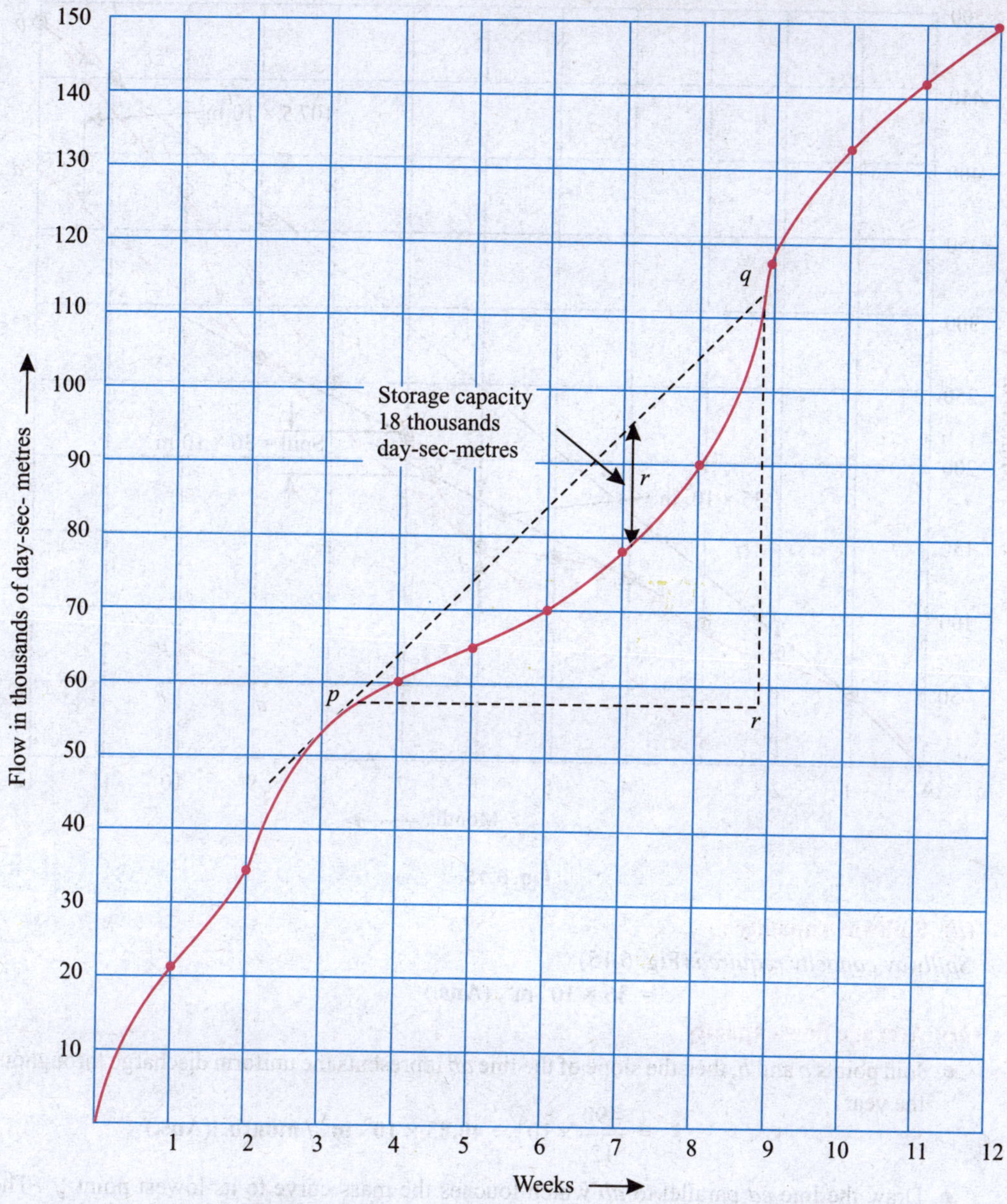

Fig. 6.14

By using the above tabulated data, the mass curve can be drawn as shown in Fig. 6.15.

***(i)* Required capacity for the uniform flow of 25 millions–m^3 per month :**

- For finding the capacity of the reservoir for uniform flow of 25 millions–m^3 per month, construct the Δpqr as shown in Fig. 6.15. *qr* represents one month and *pr* represents 25 millions–m^3.
- Now draw the parallel lines to the line *pq* through the points *e* and *g* which are apex of mass curve. The greatest departure of the mass curve from these lines represents the storage capacity.

$\therefore$ *Storage capacity* = **35 × 10^6 m^3. (Ans.)**

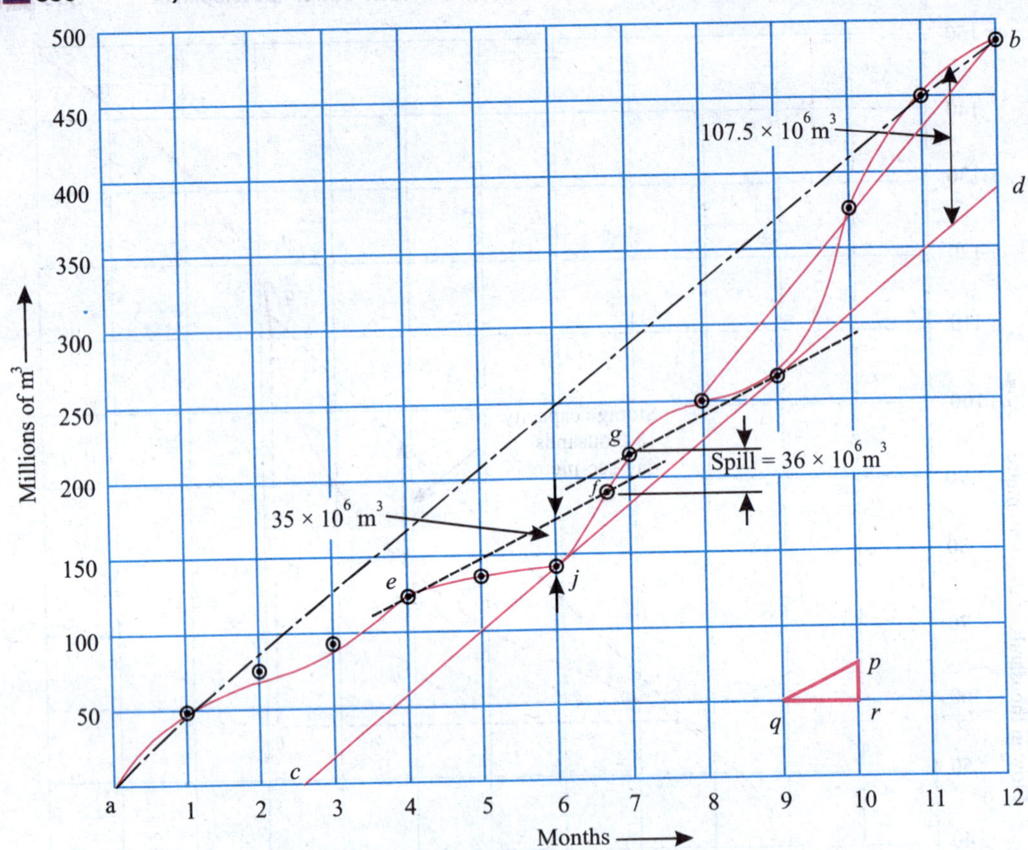

Fig. 6.15

***(ii)* Spillway capacity :**

Spillway capacity required (Fig. 6.15)

$$= \mathbf{36 \times 10^3\ m^3}. \textbf{ (Ans.)}$$

***(iii)* Average flow capacity**

- Join points a and b, then the slope of the line ab represents the uniform discharge throughout the year

$$= \frac{490}{12} \times 10^6 = \mathbf{40.83 \times 10^6\ m^3 / month.} \textbf{ (Ans.)}$$

- Draw the line cd parallel to ab which touches the mass curve to its lowest point 'j'. The *maximum departure of the line cd from the mass curve represents the required* ***storage capacity*** *for the uniform supply of 40.83 × 10⁶ m³/month.* In this case, storage capacity required

$$= \mathbf{107.5 \times 10^6\ m^3}. \textbf{ (Ans.)}$$

HIGHLIGHTS

1. A dam is a barrier to confine or rise water for storage or diversion to create a hydraulic head.
2. A canal is an open waterway excavated in natural ground. A flume is an open channel excavated on the surface or supported above ground on a trestle. A tunnel is a closed channel excavated through a natural obstruction such as ridge of higher land between the dam and the power house.

3. A surge tank is a small reservoir or tank in which the water level rises or falls to reduce the pressure swing so that they are not transmitted in full to a closed curcuit.
4. A draft tube serves the following two purposes :
   (*i*) It allows the turbine to set above tail-water level without loss of head, to facilitate inspection and maintenance.
   (*ii*) It regains, by diffuser action, the major portion of the kinetic energy delivered to it from the runner.
5. The plants which cater for the base load of the system are called 'base load plants' whereas the plants which can supply the power during peak loads are known as 'peak load plants'.
6. Microhydel plants (microstations) make use of standardized bulb sets with unit output rainging from 100 to 1000 kW working under heads between 1.5 to 10 metres.
7. The *specific speed* of a turbine is defined as the speed of a geometrically similar turbine that would develop one brake power under a head of one metre.
8. The Pelton turbine is a tangential flow impulse turbine. The pressure over the pelton wheel is constant and equal to atmosphere, so that energy transfer occurs due to purely impulse action.
9. The modern Francis water turbine is an inward mixed flow reaction turbine. It operates under medium heads and also requires medium quantity of water.
10. In the propeller turbine the runner blades are fixed and not-adjustable. In Kaplan turbine, which is a modification of propeller turbine the runner blades are adjustable and can be rotated about the pivots fixed to the boss of runner.
11. *Cavitation* may be defined as the phenomenon which manifests itself in the pitting of the metallic surfaces of turbine parts because of formation of cavitites.
12. *Hydrology* may be defined as the science which deal with the depletion and replenishment of water resources.
13. Run-off includes all the water flowing in the stream channel at any given section. It can be measured by the following methods:
   (*i*) From rainfall records
   (*ii*) Empirical formulae
   (*iii*) Run-off curve and tables
   (*iv*) Discharge observation method.
14. *Hydrograph* is defined as a graph showing discharge (run-off) of flowing water with respect to time for a specified time. It indicates the power available from the stream at different time of day, week or year.
15. *Flow duration curve* represents the run-off data for the given time. It is plotted between flow available during a period versus the fraction of time.

THEORETICAL QUESTIONS

1. Define hydrology.
2. Draw and explain the hydrologic cycle.
3. Define run-off. How is it measured ?
4. List the factors which affect run-off.
5. What is a hydrograph ?
6. What is a unit hydrograph ? What are the limitations to the use of unit hydrograph ?
7. What is a flow duration curve ?
8. What is a mass curve ?
9. Write a short note on hydropower development in India.

UNSOLVED EXAMPLES

1. The following data is available for a hydropower plant:

   Available head = 130 m; catchment area = 220 sq. km; annual average rainfall = 150 cm ; turbine efficiency = 86%; generator efficiency = 91% ; percolation and evaporation losses = 18%.

   Determine power developed in MW taking load factor as unity. **[Ans.** 8.563 MW]

2. From the investigation of a hydrosite the following data is available :

   Available head = 50 m; catchment area = 50 sq. km; rainfall = 150 cm per year; 70% of rainfall can be utilised; turbine efficiency 80%, generater efficiency = 91%; penstock efficiency = 75%; load factor = 60%.

   Determine the suitable capacity of a turbo-generator.

   [Ans. 750 kW (maximum rating). Francis or Kaplan turbine]

3. At a particular site the mean discharge (in millions of m^3) of a river in 12 months from January to December is respectively 80, 50, 40, 20, 0, 100, 150, 200, 220, 120, 100, 80.

   (*i*) Draw a hydrograph and find the mean flow.

   (*ii*) Also draw the flow duration curve.

   (*iii*) Find the power in MW available at mean flow if the head available is 100 m and overall efficiency of generation is 100 m and overall efficiency of generation is 80%.

   Take the each month of 30 days.

   [Ans. (*i*) 96.67 millions of m (*ii*) 29.2 MW]

Chapter 7

FLUIDICS

7.1. INTRODUCTION

The term **fluidics** relates to the combination of the two functions namely *fluid amplification and fluid logic.*

Fluidics is *defined as a control technology which makes use of fluids interaction to produce useful signals.*

- The *"field of fluidics" is the study of the performance and response characteristics of control systems, computing devices and logical switch gears based on the fluidic elements.*
- In finer control engineering, non-moving logic elements find a prominent place. *Irrespective of the development of electronics, low pressure pneumatics and fluidic elements have certain specific characteristics which put them at par with electronic controls even for modern sophisticated machines.* Various fluidic elements have been developed conforming to the need of logic functions in the industrial automation.

— The *basic principle is derived from the "Tesla's fluid-diode" and theory of "wall attachment" discovered by Coanda.*

— More and more fluid-logic elements in the form of logic gates like OR, NOR, etc. are being used along with power pneumatic circuits to offer better control and feedback to the pneumatic system. One of the major areas of their application is in the *field of sensors*. The present day state of art of pneumatic sensors is quite competitive with other form of sensors *e.g.,* fine mechanical, opto-electrical, inductive, hydraulic, ultra sound and magnetic devices etc. and are therefore widely used in various engineering tools and instruments.

History of fluidics :

- 1904 : *L. Prandtl* (German aerodynamist) suggested that in a wide angled diffuser the flow separation could be controlled by applying suction to the boundary layer.
- 1916 : — Nikola *Tesla* filed a patent for a "Valvular Conduit" for fluids in which there was an easy direction of flow and a difficult direction, owing to the interference caused by the divided branch flow opposing the intended flow-direction as can be seen in Fig. 7.1 which was granted in 1920.

— Whill Tesla claimed to have invented the first fluid device having no moving parts.

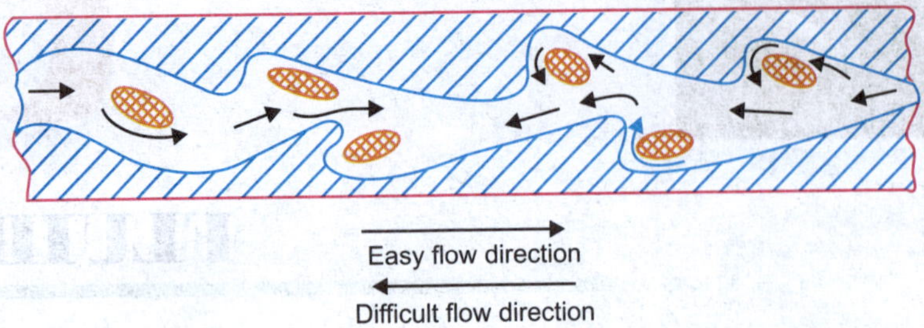

Fig. 7.1. Tesla's tube.

- 1930 : Henri *Coanda* (a Rumanian engineer) discovered that a free jet would follow an adjacent curved or inclined surface. Coanda's theory of "*wall attachment*" was a major stepping stone in the development of this field. Later this effect was used for the development of fluidic logic components.
- 1958 : Moore and Klive, working with wide angled diffusers, discovered that a jet could have two or three stable stages, depending on the angle of diffuser through which the jet was flowing.
- 1962 : Ray Auger discovered a fluidic logic element called "*Turbulence amplifier*".

7.2. ADVANTAGES, DISADVANTAGES AND APPLICATIONS OF FLUIDIC DEVICES/FLUIDICS

During the last few decades, the technology of electronic control system has leapt through milestones of innovations with various forms of transistors, ICs and the like, production costs have drastically reduced, and reliability has increased many a fold. Inspite of these developments, fluidic devices are finding wide applications and claim the following *advantages* :

Advantages :

1. More reliable.
2. Simpler in construction.
3. Smaller in size, mass or weight.
4. Offer exceptional physical and thermal stability and ruggedness.
5. Noise free.
6. Hazard free.
7. Mode of energy feeding to a fluidic system is very simple.
8. Fuidic elements are easily adaptable to logic functions in engineering applications.
9. Have good response and performance characterisitics (0.001 second).
10. Since they contain no moving parts, not much maintenance problems are encountered.

Disadvantages :

1. Unsuitable for incompressible fluids.
2. Slow speeds and low power outputs.
3. Complex systems are impracticable.
4. Inefficient in operation; they cannot be used in high-speed switching operations.
5. Not suitable for intermittent operation control systems.
6. Limited development of the field.

- The working pressure used for pneumatic fluid devices is very less within 0.05 to 0.1 bar but need not always be so. They are quite sensitive to load and position which can be easily sensed with the help of instrumentation.

Applications : The fluidic devices are used :

1. To measure flow rates;
2. To provide on-off controls;
3. To check weights;
4. To operate various types of machinery, etc.

- If used with a chemically inert gas such as nitrogen or helium, these devices are especially advantageous in cases where fire or electrical hazards are present, as in the manufacture of explosives.

7.3. FLUIDIC (OR FLUID LOGIC) ELEMENTS

7.3.1. General Aspects

The fluidic elements (also called *fuid logic elements*) were developed in the early sixties. The primary merit of these elements over all other forms of control elements is that they have a *minimum number of mechanical moving parts;* because of this, these elements are also known as "*non-moving logic*" controllers.

These element claim the following *advantages :*

1. Quite insensitive to temperature, vibration, shock, electric noise and radiation.
2. Insensitive to electromagnetic interferences etc.
3. Need no actuating force.
4. No wear and tear of elements.
5. Need very little space for mounting.

7.3.2. Coanda Effect

Physically a fluidic device is a block of material having an internal network of passages. One of the basic underlying principle of the functioning of these flow passages was given by Henri Marie Coanda (a Ramanian engineer) and is known as *"Coanda effect"* or *"Wall attachment effect"*.

This effect is explained below : Refer to Fig. 7.2.

- **Fig. 7.2. (*i*) :** A free jet of air is emitted into a confined region or orifice at a velocity high enough to *produce turbulent flow.*

  Fig. 7.2. (*ii*) : The free jet of air will continue in a given direction, pulling in with it the available air from its surroundings as it leaves the orifice. If there is greater availability of this entraining air from one side, a small vortex area (low pressure area/region) is created near the nozzle exit. This low pressure area then tends to attract the free jet, distorting it and pulling it towards the wall, because the atmospheric pressure on the other side forces the jet to cling to the surface.

  Fig. 7.3. (*iii*) : The free jet attachment continues until a small air supply is fed to the low pressure area, thus relieving the attraction of the jet to the wall. When this signal is injected, the free jet then detaches itself from the wall and resumes its normal uninterrupted flow path.

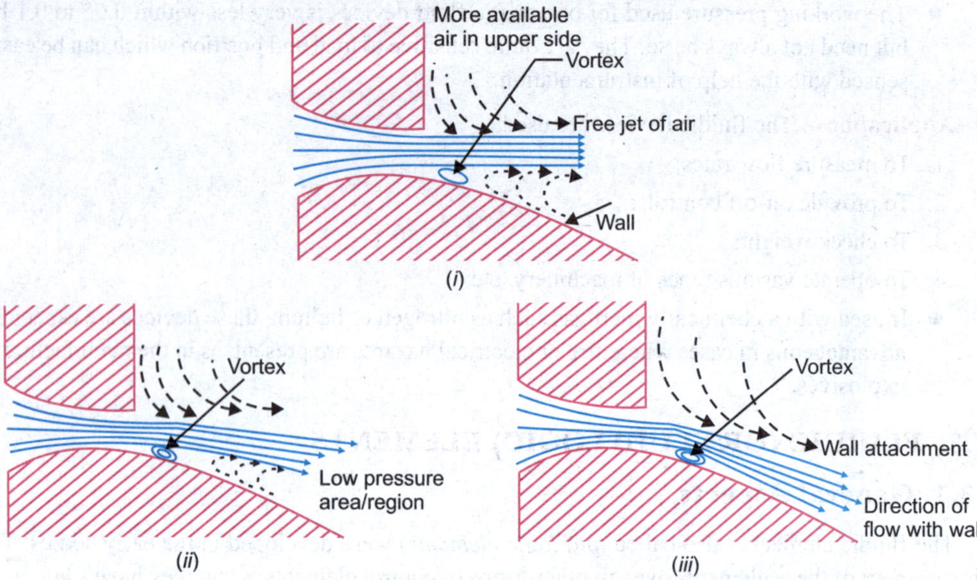

Fig. 7.2. Coanda/Wall attachment effect.

7.3.3 Classification of Fluidic Devices

Fluidic devices may be *classified* as follows :

A. **1. Digital fluidic devices :**

(*i*) Wall attachment. (*ii*) Turbulence.

(*iii*) Vortex feedback.

- A *digital element* has two outputs, and flow takes place from one or other output depending upon the presence or absence of control signal. The flow never splits between two outputs; it is either fully out of one or other. It can be compared to a simple "ON-OFF" switch.
- The digital devices are of the following two categories :

(*a*) *Monostable :*

— In the absence of a signal the device always selects one of its two output states. The other state can only be selected by applying and maintaining the signal *i.e.,* when the applied signal is removed the device returns to its former state.

These devices do *not* have any *memory*.

Example : 3/2 normally operated, spring return D.C. valve.

(*b*) *Bistable :*

— The output of this device in any direction is stable, irrespective of the fact whether applied signal is present or absent.

— These devices are *equipped with memory.*

Example : 4/2 pilot operated D.C. valve.

Examples of digital fluidic devices :

(*i*) Bistable flip-flops; (*ii*) Logic gates, etc.

2. Analogue fluidic devices :

(*i*) Stream or beam deflection. (*ii*) Impact modulator.

(*iii*) Vortex.

- An "*Analogue element*" varies its output *continuously* as a function of the control output signal and gives *proportional control* (similar to an accelerator pedal of a car)

  Examples of analogue fluidic devices :

  (*i*) Fluidic position sensors; (*ii*) Fluidic vortex amplifier;
  (*iii*) Wall attachment amplifier; (*iv*) Fluidic oscillator, etc.

B. 1. Active fluidic devices :

— In an active fluidic device, the power nozzle of the device is *continuously supplied with an air pressure source.*

2. Passive fluidic devices :

— These devices intermittently receive pressure signals at the power nozzle, and this signal usually comes from the *output of preceding element in the circuit.* In other words the air only passes momentarily through the device.

C. 1. Fluid logic devices :

(*i*) Bi-stable flip-flop; (*ii*) AND gate;
(*iii*) OR-NOR gate, etc.

2. Fluidic sensors :

(*i*) Interruptible jet sensor; (*ii*) Reflex sensor;
(*iii*) Back pressure sensor.

3. Fluidic amplifiers :

(*i*) Turbulence amplifier; (*ii*) Vortex amplifiers.

7.3.4. Fluid Logic Devices

The following terms are frequently used in connection with fluid logic devices :

1. ***Logic function or gate.*** *It is defined as an assembly of one or more logic elements which produce a desired effect on satisfying certain conditions, i.e., if we apply correct input signals, we will get the required output.*
2. ***Input signal.*** *It is the pressure or flow which is directed into the input part to control the logic function or an element.*
3. ***Output signal.*** *It is the pressure or flow which is leaving the output part of a logic function or an element.*

7.3.4.1. Bi-stable flip-flop

A bi-stable flip-flop is the most common fluid logic device. It works on the principle of *Coanda effect.*

Fig. 7.3. shows this device :

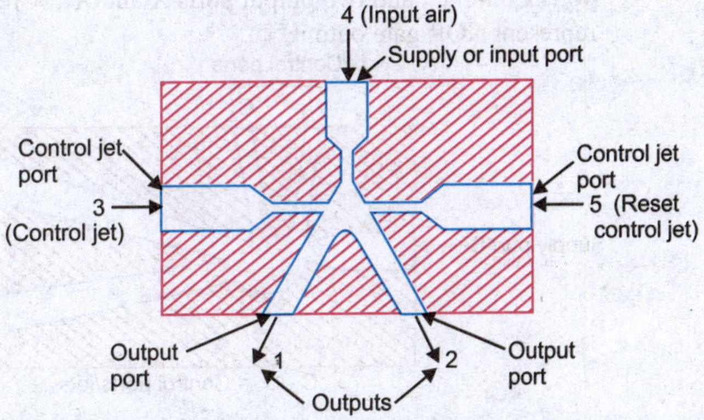

Fig. 7.3. Bi-stable flip-flop.

— In general, it consist of five ports, out of which one is supply port; two control jet ports and two output ports.

— Supply is always present on the input port (4) whereas output is dependent on the control jets (3, 5), *i.e.,* when fluid is passed from control jet 3 there will be an output – 1 and similarly when there is a supply at control jet–5 the output will be an output – 2.

- The device is suited for binary logic functions, as the output either exists or does not. The output from this device can be used as "pilot signal" for actuating various valves with low pressure actuating element.
- This device can be manufactured easily out of glass or plastic, not much larger than the size of a coin, the orifice size in such devices is of the order of 0.25 mm in diameter and the working pressure is about 0.05 to 0.1 bar. The operational speed is about 1000 cycles per second.

7.3.4.2. AND gate

Fig. 7.4. shows a fluidic AND gate/element; its operation is based on Coanda's wall attachment theory.

(*i*) Under no-control condition, output is at 0 (output–0).

(*ii*) When control jet C_1 is on, output is at 1 (output–1).

(*iii*) *When both the control jets C_1 and C_2 are on, only then there is an output at the AND gate, i.e.,* output is at 2 (output–2).

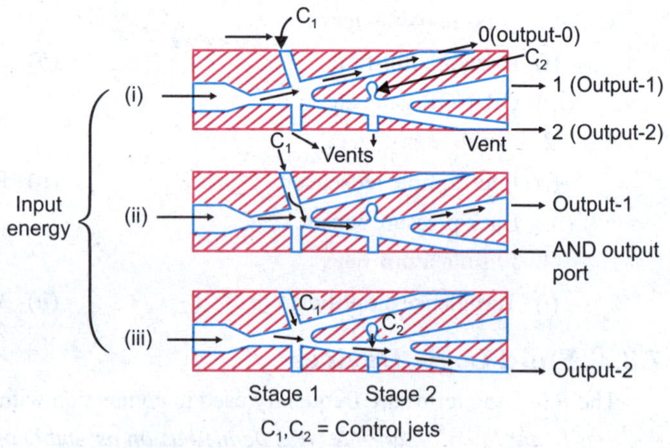

Fig. 7.4. AND gate.

7.3.4.3. OR–NOR gate

Fig. 7.5 shows fluidic OR–NOR gate.

— It consists of supply port S from which the fluid is supplied all the time. There are two control ports C_1 and C_2 and two output ports X and X_1. X represents the OR gate output while X_1 represent NOR gate output.

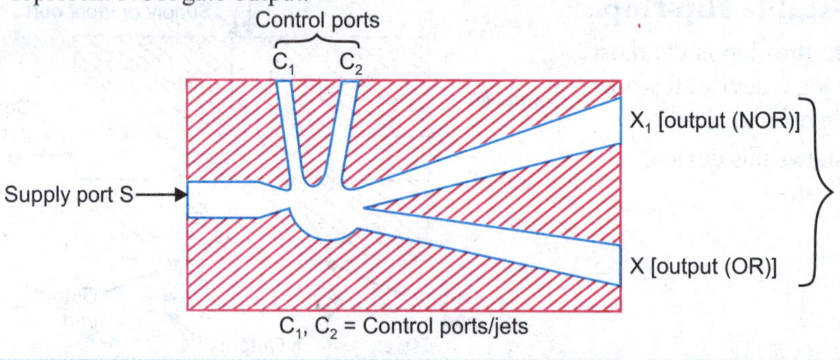

Fig. 7.5. OR–NOR gate.

— When the control jets C_1 and C_2 are absent, supply S is passed through the gate via output X_1. It implies that when there is no control input, there is an output. This is the characteristics of NOR gate and X_1 is considered as NOR output.

— When fluid stream is present at either of control jet C_1 or C_2 or both at C_1 and C_2, there will be output at X. It implies that even with one control input there is an output. This is the logic characteristic of OR gate and X output is considered as OR output.

The complete device is called as OR–NOR gate.

- Fig. 7.6 shows a circuit diagram illustrating the application of fluidic pressure for actuating a low pressure diaphragm valve.

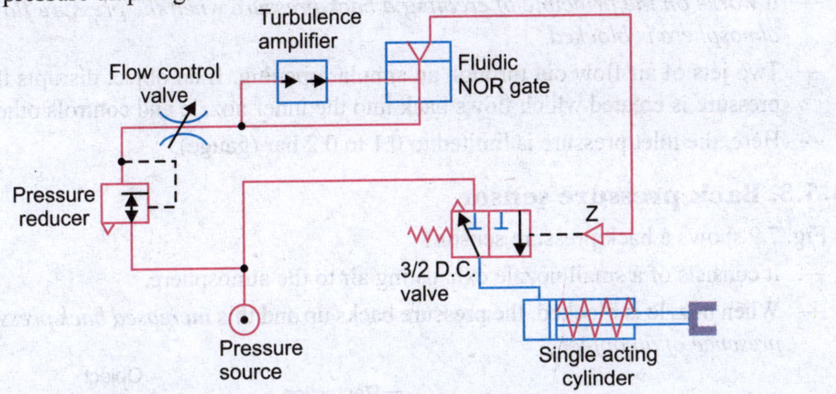

Fig. 7.6. Application of fluidic elements in pneumatic circuit.

7.3.5. Fluidic Sensors

These sensors (generally of very small size) are mainly used *to detect the presence of objects.* They are *designed to provide a signal in the form of fluid jet, to indicate the presence of an object.*

- The output of these sensors can be directly used as a *control signal for pneumatic logic circuits.* These control signals, however, need to be amplified when employed in pneumatic and hydraulic devices.

A few important fluidic sensors are described in the following subarticles.

7.3.5.1. Interruptible jet sensor

Fig. 7.7 shows an interruptible jet sensor :

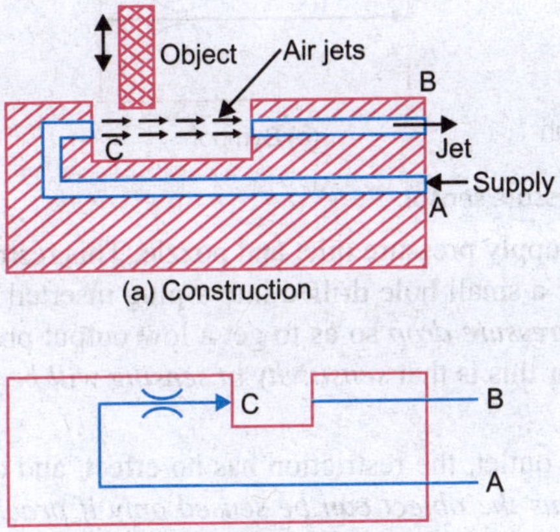

Fig. 7.7. Interruptible jet sensor.

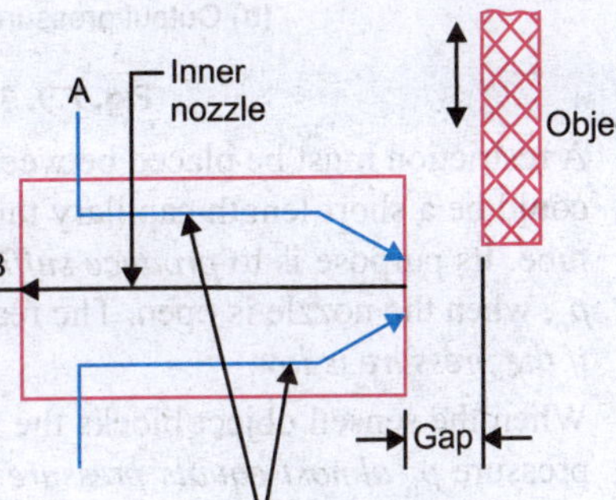

Fig. 7.8. Reflex sensor-symbol.

— Low pressure air is permitted to pass from A to B uninterrupted. But if a mechanical object comes between C and B, the air jets are blocked after leaving the port C and the signal at B disappears.

— The pressure at A ranges between 0.1 bar to 2 bar, but normal air-line pressure may also be used. However, in that case, it is better to throttle the air before it enters at inlet A.

— Sensing gap is limited to 5 mm.

7.3.5.2. Reflex sensor

Fig. 7.8. shows the symbol of a reflex sensor :

— It *works on the principle of creating a back pressure when the pressure flowing out to the atmosphere is blocked.*

— Two jets of air flow out through an annular opening. If an object disrupts the jets, a back pressure is created which flows back into the inner nozzle and controls other valves.

— Here, the inlet pressure is limited to 0.1 to 0.2 bar (gauge).

7.3.5.3. Back pressure sensor

Fig. 7.9 shows a back pressure sensor :

— It consists of a small nozzle exhausting air to the atmosphere.

— When nozzle is blocked, the pressure backs up and this *increased back pressure signals the presence of an object.*

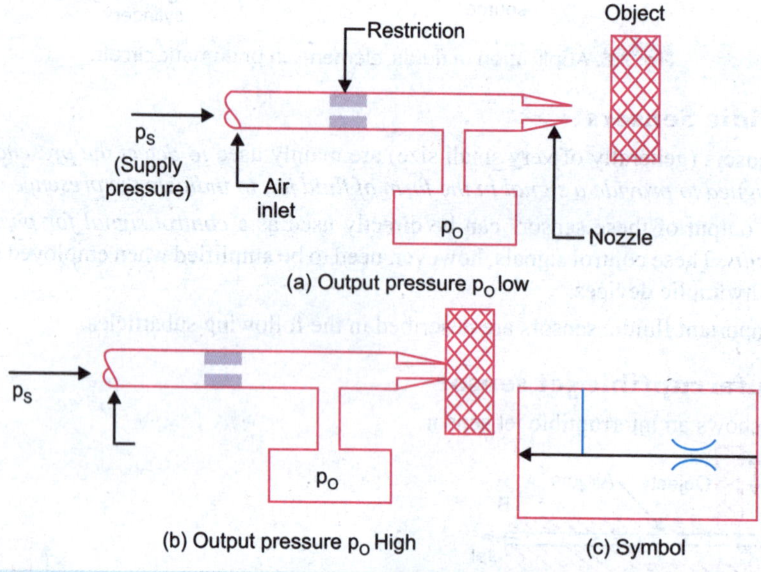

Fig. 7.9. Back pressure sensor.

— A restriction must be placed between the supply pressure inlet and nozzle. This restriction could be a short length capillary tubing or a small hole drilled into a plug inserted in the tube. Its purpose is to *produce sufficient pressure drop* so as to get a low output pressure p_o, when the nozzle is open. The reason for this is that *sensitivity of sensing will be better if the pressure is low.*

— When the sensed object blocks the nozzle outlet, the restriction has no effect, and output pressure p_o *almost equals pressure* $\mathbf{p_s}$. *Thus the object can be sensed only if practically touches the nozzle.*

7.3.6. Fluidic amplifiers

If a device gives a large change in output either pressure, flow or both, as a result of a small change in control input, the device is said to have "gain" or in other words it is an ***amplifier****.*

An amplifier can be either electric, hydraulic, pneumatic, or fluidic.

- *"Fluidic amplifiers" not only control but also provide certain amplification to the fluid signals.*

7.3.6.1. Turbulence amplifier

Fig. 7.10. Illustrates a turbulence amplifier (T.A.)

— Low pressure fluid is conducted through a long, small bore pipe to achieve *laminar flow.*

— This laminar flow issued from the inlet pipe transits the space of about 20 mm and is then collected by the outlet pipe. The space between input pipe and output pipe is protected by a cylindrical shield of diameter 20 to 30 times that of input/output pipe. This shield houses the *'control inputs'* and it is vented to atmosphere as shown in the figure.

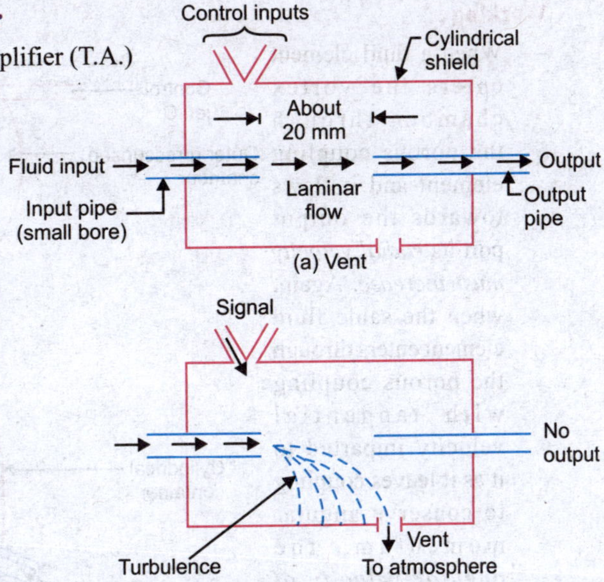

Fig. 7.10. Turbulence amplifier.

— When there is no control input or signal, the laminar flow proceeds from input to output through the open space within the turbulence amplifier. However, if there is an input control signal, this will create a turbulence between the input and output pipes; thus the output *will not collect* any flow, hence there is *No output.*

— When the control signal is *removed,* the laminar flow is reestablished and *there will be an output.*

— There can be several control inputs and the logic function is NOR.

In these devices, the switch off time is about 4 ms (milli seconds) and the total cycle time may not exceed 6 to 7 ms.

- Following are the examples of the application of the "turbulence amplifier" to some industrial problems :

(*i*) Sensing; (*ii*) Counting;
(*iii*) Discrimination; (*iv*) Timing;
(*v*) Dispensing; (*vi*) Liquid level sensing.

7.3.6.2. Vortex amplifier

This fluidic device is used to regulate the flow of fluid by *utilising the properties of a vortex.*

Fig. 7.11. shows a vortex amplifier :

— It consists of a cylindrical disc like container, which is divided by a *cylindrical porous element* into two chambers : *Outer* chamber and *vortex* chamber.

— Supply port S is provided for fluid inlet to the cylindrical disc.

— Control jet C is provided to generate vortex in the vortex chamber. Several control jets along the circumference can be provided depending upon capacity, which throw streams of fluid in *tangential direction* and generate vortex motion in fluid.

— At the centre there is an output port X from where signal is transmitted.

Working :

— When a fluid element enters the vortex chamber through the porous coupling element and it flows towards the output port its *radial velocity must increase.* Again, when the same fluid element enters through the porous coupling with tangential velocity imparted to it as it leaves coupling to conserve angular momentum, the *angular velocity of fluid must increase* as it approaches the output port X. Thus there are two amplifying properties of vortex amplifier : (*i*) Increase in radial velocity; (*ii*) Increase in angular velocity.

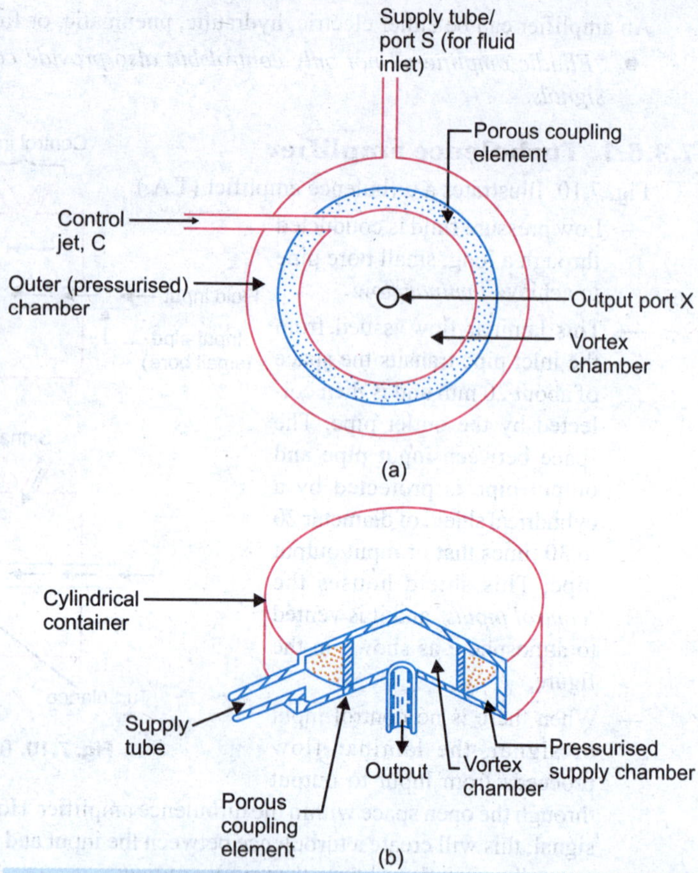

Fig. 7.11. Vortex amplifier.

— When the supply pressure is regulated, if the vortex motion of the fluid produces a centrifugal pressure drop across the vortex chamber, there is less pressure at the exit to expel the liquid, thus *less fluid flows from the exit.* The introduction of vortex motion into the chamber effectively throttles the fluid through the chamber.

It has been observed from the curves of output flow as function of the control pressure applied to vortex chamber that, the *flow is reduced with the increase of control pressure.*

— *When the control pressures are sufficiently high, the flow from supply is zero or even slightly reversed.* Centrifugal pressure increases the pressure at the inside surface of the porous coupling element so that the total pressure is *equal to or slightly higher than the supply pressure.* As a result, the control flow *can cut-off completely the flow from the supply chamber.*

The flow versus control pessure curves, for high control pressures, become tangent to the time of zero supply flow.

- *In any case, the control pressure should not be less than the supply pressure.*

7.4. COMPARISON AMONG DIFFERENT SWITCHING ELEMENTS

The comparison among different switching elements (Pneumatic valves, moving part logic elements and fluidics) is given in tabular form on the next page :

S.No.	Aspects	Pneumatic valves	Moving part logic elements	Fluidics
1.	*Air consumption*	Low	Low	High
2.	*Supply air pressure*	High	Intermediate to high	Low
3.	*Relative size*	Large	Intermediate	Intermediate
4.	*Response time*	10 to 20 ms	5 to 10 ms	1-2 ms
5.	*Sensitivity to dirt and conversion*	Good	Good	Poor
6.	*Sensitivity to shock and vibration*	Excellent	Intermediate	Excellent
7.	*Sensitivity to electric noise and radiation*	Excellent	Excellent	Excellent
8.	*Sensitivity to high temperatures*	Good	Good	Excellent
9.	*Life expectancy*	10^7 to 10^8 cycles	10^7 to 10^8 cycles	Unlimited (with clean air)

HIGHLIGHTS

1. *Fluidics* is defined as a control technology which makes use of fluids interaction to produce useful signals.
2. Fluid devices may be of following types :
   (*i*) Digital; Analogue; (*ii*) Active; passive; (*iii*) Logic devices; sensor; amplifiers.
3. Fluidics amplifiers not only control but also provide certain amplification to the fluid signals.

OBJECTIVE TYPE QUESTIONS

Fill in the Blank or Say "Yes" or "No".

1. 'Valvular conduit' was patented by ....
2. The control circuit element has a limited life.
3. A digital element has one output.
4. .... element varies its output continuously as a function of the control output signal and gives proportional control.
5. The digital devices are least important for industrial applications.
6. A device having bistable properties is also said to have ....
7. In .... device the air only passes momentarily.
8. A logic gate is a device which has inputs and outputs.
9. .... devised the first NOT gate.
10. Coanda effect was discovered in 1930 by ....
11. Bi-stable flip-flop works on the principle of .... effect.
12. The output from Bi-stable flip-flop can be used as a pilot signal for actuating various valves with low pressure actuating element.
13. Fluidic sensors are primarily used for detecting the presence of objects.

14. Vortex amplifier is used to regulate the flow of fluid by utilizing properties of a vortex.
15. The power required to turn off the turbulence amplifier is much more than the power in the output

ANSWERS

1. Nikola Tesla	2. No	3. No	4. Analogue
5. No 6. Memory	7. passive	8. Yes	9. Prandtl
10. Henri Coanda	11. Coanda	12. Yes	13. Yes
14. Yes. 15. No			

THEORETICAL QUESTIONS

1. Define the term 'Fluidics'.
2. Enlist the advantages, disadvantages and applications of fluidic devices/fluidics.
3. Explain briefly "Coanda effect".
4. Give the classification of fluidic devices.
5. Differentiate between the following fluidic devices :
   (*i*) Digital and analogue. (*ii*) Active and passive.
   (*iii*) Mono-stable and bi-stable.
6. Define the following terms :
   (*i*) Logic function or gate. (*ii*) Input signal.
   (*iii*) Output signal.
7. Explain briefly any two of the following fluid logic devices :
   (*i*) Bi-stable flip-flop. (*ii*) AND gate.
   (*iii*) OR-NOR gate.
8. Describe briefly the following fluidic sensors :
   (*i*) Interruptible jet sensor. (*ii*) Back pressure sensor.
9. Explain briefly air following fluidic amplifiers :
   (*i*) Turbulence amplifier. (*ii*) Vortex amplifier.
10. Give the comparison among pneumatic valves, moving part logic elements and fluidics.

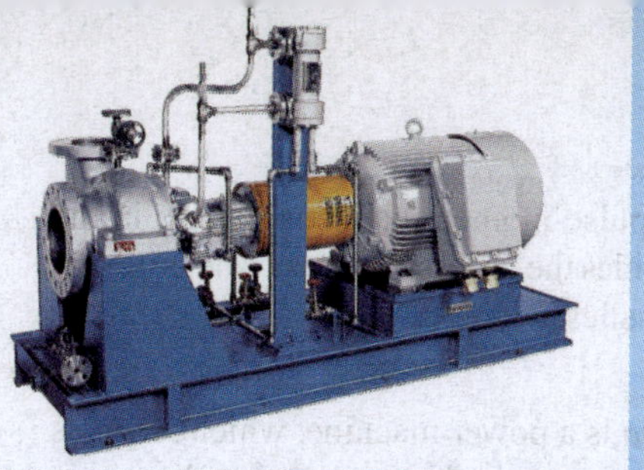

Chapter 8

UNIVERSITIES' QUESTIONS (LATEST) WITH "SOLUTIONS"

SECTION A: SHORT ANSWER QUESTIONS

Q. 1. *What is a "fluid jet"?*

Ans. A ***fluid jet*** is a stream of fluid issuing from a nozzle with a high velocity and hence a high kinetic energy. When it impinges on a plate or vane, it exerts a force on it (due to change in momentum). This force (hydrodynamic) can be evaluated by using Impulse-momentum principle.

Q. 2. *Write down the formulae for the force exerted by a jet of water at a stationary plate (F_x) in the following cases: (i) Vertical plates; (ii) Inclined plate; (iii) Curved plate and jet strikes at one of tips of the jet.*

Ans. (*i*) $F_x = \rho a V^2$; (*ii*) $\rho a V^2 \sin\theta$; (*iii*) $2\rho a V^2 \cos\theta$

where, V = Velocity of the jet; θ = Angle between the jet and plate for inclined plate, and angle made by the jet with the direction of for curved plate.

Q. 3. *In case of jet propulsion of ships, what is the efficiency of propulsion when the inlet orifices face the direction of motion of the ship?*

Ans. Efficiency of propulsion, $\eta = \dfrac{2u}{V + 2u}$

where, u = Velocity of the moving ship; V = Absolute velocity of the moving jet.

Q. 4. *Point out the significance of word 'Free' in impact of free jets.*

Ans. 'Free' in impact of free jets means 'constant pressure throughout' when the jet impinges upon stationary or moving objects such as flat plates and vanes of different shapes and orientations in the study of 'impact of jets'.

Q. 5. *Explain impulse-momentum equation.*

Ans. When a force (push or pull) is applied on body it tries to change the state of rest or state of motion of that body. The amount of force applied is equal to the rate of change of momentum, where momentum is the product of mass and velocity.

Mathematically, $$F = ma = m\left(\frac{dv}{dt}\right) = \frac{d}{dt}(mv)$$

or, $$F = \frac{m}{t}(v_2 - v_1)$$

or, $$Ft = m(v_2 - v_1)$$

where, product Ft is the 'impulse' and is equal to the change in momentum.

Q. 6. *What do you mean by 'jet propulsion'?*

Ans. ***Jet propulsion*** is one of the applications of the 'impulse-moment equation' where in the reaction of high velocity jet issuing from a nozzle provides the necessary thrust. The principle is employed in propelling the ships, aircrafts and missiles.

Q. 7. *What is a dynamic machine?*

Ans. The term dynamic means power. A ***dynamic machine*** is a power machine, which receives energy from the flowing fluid in the form of momentum and converts the change in momentum into useful work.

Q. 8. *What is an impulse turbine?*

Ans. In 'impulse turbine' a high velocity jet issued from nozzle strikes a series of suitably shaped buckets fixed on the periphery of a wheel. The wheel gets resulting momentum and it gets rotated and thus we get the mechanical energy from the turbine.

Q. 9. *Classify turbines on the basis of direction of flow.*

Ans. The turbines are classified on the basis of direction of flow through the runner as follows:

(*i*) Tangential flow turbine;

(*ii*) Radial flow turbine;

(*iii*) Axial flow turbine;

(*iv*) Mixed flow turbine.

Q. 10. *What is 'scale effect'?*

Ans. However a smooth a model is made, the geometric similarity between the prototype and model cannot be extended to surface roughness. This variation of surfaceness with respect to the size of turbine will cause a small but appreciable variation in the proportion of the effective head lost due to hydraulic friction. Thus the efficiency of prototype will be different from the corresponding model efficiency. This aspect is referred to as ***scale effect***.

Q. 11. *List down some advantages of centrifugal pump over displacement pump.*

Ans. Some of the advantages claimed by centrifugal pump over displacement (reciprocating) pump are:

(*i*) The cost of a centrifugal pump is less as it has fewer parts.

(*ii*) Installation and maintenance are easier and cheaper.

(*iii*) Its discharging capacity is much greater than that of a reciprocating pump.

(*iv*) Its performance characteristics are superior.

(*v*) It can be directly coupled to an electric motor or an oil engine.

- However, because of higher efficiency the reciprocating pumps are still employed for high heads and small discharges. A reciprocating pump can build up very high pressures (as high as 700 bar or even more) and as such these pumps are made for lifting oils from very deep oil wells.

Q. 12. *What do you understand by 'specific speed' of a centrifugal pump?*

Ans. The 'specific speed' of a centrifugal pump is defined as the speed of a geometrically similar pump which would deliver unit quantity (one cubic metre of liquid per second) against a unit head (one metre).

It is denoted by N_s. The specific speed of is a characteristic of pumps which can be used as a basis for comparing the performance of different pumps.

Q. 13. *Explain the term 'negative slip' as referred to reciprocating pumps.*

Ans. In most of the reciprocating pumps $Q_{act.}$ (actual discharge) is less than $Q_{th.}$ (theoretical discharge); in such a case the value of C_d (coefficient of discharge) is less than unity and the slip of the pump is '*positive*'. However, in some cases Q_{act} may be 'more' than Q_{th} ; in such a case C_d is more than unity and the slip will be '*negative*'. The slip will be negative when there is direct connection between the suction and delivery sides before the end of the suction stroke. This happens if the momentum of liquid in the suction is large enough to open the delivery valve before the beginning of delivery stroke. The negative slip is possible in case of pumps having *long suction pipe and a short delivery pipe, especially when these are operating at high speeds*.

Q. 14. *Explain the term Net Positive Suction Head (NPSH).*

Ans. NPSH is defined as the absolute pressure head at the inlet to the pump, minus the vapour pressure head (in absolute units) plus the velocity head.

or, NPSH = Absolute pressure head at the inlet of the pump – vapour pressure head (absolute units) + velocity head.

This term is frequently used in pump industry and has significance only when cavitating liquids are used. NPSH is a parameter (dimensional) that can be used to check cavitation in pump.

The minimum NPSH depends upon the pump design, its speed and the discharge.

Q. 15. *List the various functions of surge tanks.*

Ans. Surge tank has the following functions:

(*i*) To control the pressure variations, due to rapid changes in the pipeline flow, thus eliminating water hammer possibilities.

(*ii*) To regulate the flow of water to the turbines.

(*iii*) To reduce the distance between the free water surface and turbine, thereby reducing the water hammer effect on penstock.

Q. 16. *What is a 'hydraulic ram'?*

Ans. ***Hydraulic ram*** is a device with which small quantities of water can be pumped to higher levels from the available large quantity of water of low head.

Q. 17. *What is the function of notch in Pelton turbine?*

Ans. A notch made near the edge of the outer rim of each bucket is carefully sharpened *to ensure a loss-free entry of the jet into the buckets i.e.* the path of the jet is not obstructed by incoming buckets.

Q. 18. *What are the materials used for the buckets of Pelton wheel?*

Ans. The buckets are the most important part of the Pelton turbine, they have to be designed to withstand the full force of the jet. Thus, they are made of special bronze or steel alloys with nickel, chromium or stainless steel.

Q. 19. *What is a 'draft tube'?*

Ans. A ***draft tube*** is an expanding device which has an expanding pressure conduit hermetically fixed at the runner outlet and having the other end below the minimum tail water level, that helps to convert the velocity head into pressure or potential head.

It is an integral part of mixed and axial flow turbines. Because of the draft tube it is possible to have the pressure at runner outlet *much below the atmospheric level*.

Q. 20. *What is 'priming' and why is it necessary?*

Ans. The operation of filling the suction pipe, casing of the pump and a portion of the delivery pipe completely from outside source with the liquid to be raised, before starting the pump, the remove any air, gas or vapour from these parts of the pump is called ***priming*** of a centrifugal pump.

If a centrifugal pump is not primed before starting, air pockets inside the impeller may give rise to vortices and cause discontinuity of flow. Further, dry running of the pump may result in rubbing and seizing of the wearing rings and cause serious damage.

Q. 21. *What is meant by 'degree of reaction'?*

Ans. The ***degree of reaction*** (R) is defined as the ratio of change of pressure energy in the runner to the change of total energy in the runner per kg of water.

i.e.
$$R = \frac{\text{Change in pressure energy}}{\text{Change in total energy}}$$

Q. 22. *Explain 'runaway speed'?*

Ans. The ***runaway speed*** is the maximum speed, governor being disengaged, at which a turbine would run when there is no external load but operating under design head and discharge. All the moving parts including the rotor of alternator should be designed for the centrifugal stresses caused by this maximum speed.

The practical values of runaway speed for various turbines with respect to their rated speed N are as follows:

Pelton wheel = 1.8 to 1.9 N; Francis turbine (mixed flow) = 2.0 to 2.2 N; Kaplan turbine (axial flow) = 2.5 to 3.0 N.

Q. 23. *What is a submersible pump?*

Ans. A ***submersible pump*** is a device which has a motor closely coupled to a pump body. The whole assembly is submerged in the fluid to be pumped. This *pump pushes fluid to the surface as opposed to jet pumps having pull fluids.*

The submersible pumps are more efficient than jet pumps.

Q. 24. *List the factors which influence the speed of reciprocating pump.*

Ans. Speed of reciprocating pump is influenced by:

(*i*) Asolute pressure inside the cylinder; (*ii*) Cavitation produced; (*iii*) Acceleration of piston; (*iv*) Friction in the pipes.

Q.25. *What is a 'fluid coupling'?*

Ans. A ***fluid coupling*** is a device which is employed for transmission of power from one shaft to another through a liquid medium. It has no mechanical connection or face to face contact. The magnitudes of input and output torques are equal.

Q. 26. *Define 'Thoma's cavitation parameter'.*

Ans. Prof. Dietrich Thomas of Munich (Germany) suggested a 'cavitation factor (sigma)' to determine the zone where turbine can work without being affected from cavitation. The critical value cavitation factor (σ_c) is given by:

$$\sigma_c = \frac{(H_a - H_v) - H_s}{H}$$

where, H_a = Atmospheric pressure head in meters of water; H_v = Vapour pressure in metres of water corresponding to the water temperature, H = Working head of turbine (difference between head race and tail race level in metres), and H_s = Suction pressure head (or height of turbine outlet above tail race level in metres).

Q. 27. *What is meant by speed ratio of a Pelton wheel?*

Ans. Speed ratio $(K_u) = \dfrac{\text{The peripheral velocity of wheel } (u)}{\text{The theoretical velocity of jet } (\sqrt{2gH})}$

Q. 28. *What is the utility of an 'Air lift pump'?*

Ans. An ***air lift pump*** is a device used to lift water from a deep well or sump by utilising the compressed air.

Q. 29. *What are the salient features of an 'Air lift pump'?*

Ans. The salient features of an air lift pump are:

(*i*) It has no moving parts below water level and consequently no wear and tear.

(*ii*) It can raise more water through a bore hole of given diameter than any other pump.

(*iii*) It can pump solids without any damage to the system.

(*iv*) It is suitable for draining water in the mines where compressor units are already installed.

Q. 30. *Define the term Net or Effective head.*

Ans. The head available at the inlet of the turbine is known as ***net or effective head.*** It is denoted by H and is given by:

$$H = H_g - h_f - h$$

where, h_f = Total loss of head between the head race and entrance of the turbine

$= \dfrac{4fLV^2}{D \times 2g}$ (f = coefficient of friction, L = length of penstock, D = diameter of penstock, V = Velocity of flow in penstock), and

h = Height of nozzle above the tail race.

Q. 31. *What is the function of scroll casing in reaction turbines?*

Ans. A ***scroll casing*** constitutes a close passage whose cross-sectional area gradually decreases along the flow direction, area is maximum at inlet and nearly zero at exit. It provides the limited area around the runner to maintain the constant velocity of water flow around the runner.

Q. 32. *What do you mean by the capacity of Hydraulic accumulator?*

Ans. ***Hydraulic accumulator*** is a device used to store the energy of liquid under pressure and make this energy available (as a quick secondary source of power) to hydraulic machines, such as presses, lifts and cranes.

The maximum amount of energy that the accumulator can store is known as the '*capacity of the accumulator*'.

Q. 33. *What are air vessels?*

Ans. An ***air vessel*** is a closed chamber containing compressed air in the upper part and liquid being pumped in the lower part. One air vessel is fixed on the suction pipe near the suction valve and one is fixed on the delivery pipe near the delivery valve.

The air vessels are used for the following *purposes*:

(*i*) To get continuous supply of liquid at a uniform rate (whatever fluctuations take place, they occur between the air vessels and the pump).

(*ii*) To save the power required to drive the pump (By the use of air vessels the acceleration and friction heads are considerably reduced, thereby the work is also reduced).

(*iii*) To run the pump at much higher speed without any danger of separation (By fitting the air vessels as close to the pump as possible, the length of the pipe in which acceleration takes place is reduced due to which acceleration head is reduced, and pump can run at a high speed without separation.

Q. 34. *What do you understand by a 'reciprocating pump'?*

Ans. A ***reciprocating pump*** is a *positive displacement pump* as it sucks and raises the liquid by actually displacing it with a piston/plunger that executes a reciprocating motion in a closely fitting cylinder. The amount of liquid pumped with disc pistons create pressures upto 25 bar and the plunger pumps built up still higher pressures. Discharge from these pumps is almost wholly dependent on the pump speed.

The total efficiency of a reciprocating pump is about 10 to 20 per cent higher than a comparable centrifugal pump.

The reciprocating is generally employed for:

(*i*) Light oil pumping; (*ii*) Feeding small boilers, and

(*iii*) Pneumatic pressure systems.

Q. 35. *Define the draft tube efficiency.*

Ans. The 'efficiency of a draft tube (η_d)' is defined as the ratio of net gain in pressure head to the velocity head at entrance of draft tube. Thus,

$$\eta_d = \frac{\text{Net gain in pressure head}}{\text{Velocity head at entrance of draft tube}}$$

$$= \frac{\left(\dfrac{V_2^2 - V_3^2}{2g} - h_f\right)}{\dfrac{V_2^2}{2g}}$$

where, V_2 = Velocity of water at inlet of the draft tube, and

V_3 = Velocity of water at outlet of the draft tube.

$$\left[h_f = \frac{V_2^2 - V_3^2}{2g} - \eta_d \times \frac{V_2^2}{2g}\right]$$

Q. 36. *Why the draft tube is not used for Pelton turbine?*

Ans. In case of Pelton all the K.E. is lost; and draft tube is not used because the pressure value is just the atmospheric so that there is no requirement of draft tube.

Q. 37. *Define the term 'Impact of jet'.*

Ans. A fluid jet is a stream of fluid obtained from nozzle. When this jet strikes on flat or curved plate the momentum is changed and a hydrodynamic force is exerted. So 'Impact of jet' terms refer to the study of the effect when a jet strikes on the plate or vane under the various conditions.

Q. 38. *What are the functions of a draft tube?*

Ans. A draft tube performs the following functions:

(*i*) It allows the turbine to be set above tail-water level without loss of head to facilitate inspection and maintenance.

(*ii*) It regains by diffusion action, the major portion of the kinetic energy delivered to it from the runner.

Q. 39. *Define the term 'Gross head'.*

Ans. The 'Gross head' or Total head is the difference between the water level at the reservoir (also known as head race) and the level at the tail race.

Q. 40. *List the advantages of Kaplan turbine over Francis turbine.*

Ans. Kaplan turbine claims the following advantages over Francis turbine:

(*i*) For the same power developed Kaplan turbine is more compact in construction and smaller in size.

(*ii*) Part-load efficiency is considerably high.

(*iii*) Low frictional losses (because of small number of blades used).

Q. 41. *What are the bases on which hydraulic turbines are classified?*

Ans. The hydraulic turbines are classified on the following bases:

(*i*) According to the head and quantity of water available.

(*ii*) According to the name of the originator.

(*iii*) According to the action of water on moving blades.

(*iv*) According to the direction of flow of water in the runner.

(*v*) According to the disposition of the turbine shaft.

(*vi*) According to the specific speed N_s.

Q. 42. *What is the principle of working of a centrifugal pump?*

Ans. A centrifugal pump works on the principle that when a certain mass of fluid is rotated by an external source, it is thrown away from the central axis of rotation and a centrifugal head is impressed which enables it to rise to a higher level.

Q. 43. *What are the functions of a multi-stage pump?*

Ans. A multi-stage centrifugal pump is one which has two or more identical impellers mounted on the same shaft or on different shafts. The important functions performed by a multi-stage pump are:

(*i*) To *produce greater heads* than that permissible with a single impeller, discharge remaining constant. The task can be achieved by '*series arrangement*' where in the impellers are mounted on the *same shaft* and enclosed in the same casing.

(*ii*) To *discharge a large quantity of liquid*, head remaining same. This task is accomplished by '*parallel arrangement*' where in impellers are mounted on *separate shafts*.

Q. 44. *What is cavitation and how can it be avoided in reaction turbines?*

Ans. The formation, growth, and collapse of vapour filled cavities or bubbles in a flowing liquid due to local fall in fluid pressure is called ***cavitation.*** When the cavities collapse (the collapsing pressure is of order of 100 times the atmospheric pressure) on the surface of a body, due to repeated 'hammering' action, the metal particles give way ultimately due to fatigue and indentations are formed; this erosion of material is called ***pitting.***

The following methods may be used to avoid cavitation:

(*i*) Runner turbine may be kept under water. But it is not advisable as the inspection and repair of the turbine is difficult. Alternatively, the runner of low specific speed may be used.

(*ii*) It is possible to reduce the cavitation effect by selecting materials which resist better the cavitation effect. The cast steel is better than cast iron and stainless steel or alloy steel is still better than cast steel.

(*iii*) The cavitation effect can be reduced by polishing the surface. That is why the cast steel runners and blades are coated with stainless steel.

(*iv*) The cavitation may be avoided by selecting a runner of proper specific speed for given head.

Q. 45. *How are hydraulic turbines classified according to specific speed?*

Ans. The specific speed of a turbine is defined as the speed of a geometrically similar turbine that would develop 1 kW under 1 m head. All geometrically similar turbines (irrespective of the sizes) will have the same specific speeds when operating under the same head.

$$\text{Specific speed, } N_s = \frac{N\sqrt{P}}{H^{5/4}}$$

where, N = The normal speed, P = Power output of the turbine, and H = The net or effective head in metres.

Turbines with low specific speeds work under high head and low discharge conditions, while *high specific speed turbines work under low head and high discharge conditions*.

Q. 46. *Distinguish between an impulse turbine and a reaction turbine.*

Ans.

S.No.	Impulse turbine	Reaction turbine
1.	The available energy is converted into K.E. by a nozzle.	The energy of the fluid is partly transformed into K.E. before it (fluid) enters the runner of the turbine.
2.	The pressure remains same (atmospheric) throughout the action of water on the runner.	After entering the runner with an excess pressure, water undergoes changes both in velocity and pressure while passing through the runner.
3.	Always installed above the tail race. No draft tube is used.	Unit may be installed above or below the tail race, use of a draft tube is made.
4.	Water may be allowed to enter a part or whole of the wheel circumference.	Water is admitted over the circumference of the wheel.

Q. 47. *What are the functions of a surge tank?*

Ans. A ***surge tank*** is a small reservoir or tank in which the water level rises or falls to reduce the pressure swings so that they are not transmitted in full to a closed circuit. In general a surge tank is employed to serve the following *purposes*:

(*i*) To reduce the distance between the free water surface and turbine thereby reducing the water hammer effect (the water hammer is defined as the change in pressure rapidly above or below normal pressure caused by sudden changes in rate of flow through the pipe according to the demand of the prime mover) on penstock and also protect upstream tunnel from high pressure rises.

(*ii*) To serve as *supply tank* to the turbine when water in the pipe is accelerating during increased load conditions and *storage tank* when the water is decreasing during reduced load conditions.

Q. 48. *What do you understand by governing of hydraulic turbines?*

Ans. Governing of hydraulic turbine means *speed regulation*. Governing of a turbine is necessary as a turbine is directly coupled to an electric generator, which is required to run at constant speed under all fluctuating load conditions. This is achieved by a *governor* called oil pressure governor.

The power produced by water turbine is directly proportional to the available head and discharge through the turbine. The quantity of water flowing can be controlled by varying the area of flow at the turbine inlet.

In Pelton turbine, the flow area is changed by moving the spear inside the nozzle and in reaction turbine, the area of flow is varied by rotating the guide vanes with the help of governor in a controlling unit.

Q. 49. *How does a Kaplan turbine differ from a propeller turbine?*

Ans. A ***propeller turbine*** is quite suitable when the *load* on the turbine *remains constant*. At part load its efficiency is very low, since the *blades are fixed*, the water enters with shock (at part load) and eddies are formed which reduce the efficiency. This defect of the propeller turbine is removed in Kaplan turbine.

In a ***Kaplan turbine*** the runner *blades are adjustable* and can be rotated about pivots fixed to the boss of the runner. The *blades are adjusted automatically* by servomechanism so that at all loads the flow enters them *without shock*. Thus, a *high efficiency* is maintained *even at part loads*. It behaves like a propeller turbine at full-load conditions.

Q. 50. *Differentiate between Francis and Kaplan turbines.*

Ans.

S. No.	Francis Turbine	Kaplan turbine
1.	Radially inward or mixed flow.	Partially axial flow.
2.	Horizontal or vertical shaft.	Only vertical shaft.
3.	Runner vanes are not adjustable (16 to 24 blades)	Runner vanes are adjustable (3 to 8 blades)
4.	Medium head: 60 m to 250 m.	Low head: upto 30 m.
5.	Medium flow rate.	Large flow rate.
6.	Specific speed: 50–250	Specific speed: 250–850

Q. 51. *Define specific speed of turbine and write down its expression.*

Ans. The ***specific speed*** of a turbine is defined as the speed of a turbine which is identical in shape, geometrical dimensions, blade angles, gate opening, etc. which would develop unit power when working under a unit head.

Specific speed (N_s) is given by:

$$N_s = \frac{N\sqrt{P}}{H^{5/4}}$$

where, N = Speed of the runner in r.p.m., H = Head of water, and P = Power developed.

Q. 52. *What is priming of centrifugal pump?*

Ans. ***Priming*** of centrifugal pump is the operation of filling the suction pipe, casing of the pump and a portion of the delivery pipe completely from outside source with the liquid to be raised, before starting the pump, to remove any air, gas or vapour from these parts of the pump.

Q. 53. *How are various types of centrifugal pumps primed?*

Ans. ***Small pumps*** are usually primed by pouring liquid into the *funnel* provided for the purpose.

Large pumps are primed by evacuating the casing and the suction by a *vacuum pump* or by an *ejector*; the liquid is thus drawn up the suction pipe from the sump and the pump is filled with liquid.

The internal construction of some pumps is such that *special arrangements* containing a supply of liquid are provided in the suction pipe due to which automatic priming of the pump occurs; such pumps are known as ***self priming pumps***.

Q. 54. *Distinguish between the positive and non-positive displacement pumps.*

Ans. ***Positive displacement pump***: It causes a liquid to move by trapping a fixed amount of it, then forcing (displacing) that trapped volume into the discharge pipe, *e.g.* tube, gear, screw pump etc.

Non-positive displacement pump (*rotodynamic type*)**:** It is pump in which the dynamic motion of a fluid is increased by pump action, *e.g.* centrifugal, turbine, propeller etc.

Q. 55. *Define slip for reciprocating pumps. When does negative slip occur?*

Ans. The difference between the theoretical discharge and actual discharge is called the ***slip*** of the pump

i.e., $$\text{Slip} = Q_{th.} - Q_{act.}$$

But the slip is oftenly expressed in percentage which is given by:

$$\%\ \text{slip} = \frac{Q_{th.} - Q_{act.}}{Q_{th.}} \times 100 = \left(1 - \frac{Q_{act.}}{Q_{th.}}\right) \times 100 = (1 - C_d) \times 100$$

where, C_d = Coefficient of discharge.

The percentage of slip for the pumps maintained in *good condition* is of the order of 2% or even less.

Negative slip. In most of the reciprocating pumps $Q_{act.}$ is *less* than $Q_{th.}$; in such a case the value of C_d is *less* than unity and the slip of the pump is 'positive'. However, in some cases $Q_{act.}$ may be *more* than $Q_{th.}$; in such a case C_d is *more* than unity and the slip will be '*negative*'. The slip will be negative when there is a *direct connection between the suction and delivery sides before the end of the suction stroke.* This happens if the momentum of liquid in the suction pipe, is large enough to open the delivery valve before the beginning of delivery stroke. The negative slip is possible in case of pumps having *long suction pipe and a short delivery pipe*, especially when these are operating at high speeds.

Q. 56. *Why is the efficiency of Kaplan turbine nearly constant irrespective of speed variation under load?*

Ans. Kaplan turbine has the concept of adjusting the runner vanes in face of changing load conditions on the turbine. With proper adjustment of blades during its running the Kaplan turbine is capable of giving a constant and high efficiency for a wide range of load conditions. The pitch of the blades is also automatically adjusted by the governor through the action of a servomotor.

Q. 57. *What is a hydraulic intensifier?*

Ans. ***Hydraulic intensifier*** is a device which increases the pressure of a given liquid with the help of low pressure liquid of large quantity.

Q. 58. *How are hydraulic pumps classified?*

Ans. Pumps may be placed in one of the two general categories:

(*i*) *Dynamic pressure pumps*: Centrifugal pump, jet pump, propeller, turbine.

(*ii*) *Positive displacement pump*: Piston plunger, gear, vane, screw pump etc.

Q. 59. *Define manometric head of a pump.*

Ans. The ***manometric head*** is defined as the head against which a centrifugal pump has to work. It is the head measured across the pump inlet and outlet flanges. It is denoted by H_{mano}.

Q. 60. *What is manometric efficiency?*

Ans. The ratio of the manometric head developed by the pump to the head imparted by the impeller to the liquid is known as '*manometric efficiency*'.

i.e.
$$\eta_{mano} = \frac{\text{Manometric head}}{\text{Head imparted by impeller to the liquid}}$$

SECTION B: QUESTIONS WITH SOLUTIONS

Q. 1. *A jet of water strikes with a velocity of 30 m/s a flat plate inclined at 45° with the axis of the jet. If the cross-sectional area of the jet is 20 cm², determine the following:*

(i) The force exerted by the jet on the plate,

(ii) The components of the force in the direction of the jet, and

(iii) The ratio in which the discharge gets divided after striking the plate.

Solution. *Given*: Velocity of the jet, $V = 30$ m/s, Inclination of the plate with the jet axis, $\theta = 45°$; Area of the jet, $a = 20 \text{ cm}^2 = 20 \times 10^{-4} \text{ m}^2 = 0.002 \text{ m}^2$.

(i) The force exerted by the jet, *F*:

$$F = \rho a V^2 \sin\theta \qquad \text{[Eqn. (1.3)]}$$
$$= 1000 \times 0.002 \times 30^2 \times \sin 45° = \mathbf{1272.8\ N\ (Ans.)}$$

(ii) The components of the force, *F*:

$$F_x = F\sin\theta = 1272.8 \times \sin 45° = \mathbf{900\ N\ (Ans.)}$$
$$F_y = F\cos\theta = 1272.8 \times \cos 45° = \mathbf{900\ N\ (Ans.)}$$

(iii) The ratio in which the discharge gets divided:

$$\frac{Q_1}{Q_2} = \frac{1+\cos\theta}{1-\cos\theta} \qquad \text{[Eqn. (1.7)]}$$

or,
$$\frac{Q_1}{Q_2} = \frac{1+\cos 45°}{1-\cos 45°} = \mathbf{5.83\ (Ans.)}$$

Q. 2. *A jet of water of 50 mm diameter strikes a curved vane at its centre with a velocity of 20 m/s. The curved vane is moving with a velocity of 5 m/s in the direction of the jet. The jet is deflected through an angle of 160°. Assuming the plate to be smooth, calculate:*

(i) Thrust on the plate in the direction of the jet,

(ii) Power of the jet; and

(iii) Efficiency of the jet.

Solution. *Given*: Diameter of the jet, $d = 50$ mm $= 0.05$ m; Velocity of the jet, $V = 20$ m/s; Velocity of the vane, $u = 5$ m/s; Angle of deflection of the jet = 160°; Angle made by the relative velocity at the outlet of the vane = 180 – 160 = 20°.

(i) Thrust on the plate:

$$\text{Area of the jet, } a = \frac{\pi}{4}d^2 = \frac{\pi}{4} \times 0.05^2 = 0.001963 \text{ m}^2$$

Thrust on the plate in the direction of the jet,

$$F_x = \rho a (V-u)^2 (1+\cos\theta)$$
$$= 1000 \times 0.001963\,(20-5)^2\,(1+\cos 20°) = \mathbf{856.7\ N\ (Ans.)}$$

(ii) Power of the jet:

Work done by the jet on the vane per second

$$= F_x \times u = 856.7 \times 5 = 4283.5 \text{ Nm/s}$$

Power of the jet $= 4283.5$ Nm/s $= 4283.5$ W or **4.283 kW (Ans.)**

(*iii*) Efficiency of the jet, η_{jet}:

$$\eta_{jet} = \frac{\text{Work done by the jet/sec.}}{\text{Kinetic energy of the jet/sec.}}$$

$$= \frac{4283.5}{\frac{1}{2}(\rho a V)\times V^2} = \frac{4283.5}{\frac{1}{2}(1000\times 0.001963\times 20)\times 20^2}$$

$= 0.5455$ or **54.55% (Ans.)**

Q. 3. *The following data relate to a Pelton wheel:*

Heat = 80 m; Speed of wheel = 280 r.p.m.; Shaft power of wheel = 130 kW; Speed ratio = 0.48; Co-efficient of velocity = 0.97; Overall efficiency = 82%.

Design the Pelton wheel.

Solution. *Given*: Effective head, $H = 80$ m; Speed of wheel, $N = 280$ r.p.m.; Shaft power, $P =$ 130 kW; Speed ratio, $K_u = 0.48$; Co-efficient of velocity, $C_v = 0.97$; Overall efficiency, $\eta_0 = 82\%$.

Design of the Pelton wheel means to find diameter of the wheel (D), diameter of the jet (d), width and depth of buckets and number of buckets on the wheel.

(*i*) Diameter of wheel, *D*:

$$\text{Velocity of jet, } V_1 = C_v\sqrt{2gH} = 0.97\sqrt{2\times 9.81\times 80}$$

$$= 38.4 \text{ m/s}$$

$\therefore$ Bucket velocity, $u(= u_1 = u_2) = K_u \times V_1 = 0.48 \times 38.4 = 18.4$ m/s

But, $$u = \frac{\pi D N}{60}, \textit{ or, } D = \frac{60u}{\pi N} = \frac{60\times 18.4}{\pi \times 280} = \mathbf{1.255\ m}\ \textbf{(Ans.)}$$

(*ii*) Diameter of the jet, *d*:

$$\text{Overall efficiency, } \eta_0 = \frac{\text{Shaft power}}{\text{Water power}} = \frac{P}{wQH}$$

or, $$0.82 = \frac{130}{9.81\times Q\times 80} \qquad (\because w = 9.81 \text{ kN/m}^3)$$

or, $$\text{Discharge, } Q = \frac{130}{0.82\times 9.81\times 80} = 0.202 \text{ m}^3/\text{s}$$

But, $$Q = \text{Area of jet} \times \text{velocity of jet}$$

$$0.202 = \frac{\pi}{4}d^2 \times 38.4$$

or, $$d = \left(\frac{0.202\times 4}{\pi\times 38.4}\right)^{1/2} = 0.0818 \text{ m or } \mathbf{81.8\ mm}\ \textbf{(Ans.)}$$

(*iii*) Size of buckets:

Width of the bucket, $B \simeq$ 3 to 4 times jet diameter (d)

$= 3.5d = 3.5 \times 81.8 =$ **286.3 mm (Ans.)**

Radial length of bucket, $L =$ 2 to 3 times jet diameter (d)

$\simeq 2.5d = 2.5 \times 81.8 =$ **204.5 mm (Ans.)**

Depth of bucket, $T =$ 0.8 to 1.2 times jet diameter (d)

$\simeq 1.0d =$ **81.8 mm (Ans.)**

Q. 4. *A conical draft tube having inlet and outlet diameters 1.1 m and 1.7 m discharges water at outlet with a velocity of 2.8 m/s. The total length of the draft tube is 7.0 m and 1.4 m of the length of draft tube is immersed in water. If the atmospheric pressure head is 10.3 m of water and loss of head due to friction in the draft tube is equal to 0.18 × velocity head at outlet of the tube.*

Determine: (i) Pressure head at inlet, and (ii) Efficiency of the draft tube.

Solution. Refer to Fig. 1. *Given*: Inlet diameter of the draft tube, $d_i = 1.1$ m; Outlet diameter, $d_o = 1.7$ m; Velocity at outlet, $V_3 = 2.8$ m/s; Total length of draft tube, $H_s + y = 7.0$ m; Length of draft tube in water, $y = 1.4$ m; Atmospheric pressure head, $\frac{p_a}{w} = 10.3$ m; Loss of head due to friction, $h_f = 0.18 \times$ velocity head at outlet $= 0.18 \times \frac{V_1^2}{2g}$

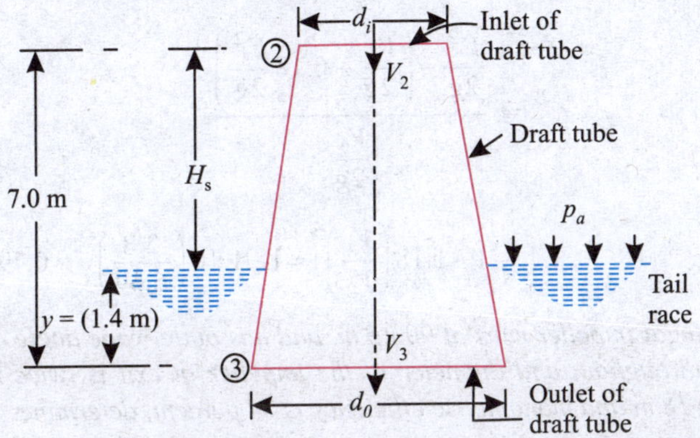

Fig. 1

(i) Pressure head at inlet, $\frac{p_2}{w}$:

Discharge through the draft tube,

$$Q = A_3 V_3 = \frac{\pi}{4} \times d_o^2 \times V_3 = \frac{\pi}{4} \times 1.7^2 \times 2.8 = 6.36 \text{ m/s}$$

$$\text{Velocity of inlet, } V_2 = \frac{Q}{A_2} = \frac{6.36}{\frac{\pi}{4} \times d_i^2} = \frac{6.36}{\frac{\pi}{4} \times 1.1^2} = 6.69 \text{ m/s}$$

We know that,

$$\frac{p_2}{w} = \frac{p_a}{w} - H_s - \left(\frac{V_2^2 - V_3^2}{2g} - h_f\right) \qquad \text{[Eqn. (2.32)]}$$

$$= \frac{p_a}{w} - H_s - \left(\frac{V_2^2 - V_3^2}{2g} - 0.18\frac{V_3^2}{2g}\right)$$

$$= 10.3 - (7.0 - 1.4) - \left(\frac{6.69^2 - 2.8^2}{2 \times 9.81} - 0.18 \times \frac{2.8^2}{2 \times 9.81}\right)$$

$$= 4.7 - (1.88 - 0.072) = \mathbf{2.892 \text{ m (abs.)}} \text{ (Ans.)}$$

(ii) Efficiency of the draft tube, η_d:

We know that,

$$\eta_d = \frac{\left(\frac{V_2^2 - V_3^2}{2g} - h_f\right)}{\frac{V_2^2}{2g}} \qquad \text{[Eqn. (2.33)]}$$

$$= \frac{\frac{V_2^2 - V_3^2}{2g} - 0.18\frac{V_3^2}{2g}}{\frac{V_2^2}{2g}}$$

$$= \frac{\frac{V_2^2}{2g} - \left(\frac{V_3^2}{2g} + 0.18\frac{V_3^2}{2g}\right)}{\frac{V_2^2}{2g}}$$

$$= 1 - 1.18\left(\frac{V_3}{V_2}\right)^2 = 1 - 1.18\left(\frac{2.8}{6.69}\right)^2 = 0.793 \text{ or } \mathbf{79.3\%} \text{ (Ans.)}$$

Q. 5. *A centrifugal impeller runs at 90 r.p.m. and has outlet vane angle of 62°. The velocity of flow is 2.7 m/s throughout and diameter of the impeller at exit is twice that at inlet. If the manometric head is 18 m and manometric efficiency is 74 percent, determine:*

(i) The diameter of the impeller at the exit, and (ii) Inlet vane angle.

Solution. Refer to Fig. 2.

Given: Speed, N = 90 r.p.m.; Outlet vane angle, ϕ = 62°; Velocity of flow, $V_{f1} = V_{f2}$ = 2.7 m/s; Manometric head, H_{mano} = 18 m; Manometric efficiency, η_{mano} = 74%. Diameter of the impeller at outlet, $D_2 = 2D_1$ (diameter at inlet).

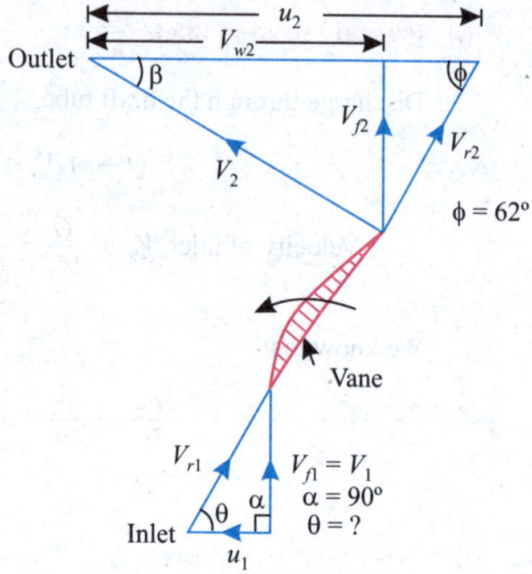

Fig. 2

(i) Diameter of the impeller at the exit, D_2:

$$\eta_{mano} = \frac{gH_{mano}}{V_{w2}u_2} \qquad \text{[Eqn. (3.9)]}$$

or, $$0.74 = \frac{9.81 \times 18}{V_{w2}u_2}$$

or, $$V_{w2}u_2 = \frac{9.81 \times 18}{0.74} = 238.6 \qquad ...(i)$$

From velocity triangle at outlet (Fig. 2), we get,

$$\tan\phi = \frac{V_{f2}}{u_2 - V_{w2}}$$

or, $$u_2 - V_{w2} = \frac{V_{f2}}{\tan\phi}, \text{ or, } V_{w2} = u_2 - \frac{V_{f2}}{\tan\phi}$$

or, $$V_{w2} = u_2 - \frac{2.7}{\tan 62°} = u_2 - 1.436$$

Inserting this value of V_{w2} in (*i*), we have:

$$(u_2 - 1.436)u_2 = 238.6$$

or, $u_2^2 - 1.436u_2 - 238.6 = 0$

or,
$$u_2 = \frac{1.436 \pm \sqrt{1.436^2 + 4 \times 238.6}}{2}$$

$$= \frac{1.436 \pm 30.927}{2} = 16.18 \text{ m/s} \qquad \text{(ignoring –ve sign)}$$

Also, tangential velocity of impeller at outlet,

$$u_2 = \frac{\pi D_2 N}{60}, \text{ or, } D_2 = \frac{60u_2}{\pi N}$$

$$\therefore \quad D_2 = \frac{60 \times 16.18}{\pi \times 90} = \mathbf{3.43\ m\ (Ans.)}$$

(*iii*) Inlet vane angle, θ:

Tangential velocity of the impeller at inlet

$$u_1 = \frac{u_2}{2} = \frac{16.18}{2} = \mathbf{8.09\ m/s\ (Ans.)} \qquad \left(\because D_1 = \frac{D_2}{2}\right)$$

From velocity triangle at *inlet*, we get,

$$\tan\theta = \frac{V_{f1}}{u_1} = \frac{2.7}{8.09} = 0.3337$$

or,
$$\theta = \tan^{-1}(0.3337) = \mathbf{18.45° \ (Ans.)}$$

Q. 6. *A centrifugal pump impeller has diameters at inlet and outlet as 350 mm and 700 mm respectively. The flow velocity at outlet is 2.3 m/s and vanes are set back at an angle of 45° at the outlet. If the manometric efficiency is 75 per cent, calculate the minimum starting speed of the pump.*

Solution. *Given*: Diameter of impeller at inlet, $D_1 = 350$ mm = 0.35 m; Diameter of impeller at outlet, $D_2 = 700$ mm = 0.7 m; The flow velocity at outlet, $V_{f2} = 2.3$ m/s; Outlet vane angle, $\phi = 45°$; Manometric efficiency, $\eta_{mano} = 75\%$.

Minimum starting speed of the pump, $N_{min.}$:

Refer to Fig. 3. From velocity triangle at *outlet*, we have:

$$\tan\phi = \frac{V_{f2}}{u_2 - V_{w2}},$$

or,
$$u_2 - V_{w2} = \frac{V_{f2}}{\tan\phi} = \frac{2.3}{\tan 45°} = 2.3 \text{ m/s}$$

or,
$$V_{w2} = u_2 - 2.3$$

But,
$$u_2 = \frac{\pi D_2 N_{min.}}{60} = \frac{\pi \times 0.7\ N_{min.}}{60}$$
$$= 0.0366\ N_{min.}$$

$$\therefore \quad V_{w2} = 0.0366\ N_{min.} - 2.3$$

For minimum speed, we have:

$$N_{min} = \frac{120 \times \eta_{mano} \times V_{w2} \times D_2}{\pi(D_2^2 - D_1^2)}$$

...[Eqn. (3.15)]

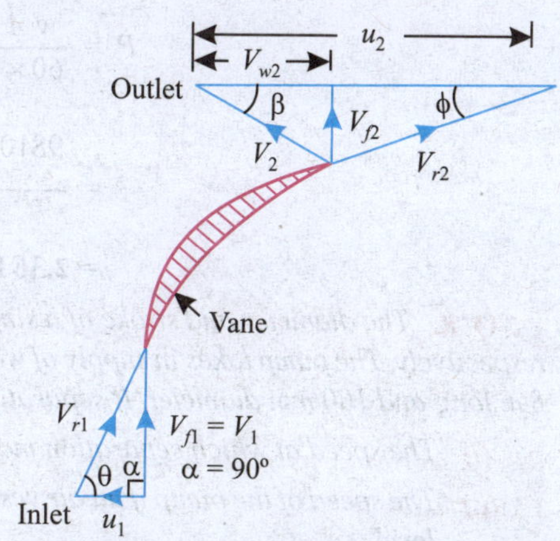

Fig. 3

$$= \frac{120 \times 0.75 \times (0.0366\, N_{min} - 2.3)}{\pi(0.7^2 - 0.35^2)}$$

or, $$N_{min.} = 77.95\,(0.0366\, N_{min} - 2.3)$$
$$= 2.853\, N_{min} - 179.285$$

or, $$1.853\, N_{min} = 179.285$$

or, $$N_{min} = \frac{179.285}{1.853} \simeq \mathbf{97\ r.p.m}$$ **(Ans.)**

Q. 7. *A single acting reciprocating pump, running at 60 r.p.m., delivers 0.01518 m^3/s of water. The diameter of the piston is 240 mm and stroke length 360 mm. The suction and delivery heads are 3.4 m and 11.4 m respectively. Determine:*

(i) Theoretical discharge, *(ii) Coefficient of discharge,*
(iii) Percentage slip of the pump, and *(iv) Power required to run the pump.*

Solution. *Given*: Speed of the pump, N = 60 r.p.m.; Actual discharge, $Q_{act.}$ = 0.01518 m^3/s; Diameter of the piston, D = 240 mm = 0.24 m; Stroke length, L = 360 mm = 0.36 m; Suction head, h_s = 3.4 m; Delivery head, h_d = 11.4 m.

(i) Theoretical discharge, Q_{th}:

$$Q_{th} = \frac{ALN}{60} = \frac{\left(\frac{\pi}{4} \times 0.24^2\right) \times 0.36 \times 60}{60} = \mathbf{0.01629\ m^3/s}$$ **(Ans.)**

(ii) Coefficient of discharge, C_d:

$$C_d = \frac{Q_{act.}}{Q_{th.}} = \frac{0.01518}{0.01629} = \mathbf{0.932}$$ **(Ans.)**

(iii) Percentage slip of the pump:

$$\%\ \text{slip} = \frac{Q_{th.} - Q_{act.}}{Q_{th.}} \times 100$$

$$= \frac{0.01629 - 0.01518}{0.01629} \times 100 = \mathbf{6.8\%}$$ **(Ans.)**

(iv) Power required to run the pump, P:

$$P = \frac{w\,A\,L\,N}{60 \times 1000}(h_s + h_d)\ \text{k.W}$$

$$= \frac{9810 \times \left(\frac{\pi}{4} \times 0.24^2\right) \times 0.36 \times 60}{60 \times 1000}(3.4 + 11.4)$$

$$= \mathbf{2.36\ kW}$$ **(Ans.)**

Q. 8. *The diameter and stroke of a single-acting reciprocating pump are 250 mm and 450 mm respectively. The pump takes in supply of water from sump 2.9 m below the pump and through a pipe 8 m long and 100 mm diameter. If separation occurs at 2.4 m of water absolute, determine:*

(i) The speed at which separation may take place at the beginning of suction of stroke, and

(ii) The speed of the pump if an air vessel is fitted on the suction side 2 m above the sump water level.

Take atmospheric pressure head = 10.3 m of water, and friction coefficient, f = 0.01.

Solution. *Given*: Diameter of piston, $D = 250$ mm = 0.25 m; Stroke length, $L = 450$ mm = 0.45 m; Crank radius, $r = \frac{0.45}{2} = 0.225$ m; Suction head, $h_s = 2.9$ m; Diameter of the suction pipe, $d_s = 200$ mm = 0.2 m; Length of suction pipe, $l_s = 8$ m; Separation head = 2.4 m of water absolute; Atmospheric pressure head, $H_{atm.} = 10.3$ m of water; Friction coefficient, $f = 0.01$.

$$\text{Area of piston, } A = \frac{\pi}{4} \times D^2 = \frac{\pi}{4} \times 0.25^2 = 0.049 \text{ m}^2$$

$$\text{Area of suction pipe, } a_s = \frac{\pi}{4} \times 0.2^2 = 0.0314 \text{ m}^2$$

(*i*) The speed at which separation may take place (no air vessel fitted), *N*:

The pressure head due to acceleration in the suction pipe, is given by:

$$h_{as} = \frac{l_s}{g} \times \frac{A}{a_s} \omega^2 r \cos\theta$$

At the beginning of the suction stroke, $\theta = 0°$, we have:

$$h_{as} = \frac{l_s}{g} \times \frac{A}{a_s} \omega^2 r = \frac{8}{9.81} \times \frac{0.049}{0.0314} \times \omega^2 \times 0.225 = 0.286\,\omega^2$$

Pressure head at the beginning of suction stroke

$$= (h_s + h_{as}) = (2.9 + 0.286\,\omega^2) \text{ below atmospheric head}$$

Limiting condition for *no separation* gives,

$$H_{atm} - (h_s + h_{as}) = h_{sep.}$$

or, $$10.3 - (2.9 + 0.286\,\omega^2) = 2.4$$

or, $$\omega = \left(\frac{10.3 - 2.4 - 2.9}{0.286}\right)^{1/2} = 4.18 \text{ rad./s}$$

But, $$\omega = \frac{2\pi N}{60}, \quad \text{or,} \quad N = \frac{60\,\omega}{2\pi} = \frac{60 \times 4.18}{2\pi} = \textbf{39.9 r.p.m} \text{ (Ans.)}$$

(*ii*) The speed of pump when an air vessel is fitted on suction side, *N*:

Since the air vessel is installed 2 m above the sump water level, therefore:

(*a*) There will be a loss of head due to friction in the suction pipe for the length of $8 \times \frac{2.4}{2.9} = 6.62$ m; (*ii*) the acceleration pressure head will be restricted in the remaining (8 – 6.62) m = 1.38 m length of suction pipe.

The pressure head due to acceleration (h_{as}) in the suction pipe at the beginning of suction stroke ($\theta = 0°$) is given by:

$$h_{as} = \frac{l_s}{g} \times \frac{A}{a_s} \times \omega^2 r = \frac{1.38}{9.81} \times \frac{0.049}{0.0314} \times \omega^2 \times 0.225 = 0.049\,\omega^2$$

The velocity of water in the suction pipe **fitted with air vessel.**

$$v_s = \frac{A}{a_s} \times \frac{\omega r}{\pi} = \frac{0.049}{0.0314} \times \frac{\omega \times 0.225}{\pi} = 0.112\,\omega$$

$$\text{Loss of head due to friction, } h_{fs} = \frac{4 f\, l_s\, v_s^2}{d_s \times 2g}$$

$$= \frac{4 \times 0.01 \times 6.62 \times (0.112\,\omega)^2}{0.2 \times 2 \times 9.81} = 8.46 \times 10^{-4}\,\omega^2$$

Limiting condition for *no separation* gives:

$$H_{atm} - (h_s + h_{as} + h_{fs}) = h_{sep.}$$

or, $$10.3 - (2.9 + 0.049\,\omega^2 + 8.46 \times 10^{-4}\,\omega^2) = 2.4$$

or, $$10.3 - 2.9 - 0.0498\,\omega^2 = 2.4$$

or, $$\omega = \left(\frac{10.3 - 2.9 - 2.4}{0.0498}\right)^{1/2} = 10.02 \text{ rad/s}$$

But, $$\omega = \frac{2\pi N}{60} \text{ or } N = \frac{60\,\omega}{2\pi} = \frac{60 \times 10.02}{2\pi} = \mathbf{95.7\ r.p.m.}\ \textbf{(Ans.)}$$

Q. 9. *A hydroelectric power plant produces 30 MW under a head of 18 m. If the overall efficiency of the plant is 75%, determine:*

(i) Type of turbine; (ii) Synchronous speed of generator.

Solution. *Given*: Power developed, $P = 30$ MW ($= 30 \times 10^3$ kW); Head, $H = 18$ m; Overall efficiency, $\eta_0 = 75\%$.

(*i*) Type of turbine:

$$P = \eta_0 \times w\,Q\,H$$

$$25 \times 10^3 = 0.75 \times 9.81 \times Q \times 18$$

$$\therefore \quad Q = \frac{30 \times 10^3}{0.75 \times 9.81 \times 18} = 226.5 \text{ m}^3/\text{s}$$

As the head is low and discharge is high so "propeller type of turbine" should be used. **(Ans.)**

(*ii*) Synchronous speed of the generator, $N_{syn.}$:

$$\text{Specific speed, } N_s = \frac{1150}{(H)^{1/4}} \text{ (approx.)}$$

$$= \frac{1150}{(18)^{1/4}} = 558.3 \text{ r.p.m.}$$

$$\text{Speed of rotation, } N = \frac{N_s \times H^{5/4}}{\sqrt{P}} \qquad \left(\because N_s = \frac{N\sqrt{P}}{H^{5/4}}\right)$$

$$= \frac{558.3 \times (18)^{5/4}}{\sqrt{30 \times 10^3}} = 119.5 \text{ r.p.m.}$$

For generator, $$N = \frac{120\,f}{p}$$

[where, f = frequency (= 50 Hz)]

$$\therefore \quad \text{Number of poles, } p = \frac{120 \times 50}{119.5} = 50\ \textbf{(Ans.)}$$

(Number of poles is necessarily an even number).

Q. 10. *A jet of water 24 mm diameter and moving at 12 m/s, strikes upon the centre of a symmetrical vane. After impingement, the jet gets deflected through 165° by the vane. Presuming vane to be smooth, determine:*

(i) The force exerted by jet on the vane, and

(ii) The ratio of velocity at outlet to that at inlet if actual reaction of the vane is 121 N.

Solution. *Given*: Diameter of the jet, $d = 24$ mm $= 0.024$ m; Velocity of jet, $V = 12$ m/s; Angle of deflection = 165°; Actual reaction of the vane = 121 N.

(i) The force exerted by the jet on the vane F:

Refer to Fig. 4.

$$165° = 180 - \theta, \quad \text{or,} \quad \theta = 180 - 165 = 15°$$

For *smooth vane,* the theoretical force (or thrust) exerted by the jet on the vane is, given by:

$$F = \rho a V^2 (1 + \cos\theta) \qquad ...(i)$$

$$= 1000 \times \left(\frac{\pi}{4} \times 0.024^2\right) \times 12^2 (1 + \cos 15°)$$

$$= \mathbf{128\ N}\ \textbf{(Ans.)}$$

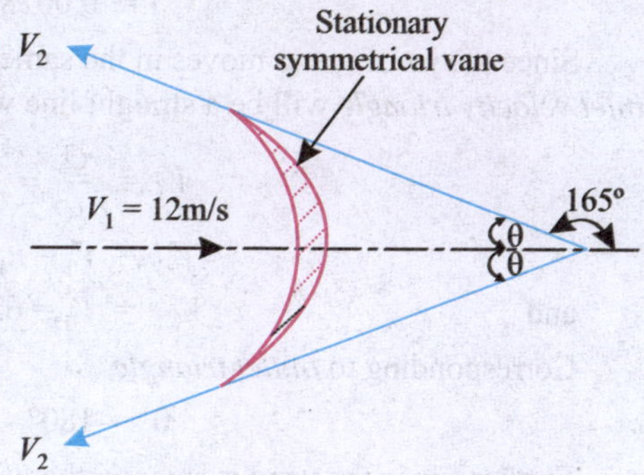

Fig. 4

(ii) $\frac{V_2}{V_1}$:

Actual reaction of the vane = 121 N (Given)

If the vane is *not smooth,* then outgoing velocity at the vane tip is less than the incoming velocity,

i.e., $\qquad \frac{V_2}{V_1} = K,\quad$ where, $\quad K < 1$. Then,

Eqn. (*i*) gets modified to:

$$F = \rho a V^2 (1 + K\cos\theta)$$

$$121 = 1000 \times \left(\frac{\pi}{4} \times 0.024^2\right) \times 12^2 (1 + K \cos 15°)$$

or
$$1 + K\cos 15° = \frac{121}{1000 \times \left(\frac{\pi}{4} \times 0.024^2\right) \times 12^2} = 1.857$$

or, ∴
$$K = \frac{1.857 - 1}{\cos 15°} = \mathbf{0.887}\ \textbf{(Ans.)}$$

Q. 11. *A jet of 60 mm diameter impinges on a curved vane and is deflected through an angle of 165°. The vane moves in the same direction as that of jet with a velocity of 25 m/s. If the rate of flow is 180 litres per second, determine the component of force on the vane in the direction of motion. How much would be the power developed by the vane and what would be the water efficiency?*

Solution. *Given*: Diameter of the jet, $d = 60$ mm $= 0.06$ m; Angle of deflection = 165°; Velocity of the vane, $u_1 = u_2 (= u)$ = 25 m/s; Rate of flow, $Q = 180$ litres p/s = 0.18 m³/s.

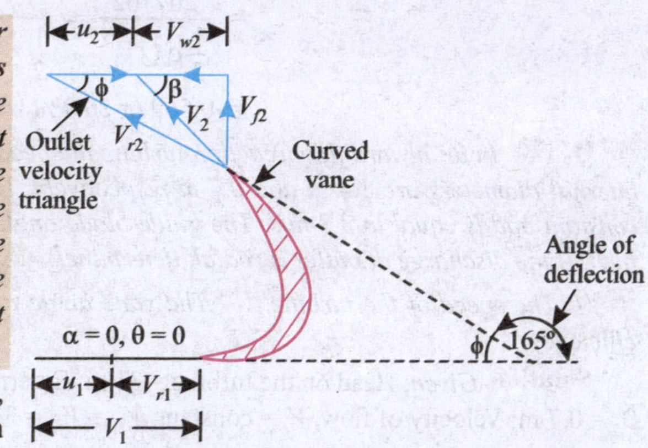

Fig. 5

$$\text{Area of the jet, } a = \frac{\pi}{4}d^2$$
$$= \frac{\pi}{4}\times 0.06^2$$
$$= 0.002827 \text{ m}^2$$

Since the jet of water moves in the same direction as that of vane, $\alpha = \theta = 0$ and, therefore, the *inlet velocity triangle* will be a straight line with

$$V_1 = \frac{Q}{a} = \frac{0.18}{0.002827} = 63.67 \text{ m/s}$$
$$V_{r1} = V_1 - u_1 = 63.67 - 25 = 38.67 \text{ m/s}$$

and $$V_{w1} = V_1 = 63.67 \text{ m/s}$$

Corresponding to *outlet triangle,*

$$\theta = 180° - 165° = 15°$$

Further, since the vane is *smooth,* therefore,

$$V_{r2} = V_{r1} = 38.67 \text{ m/s}$$
$$V_{w2} = V_{r2}\cos\phi - u_2$$
$$= 38.67\cos 15° - 25 = 12.35 \text{ m/s} \quad (\because u_1 = u_2 = 25 \text{ m/s})$$

Power developed by the vane:

Force exerted by the jet on the vane in the direction of motion,

$$F = \rho\, a\, V_{r1}(V_{w1} + V_{w2})$$
$$= 1000 \times 0.002827 \times 38.67 \times (63.67 + 12.35) = 8310.5 \text{ N}$$
$$\text{Work done} = \text{Force} \times \text{Velocity}$$
$$= 8310.3 \times 25 = 207762 \text{ Nm/s or J/s or W}$$

Hence, power developed = 207762 W or **207.8 kW (Ans.)**

Water efficiency (Efficiency of vane):

Efficiency of vane (water efficiency)

$$= \frac{\text{Work done on the vane}}{\text{Kinetic energy supplied by the jet}}$$
$$= \frac{207762}{\frac{1}{2}\rho Q V_1^2} = \frac{207762}{\frac{1}{2}\times 1000 \times 0.18 \times 63.67^2}$$
$$= 0.569 \text{ or } \mathbf{56.9\%} \textbf{ (Ans.)}$$

Q. 12. *In an inward flow reaction turbine the head on the turbine is 31 m. The external and internal diameters are 1.4 m and 0.7 m respectively. The velocity of flow through the remner is constant and is equal to 3.3 m/s. The guide blade angle is 11° and the runner vanes are rigid at inlet. If the discharge at outlet is rodial, determine:*

(i) The speed of the turbine, (ii) The vane angle at outlet of the runner, and (iii) Hydraulic efficiency.

Solution. *Given*: Head on the turbine = 31 m; External diameter, D_1 = 1.4 m; Internal diameter, D_2 = 0.7 m; Velocity of flow, V_f = constant, $V_{f1} = V_{f2}$ = 3.3 m/s; Guide blade angle = 11°.

Refer to Fig. 6.

Runner vanes are *radial at inlet,*

$$\therefore \quad \theta = 90°; \quad V_{w1} = u_1$$

Discharge is radial,

$$\therefore \quad V_{w2} = 0, \quad V_2 = V_{f2} = 3.3 \text{ m/s}$$

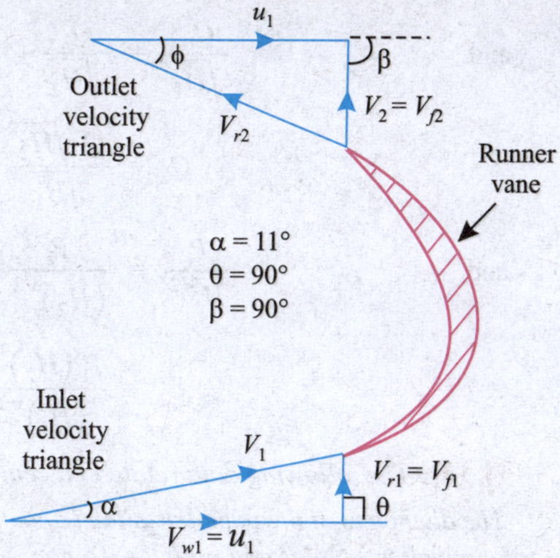

Fig. 6

(i) The speed of the turbine, *N*:

From inlet velocity triangle, we have:

$$\tan \alpha = \frac{V_{f1}}{u_1}$$

$$\text{or,} \quad u_1 = \frac{V_{f_1}}{\tan \alpha} = \frac{3.3}{\tan 11°}$$

$$\text{or,} \quad u_1 = 16.98 \text{ m/s}$$

$$\text{Also,} \quad u_1 = \frac{\pi D_1 N}{60}$$

$$\text{or,} \quad N = \frac{60\, u_1}{\pi D_1}$$

$$\text{or,} \quad N = \frac{60 \times 16.98}{\pi \times 1.4} = \mathbf{231.6\ r.p.m.}\ \text{(Ans.)}$$

(ii) The vane angle at the outlet of the runner, ϕ:

$$u_2 = \frac{\pi D_2 N}{60} = \frac{\pi \times 0.7 \times 231.6}{60} = 8.49 \text{ m/s}$$

From *outlet velocity triangle,* we have:

$$\tan \phi = \frac{V_{f2}}{u_2} = \frac{3.3}{8.49} = 0.3887$$

$$\phi = \tan^{-1}(0.3887) = \mathbf{21.2°}\ \text{(Ans.)}$$

(iii) Hydraulic efficiency, η_{HG}:

$$\eta_h = \frac{V_{w1}\, u_1}{g\, H} \qquad (V_{w2} = 0, \text{ the discharge being radial at outlet})$$

$$= \frac{16.98 \times 16.98}{9.81 \times 31} = 0.948 \text{ or } \mathbf{94.8\%}\ \text{(Ans.)}$$

Q. 13. *A turbine is to operate under a head of 24 m at 180 r.p.m. The discharge is 8.5 m³/s. If the efficiency is 92 per cent, determine the performance of turbine under a head of 18 m.*

Solution. *Given*: Head under which turbine works, $H_1 = 24$ m; speed of the turbine, $N_1 = 180$ r.p.m; Discharge through the turbine, $Q_1 = 8.5$ m^3/s; Efficiency (overall), $\eta_0 = 91\%$.

Performance of the turbine under a head, $H_2 = 18$ m means to find speed (N_2), discharge (Q_2) and power generated (P_2) by the turbine when working under a head of 18 m.

$$\text{Overall efficiency, } \eta_0 = \frac{\text{Shaft power}}{\text{Water power}} = \frac{P}{w\,Q\,H} = \frac{P_1}{w\,Q_1 H_1}$$

$$P_1 = \eta_0 \times w\, Q_1\, H_1 = 0.92 \times 9.81 \times 8.5 \times 24 = 1841 \text{ kW}$$

$$\text{Now,} \quad \frac{N_1}{\sqrt{H_1}} = \frac{N_2}{\sqrt{H_2}}$$

$$\therefore \quad N_2 = \frac{N\sqrt{H_2}}{\sqrt{H_1}} = \frac{180 \times \sqrt{18}}{\sqrt{24}} = \mathbf{155.88\ r.p.m.}\ \text{(Ans.)}$$

and,
$$\frac{Q_1}{\sqrt{H_1}} = \frac{Q_2}{\sqrt{H_2}}$$

$$\therefore \quad Q_2 = \frac{Q_1\sqrt{H_2}}{\sqrt{H_1}} = \frac{8.5\times\sqrt{18}}{\sqrt{24}} = \mathbf{7.36\ m^3/s}\ \text{(Ans.)}$$

and,
$$\frac{P_1}{H_1^{3/2}} = \frac{P_2}{(H_2)^{3/2}}$$

$$\therefore \quad P_2 = \frac{P_1(H_2)^{3/2}}{(H_1)^{3/2}} = \frac{1841\times(18)^{3/2}}{(24)^{3/2}} = \mathbf{1195.8\ kW}\ \text{(Ans.)}$$

Q. 14. *The following data relate to a centrifugal pump:*

The diameters of the impeller at inlet and outlet = 170 mm and 340 mm respectively; The width of the impeller at inlet and outlet = 14 mm and 7 mm respectively; The rate of flow through the pump = 16.9 litres/sec; Speed of the impeller = 1400 r.p.m. Vane angle at the outlet = 45°.

The water enters the impeller radially at inlet.

Neglecting losses through the impeller, find the pressure rise in the impeller.

Solution. *Given*: D_1 = 170 mm = 0.17 m; D_2 = 340 mm = 0.34 m; B_1 = 14 mm = 0.014 m; B_2 = 7 mm = 0.007 m; Q = 16.9 litres/s = 0.0169 m^3/s; N = 1400 r.p.m; ϕ = 45°.

Pressure rise in the impeller:

Velocity of flow at inlet,

$$V_{f1} = \frac{Q}{\pi D_1 B_1}$$

$$= \frac{0.0169}{\pi\times 0.17\times 0.014} = 2.26\ \text{m/s}$$

Velocity of flow at outlet,

$$V_{f2} = \frac{Q}{\pi D_2 B_2}$$

$$= \frac{0.0169}{\pi\times 0.34\times 0.007} = 2.26\ \text{m/s}$$

Tangential velocity of impeller at outlet,

$$u_2 = \frac{\pi D_2 N}{60} = \frac{\pi\times 0.34\times 1400}{60} = 24.92\ \text{m/s}$$

Pressure rise in the impeller

$$= \frac{1}{2g}\left(V_{f1}^2 + u_2^2 - V_{f2}^2\ \text{cosec}^2\ \phi\right) \quad \text{...[Eqn. (3.165)]}$$

$$= \frac{1}{2\times 9.81}(2.26^2 + 24.92^2 - 2.26^2 \times \text{cosec}^2\ 45°)$$

$$= \frac{1}{2\times 9.81}(5.108 + 621.006 - 10.215) = \mathbf{31.39\ m}\ \text{(Ans.)}$$

Q. 15. *Tests on a pump model indicate a cavitation parameter σ_c = 0.11. A homologous unit is to be installed at a location when atmospheric pressure, p_a = 0.89 bar and vapour pressure p_v = 0.034 bar absolute and is to pump water against a head of 22 m.*

Calculate the maximum permissible suction head.

Solution. *Given*: Cavitation parameter, $\sigma_c = 0.11$; Atmospheric pressure, $p_a = 0.89$ bar or $H_a = \frac{0.89 \times 10^5}{9810} = 9.07$ m of water; Vapour pressure, $p_u = 0.034$ bar or $H_v = \frac{0.034 \times 10^5}{9810} = 0.346$ m of water; Manometric head, $H_{mano} = 22$ m.

Maximum permissible suction head, h_s:

We know that, $$\sigma = \frac{H_a - H_s - H_v}{H_{mano.}}$$

or, $$\sigma_c = \frac{H_c - h_s - H_v}{mano.}$$ neglecting head lost due to friction

$$\left(\because H_s = h_s + h_{fs} + \frac{V_2^2}{2g}\right)$$

or, $$0.11 = \frac{9.07 - h_s - 0.346}{22}$$

or, $$0.11 \times 22 = 8.724 - h_s$$

or, $$h_s = \mathbf{6.304\ m}\ \text{(Ans.)}$$

Q. 16. *A single-acting reciprocating pump has a diameter (piston) of 120 mm and stroke length 240 mm. The length and diameter of the suction pipe are 7.0 m and 60 mm respectively. If the suction lift of the pump is 3.3 m and separation occurs when pressure in the pump falls below 2.6 m of water absolute and the manometer reads 762 mm of mercury, determine the maximum speed at which pump can be run without separation in the suction pipe.*

Solution. *Given:* Piston diameter, $D = 120$ mm $= 0.12$ m, $\therefore$ Area, $A = \frac{\pi}{4} \times 0.12^2 = 0.01131\ m^2$; Stroke length, $L = 240$ mm $= 0.24$ m; Crank radius, $r = \frac{L}{2} = 0.24/2 = 0.12$ m; Length of suction pipe, $l_s = 7.0$ m; Diameter of suction pipe, $d_s = 60$ mm $= 0.06$ m, $\therefore$ Area of suction pipe, $a_s = \frac{\pi}{4} \times 0.06^2 = 0.002827\ m^2$; The suction lift of the pump, $h_s = 3.3$ m; Separation pressure head, $h_{sep.} = 2.6$ m.

Maximum speed at which pump can run without separation, N:

Atmospheric head, $H_{atm.} = \frac{762}{1000} \times 13.6 = 10.363$ m of water during suction stroke, the possibility of separation is only at the *beginning of the stroke.* At the beginning of suction stroke, $\theta = 0°$ and $\cos\theta = 1$, that gives:

$$h_{as} = \frac{l_s}{g} \times \frac{A}{a_s} \times \omega^2 r = \frac{7.0}{9.81} \times \frac{0.01131}{0.002827} \times \omega^2 \times 0.12 = 0.342\ \omega^2$$

Pressure head in the cylinder at the beginning of suction stroke

$$= (h_s + h_{as})\ \text{vacuum}$$
$$= H_{atm.} - (h_s + h_{as})\ \text{absolute}$$

This absolute pressure (at the beginning of suction stroke) *should not fall below the vapour pressure head ($h_{sep.}$) to avoid separation,* thus in the *limiting condition,*

$$H_{atm.} - (h_s + h_{as}) = h_{sep.}$$

or $$10.363 - (3.3 + 0.342\ \omega^2) = 2.6$$

or $$\omega^2 = \frac{10.363 - 2.6 - 3.3}{0.342}$$

or $$\text{Angular velocity, } \omega = 3.6\ \text{rad/s}$$

But, $\omega = \frac{2\pi N}{60}$ or $N = \frac{60\,\omega}{2\pi} = \frac{60 \times 3.6}{2 \times \pi} = \mathbf{34.38\ r.p.m.}$ **(Ans.)**

Q. 17. *The diameters of ram and plunger of a hydraulic press are 120 mm and 15 mm respectively.*

(i) Find the force required to be applied on the plunger to raise a load of 30 kN on the ram.

(ii) If the plunger has a stroke of 220 mm, how many strokes will be required to lift the load by 450 mm. Also calculate the volume of additional liquid required.

(iii) Further, if the time taken to lift the load is 11 minutes, what will be power required to drive the plunger.

Solution. *Given*: Diameter of the ram, D = 120 mm = 0.12 m, ∴ Area of the ram = $\frac{\pi}{4} \times 0.12^2$ = 0.01131 m^2; Diameter of the plunger, d = 15 mm = 0.015, ∴ Area of the plunger = $\frac{\pi}{4} \times 0.015^2$ = 0.0001767 m^2; Load to be raised, W = 30 kN; Stroke of the plunger = 220 mm = 0.22 m; Distance through which load is to be lifted = 450 mm = 0.45 m; Time taken to lift the load = 11 minutes.

(i) Force required to raise a load of 30 kN, *F*:

Since pressure intensity is same throughout a static mass of fluid, therefore,

$$\frac{F}{a} = \frac{W}{A} \quad \text{or} \quad \frac{F}{0.0001767} = \frac{30}{0.01131}$$

or, $$F = \frac{30 \times 0.0001767}{0.01131} = \mathbf{0.468\ kN}\ \textbf{(Ans.)}$$

(ii) Number of strokes required, *n*:

Number of strokes required to lift the load by 0.45 m,

$$n = \frac{\text{Total volume of liquid to be displaced}}{\text{Volume of liquid displaced in one stroke of the plunger}}$$

$$= \frac{\frac{\pi}{4} \times (0.12)^2 \times 0.45}{\frac{\pi}{4} \times (0.015)^2 \times 0.22} = \mathbf{131}\ \textbf{(Ans.)}$$

$$\textit{Volume of additional liquid} = \frac{\pi}{4} \times 0.12^2 \times 0.45 = \mathbf{0.00509\ m^3}\ \textbf{(Ans.)}$$

(iii) Power required to drive motor, *P*:

$$\text{Work done by the press} = 30 \times 0.45 = 13.5 \text{ kNm (in 11 minutes)}$$

$$\text{Work done per second} = \frac{13.5}{11 \times 60} = 0.02045 \text{ kNm/s}$$

$$\therefore \quad \text{Power required, } P = 0.02045 \text{ kW} = \mathbf{20.45\ W}\ \textbf{(Ans.)}$$

Q. 18. *A 80 mm diameter jet having a velocity of 28 m/s strikes a flat plate, the normal of which is inclined at 40° to the axis of the jet. Determine the normal force on the plate:*

(i) When the plate is stationary; (ii) When the plate is moving with a velocity of 12 m/s in the direction of jet, away from the jet.

What is the power and efficiency of the jet when the plate is moving?

Solution. Diameter of the jet, d = 80 mm = 0.08 m, ∴ Area of the jet, $a = \frac{\pi}{4} \times 0.08^2 = 0.00503$ m^2; Angle between the jet and the plate, θ = 90° – 40° = 50°; Velocity of the jet, V = 28 m/s; Velocity of the plate, u = 12 m/s.

Normal force on the plate:

(*i*) *When the plate is stationary,* the normal force on the plate is given by:

$$F_x = \rho\, a\, V^2 \sin\theta \qquad \text{...[Eqn. (3.2)]}$$

$$= 1000 \times 0.00503 \times 28^2 \times \sin 50° = \mathbf{3020.9\ N}$$ **(Ans.)**

(*ii*) *When the plate is moving with a velocity of 12 m/s and moving away from the jet,* the normal force on the plate is given by the relation:

$$F_n = \rho\, a\, (v - u)^2 \sin\theta$$

$$= 1000 \times 0.00503 \times (28 - 12)^2 \sin 50° = \mathbf{986.4\ N}$$ **(Ans.)**

Power and efficiency of the jet when the plate is moving:

Work done per second by the jet

= Force in the direction of the jet × distance moved by the plate in the direction of the jet/sec.

$$= F_x \times u$$

where, $F_x = F_n \times \sin\theta = 986.4 \times \sin 50° = \mathbf{755.6\ N}$ **(Ans.)**

∴ Work done $= 755.6 \times 12 = 9067.2$ Nm/s

Hence, *Power of the jet* $= 9067.2$ J/s $= 9067.2$ W $=$ **9.067 kW (Ans.)**

$$\textit{Efficiency of the jet} = \frac{\text{Work done on the plate}}{\text{Kinetic energy supplied by the jet}}$$

$$= \frac{9067.2}{\frac{1}{2}(\rho a V) \times V^2} = \frac{9067.2}{\frac{1}{2} \times (1000 \times 0.00503 \times 28) \times 28^2}$$

$$= 0.1642 \text{ or } \mathbf{16.42\%}$$ **(Ans.)**

Q. 19. *A single jet Pelton wheel runs at 280 r.p.m under a head of 480 m. The jet diameter is 190 mm, its deflection inside the bucket is 160° and its relative velocity is reduced by 12 percent due to friction. Determine:*

(i) Water power; (ii) Resultant force on the bucket; and (iii) Overall efficiency.

Assume: Mechanical losses = 2.6 per cent, coefficient of velocity = 0.97; and speed ratio = 0.45.

Solution. *Given*: Speed of the wheel, $N = 280$ r.p.m.; Diameter of the jet; $d = 190$ mm $= 0.19$ m; Net head, $H = 480$ m; Angle of deflection of jet = 160°; Reduction of relative velocity due to friction = 12%; Mechanical losses = 2.6%; Coefficient of velocity, $C_v = 0.97$; Speed ratio, $K_u = 0.45$.

(*i*) Water power:

$$\text{Velocity of jet, } V_1 = C_v \sqrt{2\,\text{gH}} = 0.97\sqrt{2 \times 9.81 \times 480} = 94 \text{ m/s}$$

Discharge through the Pelton wheel,

$$Q = \text{Area of jet } (a) \times \text{velocity } (V_1)$$

$$= \frac{\pi}{4} \times 0.19^2 \times 94 = 2.66 \text{ m}^3/\text{s}$$

$$\text{Water power} = w\,Q\,H = 9.81 \times 2.66 \times 480 = \mathbf{12525\ kW}$$ **(Ans.)**

(*ii*) Resultant force on the bucket:

Peripheral speed of the wheel, $u = K_u \sqrt{2\,\text{gH}} = 0.45\sqrt{2 \times 9.81 \times 480} = 43.67$ m/s

Refer to Fig. 7.

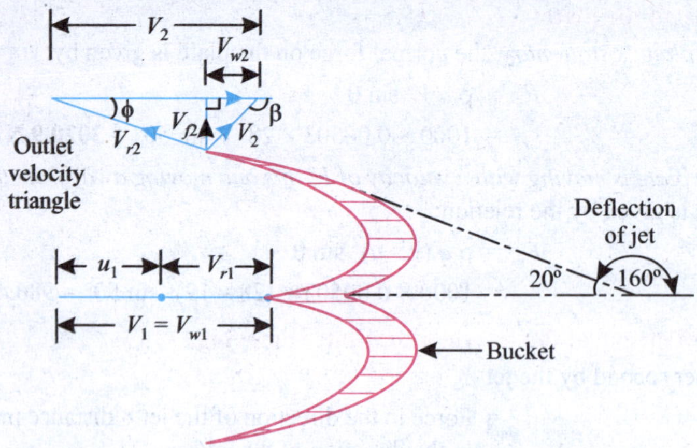

Fig. 7

At inlet to turbine:

$$V_{w1} = V_1 = 94 \text{ m/s}$$
$$V_{r1} = (V_1 - u_1) = 94 - 43.67 = 50.33 \text{ m/s}$$

At exit from the turbine:

The blade angle at exit,

$$\phi = 180° - 160° = 20°$$
$$V_{r2} = \left(100 - \frac{12}{100}\right) \times V_{r1}$$

or,
$$V_{r2} = 0.88 \times 50.33 = 44.3 \text{ m/s}$$

As $V_{r2} \cos \theta$ is less than blade speed u, the velocity triangle at outlet will be as shown in Fig. 7. ($\beta > 90°$)

$$V_{w2} = u_2 - V_{r2} \cos \phi = 43.67 - 44.3 \cos 20° = 2.04 \text{ m/s}$$
$$(\because u_1 = u_2 = u)$$

Resultant force on the bucket,

$$F = \rho Q (V_{w1} - V_{w2})$$
$$= 1000 \times 2.66 (94 - 2.04) = \mathbf{244614 \text{ N}} \textbf{ (Ans.)}$$

***(iii)* Brake power, *P*:**

Power developed by the wheel

$$= F \times u$$
$$= 244614 \times 43.67 \text{ Nm/s or W}$$
$$= 244614 \times 43.67 \times 10^{-3} \text{ kW}$$
$$= \mathbf{10682 \text{ kW}} \textbf{ (Ans.)}$$

***(iv)* Overall efficiency, η_0:**

$$\eta_0 = \frac{\text{Brake power}}{\text{Water power}}$$
$$= \frac{10682}{12525} = 0.853 \text{ or } \mathbf{85.3\%} \textbf{ (Ans.)}$$

Q. 20. *A Kaplan turbine develops 20 MW at average head of 32 m. Assuming a speed ratio of 2, flow ratio of 0.58, diameter of the boss equal to 0.34 times the diameter of the runner and an overall efficiency of 85 per cent, determine the diameter, speed and specific speed of the turbine.*

Solution. *Given*: Shaft power, $P = 20$ MW $= 20 \times 10^3 = 20{,}000$ kW; Head, $H = 32$ m; Speed ratio, $K_u = 2$, Flow ratio $K_f = 0.58$; Diameter of boss $(D_b) = 0.34 \times$ diameter of the runner (D_0), *i.e.*, $D_b = 0.34\, D_0$; Overall efficiency, $\eta_0 = 85\%$.

Refer to Fig. 8.

Diameter of the runner, D_0:

$$K_u = \frac{u_1}{\sqrt{2gH}} = 2, \quad u_1 = 2 \times \sqrt{2gH} = 2 \times \sqrt{2 \times 9.81 \times 32} = 50.11 \text{ m/s}$$

$$K_f = \frac{V_{f1}}{\sqrt{2gH}} = 0.58$$

$$\text{or,} \quad V_{f1} = 0.58 \times \sqrt{2gH} = 0.58 \times \sqrt{2 \times 9.81 \times 32} = 14.5 \text{ m/s}$$

Overall efficiency,

$$\eta_0 = \frac{\text{Shafter power } (P)}{\text{Water power}} = \frac{20000}{w\,Q\,H}$$

$$0.85 = \frac{20000}{9.81 \times Q \times 32}$$

$$\text{or,} \quad Q = \frac{20000}{0.85 \times 9.81 \times 32} = 74.9 \text{ m/s}$$

Also, Q = Area of flow × velocity of flow

$$74.9 = \frac{\pi}{4}\left(D_0^2 - D_b^2\right) \times V_{f1}$$

$$\text{or, } 74.9 = \frac{\pi}{4}\left[D_0^2 - (0.34\, D_0)^2\right] \times 14.5$$

$$= \frac{\pi}{4} D_0^2 \left[1 - (0.34)^2\right] \times 14.5 = 10.07\, D_0^2$$

$$\text{or,} \quad D_0 = \left(\frac{74.9}{10.07}\right)^{1/2} = \mathbf{2.73 \text{ m}} \text{ (Ans.)}$$

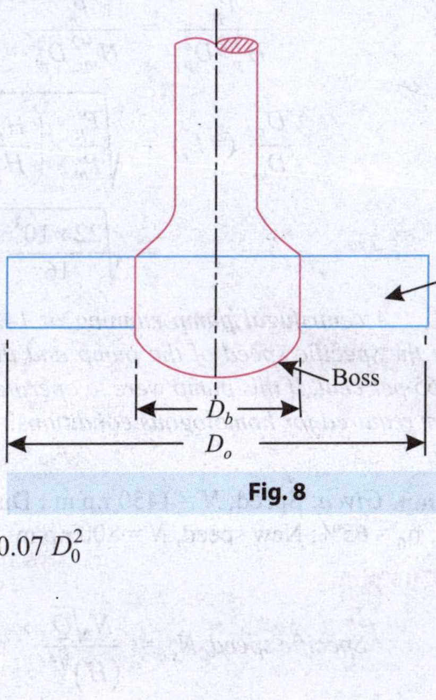

Fig. 8

Speed of the turbine, N:

$$u_1 = \frac{\pi D_0 N}{60}, \quad \text{or,} \quad N = \frac{60\, u_1}{\pi D_0} - \frac{60 \times 50.11}{\pi \times 2.73} = \mathbf{350.6 \text{ r.p.m.}} \text{ (Ans.)}$$

Specific speed of the turbine, N_s:

$$N_s = \frac{N\sqrt{P}}{H^{5/4}} = \frac{350.6 \times \sqrt{20000}}{(32)^{5/4}} = \mathbf{651.5} \text{ (Ans.)}$$

Q. 21. *A hydro-turbine is required to give 22 MW at 45 m head and 90 r.p.m runner speed. The laboratory facilities available permit testing of 16 kW model at 4.5 m head. What should be model runner speed and model to prototype scale ratio?*

Solution. *Given*: $P_p = 22$ MW; $H_p = 45$ m; $N_p = 90$ r.p.m.; $P_m = 16$ kW; $H_m = 4.5$ m.

N_m; $\dfrac{D_p}{D_m}$ $(= L_r)$:

Prototype specific speed, $(N_s)_p = \dfrac{N_p\sqrt{P_p}}{(H_p)^{5/4}}$ (where P is in kW)

$$= \frac{90 \times \sqrt{22 \times 10^3}}{(45)^{5/4}} = 114.5$$

For model, $114.5 = \dfrac{N_m \times \sqrt{P_m}}{(H_m)^{5/4}}$ $[\because (N_s)_p = (N_s)_m]$

or, $N_m = \dfrac{114.5 \times (H_m)^{5/4}}{\sqrt{P_m}} = \dfrac{114.5 \times (4.5)^{5/4}}{\sqrt{16}}$ = **187.6 r.p.m. (Ans.)**

For similar turbines $\dfrac{P}{H^{3/2}D^2}$ should be *equal*.

Then, $\dfrac{P_p}{H_p^{3/2}D_p^2} = \dfrac{P_m}{H_m^{3/2}D_m^2}$

or, $\dfrac{D_p}{D_m} (= L_r) = \sqrt{\dfrac{P_p}{P_m} \times \left(\dfrac{H_m}{H_p}\right)^{3/2}}$

$$= \sqrt{\frac{22 \times 10^3}{16} \times \left(\frac{4.5}{45}\right)^{3/2}} = \mathbf{6.594}$$ **(Ans.)**

Q. 22. *A centrifugal pump running at 1450 r.p.m. delivers 0.18 m³/s at a head of 12 m. Calculate the specific speed of the pump and the power input. Assume overall efficiency of the pump as 65 per cent. If this pump were to operate at 800 r.p.m. what would be the head, discharge and power required for homologous conditions? Assume overall efficiency remains unchanged at new r.p.m.*

Solution. *Given*: Speed, $N = 1450$ r.p.m.; Dischange, $Q = 0.18$ m^3/s; Head, $H = 12$ m; Overall efficiency, $\eta_0 = 65\%$; New speed, $N = 800$ r.p.m.

At 1450 r.p.m:

Specific speed, $N_s = \dfrac{N\sqrt{Q}}{(H)^{3/4}}$ ...[Eqn. (3.25)]

$$= \frac{1450 \times \sqrt{0.18}}{(12)^{3/4}} = \mathbf{95.4}$$ **(Ans.)**

Power input $= \dfrac{wQH}{\eta_0} = \dfrac{9.81 \times 0.18 \times 12}{0.65}$ = **32.6 kW (Ans.)**

At 800 r.p.m:

$$N_u = \frac{N}{\sqrt{H}}$$

i.e., $\dfrac{800}{\sqrt{H}} = \dfrac{1450}{\sqrt{12}}$

or, $$\text{Head, } H = \left(\frac{800}{1450}\right)^2 \times 12 = \mathbf{3.65\ m}\ \textbf{(Ans.)}$$

$$Q_u = \frac{Q}{\sqrt{H}}$$

i.e., $$\frac{Q}{\sqrt{3.65}} = \frac{0.18}{\sqrt{12}}$$

or, $$\text{Discharge, } Q = \sqrt{\frac{3.65}{12}} \times 0.18 = \mathbf{0.09927\ m^3/s}\ \textbf{(Ans.)}$$

$$P_u = \frac{P}{(H)^{3/2}}$$

i.e., $$\frac{P}{(3.65)^{3/2}} = \frac{32.6}{(12)^{3/2}}$$

or $$\text{Power input, } P = \left(\frac{3.65}{12}\right)^{3/2} \times 32.6 = \mathbf{5.469\ kW}\ \textbf{(Ans.)}$$

Q. 23. *A single-stage centrifugal pump runs at 550 r.p.m. and delivers 290 m^3/min of water against a head of 135 m. The pump impeller is 2.2 m in diameter and it has a positive suction lift (including the velocity head and friction) of 3.3 m. Laboratory tests are to be conducted on a model with 0.5 m diameter impeller and on a reduced head of 105 m. Determine the speed, discharge and suction lift for the laboratory tests.*

Assume atmospheric head = 10.15 m of water, and vapour head = 0.3 m of water.

Solution.

Prototype pump:	*Model pump:*
Speed, N_p = 550 r.p.m	Speed, N_m = ?
Discharge, Q_p = 290 m^3/s	Discharge, Q_m = ?
Manometric head, $(H_{mano})_p$ = 135 m	Manometric head, $(H_{mano})_m$ = 105 m
Diameter of impeller, D_p = 2.2 m	Diameter of impeller, D_m = 0.5 m
Positive suction lift = 3.3 m	Positive suction lift, h_s = ?
Vapour head, H_v = 0.3 m of water	Atmospheric head, H_a = 10.15 m

(*i*) Speed of the model pump, N_m:

We know that: $$\left(\frac{\sqrt{H_{mano}}}{DN}\right)_m = \left(\frac{\sqrt{H_{mano}}}{DN}\right)_p \qquad \text{...[Eqn. (3.27)]}$$

$$\frac{\sqrt{(H_{mano})_m}}{D_m\, N_m} = \frac{\sqrt{(H_{mano})_p}}{D_p\, N_p}$$

or, $$N_m = \frac{\sqrt{(H_{mano})_m}}{\sqrt{(H_{mano})_p}} \times \frac{D_p}{D_m} \times N_p$$

or, $$N_m = \sqrt{\frac{105}{135}} \times \frac{2.2}{0.5} \times 550 = \mathbf{2134\ r.p.m.}\ \textbf{(Ans.)}$$

(*ii*) Discharge for the model pump, Q_m:

We know that: $$\left(\frac{Q}{D^3 N}\right)_m = \left(\frac{Q}{D^3 N}\right)_p \qquad \text{...Eqn. (3.28)}$$

i.e.,
$$\frac{Q_m}{D_m^3 N_m} = \frac{Q_p}{D_p^3 N_p}$$

or,
$$Q_m = Q_p \left(\frac{D_m}{D_p}\right)^3 \times \frac{N_m}{N_p}$$

or,
$$Q_m = 290\left(\frac{0.5}{2.2}\right)^3 \times \frac{2134}{550} = \mathbf{13.21\ m^3/min.\ (Ans.)}$$

(*iii*) Positive suction lift with which model should be tested, h_s:

Cavitation factor for the prototype,

$$\sigma_p = \frac{H_a - H_s - H_v}{(H_{mano})_p} = \frac{10.15 - 3.3 - 0.3}{135} = 0.0485$$

For cavitation similarity, $\sigma_m = \sigma_p$

$$\sigma_m = \frac{10.15 - H_s - 0.3}{105} = 0.0485$$

or,
$$H_s = (10.15 - 0.3) - 105 \times 0.0485$$
$$= \mathbf{4.75\ m\ (including\ velocity\ head\ and\ friction)\ (Ans.)}$$

Q. 24. *The plunger diameter and stroke length of a single-acting reciprocating pump are 250 mm and 420 mm respectively. The speed of the pump is 60 r.p.m. The diameter and length of delivery pipe are 125 mm and 50 m respectively. If the pump is equipped with an air vessel on the delivery side at the centre line of the pump, find the power saved in overcoming friction in the delivery pipe. Assume coefficient of friction, f = 0.01.*

Solution. Diameter of plunger, $D = 250$ mm $= 0.25$ m, ∴ Area $A = \frac{\pi}{4} \times 0.25^2 = 0.04909$ m^2; Stroke length, $L = 420$ mm $= 0.42$ m; Crank radius, $r = \frac{0.42}{2} = 0.21$ m; Speed of the pump, $N = 60$ r.p.m, ∴ Angular velocity, $\omega = \frac{2\pi N}{60} = \frac{2\pi \times 60}{60} = 6.283$ rad/s; Diameter of delivery pipe, $d_d = 125$ mm $= 0.125$ m, ∴ Area of delivery pipe, $a_d = \frac{\pi}{4} \times 0.125^2 = 0.01227$ m^2; Length of delivery pipe, $l_d = 50$ m; Coefficient of friction, $f = 0.01$.

Power saved in overcoming friction in the delivery pipe:

Maximum velocity of water in delivery pipe,

$$v_d = \frac{A}{a_d}\omega r = \frac{0.04909}{0.01227} \times 6.283 \times 0.21 = 5.28 \text{ m/s}$$

Maximum loss of head due to friction,

$$h_{fd} = \frac{4 f\, l_d\, v_d}{2g \times d_d} = \frac{4 \times 0.01 \times 50 \times (5.28)^2}{2 \times 9.81 \times 0.125} = 22.7 \text{ m}$$

Power required to overcome friction

$$= \frac{\omega A L N}{60} \times \left(\frac{2}{3} h_{fd}\right) = \frac{9810 \times 0.04909 \times 0.42 \times 60}{60} \times \left(\frac{2}{3} \times 22.7\right)$$

$$= 3060 \text{ W or } 3.06 \text{ kW}$$

"With air vessel fitted", the velocity in the delivery pipe becomes constant and is given by:

$$v_d = \frac{A}{a_d} \times \frac{\omega r}{\pi} = \frac{0.04909}{0.01227} \times \frac{6.283 \times 0.21}{\pi} = 1.68 \text{ m/s.}$$

Loss of head due to friction,

$$h_{fd} = \frac{4 f\, l_d\, v_d^2}{2g \times d_d} = \frac{4 \times 0.01 \times 50 \times 1.68^2}{2 \times 9.81 \times 0.125} = 2.3 \text{ m}$$

Power required to overcome friction

$$= \frac{\omega\, A\, L\, N}{60} \times h_{fd} = \frac{9810 \times 0.04909 \times 0.42 \times 60}{60} \times 2.3 = 465 \text{ W or } 0.465 \text{ kW}$$

Hence power saved by fitting an air vessel

$= 3.06 - 0.465 =$ **2.595 kW (Ans.)**

Q. 25. *A hydraulic ram is being supplied water at the rate of 0.045 m³/s from a height of 4.9 m and it raises 0.0045 m³/s to a height of 32 m from the ram. The length and diameter of the pipe are 110 m and 65 mm respectively. If the coefficient of friction is 0.01, calculate D'Aubuisson's and Rankine's efficiencies.*

Solution. *Given*: Disharge through the supply pipe, $Q = 0.045$ m^3/s; Supply head, $h = 4.9$ m; Discharge raised, $q = 0.0045$ m^3/s; Height of water raised from hydraulic ram, $H = 32$ m; Length of pipe, $l = 110$ m; Diameter of the pipe, $d = 65$ mm $= 0.065$ m; Coefficient of friction, $f = 0.01$.

Efficiency of the ram:

Head lost due to friction in the delivery pipe,

$$h_f = \frac{4 f\, l\, V^2}{d \times 2g} = \frac{4 \times 0.01 \times 110 \times V^2}{0.065 \times 2 \times 9.81} = 3.45\, V^2$$

But, V = Velocity of water in delivery pipe

$$= \frac{q}{\frac{\pi}{4} d^2} = \frac{0.0045}{\frac{\pi}{4} \times 0.065^2} = 1.356 \text{ m/s}$$

$\therefore \quad h_f = 3.45\, V^2 = 3.45 \times 1.356^2 = 6.34$ m

Effective head developed by the ram,

$$H_e = H + h_f = 32 + 6.34 = 38.34 \text{ m}$$

$$D'Aubuisson's\ efficiency = \frac{q \times H_e}{Q \times h} = \frac{0.0045 \times 38.34}{0.045 \times 4.9} = 0.7824 \text{ or } \mathbf{78.24\%} \text{ (Ans.)}$$

$$Rankine's\ efficiency = \frac{q\,(H_e - h)}{(Q - q)\,h}$$

$$= \frac{0.0045\,(38.34 - 4.9)}{(0.045 - 0.0045) \times 4.9} = 0.758 \text{ or } \mathbf{75.8\%} \text{ (Ans.)}$$

GATE and UPSC Examinations' Questions with Answers/Solutions

A. Conventional Questions with Answers/Solutions

B. Multiple-Choice Questions with Answers and "Explanations"

Chapter 9

A. CONVENTIONAL QUESTIONS WITH ANSWERS/SOLUTIONS

Q. 9.1. ***(Centrifugal pumps)*** *Explain the effect of vane angle on manometric efficiency in a centrifugal pump. Why is vane angle in a centrifugal pump not kept at values below 20°.* **(GATE)**

Ans. Effect of outlet angle (ϕ) on manometric efficiency.

The total energy of liquid before entering the impeller, with reference to the centre of pump, is taken as zero. After leaving the impeller, the liquid has a pressure energy (H_{mano}) and kinetic energy $\left(\frac{V_2^2}{2g}\right)$. The energy supplied to the impeller is $\frac{V_{w2}u_2}{g}$. Neglecting the losses in the pump and equating the energy given to the impeller to the increase in total energy, we have:

$$\frac{V_{w2}u_2}{g} = H_{mano} + \frac{V_2^2}{2g}$$

or,
$$H_{mano} = \frac{V_{w2}u_2}{g} - \frac{V_2^2}{2g} \quad ...(i)$$

From velocity triangle at outlet shown in Fig. 9.1, we have:

$$V_2^2 = V_{w2}^2 + V_{f2}^2, \text{ and } V_{w2} = u_2 - \frac{V_{f2}}{\tan\phi} = u_2 - V_{f2}\cot\phi$$

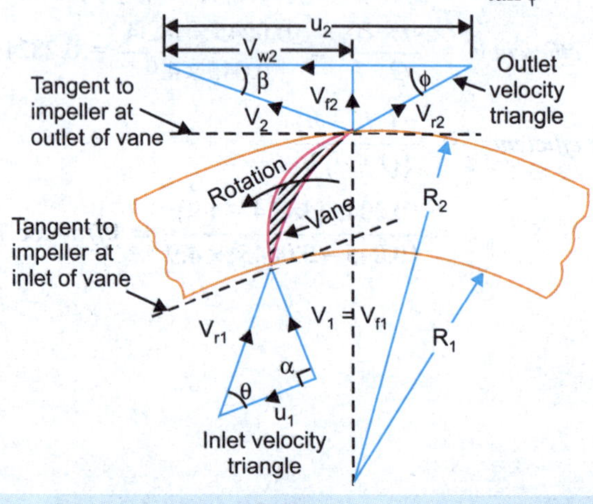

Fig. 9.1 Velocity triangles for an impeller vane.

Then, $$H_{mano} = \frac{(u_2 - V_{f2}\cot\phi)\times u_2}{g} - \frac{(u_2 - V_{f2}\cot\phi)^2 \times V_{f2}^2}{2g}$$

$$= \frac{2(u_2^2 - u_2V_{f2}\cot\phi) - [(u_2^2 + V_{f2}^2\cot^2\phi - 2u_2V_{f2}\cot\phi) + V_{f2}^2]}{2g}$$

$$= \frac{2u_2^2 - 2u_2V_{f2}\cot\phi - u_2^2 - V_{f2}^2\cot^2\phi + 2u_2V_{f2}\cot\phi - V_{f2}^2}{}$$

$$= \frac{u_2^2 - V_{f2}^2(1+\cot^2\phi)}{2g} = \frac{u_2^2 - V_{f2}^2\operatorname{cosec}^2\theta}{2g} \quad \text{... (ii)}$$

The manometric efficiency of the pump, under the ideal condition assumed above, will become:

$$\eta_{mano} = \frac{H_{mano}}{(V_{w2}u_2/g)} = \frac{gH_{mano}}{V_{w2}u_2} = \frac{u_2^2 - V_{f2}^2\operatorname{cosec}^2\theta}{2u_2(u_2 - V_{f2}\cot\phi)}$$

Let us assume the flow ratio $K_f \left[= \frac{V_{f2}}{\sqrt{2gH_{mano}}} \right] = 0.25$; computing the value of u_2 in terms of H_{mano} from eqn. (*ii*) for different values of ϕ, it will be observed that as the value of ϕ varies from 90° to 20° the value of η_{mano} increases from 0.47 to 0.73. A further decrease in the angle ϕ will increase the efficiency, but it is impracticable to have the angle ϕ less than 20°, as it would *result in long and narrow blades, with very high frictional losses.* As such the **minimum value** *of* ϕ is 20°.

Q. 9.2. ***(Hydraulic lift)*** *A hydraulic lift of the type commonly used for greasing automobiles consists of a 280 mm diameter ram that slides in a 280.18 mm cylinder. The annual space between the ram and cylinder is filled with oil having a kinematic viscosity of 0.00042 m²/s and specific gravity of 0.86. If the rate of travel of the ram is 0.22 m/s, find the frictional resistance when 2 m of the ram is engaged in the cylinder.* **(IES)**

Solution. *Given:* $d_r = 280$ mm, or 0.28 m; $d_{cy} = 280.18$ mm; $\nu = 0.00042$ m²/s; Specific gravity = 0.86; V = 0.22 m/s; Ram engagement = 2 m.

Now, $\nu = \frac{\mu}{\rho}$ or $\mu = \nu\rho = 0.00042 \times (0.86 \times 1000) = 0.3612$

Frictional resistance, *F*:

We know that: $$\frac{F}{A} = \mu\frac{du}{dy}$$

or, $$F = \frac{\mu AV}{dy} = \frac{0.3612 \times (\pi \times 0.28 \times 2) \times 0.22}{\left(\frac{280.18 - 280}{2}\right) \times 10^{-3}}$$

$$= \mathbf{1553.3\ N} \quad \textbf{(Ans.)}$$

Q. 9.3. ***(Pelton turbine)*** *A three jet pelton turbine is required to generate 10000 kW under a net head of 400 m. The blade angle at the outlet is 15° and the decrease in the relative velocity while passing over the blade is 5%. Determine:*

(i) The diameter of the jet.

(ii) The force exerted by a single jet on the buckets.

Given, overall efficiency of turbine = 80%

co-effecient of velocity = 0.98

speed ratio = 0.46

Further, if the jet ratio is not to be less than 10, calculate

(iii) Speed of the wheel for a frequency of 50 hertz/sec and corresponding wheel diameter. **(IES)**

Solution. *Given*: No. of jets = 3; Total power, $P = 10000$ kW; Net head, H = 400 m; Blade angle at the outlet, $\phi = 15°$; Relative velocity at the outlet, $V_{r2} = 0.95\ V_{r1}$; Overall efficiency, $\eta_0 = 80\%$, Co-efficient of velocity, $C_v = 0.98$; Speed ratio, $K_u = 0.46$; $f = 50$ hertz.

(i) Diameter of the jet, *d*:

Now, $$\eta_0 = \frac{P}{wQH}$$

or, $$0.8 = \frac{10000}{9.81 \times Q \times 400}$$

or, $$Q = \frac{10000}{0.8 \times 9.81 \times 400} = 3.18 \text{ m}^3/\text{s}$$

$$\therefore \text{ Discharge through one nozzle} = \frac{3.18}{3} = 1.06 \text{ m}^3/\text{s}$$

Also, $$1.06 = \text{Area of jet} \times \text{velocity of jet}$$

or, $$1.06 = \frac{\pi}{4} d^2 \times \left[C_v \times \sqrt{2gH}\right]$$

or, $$= \frac{\pi}{4} d^2 \times 0.98 \times \sqrt{2 \times 9.81 \times 400}$$

$$\therefore \quad d = \left[\frac{1.06 \times 4}{\pi \times 0.98 \times \sqrt{2 \times 9.81 \times 400}}\right]^{1/2} = 0.125 \text{ m or } 125 \text{ mm} \quad \textbf{(Ans.)}$$

(ii) Force exerted by a single jet on the buckets:

Refer to Fig. 2.5.

$$u_1 = K_u \times V_1 = K_u \times \sqrt{2gH} = 0.46 \times \sqrt{2 \times 9.81 \times 400} = 40.75 \text{ m/s}$$

Now, $$V_{r1} = V_1 - u_1 = C_r \sqrt{2gh} - 40.75 = 0.98\sqrt{2 \times 9.81 \times 400} - 40.75 = 46.07 \text{m/s}$$

and, $$V_{r2} = 0.95\ V_{r1} = 0.95 \times 46.07 = 43.77 \text{ m/s}$$

$$V_{w1} = V_1 = 0.98\sqrt{2 \times 9.81 \times 400} = 86.82 \text{ m/s}$$

$$V_{w2} = V_{r2} \cos\phi - u_2 = 43.77 \times \cos 15° - 40.75 = 1.53 \text{ m/s}$$

$(\because u_1 = u_2 = 40.75 \text{ m/s})$

Force exerted by a single jet on the buckets

$= \rho \times \text{discharge through one jet} \times (V_{w1} + V_{w2})$

$= 1000 \times 1.06\ (86.82 + 1.53) =$ **93651 N or 93.651 kN (Ans.)**

(iii) Speed of the wheel for a frequency of 50 hertz/sec and corresponding wheel diameter:

Jet ratio = 10 or $\frac{D}{d} = 10$, $\quad \therefore D = 10d = 10 \times 0.125 = 1.25$ m

Also, $$u_1 = \frac{\pi DN}{60}, \text{ or } N = \frac{60\mu_1}{\pi D} = \frac{60 \times 40.75}{\pi \times 1.25} = 622.6 \text{ rpm}$$

Now, frequency, $$f = \frac{Np}{60} \quad (N = \text{Speed}, p = \text{no. of poles})$$

$$p = \frac{60f}{N} = \frac{60 \times 50}{622.6} = 4.82 \text{ say } 5$$

Now, for $p = 5, N = \frac{60f}{p} = \frac{60 \times 50}{5} = \textbf{600 rpm}$ **(Ans.)**

Again, $u = \frac{\pi DN}{60}$ or $D = \frac{60u}{\pi N} = \frac{60 \times 40.75}{\pi \times 600} = \textbf{1.3 m}$ **(Ans.)**

$\therefore$ Jet ratio becomes: $\frac{D}{d} = \frac{1.3}{0.125} = 10$

Hence the given condition is *satisfied.*

Q. 9.4. *(Centrifugal pump)* *A centrifugal pump is to deliver 4.5 m³/s when running at 750 rpm. The diameter of the impeller at inlet is 53 cm and at outlet is 76 cm. It may be assumed that the air enters radially with a speed of 15 m/s. The vanes are set backwards at oultet at 70° to the tangent, and width at outlet is 10 cm. The volute casing gives at 30% recovery of the outlet velocity head. The losses in the impeller may be taken as equivalent to 25% of the outlet velocity head. Blade thickness effects may be naglected. Determine :*

(i) The manometric efficiency, and

(ii) The pressure at the discharge. **(IES)**

Solution. Refer to Fig. 9.2

Given: $Q = 4.5$ m³/s; $N = 750$ rpm; $D_1 = 53$ cm $= 0.53$ m; $D_2 = 76$ cm $= 0.76$ m; $V_1 = V_{f1} = 15$ m/s; $\phi = 70°$; $B_2 = 10$ cm $= 0.1$ m; Volume casing recovery = 30% of V_2; Losses in impeller = 25% of the outlet velocity head $= 0.25 \times \frac{V_2^2}{2g}$.

(*i*) **Manometric efficiency, η_{mano} :**

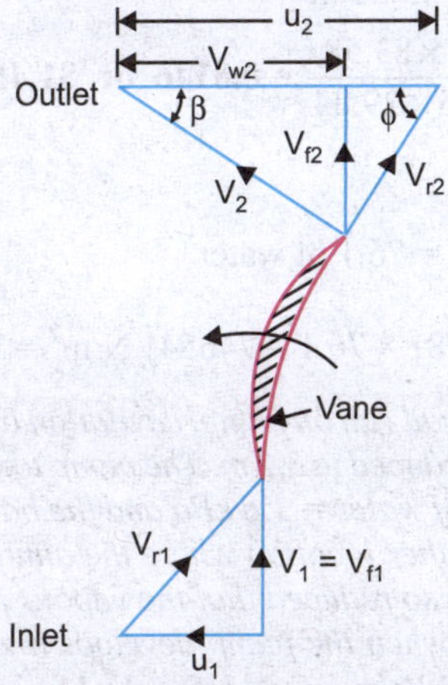

Fig. 9.2

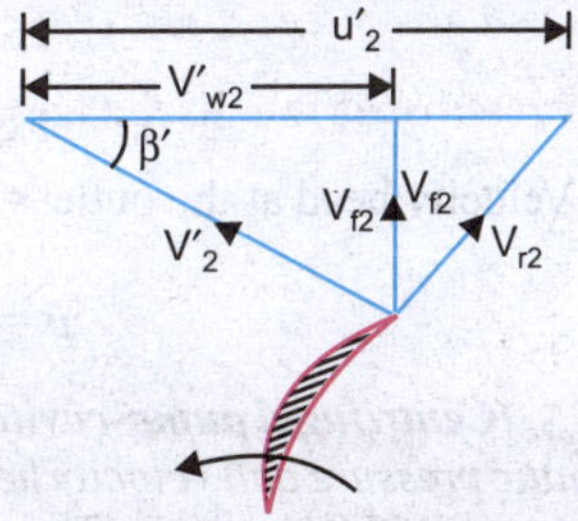

Fig. 9.3

$$u_2 = \frac{\pi D_2 N}{60} = \frac{\pi \times 0.76 \times 750}{60} = 29.84 \text{ m/s}$$

$$Q = \pi D_2 B_2 V_{f2} \quad \text{or} \quad 4.5 = \pi \times 0.76 \times 0.1 \times V_{f2}$$

or,
$$V_{f2} = \frac{4.5}{\pi \times 0.76 \times 0.1} = 18.85 \text{ m/s}$$

From velocity triangle at outlet, we have :

$$\sin\phi = \frac{V_{f2}}{V_{r2}}, \text{ or } \sin 70° = \frac{18.85}{V_{r2}}$$

or, $$V_{r2} = \frac{18.85}{\sin 70°} = 20.06 \text{ m/s}$$

and, $$V_{w2} = u_2 - V_{r2}\cos\phi$$
$$= 29.84 - 20.06 \times \cos 70° = 22.98 \text{ m/s}$$

Now, $$\tan\beta = \frac{V_{f2}}{V_{w2}} = \frac{18.85}{22.98} \text{ or } \beta = 39.36°$$

Also, $$\cos\beta = \frac{V_{w2}}{V_2} \text{ or } V_2 = \frac{V_{w2}}{\cos\beta} = \frac{22.98}{\cos 39.36°} = 29.72 \text{ m/s}$$

Refer to Fig 9.3.

Since volute casing gives 30% recovery of the outlet velocity head (given), therefore,

$$V_2 = 1.3 \times 29.72 = 38.64 \text{ m}$$

$\therefore$ $$\sin\beta' = \frac{18.85}{38.64} = 0.4878, \text{ or } \beta' = 29.2°$$

Now, $$V'_{w2} = V'_2 \cos\beta' = 38.64 \times \cos 29.2° = 33.73 \text{ m/s}$$

Manometric head, $$H_{mano} = \frac{V_{w2}u_2}{g} - 0.25 \times \frac{V_2'^2}{2g}$$
$$= \frac{33.73 \times 29.84}{9.81} - 0.25 \times \frac{38.64^2}{2 \times 9.81}$$
$$= 102.6 - 19.02 = 83.58 \text{ m}$$

Hence, $$\eta_{mano} = \frac{gH_{mano}}{V'_{w2}u_2} = \frac{9.81 \times 83.58}{33.73 \times 29.84} = \mathbf{0.8146 \text{ or } 81.46\%} \quad \textbf{(Ans.)}$$

(ii) **Pressure at the discharge *p*:**

Velocity head at the outlet $= \dfrac{V_2'^2}{2g} = \dfrac{38.64^2}{2 \times 9.81} = 76.1$ of water

$\therefore$ $$p = \rho gh = 1000 \times 9.81 \times 76.1 = 746541 \text{ N/m}^2 = \mathbf{7.46 \text{ bar}} \quad \textbf{(Ans.)}$$

Q. 9.5. ***(Centrifugal pump-cavitation)*** *During a trail run on pump, cavitation occurred when the sum of static pressure and velocity head at inlet was reduced to 3.2 m. The pump total head was 35 m at a discharge of 0.045 m³/sec. The vapour pressure of water = 1.8 kPa and the barometer pressure = 750 mm of Hg. If the pump is to operate at some other location where the atmospheric pressure was reduced to 620 mm of Hg and the temperature is so reduced that the vapour pressure of water is 830 Pa, what is the value of caviation parameter when the pump develops the same total head and discharge? Is it necessary to reduce the height of the pump and if so, by how much?* **(GATE)**

Solution. *Given:* Total suction head, $H_s = 32$ m; Pump total head, $H_{mano} = 35$ m; $Q = 0.045$ m³/s; Vapour pressure, $H_v = 1.8 \text{ kPa} = \dfrac{1.8 \times 1000}{1000 \times 9.81} = 0.183$ m or water; Atmospheric pressure, $H_a = 750$ mm of $H_g = 750 \times 13.6 \times 10^{-3} = 10.2$ m of water.

Thoma's cavitation factor, $$\sigma = \frac{H_a - H_s - H_a}{H_{mano}}$$
$$= \frac{10.2 - 3.2 - 0.183}{35} = 0.1948 = \sigma_c$$

This is called *critical* cavitation factor, since cavitation occurs at this location and by this arrangement at '*new location*':

$$H_a = 620 \text{ mm of Hg} = 0.62 \times 13.6 = 8.432 \text{ m of water}$$

$$H_v = \frac{830}{1000 \times 9.81} = 0.0846 \text{ m of water}$$

Again, $H_{mano} = 35$ m and $H_s = 3.2$ m ... remain the same

$$\therefore \text{ New cavitation factor, } \sigma_{new} = \frac{H_a - H_s - H_v}{H_{mano}}$$

$$= \frac{8.432 - 3.2 - 0.0846}{35} = 0.147$$

Thus, $\sigma_{new} < \sigma_c$ *i.e.*, $0.147 < 0.1948$ and as such cavitation occurs.

Therefore, it is necessary to '*reduce the height of the pump*'.

If 'x' is the new height, then we have:

$$\frac{8.432 - x - 0.0846}{35} = 0.1948$$

or, $8.3474 - x = 6.818$ or $x = 1.5294$ m

Hence, H_s must be *reduced* from 3.2 m to **1.5294** m to *avoid cavitation* **(Ans.)**

Q. 9.6. ***(Centrifugal pump)*** *A centrifugal pump is required to lift 9100 lit/s of water against a head of 6 m while running at 500 r.p.m. The velocity of flow through the wheel is 2 m/s and the manometric efficiency is 60%. The angle of vane tip with the direction of a motion is 150°. Work out the following:*

(i) Sketch an impeller of centrifugal pump and draw the velocity triangle.

(ii) The diameter of impeller.

(iii) The width of impeller. **(GATE)**

Solution. *Given:* $Q = 9100$ lit/s $= 9.1$ m^3/s; $H_{mano} = 6$ m; $N = 500$ r.p.m., $V_{f1} = V_{f2} = 2$ m/s; $\eta_{mano} = 60\%$; $\phi = 180° - 150° = 30°$.

(*i*) Sketch of the impeller–velocity triangle (Refer to Fig. 9.4.)

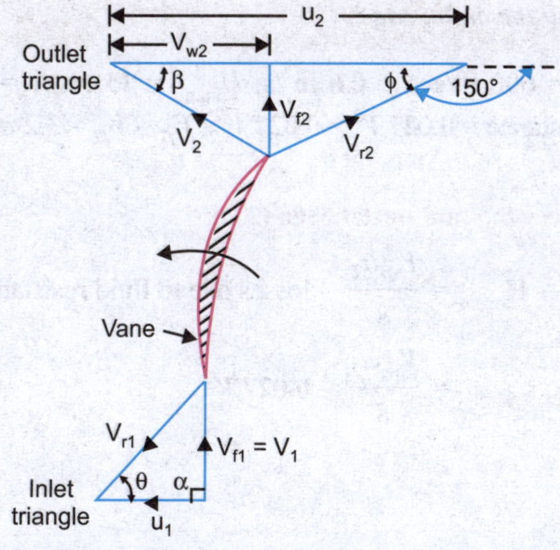

Fig. 9.4

(ii) Diameter of impeller, D_2:

$$\eta_{mano} = \frac{gH_{mano}}{V_{w2}u_2} \quad \text{... [Eqn. (3.9)]}$$

$$0.6 = \frac{9.81\times 6}{V_{w2}u_2} \text{ or } V_{w2}u_2 = \frac{9.81\times 6}{0.6} = 98.1 \quad \text{...}(i)$$

From velocity triangle at the outlet, we have:

$$\tan\phi = \frac{V_{f2}}{(u_2 - V_{w2})} \text{ or } u_2 - V_{w2} = \frac{V_{f2}}{\tan\phi} = \frac{2}{\tan 30°} = 3.464$$

From (*i*) and (*ii*), we get:

$$u_2\,(u_2 - 30.464) = 98.1$$

or,
$$u_2^2 - 3.464\,u - 98.1 = 0$$

or,
$$u_2 = \frac{3.464 \pm \sqrt{3.464^2 + (4\times 98.1)}}{2} = 11.787 \text{ m/s}$$

But,
$$u_2 = \frac{\pi D_2 N}{60} \text{ or } 11.787 = \frac{\pi\, D_2 \times 500}{60}$$

or,
$$D_2 = \frac{11.787\times 60}{\pi\times 500} = \mathbf{0.45\ m} \quad \textbf{(Ans.)}$$

(iii) Width of impeller, B_2:

$$Q = \pi D_2 B_2 V_{f2} = \text{or } 9.1 = \pi \times 0.45 \times B_2 \times 2$$

or,
$$B_2 = \frac{9.1}{\pi\times 0.45\times 2} = \mathbf{3.218\ m} \quad \textbf{(Ans.)}$$

Q. 9.7. (Centrifugal pump) *A centrifugal pump is required to discharge 600 litres of water per second and develop a head of 15 m when the impeller rotates at 750 rpm. The manometric efficiency is 80%. The loss of head in the pump due to fluid resistance being assumed to be 0.027 V^2 m of water, where V is the velocity with which water leaves the impeller. Water enters the impeller without shock or whirl and the velocity of flow is 3.2 m/sec. Determine:*

(i) The impeller diameter and outlet area.

(ii) The blade angle at the outlet edge. **(IES)**

Solution. *Given:* Q = 600 lit/sec = 0.6 m^3/s; H_{mano} = 15 m; N = 750 rpm; H_{mano} = 80%; Head loss due to fluid resistance = 0.027 V^2 = 0.027 V_2^2; $V_{f1} = V_{f2}$ = 3.2 m/s.

Refer to Fig. 9.5.

(i) Impeller diameter (D_2) and outlet area (A_2):

$$H_{mano} = \frac{V_{w2}u_2}{g} - \text{losses due to fluid resistance}$$

$$= \frac{V_{w2}u_2}{g} - 0.027\,V^2$$

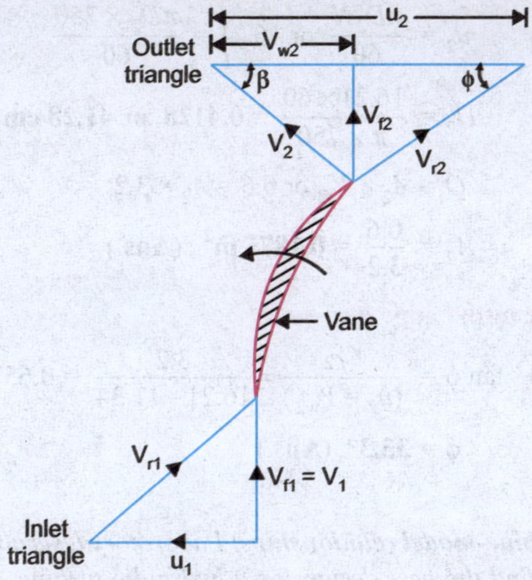

Fig. 9.5

Now, $\eta_{mano} = \dfrac{H_{mano}}{(V_{w2}u_2/g)}$ or, $\dfrac{g(H_{mano})}{V_{w2}u_2} = \dfrac{g\left[\dfrac{V_{w2}u_2}{g} - 0.027\,V_2^2\right]}{V_{w2}u_2}$

or, $0.8 = \dfrac{V_{w2}u_2 - g \times 0.027\,V^2}{V_{w2}u_2} = 1 - \dfrac{g \times 0.027 V_2^2}{V_{w2}u_2}$

or, $\dfrac{g \times 0.027 V_2^2}{V_{w_2}u_2} = 0.2$ or $V_{w2}u_2 = \dfrac{0.027V^2 g}{0.2}$ ... (*i*)

Again, $H_{mano} = \dfrac{V_{w2}u_2}{g}$ – losses

or, $15 + \text{losses} = \dfrac{V_{w2}u_2}{g}$ ... (*ii*)

$$15 + 0.027\,V_2^2 = \frac{V_{w2}u_2}{g} = \frac{0.027V^2 g}{0.2 \times g} = 0.135\,V^2$$

or, $V_2^2\,(0.135 - 0.027) = 15$ or $V_2 = 11.78$ m/s

From velocity triangle at outlet, we have:

$$\sin\beta = \frac{V_{f2}}{V_2} = \frac{3.2}{11.78} \text{ or } \beta = 15.76°$$

and $\cos\beta = \dfrac{V_{w2}}{V_2}$ or $\cos 15.76° = \dfrac{V_{w2}}{11.78}$

or, $V_{w2} = \cos 15.76° \times 11.78 = 11.34$ m/s

From (*ii*), we get: $V_{w2}u_2 = (15 + 0.027V_2{}^2)g$

$= [15 + (0.027 \times 11.78^2)] \times 9.81 = 183.9$

or, $u_2 = \dfrac{183.9}{V_{w2}} = \dfrac{183.9}{11.34} = 16.21$ m/s

Also, $u_2 = \frac{\pi D_2 N}{60}$ or $16.21 = \frac{\pi D_2 \times 750}{60}$

or, $D_2 = \frac{16.21 \times 60}{\pi \times 750} = \mathbf{0.4128 \text{ or } 41.28 \text{ cm}}$ **(Ans.)**

We know that: $Q = A_2 \times V_{f2}$ or $0.6 = A_2 \times 3.2$

or, $A_2 = \frac{0.6}{3.2} = \mathbf{0.1875 \text{ m}^2}$ **(Ans.)**

(*ii*) **Blade angle at the outlet edge ϕ :**

$$\tan \phi = \frac{V_{f2}}{(u_2 - V_{w2})} = \frac{3.2}{16.21 - 11.34} = 0.657$$

or, $\phi = \mathbf{33.3°}$ **(Ans.)**

Q. 9.8. ***(Hydraulic turbine-model relationships)*** *Laboratory tests are required to be performed on a 1/10 scale model to find the best design for a hydraulic turbine. The turbine is required to develop 30,000 kW under a net head of 50 m running at 300 rpm. The head available at the laboratory is 12 m. Determine:*

(i) The corresponding speed,
(ii) Power, and
(iii) Discharge.
The overall efficiency is 88%. **(GATE)**

Solution. *Given:* Scale model $= \frac{D_m}{D_p} = \frac{1}{10}$; Power, P_p = 30000 kW; Head, H_p = 50 m; N_p = 300 rpm; H_m = 12 m; η_0 = 88%.

(*i*) **Speed N_m :**

We know that: $\frac{H}{N^2 D^2}$ = constant ... [Eqn. (2.43)]

$$\therefore \quad \frac{H_p}{N_p^2 \times D_p^2} = \frac{H_m}{N_m^2 \times D_m^2}$$

or, $N_m{}^2 = \left(\frac{H_m}{H_p}\right) \times \left(\frac{D_p^2}{D_m^2}\right) \times N_p^2 = \frac{12}{50} \times 10^2 \times 300^2 = 2160000$

or, $N_m = 1469.7$ **rpm (Ans.)**

(*ii*) **Power, P_m :**

Also, $\frac{P}{N^3 D^5}$ = constant ... [Eqn (2.45)]

$$\therefore \quad \frac{P_p}{N_p^3 D_p^5} = \frac{P_m}{N_m^3 \times D_m^5}$$

or, $P_m = P_p \times \left(\frac{N_m}{N_p}\right)^3 \times \left(\frac{D_m}{D_p}\right)^5 = 30000 \times \left(\frac{1469.7}{300}\right)^3 \times \left(\frac{1}{10}\right)^5$

$= \mathbf{35.273 \text{ kW}}$ **(Ans.)**

(iii) **Discharges Q_m, Q_p :**

Overall efficiency, $$\eta_0 = \frac{P_m}{\rho g Q_m H_m}$$

$$0.88 = \frac{35.273 \times 10^3}{1000 \times 9.81 \times Q_m \times 12}$$

$$\therefore \quad Q_m = \frac{35.273 \times 10^3}{0.88 \times 1000 \times 9.81 \times 12} = \mathbf{0.34\ m^3/s} \quad \textbf{(Ans.)}$$

We know that: $$\frac{Q}{ND^3} = \text{constant} \quad \text{... [Eqn. (2.44)]}$$

$$\therefore \quad \frac{Q_p}{N_p D_p^3} = \frac{Q_m}{N_m D_m^3}$$

or, $$Q_p = Q_m \times \left(\frac{N_p}{N_m}\right) \times \left(\frac{D_p}{D_m}\right)^3$$

$$= 0.34 \times \left(\frac{300}{1469.7}\right) \times \left(\frac{10}{1}\right)^3 = \mathbf{69.4\ m^3/s} \quad \textbf{(Ans.)}$$

Q. 9.9. ***(Pelton turbine)*** *A pelton turbine is to work at the foot of dam whose reservoir level is 220 m. The head at the full opening at the turbine nozzle is 200 m and coefficient of velocity is 0.98. The turbine is to operate at 200 rpm and develop a power of 3.7 MW. Assuming blade to jet speed ratio as 0.46, estimate the wheel diameter. If the blade outlet angle is 16°, determine the blade and hydraulic efficiencies. Neglecting frictional losses.* **(IES)**

Solution. *Given:* Gross head, H_g = 220 m; Net or effective head, H = 200 m; Co-efficient of velocity, C_v = 0.98; Speed, N = 200 rpm; Power = 3.7 MW; Speed ratio, $K_u = \frac{u}{V_1} = 0.4$. Blade outlet angle, ϕ = 16°.

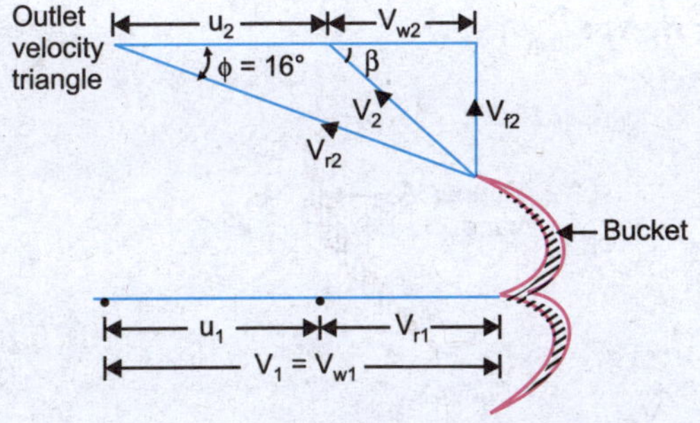

Fig. 9.6

Refer to Fig. 9.6.

Wheel diameter, *D*;

$$V_1 = C_v\sqrt{2gH} = 0.98 \times \sqrt{2 \times 9.81 \times 200} = 61.4 \text{ m/s}$$

Speed of the wheel, $u\ (=u_1 = u_2) = 0.46 \times V_1$

$$= 0.46 \times 61.4 = 28.24 \text{ m/s}$$

Also, $$u = \frac{\pi DN}{60} \text{ or } 28.24 = \frac{\pi \times D \times 200}{60}$$

or, $$D = \frac{28.24 \times 60}{\pi \times 200} = \mathbf{2.697\ m/s} \quad \textbf{(Ans.)}$$

Blade efficiency, η_{bl}:

$$V_{r2} = V_{r1} = V_1 - u = 61.4 - 28.24 = 33.16 \text{ m/s}$$

$$V_{w2} = V_{r2} \cos 16° - u = 33.16 \cos 16° - 28.24 = 3.635 \text{ m/s}$$

We know that: $$\eta_{bl} = \frac{2(V_{w1} + V_{w2}) \times u}{V_1^2} = \frac{2(61.4 + 3.635) \times 28.24}{(61.4)^2}$$

$$= \mathbf{0.974 \text{ or } 97.4\%} \quad \textbf{(Ans.)}$$

Hydraulic efficiency, η_h:

$$\eta_h = \frac{(V_{w1} + V_{w2})u}{gH} \quad \text{... [Eqn. (2.7)]}$$

$$= \frac{(61.4 + 3.635) \times 28.24}{9.81 \times 200} = \mathbf{0.936 \text{ or } 93.6\%} \quad \textbf{Ans.}$$

Q. 9.10. ***(Reaction turbines)*** *The inward flow reaction turbines have the same diameter viz 0.7 m and the same efficiency. Both the runners work under the same head and they have the same velocity of flow, viz 5.8 m/s. One of the runner, A revolves at 500 rpm and the inlet blade angle is 65°. If the other runner, B has an inlet blade angle of 110°, at what speed should it run?*

Solution. *Given:* $D_A = D_B = 0.7$ m; $\eta_A = \eta_B$; $V_{fA} = V_{fB} = 5.8$ m/s; $N_A = 500$ rpm; Inlet blade angles: $\theta_A = 65°$, $\theta_B = 110°$.

Speed of the runner B, N_B:

Refer to Fig. 9.7.

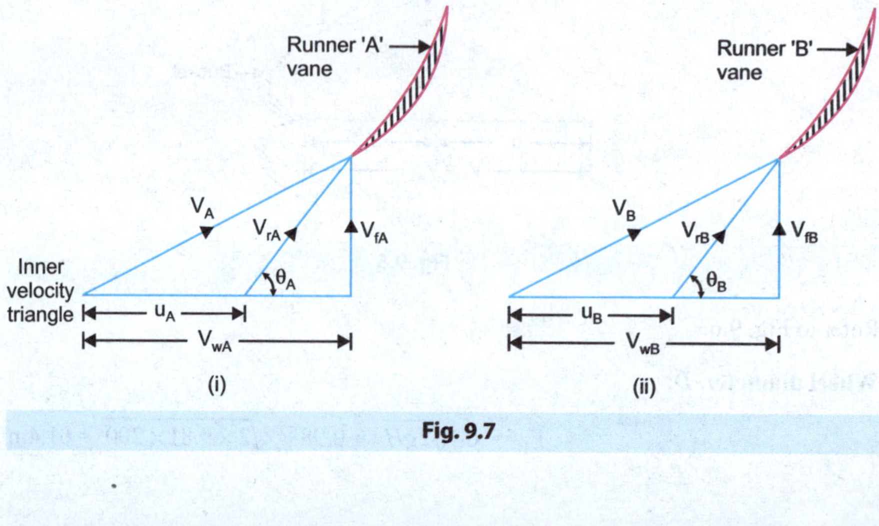

Fig. 9.7

From inner velocity triangles, we have :

$$\tan\theta_A = \frac{V_{fA}}{V_{wA} - u_A} \quad \text{... (i)}$$

$$\tan\theta_B = \frac{V_{fB}}{V_{wB} - u_B} \quad \text{... (ii)}$$

From eqns. (*i*) and (*ii*), we get:

$$\frac{\tan 65°}{\tan 110°} = \frac{V_{fA}}{V_{wA} - u_A} \times \frac{V_{wB} - u_B}{V_{fB}} = \frac{V_{wB} - u_B}{V_{wA} - u_A}$$

or, $$-0.7805 = \frac{V_{wB} - u_B}{V_{wA} - u_A}$$

or, $$0.7805 = \frac{u_B - V_{wB}}{V_{wA} - u_A} \quad \text{... (iii)}$$

Also, $\eta_A = \eta_B$ (given)

$$\therefore \quad \frac{V_{wA}\, u_A}{gH_A} = \frac{V_{wB}\, u_B}{gH_B}$$

or, $$V_{wB} = \frac{V_{wA} u_A}{u_B} \quad (\because H_A = H_B) \quad \text{... (iv)}$$

Substituting (*iv*) in (*iii*), we get:

$$0.7805 = \frac{u_B - \dfrac{V_{wA}.u_A}{u_A}}{V_{wA} - u_A} = \frac{u_B^2 - V_{wA} u_A}{u_B(V_{wA} - u_A)} \quad \text{... (v)}$$

From (*i*), $$\tan 65° = \frac{V_{fA}}{V_{wA} - u_A} = \frac{5.8}{V_{wA} - \left(\dfrac{\pi \times 0.7 \times 500}{60}\right)} = \frac{5.8}{V_{wA} - 18.32}$$

or, $$V_{wA} = \frac{5.8}{\tan 65°} + 18.32 = 21.02 \text{ m/s}$$

Inserting the various values in (*v*), we get:

$$\frac{u_B^2 - 21.02 \times 18.32}{u_B(21.02 - 18.32)} = 0.7805$$

or, $$u_B^{\,2} - 385.09 = 2.107\, u_B$$

or, $$u_B^{\,2} - 2.107\, u_B - 3.850.9 = 0$$

or, $$u_B = \frac{2.107 + \sqrt{2.107^2 + (4 \times 385.09)}}{2} = 20.7 \text{ m/s}$$

Also, $$u_B = \frac{\pi D_B N_B}{60} = 20.7$$

or, $$N_B = \frac{20.7 \times 60}{\pi \times 0.7} = \mathbf{564.8 \text{ rpm}} \quad \textbf{(Ans.)}$$

Q. 9.11. ***(Centrifugal pump)*** *A centrifugal pump is to discharge 0.2 m³/s at a speed of 1400 rpm against a head of 25 m. The impeller diameter is 0.3 m, its width at outlet is 0.05 m and the manometric efficiency is 80%. Determine the vane angles at the outer periphery of the impeller.* **(GATE)**

Solution. *Given:* $Q = 0.2\ \text{m}^3/\text{s}$; $N = 1400$ rpm; $H_{mano} = 25$ m; $D_2 = 0.3$ m; $B_2 = 0.05$ m; $\eta = 80\%$.

Refer to Fig. 9.8.

Vane angle at the outlet :

Tangential velocity of impeller at outlet,

$$u_2 = \frac{\pi D_2 N}{60} = \frac{\pi \times 0.3 \times 1400}{60} = 22 \text{ m/s}$$

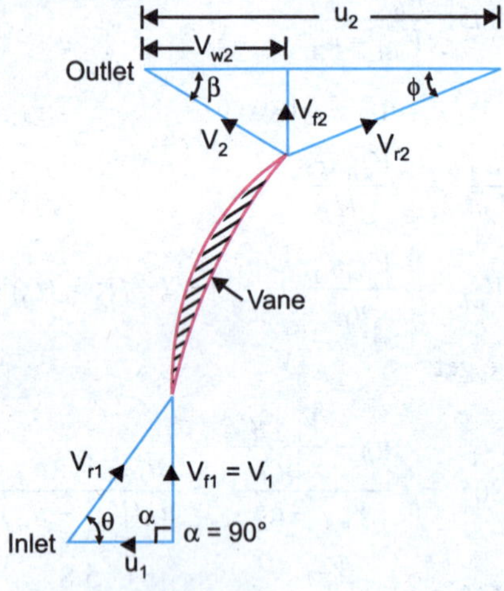

Fig. 9.8

Now, $$Q = \pi D_2 B_2 \times V_{f2}$$

or, $$V_{f2} = \frac{Q}{\pi D_2 B_2} = \frac{0.2}{\pi \times 0.3 \times 0.5} = 4.24 \text{ m/s}$$

Manometric efficiency, $$\eta_{mano} = \frac{gH_{mano}}{V_{w2} u_2}$$

or, $$V_{w2} = \frac{gH_{mano}}{\eta_{mano} \times u_2} = \frac{9.81 \times 25}{0.8 \times 22} = 13.93 \text{ m/s}$$

$$\tan \beta = \frac{V_{f2}}{V_{w2}} = \frac{4.24}{13.93} = 0.3044$$

or, $$\beta = \tan^{-1} (0.3044) = \mathbf{16.93°} \quad \textbf{(Ans.)}$$

Again, $$\tan \phi = \frac{V_{f2}}{u_2 - V_{w2}} = \frac{4.24}{22 - 13.93} = 0.5254$$

or, $$\phi = \tan^{-1} (0.5254) = \mathbf{22.71°} \quad \textbf{(Ans.)}$$

Q. 9.12. *Explain fluid coupling and torque converter and their areas of application.* **(GATE)**

Ans. Refer to Articles 5.8 and 5.9.

Q. 9.13. *Do you need priming in a gear pump? Give the reasons for your answer.* **(IES)**

Ans. In a "gear pump", *priming* is *'not'* required. The primary purpose of priming is to *remove air* from the suction pipe, casing of the pump and the delivery pipe by filling these parts with the liquid to be raised by the pump. Since the head generated by the pump is $(V_{w2}u_2/g)$ which is independent of the density of liquid, and the *density of air is very low*, the head generated will be also *very low*. In case of "gear pump" *liquid cannot slip back* into the inlet side due to the *meshing of the gear* and as such there is *no possibility for the air to get entrapped in the pump.*

Q. 9.14. *What is the necessity of governing? Explain with sketch a simple governing mechanism for a turbine.* **(GATE)**

Ans. Refer to Article 2.14.

Q. 9.15. *What is meant by cavitation and what are the methods avoiding it?* **(GATE)**

Ans. Refer to Article 2.15.

Q. 9.16. *What is the purpose of draft tube in a turbine?* **(GATE)**

Ans. Refer to Article 2.8.

B. MULTIPLE-CHOICE QUESTIONS WITH ANSWERS AND *"EXPLANATIONS"*

Q. 9.1. The specific speed of an hydraulic pump is the speed of geometrically similar pump working against a unit head and

(*a*) delivering unit quantity of water (*b*) consuming unit power

(*c*) having unit velocity of flow (*d*) having unit radial velocity.

Ans. (*a*)

Explanation:

(*a*) is the correct choice, because the specific speed of hydraulic pump is the speed of geometrically similar pump working against a *unit head* and *delivering quantity of water*. **(Ans.)**

Q. 9.2. A jet of water issues from a nozzle with a velocity of 20 m/s and it impinges normally on a flat plate moving away from it at 10 m/s. If the cross-sectional area of the jet is 0.02 m^2 and the density of water is taken as 1000 kg/m^3, then the force developed on the plate will be

(*a*) 10 N (*b*) 100 N (*c*) 1000 N (*d*) 2000 N.

Ans. (*d*)

Explanation:

$V = 20$ m/s, $u = 10$ m/s, $a = 0.02$ m^2, $\rho = 1000$ kg/m^3

$$\text{Force on the plate} = \rho a (V - u) \times (V - u) = 1000 \times 0.02 \times (20 - 10) \times (20 - 10)$$

$$= 2000 \text{ N } \textbf{(Ans.)}$$

Q. 9.3. In the statement, "in a reaction turbine installation, the head of water is decreased and the speed (r.p.m.) is also decreased at a certain condition of working. The effect of each of these changes will be to X power delivered due to decrease in head and to Y power delivered due to decrease in rpm", $\frac{N\sqrt{Q}}{(H)^{5/4}}$, X and Y stand respectively for

(*a*) decrease and increase (*b*) increase and increase
(*c*) decrease and decrease (*d*) increase and decrease.

Ans. (*a*)

EXPLANATION:

In case of hydraulic reaction turbines, $P \propto H^{3/2}$, it indicates that *with the decrease in head,* there will be decrease in power. Also speed $N \propto \frac{1}{\sqrt{P}}$ or $P \propto \frac{1}{N^2}$, which means that with *decrease in speed* there will be *increase in power*. Thus (*a*) is the correct choice. **(Ans.)**

Q. 9.4. Water, having a density of 1000 kg/m^3, issues from a nozzle with a velocity of 10 m/s and the jet strikes a bucket mounted on a Pelton wheel. The wheel rotates a 10 rad/s. The mean diameter of the wheel is 1 m. The jet is split into two equal streams by the bucket, such that each stream is deflected by 120° as shown in Fig. 9.9. Friction in the bucked may be neglected. Magnitude of the torque exerted by the water on the wheel, per unit mass flow rate of the incoming jet, is

(*a*) 0 (N-m)/(kg/s) (*b*) 1.25 (N-m)/(kg/s)
(*c*) 2.5 (N-m)/(kg/s) (*d*) 7.5 (N-m)/(kg/s).

Ans. (*d*)

EXPLANATION: Refer to Fig. 9.9

Torque = Force × radius

$$T = \rho QV(1 + \cos\theta) \times R$$

Torque per unit mass flow rate

$$= V(1 + \cos\theta)R$$
$$= 10\,(1 + \cos 60°)\,0.5$$
$$= 7.5 \text{ N-m/kg/s } \textbf{(Ans.)}$$

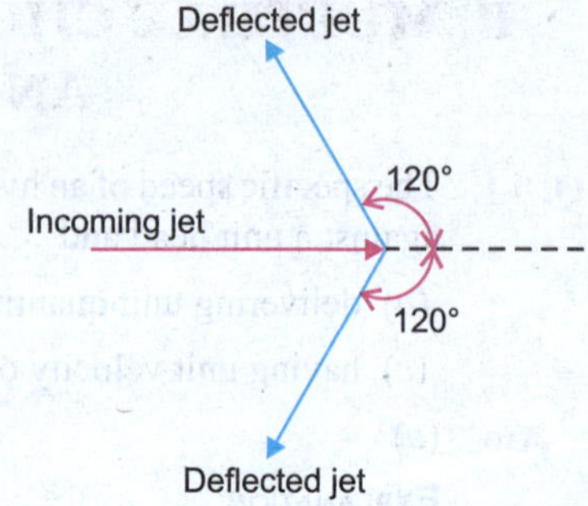

Fig. 9.9

Q. 9.5. In a Pelton wheel, the bucket peripheral speed is 10 m/s, the water jet velocity is 25 m/s and volumetric flow rate of the jet is 0.1 m^3/s. If the jet deflection angle is 120°C and the flow is ideal, the power developed is

(*a*) 7.5 kW (*b*) 15.0 kW (*c*) 22.5 kW (*d*) 37.5 kW.

Ans. (*c*)

EXPLANATION:

$$\phi = 180° - 120° = 60°$$

Power developed by pelton wheel

$$= \rho Q(V - u)\,(1 + \cos\phi).u$$
$$= 1000 \times 0.1\,(25 - 10)\,(1 + \cos 60°) \times 10$$
$$= 22500 \text{ W} = 22.5 \text{ kW } \textbf{(Ans.)}$$

Q. 9.6. The intel angle of runner blades of a Francis turbine is 90°. The blades are so shaped that the tangential component of velocity at blade outlet is zero. The flow velocity remains constant throughout the blade passage and is equal to half of the blade velocity at runner inlet. The blade efficiency of the runner is

(*a*) 25% (*b*) 50% (*c*) 80% (*d*) 89%.

Ans. (*c*)

EXPLANATION: Refer to Fig. 9.10.

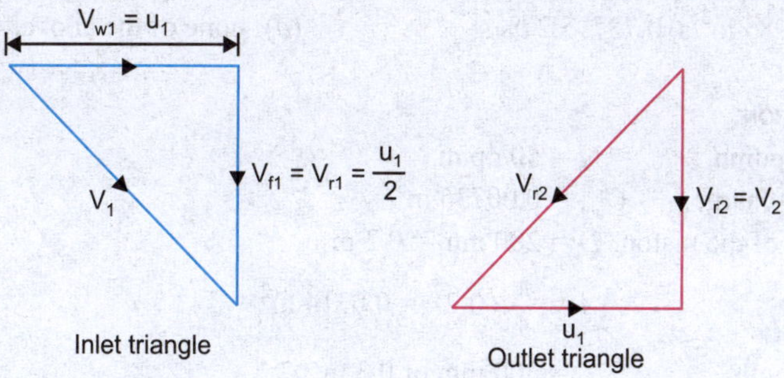

Fig. 9.10

Given: $$V_{f1} = V_{f2} = \frac{u_1}{2}$$

$$V_2 = V_{f2} = \frac{u_1}{2}$$

$$V_1^2 = (V_{w1})^2 + (V_{f1})^2$$

$$= u_1^2 + \left(\frac{u_1}{2}\right)^2 = \frac{5}{4}u_1^2$$

$$\text{Blade efficiency} = \frac{V_1^2 - V_2^2}{V_1^2} \times 100$$

$$= \frac{\frac{5}{4}u_1^2 - \frac{u_1^2}{4}}{\frac{5}{4}u_1^2} \times 100 = \frac{4}{5} \times 100 = 80\% \textbf{ (Ans.)}$$

Q. 9.7. A large hydraulic turbine is to generate 300 kW at 1000 rpm under a head of 40 m. For initial testing, a 1 : 4 scale model of the turbine operates under a head of 10 m. The power generated by the model (in kW) will be

(*a*) 2.34 (*b*) 4.68 (*c*) 9.38 (*d*) 18.75.

Ans. (*a*)

EXPLANATION:

For similar turbines specific power will be same.

i.e. $$\frac{P_1}{d_1^2 H_1^{3/2}} = \frac{P_2}{d_2^2 H_2^{3/2}}$$

$\therefore$ $$P_2 = P_1 \times \left(\frac{d_2}{d_1}\right)^2 \times \left(\frac{H_2}{H_1}\right)^{3/2}$$

$$= 300 \times \left(\frac{1}{4}\right)^2 \times \left(\frac{10}{40}\right)^{3/2}$$

i.e. $$P_2 = 2.34 \text{ kW } \textbf{(Ans.)}$$

Q. 9.8. A single-acting reciprocating pump, running at 50 r.p.m. delivers 0.00736 m^3/s of water. The diameter of the piston is 200 mm and stroke length 300 mm. The suction and delivery heads are 3.5 m and 11.5 m respectively. Then,

(*i*) Theoretical discharge, (*ii*) Co-efficient of discharge, and (*iii*) Percentage slip of the pump are respectively.

(*a*) 0.00785 m^3/s, 0.937, 6.24%
(*b*) 0.00885 m^3/s, 0.837, 3.24%
(*c*) 0.00985 m^3/s, 0.737, 5.27%
(*d*) none of the above.

Ans. (*a*)

EXPLANATION:

Speed of pump, N = 50 r.p.m.
Actual discharge, $Q_{act} = 0.00736\ m^3/s$
Diameter of the piston, $D = 200$ mm = 0.2 m

$\therefore$ Area, $A = \frac{\pi}{4} \times 0.2^2 = 0.0314\ m^2$

Stroke length, $L = 300$ mm or 0.3 m
Suction head, $h_s = 3.5$ m
Delivery head, $h_d = 11.5$ m

(*i*) *Theoretical discharge, Q_{th}:*

$$Q_{th} = \frac{ALN}{60} = \frac{0.0314 \times 0.3 \times 50}{60} = 0.00785\ m^3/s\ \textbf{(Ans.)}$$

(*ii*) *Co-efficient of discharge, C_d:*

$$C_d = \frac{Q_{act.}}{Q_{th.}} = \frac{0.00736}{0.00785} = 0.937\ \textbf{(Ans.)}$$

(*iii*) *Percentage slip of the pump:*

$$\%\ \text{slip} = \frac{Q_{th.} - Q_{act.}}{Q_{th.}} \times 100 = \frac{0.00785 - 0.00736}{0.00785} \times 100 = 6.24\%\ \textbf{(Ans.)}$$

Q. 9.9. A hydraulic intensifier gets the low pressure liquid at a pressure of 40 bar and delivers it to a machine at a pressure of 160 bar. If the intensifier has a capacity of 0.021 m^3 and stroke 1.2 m, calculate the diameters of the fixed ram and the sliding cylinder to be used for this intensifier.

(*a*) 130 mm, 260 mm
(*b*) 140 mm, 280 mm
(*c*) 149 mm, 298 mm
(*d*) none of the above.

Ans. (*c*)

EXPLANATION:

Intensity of pressure of low pressure liquid, $p_1 = 40$ bar $= 40 \times 10^5\ N/m^2$
Intensity of pressure of high pressure liquid, $p_2 = 160$ bar $= 160 \times 10^5\ N/m^2$

Capacity of intensitier = 0.021 m^3

Stroke length = 1.2 m

Diameter of fixed ram (D_2) and the sliding cylinder (D_1):

Capacity of intensifier = Area of fixed ram × stroke length

$$0.021 = A_2 \times 1.2 \text{ or } A_2 = \frac{0.021}{1.2} = 0.0175\ m^2$$

$$\therefore \quad \frac{\pi}{4} D_2^2 = 0.0175 \text{ or } D_2 = \left(\frac{4 \times 0.0175}{\pi}\right)^{1/2} = 0.149 \text{ m or } 149 \text{ mm}\ \textbf{(Ans.)}$$

Considering equilibrium of the sliding cylinder (neglecting friction effects), we have:

$$p_1A_1 = p_2A_2 \text{ or } 40 \times 10^5 \times A_1 = 160 \times 10^5 \times 0.0175 \text{ or } A_1 = 0.07 \text{ m}^2$$

$$\therefore \quad \frac{\pi}{4}D_1^2 = 0.07, \text{ or } D_1 = \left(\frac{4 \times 0.07}{\pi}\right)^{1/2} = 0.298 \text{ m or } \mathbf{298 \text{ mm (Ans.)}}$$

Q. 9.10. A hydraulic ram is receiving water at the rate of 0.022 m^3/s from a height of 3.3 m, and it raises 0.0022 m^3/s to a height of 21 m from the ram. Determine D'Aubuisson's and Rankine's efficiencies of the hydraulic ram.

(*a*) 53.64%, 49.6% (*b*) 63.64%, 59.6%

(*c*) 73.64%, 69.6% (*d*) none of the above.

Ans. (*b*)

EXPLANATION:

Discharge through the supply pipe $Q = 0.022$ m^3/s

Supply head, $h = 3.3$ m

Discharge raised, $q = 0.0022$ m^3/s

Height of water raised from hydraulic ram, $H = 21$ m

D'Aubuisson's efficiency:

$$\text{D'Aubuisson's efficiency} = \frac{qH}{Qh} = \frac{0.0022 \times 21}{0.002 \times 3.3} = 0.6364 \text{ or } 63.64\% \textbf{ (Ans.)}$$

Rankine's efficiency:

$$\text{Rankine's efficiency}: \frac{q(H-h)}{(Q-q)h} = \frac{0.0022(21-3.3)}{(0.022-0.0022) \times 3.3} = 0.596 \text{ or } 59.6\% \textbf{ (Ans.)}$$

Q. 9.11. A centrifugal pump is started with its delivery valve kept

(*a*) fully open (*b*) fully closed

(*c*) partially open (*d*) 50% open.

Ans. (*b*)

EXPLANATION:

A centrifugal pump is started with its valve fully closed. **(Ans.)**

Q. 9.12. As water flows through the runner of a reaction turbine, pressure acting on it would vary from

(*a*) more than atmospheric pressure to vacuum

(*b*) less than atmospheric pressure to zero gauge pressure

(*c*) atmospheric pressure to more than atmospheric pressure

(*d*) atmospheric pressure to vacuum.

Ans. (*a*)

EXPLANATION:

The pressure at 'runner inlet' of a reaction in drive is *more* than atmospheric pressure and at a 'runner outlet' it is *less* than atmospheric pressure (*i.e.,* vacuum) **(Ans.)**

Q. 9.13. Given that N = speed, P = power, H = head,

The specific speed of a hydraulic turbine is given by

(*a*) $\frac{N\sqrt{P}}{H^{4/5}}$ (*b*) $\frac{N\sqrt{P}}{H^{5/4}}$

(*c*) $\frac{P\sqrt{N}}{H^{4/5}}$ (*d*) $\frac{P\sqrt{N}}{H^{5/4}}$.

Ans. (*b*)

EXPLANATION:

The *specific speed* of a turbine is defined as the speed of the turbine which is identical in shape, geometrical in dimensions, blade angles, gate opening, etc. which would develop unit power when working under a unit head. It is given by:

$$N_B(\text{Specific speed}) = \frac{N\sqrt{P}}{H^{5/4}} \quad \textbf{(Ans.)}$$

Q. 9.14. Which one of the following turbines is used in underwater power stations?

(*a*) Pelton turbine (*b*) Deriaz turbine

(*c*) Tubular turbine (*d*) Turgo-impulse turbine.

Ans. (*c*)

EXPLANATION:

For underwater power stations, *tubular or bulb turbine* is used. A large bulb is centred in the water pipe which holds the generator, wiket gate and runner.

A tubular or bulb turbine has the following features:

- It is axial flow turbine with either adjustable or non-adjustable runner vanes (and hence similar to kaplan or propeller turbines).
- In such a turbine the scroll casing is *not* provided but the runner is placed in a tube extending from head water to the tail water (and hence it is called a tubular turbine).
- It is a low head turbine and is employed for heads varying from 3 m to 15 m.
- The disposition of the shaft may be vertical, inclined or horizontal.

Q. 9.15. A Pelton wheel is ideally suited for

(*a*) high head and low discharge (*b*) high head and high discharge

(*c*) low head and low discharge (*d*) medium head and medium discharge.

Ans. (*a*)

EXPLANATION:

The specific speed of an hydraulic turbine is given by $N_s\text{(specific speed)} = \frac{N\sqrt{P}}{H^{5/4}}$

(N = speed; P = power; H = head)

The specific speed for high head will be low and thus Pelton wheel/turbine will be suitable because N_s is the speed with which the turbine turns for a particular discharge Q, with unit head and thereby is able to produce unit power, thus for *low* N_s, *Q must be low.* **(Ans.)**

Q. 9.16. Consider the following turbines:

1. Kaplan
2. Pelton wheel
3. Francis

The correct sequence in increasing order of the specific speeds of these turbines is

(*a*) 2, 3, 1 (*b*) 2, 1, 3

(*c*) 3, 1, 2 (*d*) 1, 2, 3.

Ans. (*a*)

EXPLANATION:

The specific speed (N_s) for various types of turbines are:

Type of turbines	Specific speed
1. Propeller and Kaplan	300 – 1000
2. Pelton	12 – 70
3. Francis	80 – 400

Thus the correct sequence in increasing order of N_s is given by 2, 3, 1 *i.e.*, (*a*) is the correct answer **(Ans.)**

Q. 9.17. If the full-scale turbine is required to work under a head of 30 m and to run at 428 rpm, then a quarter-scale turbine model tested under a head of 10 m must run at

(*a*) 143 rpm (*b*) 341 rpm
(*c*) 428 rpm (*d*) 988 rpm.

Ans. (*d*)

EXPLANATION:

We know that: $$\left(\frac{\sqrt{H}}{DN}\right)_m = \left(\frac{\sqrt{H}}{DN}\right)_p \quad \text{... [Eqn. (3.27)]}$$

or, $$\frac{\sqrt{H_m}}{D_m N_m} = \frac{\sqrt{H_p}}{D_p N_p}$$

or, $$N_m = N_p \left(\frac{D_p}{D_m}\right)\sqrt{\frac{H_m}{H_p}}$$

$$= 428 \times (4) \times \sqrt{\frac{10}{30}} = 988 \text{ rpm} \quad \textbf{(Ans.)}$$

Q. 9.18. Consider the following turbines/wheels:

1. Francis turbine
2. Pelton wheel with two or more jets
3. Pelton wheel with a single jet
4. Kaplan turbines

The correct sequence of these turbines/wheels in increasing order of their specific speeds is

(*a*) 2, 3, 1, 4 (*b*) 3, 2, 1, 4
(*c*) 2, 3, 4, 1 (*d*) 3, 2, 4, 1.

Ans. (*b*)

EXPLANATION:

The specific speed of pelton wheel with single jet is less than that of Pelton wheel with two or more jets. Furthermore, the specific speed of Francis turbines is more than Pelton wheel and less than Kaplan turbine, hence *b* is the correct answer. **(Ans.)**

Q. 9.19. Consider the following statements pertaining to a centrifugal pump:

1. The manometric head is the head developed by pump.
2. The suction pipe has, generally, a large diameter as compared to the discharge pipe.
3. The suction pipe is provided with a foot valve and a strainer.
4. The delivery pipe is provided with a foot valve and a strainer.

Which of these statements are correct?

(*a*) 1, 2, 3, and 4 (*b*) 1 and 2
(*c*) 2 and 3 (*d*) 1 and 3.

Ans. (*c*)

EXPLANATION:

The head against which a centrifugal pump has to work is known as manometric head (H_{mano}). It is given as:

$$H_{mano} - \frac{V_{w2}u_2}{g} - \text{loss of head in the pump (}i.e.\text{, impeller and casing)}$$

($\frac{V_{w2}u_2}{g}$ = work done per second and per unit weight, assuming flow at inlet to be radial). **(Ans.)**

Q. 9.20. If a reciprocating pump having a mechanical efficiency of 80% delivers water at the rate of 80 kg/s with a head of 30 m, the brake power of the pump is

(*a*) 29.4 kW (*b*) 20.8 kW

(*c*) 15.4 kW (*d*) 10.8 kW.

Ans. (*a*)

EXPLANATION:

Power (BP) required to drive the reciprocating pump,

$$\text{P (brake)} = \frac{wQ(h_s + h_d)}{\eta} = \frac{mgH}{\eta} = \frac{80 \times 9.81 \times 30}{0.8 \times 1000}$$

$$= 29430 \text{ W} = 29.43 \text{ kW} \quad \textbf{(Ans.)}$$

Q. 9.21. The maximum number of jets generally employed in an impulse turbine without jet interference are

(*a*) 4 (*b*) 6

(*c*) 8 (*d*) 12.

Ans. (*b*)

EXPLANATION:

Normally a Pelton wheel has one nozzle or one jet. However, a number of nozzles may be employed when more power is to be produced with the same wheel. Theoretically *six nozzles* can be used on Pelton wheel. However, practical considerations limit the use of not more than two jets per runner for a vertical runner and not more than four jets per runner if it is of horizontal configuration. **(Ans.)**

Q. 9.22. For a water turbine, running at constant head and speed, the operating characteristic curves in the given figure 9.11 show that upto a certain discharge '*q*' both output power and efficiency remain zero.

The discharge '*q*' is required to

(*a*) overcome initial inertia (*b*) overcome initial friction

(*c*) keep the hydraulic circuit full (*d*) keep the turbine running at no load.

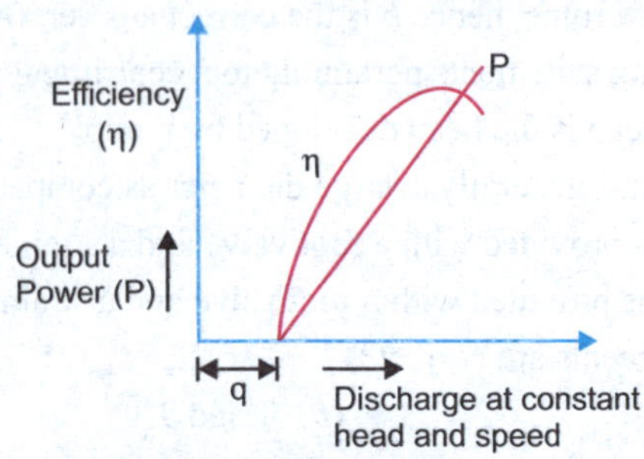

Fig. 9.11

Ans. (*b*)

EXPLANATION:

The discharge q (no load) is required to keep the turbine running under *no-load condition,* to overcome initial friction (*i.e.,* frictional losses) **(Ans.)**

Q. 9.23. The gross head on a turbine is 300 m. The length of penstock supplying water from reservoir to the turbine is 400 m. The diameter of the penstock is 1 m and velocity of water through penstock is 5 m/s. If coefficient of friction is 0.0098, the net head on the turbine would be, nearly

(*a*) 310 m (*b*) 295 m

(*c*) 200 m (*d*) 150 m.

Ans. (*b*)

EXPLANATION:

Given: Gross head, $H_g = 300$ m; L = 400 m; D = 1 m; V = 5 m/s; $f = 0.0098$

Now, Loss of head in the penstock, $h_f = \dfrac{fLV^2}{D \times 2g} = \dfrac{0.0098 \times 400 \times (5)^2}{1 \times 2 \times 9.81} = 5$ m

Hence, net head, $H = H_g - H_f = 300 - 5 = 295$ m **(Ans)**

Q. 9.24. A pump running at 1000 rpm consumes 1 kW and generates head of 10 m of water. When it is operated at 2000 rpm, its power consumption and head generated would be

(*a*) 4 kW, 50 m of water (*b*) 6 kW, 20 m of water

(*c*) 3 kW, 30 m of water (*d*) 8 kW, 40 m of water.

Ans. (*d*)

EXPLANATION:

When the diameter is constant, we have the following realtions

$$\frac{H_2}{H_1} = \left(\frac{N_2}{N_1}\right)^2 \text{ or } H_2 = H_1 \times \left(\frac{N_2}{N_1}\right)^2 = 10 \times \left(\frac{2000}{1000}\right)^2 = 40 \text{ m of water } \textbf{(Ans.)}$$

$$\text{and } \frac{P_2}{P_1} = \left(\frac{N_2}{N_1}\right)^3 \text{ or } P_2 = P_1 \times \left(\frac{N_2}{N_1}\right)^3 = 1 \times \left(\frac{2000}{1000}\right)^2 = 8 \text{ kW } \textbf{(Ans.)}$$

Q. 9.25. A centrifugal pump gives maximum efficiency when its blades are

(*a*) bent forward (*b*) bent backward

(*c*) straight (*d*) wave shaped.

Ans. (*b*)

EXPLANATION:

A centrifugal pump with its blades bent *backward* gives maximum efficiency as compared to blades with other shapes. **(Ans.)**

Q. 9.26. In utilizing scaled models in the designing of turbo-machines, which of the following relationship must be satisfied?

(*a*) $\dfrac{H}{ND^3} = \text{constant}; \dfrac{Q}{N^2D^2} = \text{constant}$

(*b*) $\dfrac{Q}{D^2\sqrt{H}} = \text{constant}; \dfrac{H}{N^3D} = \text{constant}$

(*c*) $\frac{P}{QH}$ = constant; $\frac{H}{N^2D^2}$ = constant

(*d*) $\frac{NQ^{1/2}}{H^{3/2}}$ = constant; $\frac{NP^{1/2}}{H^{3/4}}$ = constant .

Ans. (*c*)

Explanation:

We know that: $P = \rho QgH$, or $\frac{P}{QH}$ = constant **(Ans.)**

and, $u = \frac{\pi DN}{60}$

Also, $N = \frac{\sqrt{H}}{D}$ or $ND = \sqrt{H}$ or $\frac{}{N^2d^2}$ = constant **(Ans.)**

(The parameter $\frac{H}{N^2D^2}$ is called *head coefficient*, C_H)

Q. 9.27. A centrifugal pump driven by a directly coupled 3 kW motor of 1450 rpm speed, is proposed to be connected to another motor of 2900 rpm speed. The power of the motor should be

(*a*) 6 kW (*b*) 12 kW

(*c*) 18 kW (*d*) 24 kW.

Ans. (*d*)

Explanation:

For a constant diameter, $P \propto N^3$, then

$$\frac{P_2}{P_1} = \left(\frac{N_2}{N_1}\right)^3 = \left(\frac{2900}{1450}\right)^3 = 8 \text{ or } P_2 = P_1 \times 8 = 3 \times 8 = 24 \text{ kW} \quad \textbf{(Ans.)}$$

Q. 9.28. A draft tube is used in a reaction turbine

(*a*) to guide water downstream without splashing

(*b*) to convert residual pressure energy into kinetic energy

(*c*) to convert residual kinetic energy into pressure energy

(*d*) to streamline the flow in the tail race.

Ans. (*c*)

Explanation:

The draft tube serves the following two *purposes*:

1. To convert residual kinetic energy into pressure energy.
2. It allows the turbine to be set above tail water level, without loss of head, to facilitate inspection and maintenance.

- Draft tube is an integral part of mixed and axial flow turbines.

Q. 9.29. Efficiency of Pelton wheel shall be maximum if the ratio of jet velocity to tangential velocity of the wheel is

(*a*) 1/2 (*b*) 1

(*c*) 2 (*d*) 4.

Ans. (*c*)

EXPLANATION:

The hydraulic efficiency of a Pelton wheel is maximum when the tangential velocity of wheel is *half* the velocity of jet of water at inlet **(Ans.)**

Q. 9.30. A hydraulic press has a ram of 20 cm diameter and a plunger of 5 cm diameter. The force required at the plunger to lift a weight of 16×10^4 N shall be

(*a*) 256×10^4 N (*b*) 64×10^4 N

(*c*) 4×10^4 N (*d*) 1×10^4 N.

Ans. (*d*)

EXPLANATION:

Area of ram, $A = \pi/4 \times (20)^2$ cm = 100 π cm^2; Area of plunger, $a = \frac{\pi}{4} \times 5^2 = \frac{25\pi}{4}$ cm^2.

Load lifted, W = 16×10^4 N; force reqd. at the plunger, F = ?

Now, $\frac{F}{a} = \frac{W}{A}$ or $F = \frac{a}{A} \times W = \frac{\frac{25\pi}{4}\pi}{100\pi} \times 16 \times 10^4 = 1 \times 10^4$ N **(Ans.)**

Q. 9.31. Priming is necessary in

(*a*) centrifugal pumps to lift water from a greater depth

(*b*) centrifugal pumps to remove air in the suction pipe and casing

(*c*) hydraulic turbine to remove air in the turbine casing

(*d*) hydraulic turbine to increase the speed of turbine and to generate more power.

Ans. (*b*)

EXPLANATION:

The operation of filling the suction pipe, casing of the pump and a portion of the delivery pipe completely from outside source with the liquid to be raised, before starting the pump, to *remove any air, gas or vapour* form these parts of the pump is called *priming* of a centrifugal pump.

If a centrifugal pump is *not* primed before starting, air pockets inside the impeller may give rise to vortices and cause discontinuity of flow. Further dry running of the pump may result in rubbing and seizing of the wearing rings and cause serious damage.

Q. 9.32. Consider the following statements regarding air vessels provided in reciprocating pump installations:

1. The air vessels are fitted both on suction and delivery sides.
2. The air vessels are fitted far from the pump cylinder.
3. The air vessels save energy by reducing the friction loss.

Which of these statements are correct?

(*a*) 1, 2 and 3 (*b*) 1 and 2

(*c*) 2 and 3 (*d*) 1 and 3.

Ans. (*d*)

EXPLANATION:

An air vessel is a closed chamber containing compressed air in the upper part and liquid being pumped in the lower part. One air vessel is fixed on the suction pipe '*just near the suction valve*' and one is *fixed on the delivery pipe 'near delivery valve'*.

The air vessels are used for the following *purposes*:

1. To get continuous supply of liquid at a uniform rate (whatever fluctuations false place, they occur between the air vessels and the pump)
2. To save the power by reducing the friction loss required to drive the pump.
3. To run the pump at much higher speed without any danger of separation. **(Ans)**

Q. 9.33. A pump is installed at a height of 5 m above the water level in the sump. Frictional loss on the suction side is 0.6 m. If the atmospheric pressure is 10.3 m of water and vapour pressure head is 0.4 m (abs), the NPSH (Net Positive Suction Head) will be

(*a*) 3.7 m (*b*) 4 m

(*c*) 4.3 m (*d*) 4.6 m.

Ans. (*c*)

EXPLANATION:

Net Positive Suction Head, NPSH = $\frac{p_0}{w}$ – (static head + frictional head in suction + vapour pressure head)

$$= 10.3 - (5 + 0.6 + 0.4) = 4.3 \text{ m} \quad \textbf{(Ans.)}$$

Q. 9.34. Two Pelton wheels A and B have the same specific speed and are working under the same head. Wheel A produces 400 kW at 1000 rpm. If B produces 100 kW, then its rpm is

(*a*) 4000 (*b*) 2000

(*c*) 1500 (*d*) 1250.

Ans. (*b*)

EXPLANATION:

Specific speed, $N_B = \frac{N\sqrt{P}}{(H)^{5/4}}$

Since the Pelton wheels are having the same specific speed and are working under the same head, therefore :

$$N\sqrt{P} = \text{constant or } N_1\sqrt{P_1} = N_2\sqrt{P_2}$$

or,
$$N_2 = N_1\sqrt{\frac{P_1}{P_2}} = 1000\sqrt{\frac{400}{100}} = 2000 \text{ rpm} \quad \textbf{(Ans.)}$$

Q. 9.35. A centrifugal pump needs 1000 W of power when operating at 1500 rpm. What is the power requirement if the speed of the pump is increased to 3000 rpm?

(*a*) 2000 W (*b*) 4000 W

(*c*) 6500 W (*d*) 8000 W.

Ans. (*d*)

EXPLANATION:

Now, $P \propto N^3$, where P and N are power and speed respectively.

$$\therefore \quad \frac{P_2}{P_1} = \left(\frac{N_2}{N_1}\right)^3, \text{ or } P_2 = P_1 \times \left(\frac{N_2}{N_1}\right)^3 = 1000 \times \left(\frac{3000}{1500}\right)^3 = 8000 \text{ W } \textbf{(Ans.)}$$

Q. 9.36. The centrifugal pumps have impellers with outlet dimensions of each equal to twice the inner dimensions. The inner diameter of the second impeller is three times the inner diameter of the first one. What is the speed ratio N_2/N_1 of pumps, if the pumps are required to develop the same manometric head to start delivery of water?

(*a*) 9 (*b*) 4
(*c*) 1/2 (*d*) 1/3.

Ans. (*d*)

EXPLANATION:

$$(D_0)_1 = 2\,(D_i)_1;\ (D_0)_2 = 2\,(D_i)_2;\ (D_i)_2 = 3\,(D_i)_1$$

$$\therefore \quad (D_0)_2 = 2 \times 3(D_i)_1 = 6(D_i)_1$$

Also, $N \propto \dfrac{1}{D}$, $\therefore \dfrac{N_2}{N_1} = \dfrac{(D_0)_1}{(D_0)_2} = \dfrac{2(D_i)_1}{6(D_i)_1} = \dfrac{1}{3}$ **(Ans.)**

Q. 9.37. A centrifugal pump operating at 1000 rpm develops a head of 30 m. If the speed is increased to 2000 rpm and the pump operates with the same efficiency, what is the head developed by the pump?

(*a*) 60 m (*b*) 90 m
(*c*) 120 m (*d*) 150 m.

Ans. (*c*)

EXPLANATION:

In case of a centrifugal pump, $H \alpha N^2$

$$\therefore \quad \frac{H_2}{H} = \left(\frac{N_2}{N_1}\right)^2 \text{ or, } H_2 = H_1 \times \left(\frac{N_2}{N_1}\right)^2 = 30 \times \left(\frac{2000}{1000}\right)^2 = 120 \text{ W } \textbf{(Ans.)}$$

Q. 9.38. Two centrifugal pumps A and B operate at their maximum efficiencies at 1000 rpm and 500 rpm respectively. Against the same delivery head, pump A discharges 1 m^3/s and pump B discharges 4 m^3/s respectively. What is the ratio of specific speeds $(N_S)_A : (N_S)_B$?

(*a*) 1 : 2 (*b*) 1 : 1
(*c*) 1 : 4 (*d*) 4 : 1.

Ans. (*b*)

EXPLANATION:

Specific speed (N_s) in case of a centrifugal pump is given by, $\dfrac{N\sqrt{Q}}{(\ \ )^{3/4}}$

$$\therefore \quad \frac{(N_s)_A}{(N_s)_B} = \frac{1000 \times \sqrt{1}}{500 \times \sqrt{4}} = 1 : 1 \text{ (H being same) } \textbf{(Ans.)}$$

Q. 9.39. Which one of the following pairs is not correctly matched?

(*a*) Centrifugal pump : Rotating blades in the motor create centrifugal head
(*b*) Reciprocating pump : Positive displacement pump
(*c*) Turbine pump : Centrifugal pump with guide vanes
(*d*) Gear pump : Gear teeth work like rotating blades to create centrifugal head.

Ans. (*d*)

EXPLANATION:

In a gear pump, the gear teeth do *not* create centrifugal head. A gear pump has the ability to pump highly viscous fluid. **(Ans.)**

Q. 9.40. An impulse turbine operating with a single nozzle has a specific speed of 5. What will be the approximate specific speed of the turbine if the turbine is operated with one more additional nozzle of the same size?

(*a*) 4 (*b*) 6

(*c*) 7 (*d*) 10.

Ans. (*c*)

EXPLANATION:

In case of an impulse turbine, $N_s \alpha \sqrt{Q}$

$$\therefore \quad N_s = 5\sqrt{2} \simeq 7 \quad \textbf{(Ans.)}$$

Q. 9.41. Why is a minimum of Net Positive Suction Head required for a hydraulic pump?

(*a*) To prevent cavitation (*b*) To increase discharge

(*c*) To increase suction head (*d*) To increase efficiency.

Ans. (*a*)

EXPLANATION:

The NPSH (Net Positive Suction Head) should always be *positive* to *prevent cavitation.*

The term NPSH has significance only when cavitating liquids are handled. **(Ans.)**

Q. 9.42. If α is the blade angle at the outlet, then the maximum hydraulic efficiency of an ideal impulse turbine is

(*a*) $\dfrac{1+\cos\alpha}{2}$ (*b*) $\dfrac{1-\cos\alpha}{2}$

(*c*) $\dfrac{1-\sin\alpha}{2}$ (*d*) $\dfrac{1+\sin\alpha}{2}$.

Ans. (*a*)

EXPLANATION:

The maximum hydraulic efficiency of an ideal impulse turbine is given by:

$$(\eta_h)_{max} = \frac{1+\cos\alpha}{2}, \text{ where } \alpha \text{ is the blade angle } \textbf{(Ans.)}$$

Q. 9.43. In an axial flow impulse turbine, energy transfer takes place due to

(*a*) change in relative kinetic energy (*b*) change in absolute kinetic energy

(*c*) change in pressure energy (*d*) change in energy because of centrifugal force.

Ans. (*b*)

EXPLANATION:

In an axial flow impulse turbine, energy transfer takes place due to *change in absolute kinetic energy.* **(Ans.)**

Q. 9.44. The overall efficiency of Pelton turbine is 70%. If the mechanical efficiency is 85%, what is its hydraulic efficiency?

(*a*) 82.4% (*b*) 59.5%

(*c*) 72.3% (*d*) 81.5%.

Ans. (*a*)

EXPLANATION:

The overall efficiency of a pelton turbine,

$$\eta_0 = \eta_h \times \eta_m, \text{ or } 0.7 = \eta_h \times 0.85, \text{ or } \eta_h = \frac{0.7}{0.85} = 0.824 \text{ or } 82.4\% \textbf{ (Ans.)}$$

Q. 9.45. If in a pump, the discharge is halved, then assuming that the speed remains unchanged, what would be the ratio of the heads H_1/H_2?

(*a*) $\sqrt{1/3}$ (*b*) $\sqrt{2/3}$

(*c*) $\sqrt[3]{0.25}$ (*d*) $\sqrt[3]{0.5}$.

Ans. (*c*)

EXPLANATION:

In a centrifugal pump, $Q = K\,\dfrac{(H)^{3/2}}{N^2}$

(where, K = constant of proportionality, H = head, and N = speed)

If the speed remains constant, then: $Q \,\alpha\, H^{3/2}$ or $H \,\alpha\, Q^{2/3}$

$$\therefore \frac{H_1}{H_2} = \left(\frac{Q_1}{Q_2}\right)^{\frac{2}{3}} = \left(\frac{1}{2}\right)^{2/3} = \sqrt[3]{0.25} \quad \textbf{(Ans.)}$$

Q. 9.46. In a hydraulic coupling, what is the ratio of speed of the turbine runner to that of the pump impeller to maintain circulatory motion of oil?

(*a*) < 1 (*b*) = 1

(*c*) > 1 (*d*) Can be any value.

Ans. (*a*)

EXPLANATION:

In order to transmit power from pump impeller to turbine runner, circulatory motion of oil is required.

The efficiency (η) of hydraulic coupling is defined as the ratio of power output to power input, *i.e.*,

$$\eta = \frac{\text{Power output } (P_o)}{\text{Power input } (P_r)} = \frac{T_t\,\omega_t}{T_p\,\omega_p} = \frac{\omega_t}{\omega_p} \qquad (\because T_p = T_t)$$

As efficiency is always less than 1, thus $\dfrac{\omega_t}{\omega_p} < 1$ **(Ans.)**

Q. 9.47. Why is an air vessel used in a reciprocating pump?

(*a*) To obtain a continuous supply of water at uniform rate

(*b*) To reduce suction head

(*c*) To increase the delivery head

(*d*) To reduce cavitation.

Ans. (*a*)

EXPLANATION:

An air vessel are used for the following purposes:

1. To get continuous supply of liquid at a uniform rate (whatever fluctuation take place, they occur between the air vessels and the pump)
2. To save the power required to drive the pump).
3. To run the pump at such higher speed without any danger of separation. **(Ans.)**

Q. 9.48. Centrifugal pumps have which of the following advantages?

1. Low initial cost.
2. Compact, occupying less floor space.
3. Easy handling of highly viscous fluids.

Selected the correct answer using the codes given below:

(*a*) 1, 2 and 3 (*b*) 1 and 2 only

(*c*) 1 and 3 only (*d*) 2 and 3 only.

Ans. (*d*)

EXPLANATION:

Centrifugal pumps have the following advantages:

1. They are compact and occupy less floor space.
2. They can easily handle highly viscous fluids.

However, they have high initial cost. **(Ans.)**

Q. 9.49. Two pumps can operate independently at heads H_1, H_2 and discharge Q_1, Q_2 respectively. If the pumps are connected in parallel, then what are the resulting discharge (Q) and head (H)?

(*a*) $Q = Q_1 + Q_2, H = H_1 + H_2$ (*b*) $Q = Q_1 - Q_2, H = H_1 - H_2$

(*c*) $Q = Q_1 = Q_2, H = H_1 = H_2$ (*d*) $Q = Q_1 + Q_2, H = H_1 = H_2$.

Ans. (*d*)

EXPLANATION:

When the pumps are connected in *parallel*

$$Q = Q_1 + Q_2, H = H_1 = H_2 \quad \textbf{(Ans.)}$$

Q. 9.50. Which one of the following is correct for hydraulic coupling?

(*a*) It connects two shafts rotating at about the same speed

(*b*) It connects two shafts running at different speeds

(*c*) It is used to augment the torque to the driven shaft

(*d*) It is used to connect the centrifugal pump and its electrical motor for efficient operation.

Ans. (*b*)

EXPLANATION:

A hydraulic coupling connects two shafts running at *different speeds.* **(Ans.)**

HYDRAULIC MACHINES LABORATORY PRACTICALS

(Experiments : 1 to 4)

LABORATORY EXPERIMENTS

HYDRAULIC MACHINES

Experiment No. 1. *To study the operation and performance of a* **Pelton wheel.**

Experiment No. 2. *To study the performance of a* **Francis turbine.**

Experiment No. 3. *To study the performance characteristics of a single-stage* **centrifugal pump.**

Experiment No. 4. *To obtain the performance characteristics of a* **reciprocating pump.**

EXPERIMENT NO. 1. To study the operation and performance of a Pelton wheel.

A. Operation of a Pelton Wheel

Object : *To operate a Pelton-wheel turbine and understand its construction and working.*

Apparatus : Pelton wheel/turbine connected to a high head water tank. A centrifugal pump to supply water to the tank.

Theory : A Pelton wheel is a special type of *axial flow impulse* type turbine. It is employed where very high head of water is available. It converts pressure energy of water into kinetic energy which further rotates the wheel/runner of turbine.

The wheel/runner essentially consists of a disc made of cast-iron or steel fitted to the shaft. On the periphery of the wheel are attached the blades or buckets. The buckets are made of cast-iron or hard bronze and are in the form of a double hemispherical cup. The water to the wheel is delivered by one or more nozzles. The water after passing through the nozzle strikes the bucket at its centre in tangential direction and flows axially in both directions over the two cups. Usually the total deflections of jet is 160°. Due to impulse of water the wheel rotates. The turbine rotates *most efficiently when runner rotates at half the velocity of jet.*

The governing of medium and large powered Pelton wheel is usually carried out by an *oil pressure governor.* The shaft type governor is restricted to turbines of relatively low output.

Operating procedure : Refer to Fig. 1.

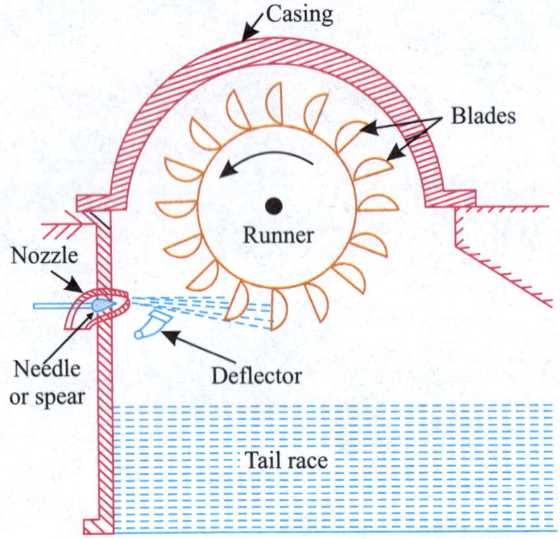

Fig. 1. Pelton wheel.

The procedure of operating a Pelton wheel includes the following steps :

1. To run the turbine; **2.** To slow down the turbine; **3.** To stop the turbine.

1. To run the turbine. (*i*) In case high head water tank receives water from a centrifugal pump, then ensure first that tank is adequately filled before running the turbine.

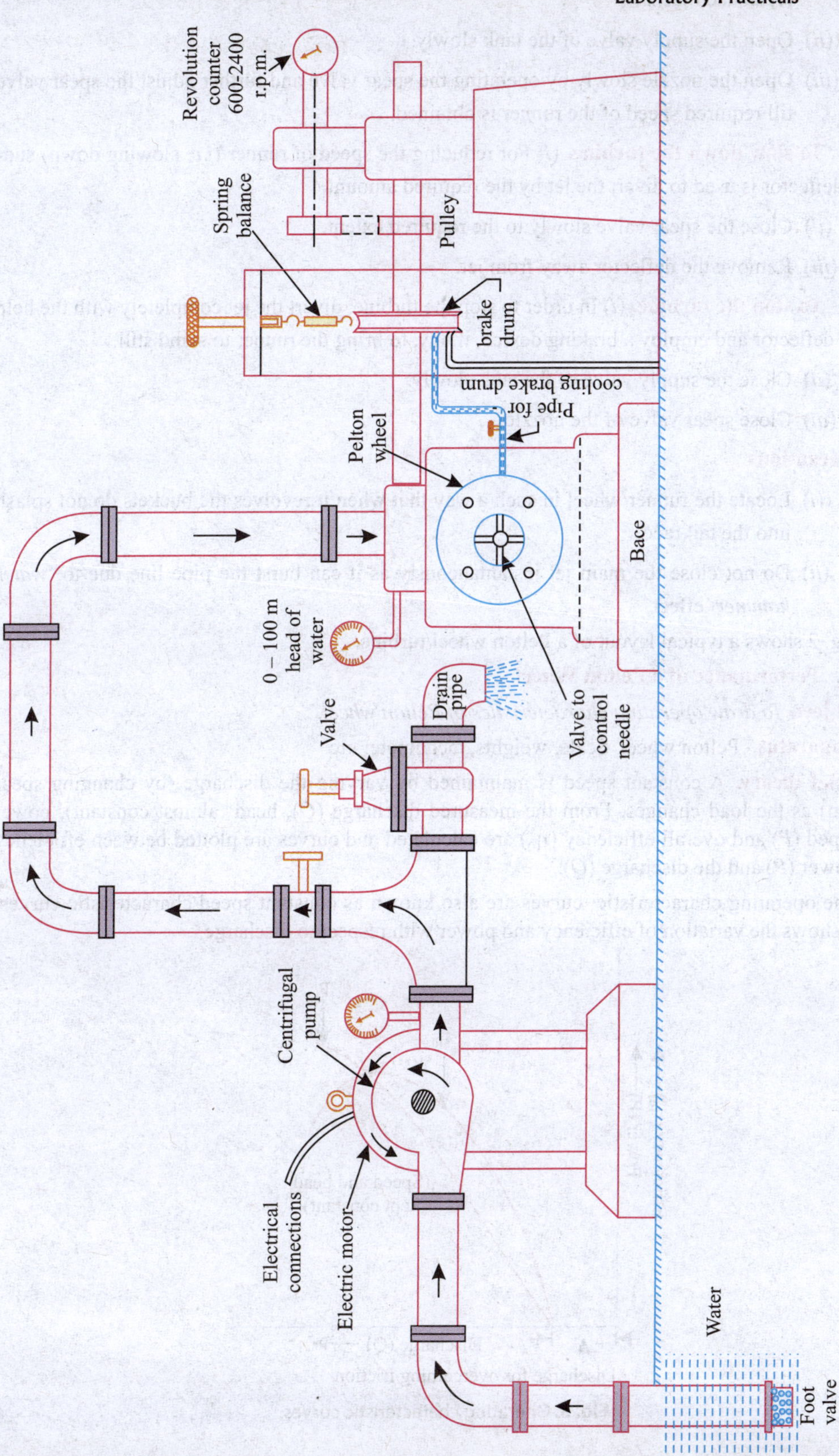

Fig. 2. Typical layout of a Pelton wheel (turbine).

(*ii*) Open the supply valve of the tank slowly.

(*iii*) Open the nozzle slowly by operating the spear valve and further adjust the spear valve till required speed of the runner is obtained.

2. To slow down the turbine. (*i*) For reducing the speed of runner (*i.e.* slowing down) suddenly, deflector is used to divert the jet by the required amount.

(*ii*) Close the spear valve slowly to the required extent.

(*iii*) Remove the deflector away from jet.

3. To stop the turbine. (*i*) In order to stop the turbine, divert the jet completely with the help of a jet deflector and employ a braking device, if any, to bring the runner to stand still.

(*ii*) Close the supply valve of the tank slowly.

(*iii*) Close spear valve of the nozzle.

Precautions :

(*i*) Locate the runner/wheel in such a way that when it revolves the buckets do not splash into the tail race.

(*ii*) Do not close the main jet instantaneously as it can burst the pipe line due to '*water hammer*' effect.

Fig. 2 shows a typical layout of a Pelton wheel/turbine.

B. Performance of a Pelton Wheel

Object. *To draw operating characteristics of Pelton wheel.*

Apparatus. Pelton wheel, Scale, weights, tachometer etc.

Brief theory. A constant speed is maintained by varying the discharge (by changing spear position) as the load changes. From the measured discharge (*Q*), head (almost constant), power developed (*P*) and overall efficiency (η_0) are calculated and curves are plotted between efficiency (η_0) power (*P*) and the discharge (*Q*).

The operating characteristic curves are also known as constant speed characteristic curves. Fig. 3 shows the variation of efficiency and power with respect to discharge.

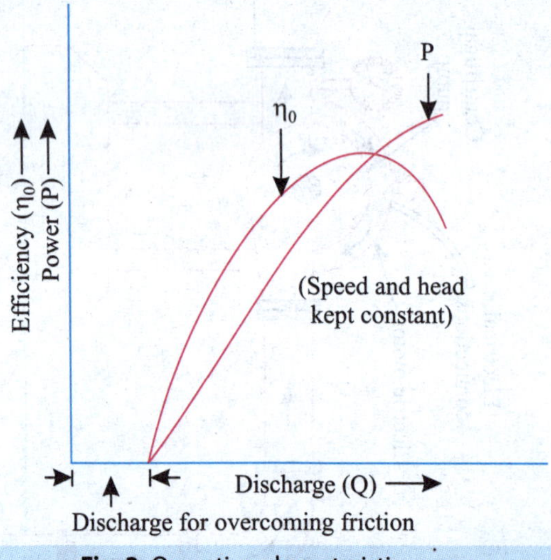

Fig. 3. Operating characteristic curves.

Formulae:

1. Discharge, $Q = 1.84\,Lh^{3/2}$ (app.)
   where, L = Length of the weir, and
   h = Head over the weir.
2. Shaft power developed (S.P.)
   $$P = (W - S) \times D/2 \times 2\,\pi\,N$$
   $$= (W - S)\,\pi\,DN \text{ watts}$$
   $$= \frac{(W - S)\,\pi\,DN}{60 \times 1000} \text{ kW}$$
   where, W = Load applied on the brake drum (N),
   S = Spring balance reading (N),
   D = Mean diameter of the brake drum (m), and
   N = Speed in r.p.m.
3. Water power (W.P.) $= \dfrac{wQH}{1000}$ kW
   where, w = Specific weight of water (= 9810 N/m^3),
   Q = Discharge in m^3/s (as calculated at 1), and
   H = Head (of water) acting on Pelton wheel, m.
4. Overall efficiency, $\eta_0 = \dfrac{S.P.}{W.P.}$

Procedure :

1. Change the load on the hanger (W) and note the reading of the spring balance (S).
2. Adjust the delivery valve of centrifugal pump to keep head (H) same and adjust spear position to keep the speed (r.p.m.) same.
3. With constant speed and almost constant head note sill level reading (hook gauge reading).
4. By using the above mentioned procedure change the loads from no load to full load and tabulate the readings as given in Table 1.
5. Plot the operating characteristics with the data detained.

Observations :

Mean diameter of the brake drum, $D (= D_d + d_r) = \ldots$ m
[D_d = diameter of the drum, d_r = dia. of rope]
Sill level (initial hook gauge reading), $h_1 = \ldots$ m
Length of the weir, $L = \ldots$ m
Speed, (r.p.m.), $N = \ldots$

Table 1. Performance of a Pelton wheel – Observations

S. No	*W*	*S*	*H*	*Final hook gauge reading (h_2)*	*h* $(= h_2 - h_1)$ *(m)*	$Q = 1{\cdot}84\,Lh^{3/2}$ *(m^3/s)*	$S.P. = \dfrac{(W-S)\pi DN}{60 \times 1000}$ kW	$W.P. = \dfrac{wQH}{1000}$ kW	$\eta_0 = \dfrac{S.P.}{W.P.}$	*Remarks*
1.										
2.										
3.										
4.										
5.										
6.										

Note : The constency of units shall be maintained carefully.

Specimen calculations : (*i*)

(*ii*)

Conclusions :

Precautions :

1. Head over the Pelton wheel should be kept constant
2. All the readings must be taken and recorded accurately.

***EXPERIMENT NO. 2.* To study the performance of a Francis turbine.**

Apparatus :

1. A Francis turbine with an arrangement for adjusting the guide vane positions (hand wheel with suitable link mechanisms).
2. Supply pump unit.
3. Flow measurement unit (*viz.* venturimeter with manometer.)
4. Tachometer.
5. Pressure gauges at the inlet and outlet of turbine.
6. Rope brake dynamometer with spring balance connected to the turbine shaft.

Theory :

Fig. 4 shows a schematic diagram of a Francis turbine.

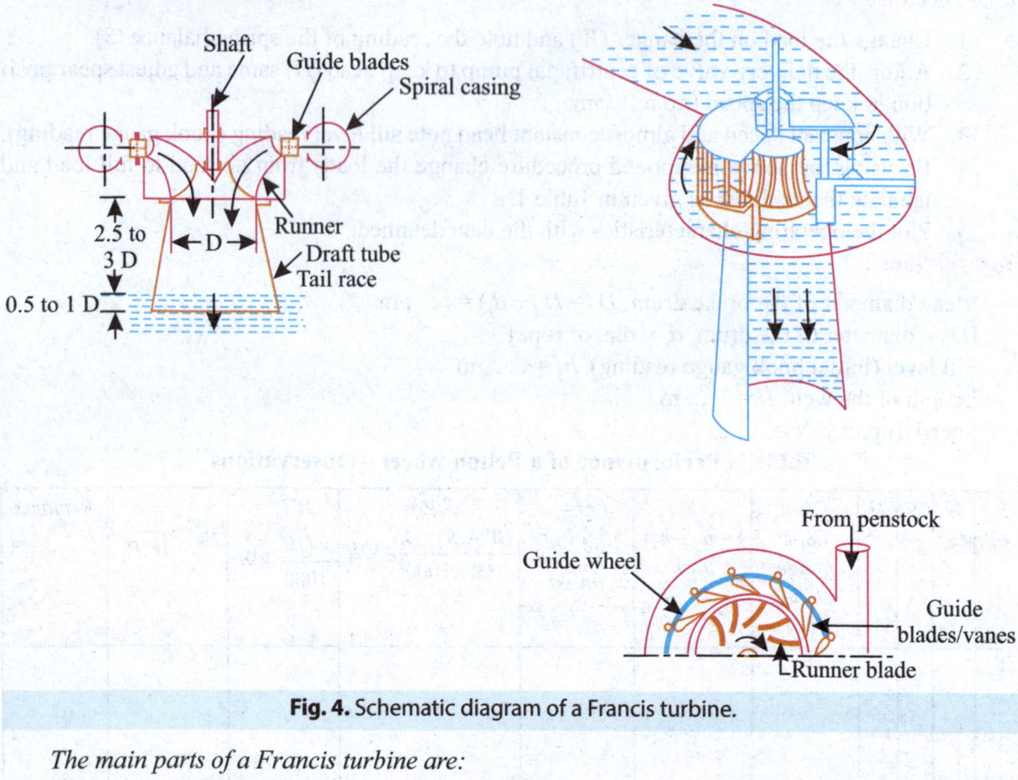

Fig. 4. Schematic diagram of a Francis turbine.

The main parts of a Francis turbine are:

1. *Penstock* ... It is a large size conduit which conveys water from the upstream of the dam/reservoir to the turbine runner.

2. *Spiral/scroll casing* ... It constitutes a closed passage whose cross-sectional area gradually decreases along the flow direction, area is maximum at inlet and nearly zero at exit.
3. *Guide vanes/wicket gates* ... These vanes direct the water onto the runner at an angle appropriate to the design. The motion to them is given by means of a hand wheel or automatically by a governor.
4. *Governing mechanism* ... It changes the position of the guide blades/vanes to affect a variation in water flow rate, when the load conditions on the turbine change.
5. *Runner and runner blades* ... The driving force on the runner is both due to impulse reactions effects;

   The number of runner blades usually varies between 16 to 24.
6. *Draft tube* ... It is a gradually expanding tube which discharges water, passing through the runner, to the tail race.

The modern Francis turbine is an *inward mixed flow reaction turbine* (in the earlier stages of development, Francis turbine had a purely radial flow runner) *i.e., water under pressure, enters the runner from the guide vanes towards the centre in radial direction and discharges out of the runner axially.* The Francis turbine operates under *medium heads* and also requires *medium quantity* of water. It is employed in the medium head power plants. This type of turbine covers a wide range of heads. Water is brought down to the turbine through a *penstock* and directed to a number of stationary orifices fixed all around the circumference of the *runner.* These stationary orifices are commonly called as guide *vanes* or *wicket gates*.

The head acting on the turbine is partly transformed into kinetic energy and the rest remains as pressure head. There is a difference of pressure between the guide vanes and the runner which is called the *reaction pressure* and is responsible for the motion of the runner. That is why a Francis turbine is also known as *reaction turbine.*

In Francis turbine *the pressure at inlet is more than that at the outlet.* This means that the water in the turbine must flow in a closed conduit. Unlike the Pelton type, where the water strikes only a few of the runner buckets at a time, in the Francis turbine the *runner is always full of water. The moment of runner is affected by the change of both the potential and kinetic energies of water.* After doing the work the water is discharged to the tail race through a closed tube of gradually enlarging section. This is known as *draft tube.* It does not allow water to fall freely to tail race level as in the Pelton turbine. The free end of the draft tube is submerged deep in tail water making, thus, the entire water passage, right from the head race upto the tail race, *totally enclosed.*

Fig. 5 shows the typical layout of a Francis turbine.

Procedure :

1. Note the inlet and outlet pipe diameters and measure the brake drum diameter and z_1 and z_2 *i.e.,* the distances of inlet and outlet pressure gauge tappings from the centreline of the turbine.
2. Start the supply pump, keeping the guide vanes completely closed.
3. Open the guide vanes partially (*e.g.* $\frac{1}{4}$ *th* or $\frac{1}{2}$ of total opening), simultaneously adjusting the load on the brake drum so that the speed of turbine is within limits.

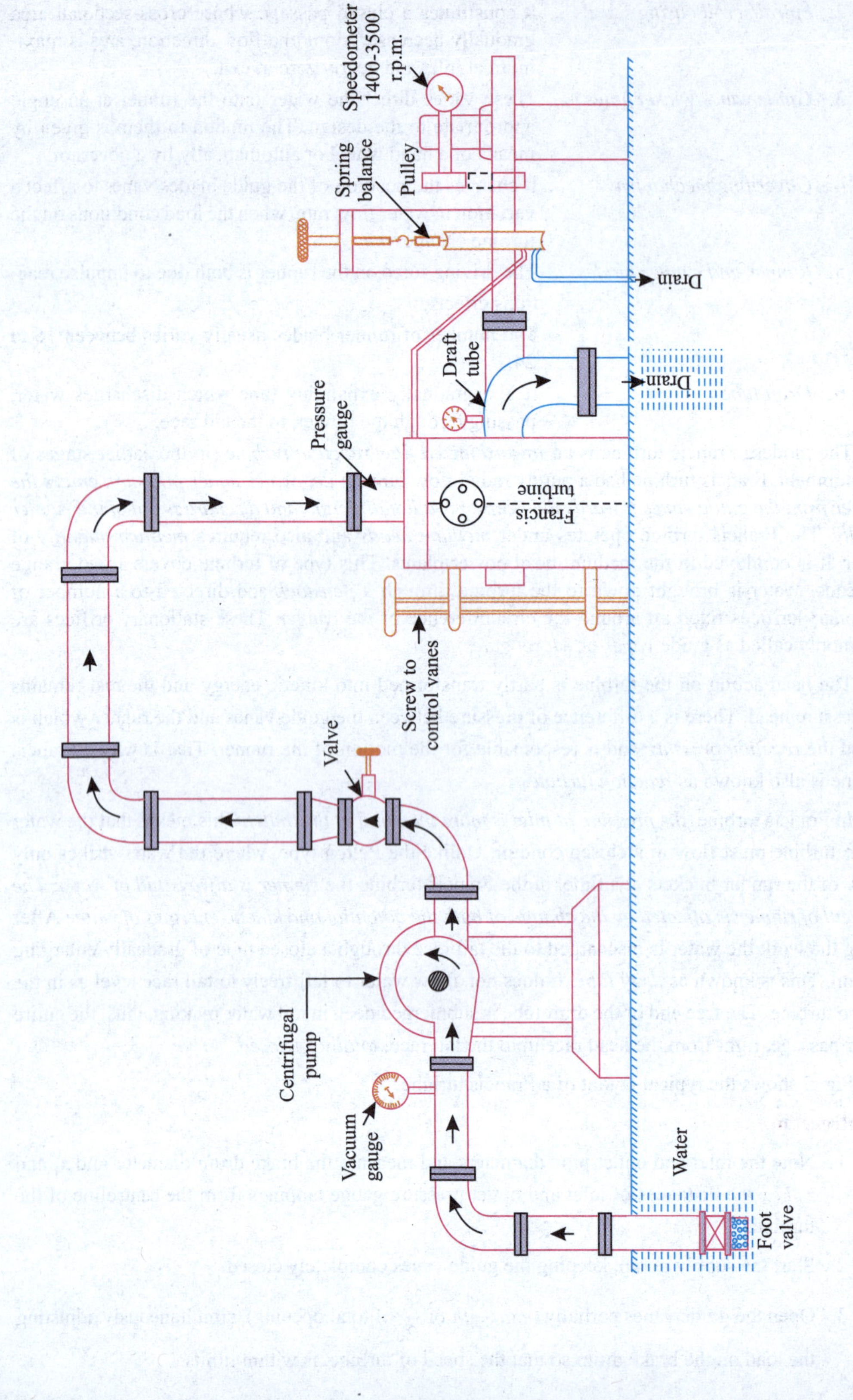

Fig. 5. Typical layout of a Francis turbine.

4. Measure the discharge (Q).
5. Note the readings of the pressure gauges (p_1, p_2).
6. Note the readings of W (load on the hanger) and S (spring balance) and the shaft speed N.
7. Vary the speed of the turbine by varying the load (*i.e.*, W and S) on the brake drum and take six to seven readings in the allowable range of speed.
8. Change the guide vane opening and repeat steps 4 to 7.

Observations :

Brake drum diameter (mean), $D = \ldots.$

Distance of inlet pressure gauge from turbine axis, $z_1 = \ldots.$

Distance of outlet pressure gauge from turbine axis, $z_2 = \ldots$

Diameter of inlet pipe, $d_1 = \ldots.$

Diameter of outlet pipe, $d_2 = \ldots.$

Table 2. Performance of a Francis turbine – Observations.

S. No.	G.V. opening	Discharge Q (m^3/s)	Pressure gauge readings		Speed	Rope tension		Input	Output	η
			p_1 (kg/cm^2)	p_2 (kg/cm^2)	N (r.p.m.)	W (kgf)	S (kgf)	P_i (H.P.)	P_o (H.P.)	%

Note : The consistency of units shall be maintained carefully.

Calculations :

For each opening of the guide vane perform the following calculations :

1. $V_1 = \dfrac{Q}{(\pi/4)\, d_1^2}$, and $V_2 = \dfrac{}{(\ \ /4)}$

2. $H = \left(\dfrac{p_1}{w} - \dfrac{p_2}{w}\right) + \left(\dfrac{V_1^2 - V_2^2}{2g}\right) + (z_1 - z_2)$

3. Input power, $P_i = \dfrac{wQH}{75}$

4. Output power (B.H.P.), $P_0 = \dfrac{(W - S)\, \pi DN}{4500}$ — MKS system (may be calculated using S.I. Units, keeping the consistency of the units)

5. $\eta = \dfrac{P_0}{P_i} \times 100$

- Plot the curves of :

  (*i*) η *vs* N ;

  (*ii*) B.H.P. *vs* N.

Result/Conclusion :

Precautions :

1. Keep the guide vanes completely closed until the supply pump develops the rated head.
2. The turbine should be loaded gradually.
3. Always keep the speed of the turbine within limits.
4. Before switching off the supply pump remove the load on the dynamometer.

***EXPERIMENT NO. 3.* To study the performance characteristics of a single stage centrifugal pump.**

Apparatus :

1. A centrifugal pump (or a working model of centrifugal pump) with all the necessary components (*e.g.,* suction pipe, delivery pipe, foot valve, strainer etc.)
2. An electronic motor coupled to the pump shaft; wattmeter.
3. Pressure gauges connected to the delivery and suction pipes as near to pump as possible.
4. Discharge measuring unit (*viz.* venturimeter with manometer).

Brief theory :

- The *centrifugal pump* is a rotodynamic machine, which increases the pressure energy of a liquid with the help of centrifugal action. In this type of pump the liquid is imparted a whirling motion due to the rotation of the impeller which creates a centrifugal head or dynamic pressure. This pressure head enables the lifting of liquid from a lower level to a higher level.

The *main parts* of a centrifugal pump are : Refer to Fig. 5.

1. **Impeller.** It is a rotating element which is provided with a number of vanes.
2. **Casing or chamber.** It surrounds the impeller and forms a passage for flow of water.
3. **Suction pipe.** It connects the inlet of the pump to the sump from which the water is to be pumped.
4. **Strainer.** It is provided at the lower end of the suction pipe which prevents the solid bodies and debris from entering the pump which if not prevented will result in damaging the impeller.
5. **Foot valve.** It is a one way valve provided above the strainer. It keeps the suction pipe filled with water when the pump is stopped. Such an arrangement helps the pump in starting which otherwise will not be possible without priming (*i.e.,* filling the pipe and pump with liquid.)
6. **Delivery pipe.** It leads water from the outlet of the pump to the desired point.
7. **Delivery valve.** It is provided on the delivery pipe just near the outlet of the pump. Its purpose is to control the flow into the delivery pipe.
8. **Prime mover.** It drives the shaft of an impeller. It may be an I.C. engine or electric motor.

Operation of a centrifugal pump :

The operation of a centrifugal pump involves the following steps :

(*i*) *Prime the pump.* Priming means filling the suction pipe, casing of the pump and a portion of the delivery pipe upto the delivery valve with the liquid to be pumped so that the air is completely driven out of these elements.

(*ii*) After priming, keeping the delivery valve still closed, start an electric motor or an engine to rotate the impeller. The rotation of the impeller inside the casing, which is full of liquid, will produce a vortex which is a responsible for imparting a centrifugal head to the water. It will also cause a *reduction of pressure* at the centre of the impeller and thus liquid will rush through the suction pipe.

(*iii*) When the impeller attains a normal speed, the delivery valve is opened to give a continuous supply of water through the delivery pipe.

- ***Troubles and their causes*** **:**

The common troubles and their causes experienced in a centrifugal pump are given as under :

1. Trouble. *Insufficient capacity or pressure and failure to deliver water/liquid.*

Causes. (*i*) Improper priming of the pump.

(*ii*) Too low a speed.

(*iii*) Too high a discharge head.

(*iv*) Too high a suction lift.

(*v*) Wrong direction of rotation.

(*vi*) Air leakage in the inlet pipe.

(*vii*) Foot valve clogged with foreign matter.

(*viii*) Mechanical defects.

(*ix*) Foot valve too small.

(*x*) Foot valve not immersed deep enough.

2. Trouble. *Pump loses water/liquid after starting.*

Causes. (*i*) Leaky suction.

(*ii*) Lift too high.

(*iii*) Excess amount of air or gases in water/liquid.

3. Trouble. *Pump takes too much power.*

Causes. (*i*) Speed too high.

(*ii*) Pumping too much water/liquid.

(*iii*) Liquids pumped of different specific gravities and viscosities than those for which the pump is designed.

(*iv*) Mechanical defects.

4. Trouble. *Pump vibrates and produces noise.*

Causes. (*i*) Misalignment.

(*ii*) Cavitation.

(*iii*) Mechanical defects (*e.g.,* bent shaft; worn out bearings etc.).

- The output power delivered by the pump is given by :

$$P_o = \frac{wQH_m}{75}$$

where, P_o = Output power in H.P.

w = Specific weight of liquid, being pumped, kg_f/m^3,

Q = Discharge of the pump, m^3/s,

H_m = Manometric head of the pump

$$= \left(\frac{p_d}{w} + \frac{V_d^2}{2g} + z_d\right) - \left(\frac{p_s}{w} + \frac{V_s^2}{2g} + z_s\right), \quad ...(1)$$

where, p_d = Pressure on the delivery side,

p_s = Pressure on the suction side,

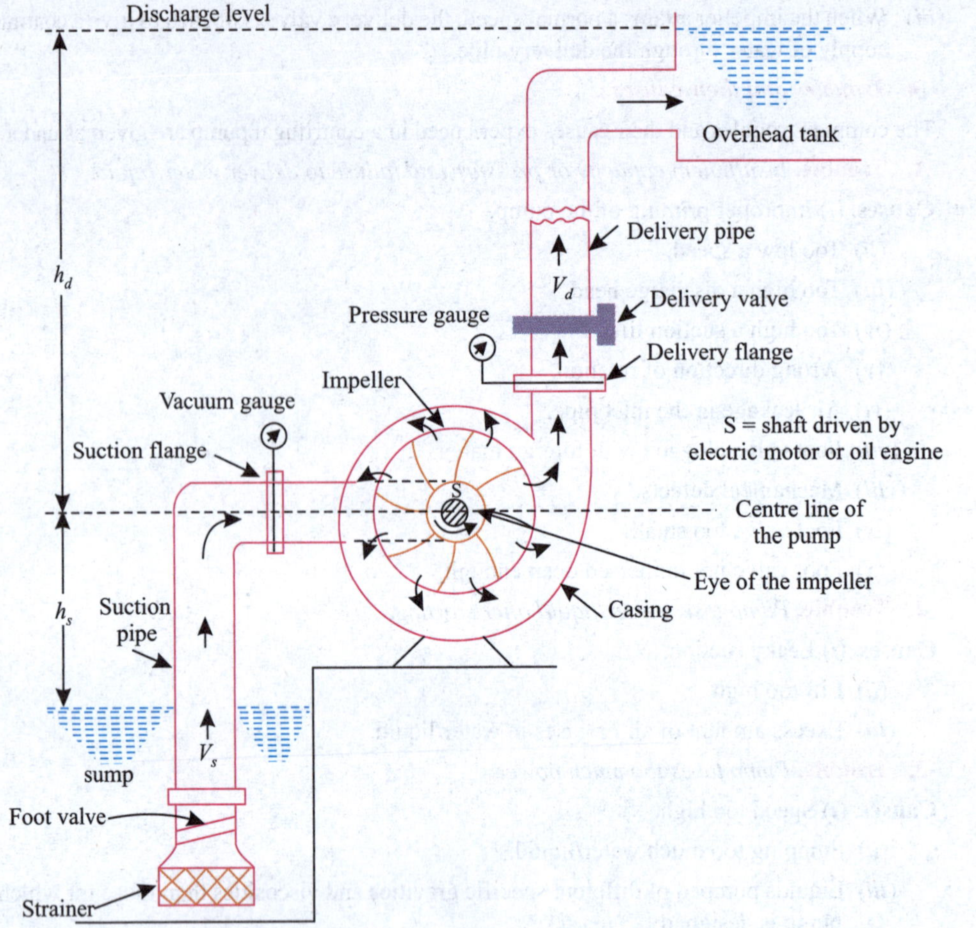

Fig. 5. Volute type centrifugal pump–component parts.

V_d = Velocity of flow on the delivery side,

V_s = Velocity of flow on the suction side,

z_d = Distance of the pressure gauge tapping on the delivery side from the pump axis, and

z_s = Distance of the pressure gauge tapping on the suction side from the pump axis.

If the net power input to the pump (*i.e.*, after taking into account all the losses) from the prime mover is P_i then, the overall efficiency is given by :

$$\% \eta = \frac{P_o}{P_i} \times 100$$

Procedure :

1. Note down the diameters of the suction and delivery pipes, wattmeter constant, overall efficiency of the prime mover, distances of the pressure gauge tappings from the pump axis.
2. Keeping the delivery valve closed prime the pump so that the suction pipe, casing and the portion of the delivery pipe upto the delivery valve are completely filled with liquid.

3. Start the motor and then open the delivery valve fully.
4. Allow the flow to stabilize and then measure the discharge (Q).
5. Note down the pressure gauge readings (p_d and p_s) and the wattmeter reading (X)
6. Change the opening of the delivery pipe.
7. Repeat steps 4 to 6 for at least ten openings of the delivery valve ranging from maximum to minimum discharge.

Observations :

Diameter of the suction pipe, d_s = ...

Diameter of the delivery pipe, d_d = ...

Distance of the pressure gauge on delivery side from the pump axis, z_d = ...

Distance of the pressure gauge on suction side from the pump axis, z_s = ...

Wattmeter constant, k = ...

Overall efficiency of the prime mover, η_p = ...

Table 3. Performance of a single stage pump – Observations

S.No.	Discharge (m^3/s)	*Pressure gauge readings*				Wattmeter readings X (watts)	Manometric head H_m (m)	Output P_0 (H.P.)	Input P_i (H.P.)	η (%)
		p_d (kg/cm^2)	H_d (m)	p_s (kg/cm^2)	H_s (m)					

Note : The consistency of units shall be maintained carefully.

Calculations : Perform the following calculations for each opening of the delivery valve.

1. $V_d = \dfrac{Q}{(\pi/4)\, d_d^2}$; $V_s = \dfrac{Q}{(\pi/4)\, d_s^2}$

2. $H_d = \dfrac{p_d}{w}$; $H_s = \dfrac{p_s}{w}$

3. H_m from eqn. (1).

4. $P_o = \dfrac{wQH_m}{75}$ H.P.

5. $P_i = \dfrac{\eta_p k X}{0{\cdot}736}$ H.P.

6. $\eta_o = \dfrac{P_o}{P_i} \times 100$

- Plot the following graphs on the same axes :

  (*i*) H_m *vs* Q;

  (*ii*) η_o *vs* Q;

  (*iii*) P_i *vs* Q.

Precautions :

1. Prime the pump to remove the air completely before starting the pump.
2. After each change in the valve opening let the flow stabilize before taking readings.

EXPERIMENT NO. 4. **To obtain the performance characteristics of a reciprocating pump.**

Apparatus :

1. A double-acting reciprocating pump with all the necessary components (*i.e.*, suction and delivery pipes, no return valves, foot valves etc.).
2. An electric motor coupled to the pump shaft; wattmeter connected to the driving motor.
3. A vacuum gauge (or U-tube manometer) and a pressure gauge connected to the suction and delivery pipes respectively as near to the pump as possible.
4. A discharge measurement unit (*viz.* venturimeter with manometer) connected to the delivery pipe.

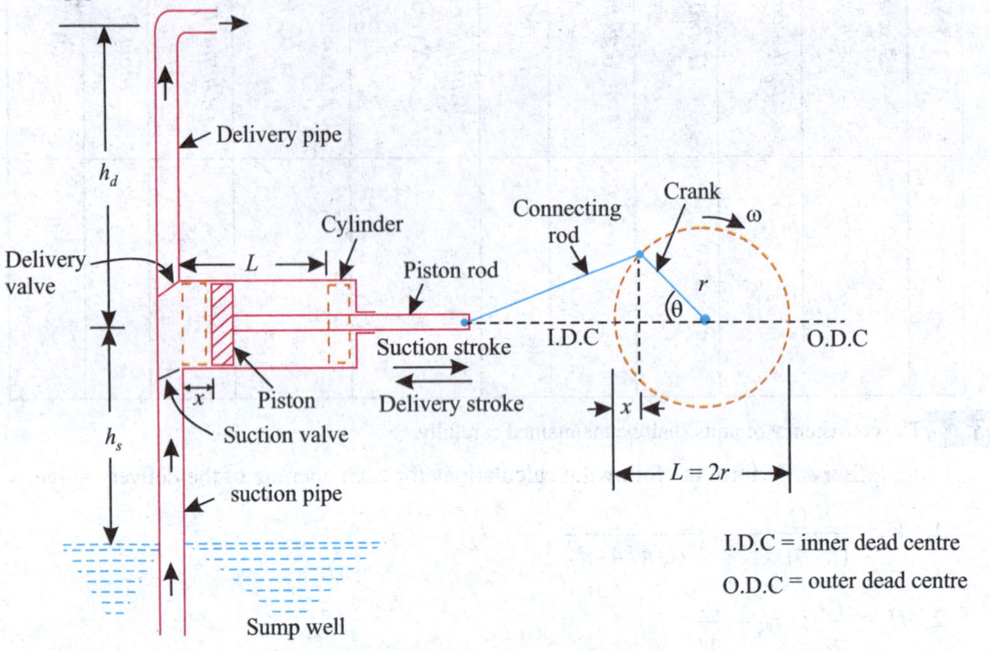

Fig. 6. Schematic view of single-acting reciprocating pump.

Theory :

- The reciprocating pump is a *positive displacement pump* as it sucks and raises the liquid by actually displacing it with the piston/plunger that executes a reciprocating motion in a closely fitting cylinder. The amount of liquids pumped is equal to the volume displaced by the piston. The total efficiency of a reciprocating pump is about 10 to 20% higher than a comparable centrifugal pump.

Refer to Fig. 6. The main parts of a reciprocating pump are :

1. Cylinder
2. Piston
3. Suction valve
4. Delivery valve
5. Suction pipe
6. Delivery pipe
7. Crank and connecting rod mechanism operated by a power source *e.g.* steam engine, internal combustion engine or an electric motor.

Working of a single-acting reciprocating pump :

As shown in Fig. 6 a single-acting reciprocating pump has one suction pipe and one delivery pipe. It is usually placed above the liquid level in the pump. When the crank rotates the piston moves backward and forward inside the cylinder. The pump operates as follows :

— Let us suppose that initially the crank is at the inner dead centre (I.D.C.) and crank rotates in the clockwise direction. As the crank rotates, the piston moves towards right and a vacuum is created on the left side of the piston. This vacuum causes suction valve to open and consequently the liquid is forced from the sump into the left side of the piston. When the crank is at the outer dead centre (O.D.C.) the suction stroke is completed and the left side of the cylinder is full of liquid.

— When the crank further turns from O.D.C. to I.D.C., the piston moves inward to the left and high pressure is built up in the cylinder. The delivery valve opens and the liquid is forced into the delivery pipe. The liquid is carried to the discharge tank through the delivery pipe. At the end of delivery stroke the crank comes to the I.D.C. and the piston is at the extreme left position.

Working of a double-acting reciprocating pump :

Refer Fig. 7. In a double-acting reciprocating pump, suction and delivery *strokes occur simultaneously.* When the crank rotates from I.D.C. in the clockwise direction, a vacuum is created on the left side of piston and the liquid is sucked in from the sump through value S_1. At the same time, the liquid on the right side of the piston is pressed and a high pressure causes the delivery valve D_2 to open and the liquid is passed on to the discharge tank. This operation continues till the crank reaches O.D.C.

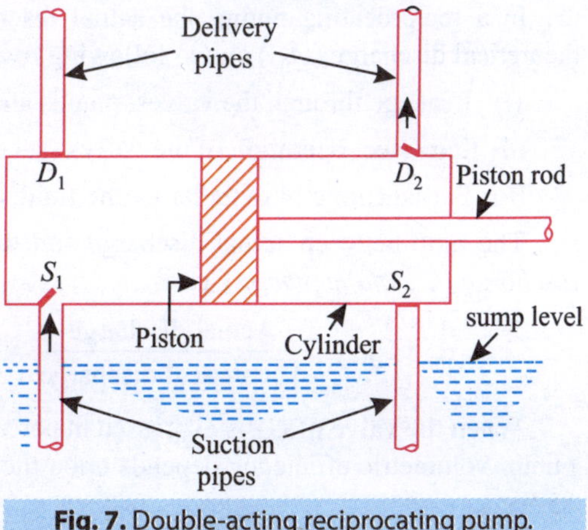

Fig. 7. Double-acting reciprocating pump.

With further rotation of the crank, the liquid is sucked in from the sump through the suction valve S_2 and is delivered to the discharge tank through the delivery valve D_1. When the crank reaches I.D.C., the piston is in the extreme left position. Thus one cycle is completed and as the crank further rotates, cycles are repeated.

Because of continuous delivery strokes, a double-acting reciprocating pump gives more uniform discharge (as compared to a single-acting pump which pumps the liquid intermittently). To get a still more uniform feed, invariably a multi-cylinder arrangement having two or more cylinders is employed.

Fig. 8 and 9 show the variations of discharge through delivery pipe (Q_d) with crank angle (θ) for single-acting and double-acting pumps respectively.

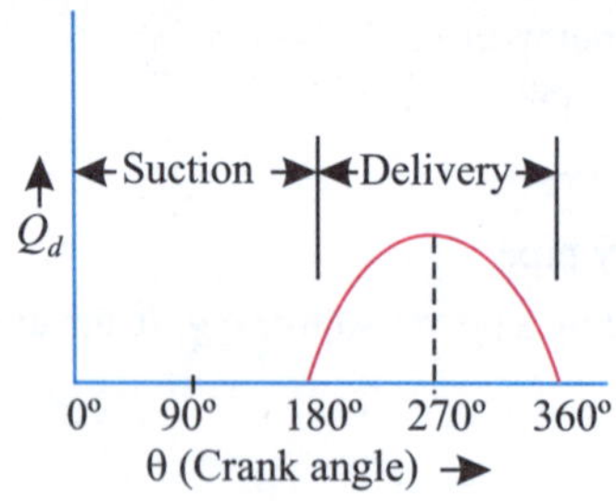

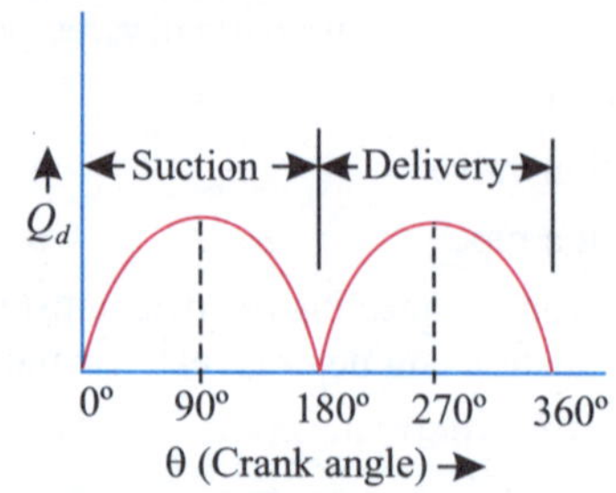

Fig. 8. Q_d v/s θ variations for single-acting pump. **Fig. 9.** Q_d v/s θ variations for double-acting pump.

Under ideal conditions the discharge in case of a double-acting reciprocating pump is given by:

$$Q_{th.} = \frac{ALN}{60} + (A - a)\frac{LN}{60}$$

$$\simeq \frac{2ALN}{60} \text{ since } a << A \quad ...(1)$$

where, $Q_{th.}$ = Theoretical discharge,

A = Cross-sectional area of the piston,

a = Cross-sectional area of the piston rod,

N = Speed of the crank pin in r.p.m., and

L = Length of stroke of the piston.

- **Co-efficient of discharge :**

In a reciprocating pump, the actual discharge ($Q_{act.}$) is always slightly different from the theoretical discharge ($Q_{th.}$) due to following *reasons :*

(*i*) Leakage through the valves, glands and piston packing,

(*ii*) Imperfect operation of the valves (suction and discharge), and

(*iii*) Partial filling of cylinder by the fluid.

The ratio between actual discharge and theoretical discharge is known as the *co-efficient of discharge (C_d) of the pump.* That is

$$C_d = \frac{\text{Actual discharge}}{\text{Theoretical discharge}} = \frac{Q_{act.}}{Q_{th.}} \quad ...(2)$$

When the valve of C_d is expressed in percentage, it is known as **'volumetric efficiency'** of the pump. Volumetric efficiency depends upon the dimensions of the pump and its value ranges from 85-98%.

- **Slip :**

The difference between the theoretical discharge and actual discharge is called the *'slip'* of the pump. That is

$$\text{Slip} = Q_{th.} - Q_{act.}$$

But the slip is often expressed in percentage which is given by,

$$\%\text{ Slip} = \frac{Q_{th.} - Q_{act.}}{Q_{th.}} \times 100 = \left(1 - \frac{Q_{act.}}{Q_{th.}}\right) \times 100 = (1 - C_d) \times 100 \quad ...(3)$$

The percentage of slip for the pumps maintained in *good condition* is of the order of 2% or even less.

- The *power generated* by the pump is given by :

$$P = \frac{w\, Q_{actual}\, H_m}{75} \quad \text{...(4)}$$

where, w = Specific weight of the liquid being pumped,

H_m = Manometric head of the pump

$$= \left(\frac{p_d}{w} + \frac{V_d^2}{2g} + z_d\right) - \left(\frac{p_s}{w} + \frac{V_s^2}{2g} + z_s\right) + h_{fs} + h_{fd} \quad \text{...(5)}$$

where, p_d = Pressure on the delivery side,

p_s = Pressure on the suction side,

V_d = Velocity of flow on the delivery side,

V_s = Velocity of flow on the suction side,

z_d = Distance of pressure gauge tapping on the delivery side from the pump axis,

z_s = Distance of pressure gauge tapping on the suction side from the pump axis,

$$h_{fd} = \text{Head loss due to friction on delivery side} = \frac{f z_d V_d^2}{d_d \times 2g} \quad \text{...(6)}$$

$$h_{fs} = \text{Head loss due to friction on suction side} = \frac{f z_s V_s^2}{d_s \times 2g} \quad \text{...(7)}$$

Here, f = Friction factor,

d_d = Diameter of delivery pipe, and

d_s = Delivery of suction pipe.

Procedure :

1. Note down the diameter (D) and stroke length (L) of the piston, wattmeter constant k, distances z_d and z_s of the pressure gauges from the pump axis.
2. Start the motor and open the delivery valve fully.
3. Measure the discharge Q_{actual}.
4. Note down the pressure gauge readings p_d and p_s and also wattmeter reading (X).
5. Measure the number of strokes occurring in a given time and hence obtain the speed of the piston in r.p.m.
6. Change the opening of the delivery valve.
7. Repeat steps 3 to 6 for at least ten different openings of the delivery valve.

Observations :

Diameter of the piston, D = ...

Stroke length of the piston, L = ...

Diameter of the suction pipe, d_s = ...

Diameter of the delivery pipe, d_d = . . .

Distance of the pressure gauge on delivery side from the pump axis, z_d = ...

Distance of the vacuum gauge on suction side from the pump axis, z_s = ...

Wattmeter constant, k = ...

Overall efficiency of the prime mover , η_p = ...

Table 4. Performance of reciprocating pump – Observations

S.No.	Discharge (m^3/s)	Pressure gauge readings				Velocities		H_m (m)	Wattmeter readings X (watts)	Output P_0 (H.P.)	Input P_i (H.P.)	Slip (%)	η (%)
		p_d (kg/cm^2)	h_{fd} (m)	p_s (kg/cm^2)	h_{fs} (m)	V_d (m/s)	V_s (m/s)						

Note : Consistency of units shall be maintained carefully.

Calculations : Perform the following calculations for each opening of the delivery valve :

1. $V_s = \dfrac{Q_{actual}}{(\pi/4)\, d_s^2}$; $V_d = \dfrac{Q_{actual}}{(\pi/4)\, d_d^2}$
2. Calculate h_{fs} and h_{fd} using eqns. (6) and (7).
3. Calculate H_m using eqn. (5).
4. $P_i = \dfrac{\eta_p\, kX}{0.736}$
5. Calculate P_o from eqn. (4).
6. $\eta = \dfrac{P_o}{P_i} \times 100$
7. Calculate $Q_{th.}$ from eqn. (1).
8. Calculate slip from eqn. (3).

- Plot the following curves:
  - (*i*) H_m vs Q
  - (*ii*) H_m vs S.H.P. (P_o)
  - (*iii*) H_m vs η.

Result/Conclusion :

Precautions :

1. Do not run the pump with the delivery valve completely closed.
2. After each change in the valve opening let the flow stabilize before taking readings.

INDEX

H

I

J

K

M

N

P

R

S

T